Peppers

There is an increased awareness on the relevance of nutraceutical and functional foods as alternatives to harmful synthetic additives used in industry. Different peppers, with an abundance of bioactive compounds, are highlighted in this book, which provides a comprehensive evaluaton of their importance as nutraceutical and functional foods to all stakeholders in the agri-food and pharmaceutical industries. ***Peppers: Biological, Health, and Postharvest Perspectives*** is a valuable addition to the existing information resource on peppers.

Key features:

- Highlights the advancements made in biodiversity, biochemistry and biosynthesis of bioactive compounds of peppers.
- Reviews the effects of processing methods on the quality of peppers to facilitate further research and development of foods having pepper as an essential nutritional component.
- Provides help in selecting better processing methods for the management of nutritional attributes and health benefits of peppers.

The book provides a blend of basic and advanced information for postgraduate students, researchers and scientists.

Functional Foods and Nutraceuticals Series

Series Editor
John Shi, Ph.D.
Guelph Food Research Center, Canada

Peppers: Biological, Health, and Postharvest Perspectives (2025)
Edited by Prasad S. Variyar, Inder Pal Singh, Vanshika Adiani, Penna Suprasanna

Anthocyanins in Sub-Tropic Fruits:
Chemical Properties, Processing, and Health Benefits (2023)
Edited by M. Selvamuthukumaran

Tea as a Food Ingredient: Properties, Processing, and Health Aspects (2022)
Edited by Junfeng Yin, Zhusheng Fu, and Yongquan Xu

Phytochemicals in Soybeans: Bioactivity and Health Benefits (2022)
Edited by Yang Li and Baokun Qi

Asian Berries: Health Benefits (2020)
Edited by Gengsheng Xiao, Yujuan Xu, and Yuanshan Yu

Phytochemicals in Goji Berries: Applications in Functional Foods (2019)
Edited by Xingqian Ye, and Yueming Jiang

Korean Functional Foods: Composition, Processing (2018)
and Health Benefits
Edited by Kun-Young Park, Dae Young Kwon, Ki Won Lee, and Sunmin Park

Phytochemicals in Citrus: Applications in Functional Foods (2017)
Xingqian Ye

Food as Medicine: Functional Food Plants of Africa (2016)
Maurice M. Iwu

Chinese Dates: A Traditional Functional Food (2016)
Edited by Dongheng Liu, Ph.D., Xingqian Ye, Ph.D., and Yueming Jiang, Ph.D.

Functional Food Ingredients and Nutraceuticals: (2015)
Processing Technologies, Second Edition
Edited by John Shi, Ph.D.

Peppers

Biological, Health, and Postharvest Perspectives

Edited by Prasad S. Variyar, Inder Pal Singh,
Vanshika Adiani, and Penna Suprasanna

CRC Press
Taylor & Francis Group
Boca Raton London New York

CRC Press is an imprint of the
Taylor & Francis Group, an **informa** business

Contents

Chapter 4 Chemical Diversity and Functionality of Capsaicinoids 40

Inder Pal Singh, Nobuyuki Mase, Ankur Kumar Tanwar,
Neha Sengar and Olivia Chatterjee

Chapter 5 Analytical Methods for Capsaicinoids and Other Bioactive Metabolites 65

Inder Pal Singh, Dharmistha Rajput and Debanjan Chatterjee

Chapter 6 Transcriptomics of Chili Pepper Fruit with Emphasis on the Carotenoid
       Biosynthetic Pathway 80

Maria Guadalupe Villa-Rivera, Octavio Martínez and Neftalí Ochoa-Alejo

Chapter 12 Post-Harvest Handling and Processing of Green and Red Indian Chillies 184

A.J. Sachin, P. Preethi, S.V.R. Reddy and P. Naresh

Chapter 13 Influence of Packaging on Postharvest Quality of Peppers 194

Komal Sonawane and Ramanand Jagtap

Chapter 14 Pepper Fixed Oil: Novel Source of Oil-Based Nutraceuticals 219

Ajay W. Tumaney, Vallamkondu Manasa, Palak Daga and
Salony Raghunath Vaishnav

Chapter 15 Effect of Newer Non-Thermal Processing 235

Admajith M. Kaimal, Amrutha M. Kaimal and Rekha S. Singhal

Preface

Peppers are an important agricultural commodity, ranked as the world's second-most-important 'fruit vegetable' after tomatoes. Peppers have a great appeal through natural colours and possess high nutritional value, health benefits and medicinal properties. Owing to the pungency principle of alkaloid capsaicin, diverse bioactivities have been known in the past few decades, making peppers a functional food and a beneficial component of the diet. Fresh pepper fruits are perishable, and hence, appropriate post-harvest processing is imperative to prevent economic losses during storage and transportation. From both a dietary and a nutritional point of view, knowledge of different processing conditions is thus of enormous significance. Newer non-thermal processing methods such as high-pressure processing, ultrasound and high-intensity pulsed electric fields etc. can aid in increasing the bioavailability of capsaicinoids, minerals, tocopherols, polyphenols, ascorbic acid, carotenoids and total antioxidant activity.

This book, with 16 chapters, presents the advancements made in chemistry, biochemistry and biosynthesis of bioactive compounds of peppers, including the impact of pre- and post-harvest processing methods. Chapter 1 is an overview of chemical constituents, quality, nutritional profiles, medicinal properties, and pre- and post-harvest factors that affect quality and novel and diverse products peppers. Chapters 2 and 3 describe the diversity of chilli (*Capsicum* spp.) genetic resources for quality traits and interspecific hybridization for obtaining plants resistant to biotic and abiotic stress conditions, respectively. Chapters 4 and 5 focus on chemical diversity and functionality of capsaicinoids and the analytical methods for estimation of capsaicinoids and other bioactive metabolites. Chapter 6 discusses the biosynthetic pathway of carotenoids of peppers. Physical, chemical and non-destructive post-harvest technologies for maintaining quality of peppers are covered in Chapter 7. The nutraceutical and medical potential of capsaicin is the focus of Chapter 8, while a detailed profile of the various secondary metabolites of Indian chilies and their medicinal properties are discussed in Chapter 9. Chapter 10 details novel pepper-based products that have improved shelf life and sensory acceptability, and Chapter 11 focusses on development of nutraceutical products of capsaicin and proposes solution for reducing capsicum and chilli waste. Post-harvest processing and handling of green and red Indian chillis and the influence of packaging on their post-harvest quality are the topics covered in Chapters 12 and 13. Chapter 14 highlights the role of pepper fixed oil as a novel source of oil-based nutraceuticals. Application of non-thermal novel technologies for post-harvest pepper processing and their efficacy in increasing shelf life of peppers are discussed in Chapter 15. Finally, the methods for detection of adulteration in peppers are detailed in Chapter 16.

This comprehensive book on peppers could facilitate further research and development of functional foods having essential nutritional components of peppers. The information will augment selection of better processing methods for the management of nutritional attributes and health benefits of peppers. The book is an excellent reference source for researchers, health professionals, students and policymakers involved in the agricultural and industrial sectors of pepper. The book is suitable for advanced undergraduate and postgraduate students specializing in agriculture, horticulture and food processing. The editors would like to place on record 'special thanks' to the contributing authors for sharing their expertise and, to the CRC team for their support during the publication of this book.

Prasad S Variyar
Vanshika Adani
I P Singh
Penna Suprasanna

The Editors

Dr. Prasad S. Variyar, formerly at Bhabha Atomic Research Centre (BARC), served as the senior scientist in the Food Technology Division, BARC, Mumbai, India, and professor at the HomiBhabha National Institute, Mumbai, India. Significant contributions include development of newer radiation-processed, minimally processed convenience foods, novel dosimeters for determining absorbed gamma radiation dose and newer biodegradable packaging for food irradiation applications. He has more than 150 publications in high-impact, international, peer-reviewed journals and has three patents and ten technology transfers to his credit.

Inder Pal Singh is a professor of natural products chemistry at National Institute of Pharmaceutical Education and Research (NIPER), SAS Nagar. His research interests include isolation of bioactive molecules from natural sources, biomimetic synthesis of bioactive natural products and their analogs for therapeutic areas such as HIV, cancer and leishmaniasis and standardization of herbal/Ayurvedic formulations using various analytical techniques. Inder Pal has been awarded Honorary Visiting Professorship of Shizuoka University, Japan, and served as Dean—Pharmacy, Maharaja Ranjit Singh Punjab Technical University, Bathinda. He has more than 165 publications and is co-author of two books.

Vanshika Adiani, PhD, is a scientist working at Food Technology Division, Bhabha Atomic Research Centre (BARC), Mumbai, India. The major focus of her work is in the fields of food processing, safety and security. She has developed rapid detection techniques for assessment of microbial quality in minimally processed fruits and colorimetric time-temperature indicator devices for easy visual determination of spoilage of minimally processed fruits. Her current research includes developing methods and prototypes for detection of spice adulteration with various synthetic non-permitted dyes using machine learning tools.

Dr. Penna Suprasanna is the director, Amity Centre for Nuclear Biotechnology, Amity University Maharashtra, Mumbai, India; former head of the Nuclear Agriculture & Biotechnology Division, Bhabha Atomic Research Centre, Mumbai, India. His research interests include agricultural biotechnology, climate-smart agriculture, molecular plant stress physiology and nuclear agriculture. He has served as an expert with International Atomic Energy Agency, Vienna, and Member, Food Security Committee, Board of Research in Nuclear Sciences, DAE. He has more than 390 research publications to his credit and several edited books.

Contributors

Satyaprakash Barik
Department of Agriculture and
    Allied Sciences
CV Raman Global University
Bhubaneswar, Odisha, India

Somnath Basak
Department of Food Engineering and
    Technology
Institute of Chemical Technology
    Mumbai, India

Debanjan Chatterjee
Department of Natural Products
National Institute of Pharmaceutical
    Education & Research (NIPER)
Punjab, India

Olivia Chatterjee
Department of Natural Products
National Institute of Pharmaceutical
    Education & Research (NIPER)
Punjab, India

Zhong-Hua Chen
School of Science
Western Sydney University
Richmond, Australia

Palak Daga
Department of Biochemistry
Council of Scientific and Industrial
    Research—Central Food Technological
    Research Institute
Mysore, India

Michelle Donovan-Mak
School of Science
Western Sydney University
Richmond, Australia

Debosree Ghosh
Department of Physiology
Government General Degree College,
    Kharagpur II
Paschim Medinipur, West Bengal, India

Ramkrishna Ghosh
Department of Botany
Government General Degree College, Kharagpur II
West Bengal, India

Talía Hernández-Pérez
Departamento de Biotecnología y Bioquímica,
Centro de Investigación y de Estudios
    Avanzados del IPN. Irapuato
Gto, México

Ramanand Jagtap
Department of Polymer and Surface
    Engineering
Institute of Chemical Technology
Mumbai, Maharashtra, India

Admajith M. Kaimal
Department of Food Engineering and
    Technology
Institute of Chemical Technology, ICT-IOC
    Campus
Bhubaneswar, India

Amrutha M. Kaimal
National Institute of Food Technology
Entrepreneurship and Management
Thanjavur, Tamil Nadu, India

Vallamkondu Manasa
Department of Biochemistry
Council of Scientific and Industrial Research—
    Central Food Technological Research Institute
Mysore, India

Octavio Martínez
Unidad de Genómica Avanzada
Centro de Investigación y de Estudios
    Avanzados del Instituto Politécnico Nacional
Irapuato, México

Nobuyuki Mase
Research Institute of Green Science and
    Technology
Shizuoka University, Johoku
Hamamatsu, Shizuoka, Japan

Kazım Mavi
Faculty of Agriculture, Department of Horticulture
Hatay Mustafa Kemal University
Hatay, Turkey

Karine Sayuri Lima Miki
Post-Graduate Program in Biotechnology
 (PPGBIOTEC)
Federal University of Amazonas
AM, Brazil

P. Naresh
Indian Institute of Horticultural Research
Bengaluru, Karnataka

Neftalí Ochoa-Alejo
Departamento de Ingeniería Genética
Unidad Irapuato, Centro de Investigación y de
 Estudios Avanzados del Instituto Politécnico
 Nacional
Irapuato, México

Sunil K. Panchal
School of Science
Western Sydney University
Richmond, Australia

Octavio Paredes-López
Departamento de Biotecnología y Bioquímica,
 Centro de Investigación y de Estudios
 Avanzados del IPN. Irapuato
Gto, México

Pavani, N.
Division of Vegetable Crops
ICAR-Indian Institute of Horticultural
 Research
Bengaluru, Karnataka, India

Naresh Ponnam
Division of Vegetable Crops
ICAR-Indian Institute of Horticultural
 Research
Bengaluru, Karnataka

P. Preethi
Indian Institute of Horticultural Research
Bengaluru, Karnataka

Dharmistha Rajput
Department of Natural Products, National
 Institute of Pharmaceutical Education &
 Research
Punjab, India

S.V.R. Reddy
Indian Institute of Horticultural Research
Bengaluru, Karnataka

Madhavi Reddy K
Division of Vegetable Crops
ICAR- Indian Institute of Horticultural Research
Bengaluru, Karnataka

Rayssa Ribeiro
Chemical Engineering Section
Laboratory of Organic Bioprocesses
Military Engineering Institute
Rio de Janeiro, Brazil

A.J. Sachin
Bihar Agricultural University
Bhagalpur, Bihar

Filipe Kayodè Felisberto dos Santos
Chemical Engineering Section
Laboratory of Organic Bioprocesses
Military Engineering Institute
Rio de Janeiro, Brazil

B. Sasikumar
Crop Improvement and Biotechnology Division
ICAR-Indian Institute of Spices Research, P O
 Marikunnu
Kozhikode, Kerala, India

Neha Sengar
Department of Natural Products
National Institute of Pharmaceutical
 Education & Research
Punjab, India

Inder Pal Singh
Department of Natural Products
National Institute of Pharmaceutical
 Education & Research
Punjab, India

Partha Sarathi Singha
Department of Chemistry
Government General Degree College
    Kharagpur II
West Bengal, India

Rekha S. Singhal
Department of Food Engineering and
    Technology
Institute of Chemical Technology
Mumbai, India

Komal Sonawane
Department of Polymer and Surface
    Engineering
Institute of Chemical Technology
Mumbai, Maharashtra, India

Penna Suprasanna
Amity Centre for Nuclear Biotechnology
Amity Institute of Biotechnology
Amity University of Maharashtra
Mumbai, India

Ankur Kumar Tanwar
Department of Natural Products
National Institute of Pharmaceutical
    Education & Research
Punjab, India

Barbara Elisabeth Teixeira-Costa
Post-Graduate Program in Biotechnology
    (PPGBIOTEC)
Federal University of Amazonas
Manaus, AM, Brazil; Department of Nutrition
    and Dietetics
Faculty of Nutrition Emília de Jesus Ferreiro,
    Federal Fluminense University—UFF,
    Niterói
Rio de Janeiro, Brazil

Ajay W. Tumaney
Department of Biochemistry, School of Life
    Sciences
University of Hyderabad
Hyderabad, India

Salony Raghunath Vaishnav
Department of Traditional Foods and Applied
    Nutrition
Council of Scientific and Industrial Research—
    Central Food Technological Research Institute
Mysore, India

Prasad S. Variyar
Food Technology Division
Bhabha Atomic Research Centre
Mumbai, India

Valdir Florêncio da Veiga Júnior
Chemical Engineering Section
Laboratory of Organic Bioprocesses
Military Engineering Institute
Rio de Janeiro, Brazil, & Post-Graduate
    Program in Biotechnology (PPGBIOTEC)
Federal University of Amazonas
Manaus, AM, Brazil

Maria Guadalupe Villa-Rivera
Departamento de Ingeniería Genética
Unidad Irapuato
Centro de Investigación y de Estudios Avanzados
    del Instituto Politécnico Nacional
Irapuato, México

Mursleen Yasin
School of Science
Western Sydney University
Richmond, Australia

1 Peppers—an Overview on The Bioactives and their Nutritional, Functional Properties and Post-Harvest Processing

Prasad S. Variyar and Penna Suprasanna

CONTENTS

1.1 INTRODUCTION

Spices have been widely used as flavouring and colouring agents and preservatives since ancient times. They exhibit numerous health benefits and are used extensively in traditional system of medicines such as Ayurveda, Unani and Chinese herbal medicines. Many compounds isolated from spices show antimicrobial activity against several common foodborne microorganisms that are known to affect food quality and shelf life (Tajkarimi et al., 2010). The FDA defines spices as an "aromatic vegetable substance in the whole, broken, or ground form, the significant function of which in food is seasoning rather than nutrition and from which no portion of any volatile oil or other flavouring principle has been removed" (Sung et al., 2012). Asia leads in the production of spices. India is a major spice-producing country, producing 75 of the 109 spices listed by the International Organization for Standardisation (ISO) (Spices Board of India, http:/ indianspices.com). The major

DOI: 10.1201/9781003378259-1

markets in the global spice trade are the USA, the European Union, Japan, Singapore, Saudi Arabia and Malaysia. The main exporting countries are China, India, Madagascar, Indonesia, Vietnam, Brazil, Spain, Guatemala and Sri Lanka (https://www.tradologie.com/lp/spice-trade.html). The global spices and seasonings market has shown significant high at USD 17.75 billion in 2021 which is expected to rise to USD 25.42 billion by 2029. North America dominated the world market at USD 7.48 billion in 2021. Europe is expected to witness substantial growth due to the inclination of consumers towards exotic and bold flavours, while the Middle East and Africa regions are also witnessing increased demand due to enhanced tourism and hospitality in these regions. The Asia Pacific market is also likely to exhibit sizable growth due to increased growth of food-processing industries in this region. India, China and Vietnam are the largest consumers and producers of spices in the world, with India being the largest producer and exporter of spice and spice products (https://www.fortunebusinessinsights.com/industry-reports/spices-and-seasonings-market-101694). For the year 2021–22, India exported spices worth USD 4.1 billion, with 51% in the total export volume contributed by value-added spice products (Spices Board of India, http://indianspices.com).

A restraining factor that can affect market growth in the coming years is the vulnerability to adulteration with other plants, artificial colours and unlisted flavourings, attributed to rising demand for exotic herbs and spices and supply constraints. Increasingly stringent quality and food safety standards have necessitated emphasis on export-oriented production in compliance with food safety and quality requirements of importing countries (https://www.fortunebusinessinsights.com/industry-reports/spices-and-seasonings-market-101694). Focusing on infrastructure investment, capacity building for quality production, ensuring food safety and demonstration of traceability can lead the global spice industry in introducing scientifically validated novel applications of spices suitable for the food and beverage, wellness, nutrition and health segments.

1.2 CAPSICUM

Pepper (*Capsicum annuum* L.) is an annual herbaceous plant belonging to the *Solanaceae* family. It is cultivated in warm-climate regions of the world such as Asia, northern America, southern and central Europe and tropical and subtropical Africa (Thampi, 2004). The genus *Capsicum* consists of approximately 22 wild species and five domesticated species, *C. annuum, C. baccatum, C. chinense, C. frutescens* and *C. pubescens*. Pepper, chili, chile, chilli, aji, paprika and *Capsicum* are used interchangeably for plants in the genus *Capsicum*. Generally, peppers are consumed raw (bell pepper) or in powdered form as a spice (chili pepper) or as a colorant (paprika) (Baenas et al., 2019). The flavour of this vegetable ranges from the sweet (non-pungent) varieties, such as paprika, to the hot species, such as chilies or cayenne (Buckenhüskes, 2003). Paprika is widely used as a table spice and as a natural colourant in the meat-processing industry, valued principally for the brilliant red colour and also for its delicate aroma. Paprika is derived from mature pods of *Capsicum annum* peppers that include sweet bell peppers, hot red peppers and other varieties of varying pungency. Chillies generally are smaller and lower in red colour and highly pungent. Some of the hot and pungent types include New Mexico, jalapeno, cayenne, Thai and habanero.

Chilli peppers enhance flavour, zest, heat, warmth, bite and colour of prepared food, and continuing affinity towards hot and spicy food is expected to boost demand for this spice. India, Pakistan, China and Turkey are the major producers of red chili, with combined share of India and Pakistan at 64%. In terms of export volume, India and Vietnam are the top two exporters, accounting for 73% of total exports globally. The global market for red chili is valued at USD 1 billion in 2021. Chilli is exported to over 144 countries (https://www.tradologie.com/lp/redchilli-trade.html). In the year 2020–2021 (Apr–Nov), India exported chilli worth of USD 78.81 million. The total volume of export in 2020–2021 (Apr–Nov) was around 42394499 Metric Tonn (MT) (Spices Board of India, http://indianspices.com). Figure 1.1 provides global production of major chilli producing countries during 2020. Among all the spices that are exported from India in 2021–22, Chilli peppers contributed to the highest export value, accounting for USD 4,464.17 million (Spices Board of India, http://indianspices.com). Peels, stalks,

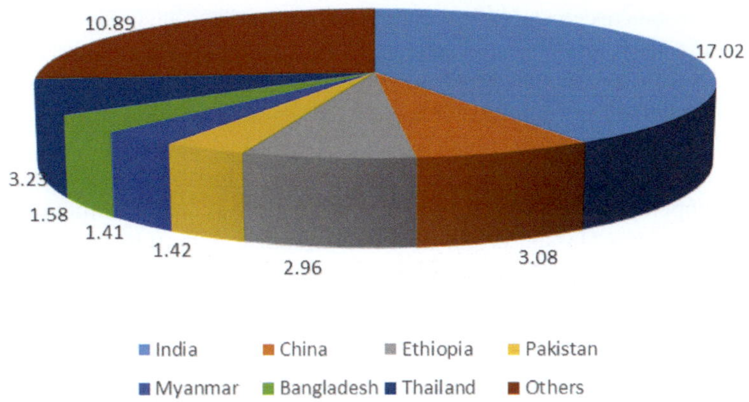

FIGURE 1.1 World production (in Lakh tonnes) of major chilli producing countries during 2020

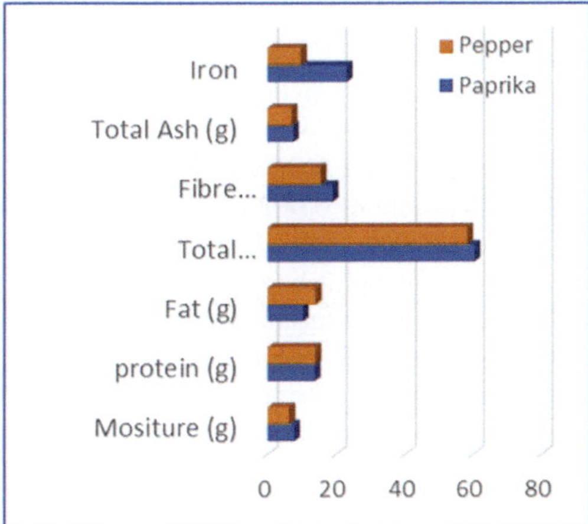

MINERALS (g)	Paprika	Pepper
Calcium	0.2	0.1
Phosphorous	0.3	0.32
Sodium	0.02	0.01
Potassium	2.4	2.1
VITAMINS (mg)		
Thiamine	0.6	0.59
Riboflavin	1.36	1.66
Niacin	15.3	14.2
Ascorbic acid	58.8	63.7
Vitamin A (IU)	4915	6165

FIGURE 1.2 General nutrient composition of paprika and chilli.

seeds and unused flesh generated by different steps of the industrial processing account for 5% to 30% of the total fruit production. Discarded as non-marketable by-products are also used as sources of nutrients and secondary metabolites, thereby providing bioprospecting value (Baenas et al., 2019).

1.3 CHEMICAL COMPOSITION

Peppers contains various chemical constituents that play an important role in their nutritional value, taste, colour and aroma. These include water, fixed (fatty) oils, steam-volatile oil, carotenoids, capsaicinoids, polyphenols, flavonoids, vitamins, resin, protein, fibre and mineral elements in varying concentration (Samarth et al., 2017) (Figure 1.2). Among these, the two most important chemical classes are the carotenoids and capsaicinoids, which contribute to the colour and pungency, respectively (Bosland and Votava, 2000; Govindarajan and Salzer, 1986). Capsaicinoids are a class of alkaloids found only in the genus *Capsicum*. The main capsaicinoids present in peppers are capsaicin (vanylamide of 8-methylnontrans-6-enoic acid) and dihydrocapsaicin (vanylamide of 8-methylnonanoic acid). They constitute more than 80% of total capsaicinoids and are the most pungent. Nordihydrocapsaicin, norcapsaicin, homocapsaicin, homodihydrocapsaicin, nornorcapsaicin, nornornorcapsaicin and nonivamide are the other minor capsaicinoids identified (Govindarajan and Salzer, 1986; Antonio et al., 2018). They are known to possess several pharmacological properties, including antioxidant anticancer, anti-inflammatory, antibacterial and analgesic activities (Szydelko et al., 2017). The non-pungent capsinoids, found only in few varieties of peppers, have a similar structure to the capsaicinoids, such as capsiate (4-hydroxy-3-methoxybenzyl (E)-8-methyl-6-nonenoate) and its derivatives dihydro-capsiate and nordihydrocapsiate, and they also occur in non-pungent red peppers, such as sweet chili pepper *Capsicum anuum* L. var. (CH-19) (Singh et al., 2009; Antonio et al., 2018).

Capsicum fruits are also rich sources of carotenoids, which are responsible for the varying colours of peppers and have been shown to act as antioxidants, immune enhancers and oxidative stress reducers (Hassan et al., 2019). Capsanthin is mainly responsible for the red colour, representing 40% to 60% of the total carotenoids in different varieties (Hassan et al., 2019; Antonio et al., 2018). Other carotenoids present are beta-carotene, alpha-carotene, β-cryptoxanthin, violaxanthin, lutein and zeaxanthin (Antonio et al., 2018). Beta-carotene, alpha-carotene and beta-cryptoxanthin are good sources of provitamin A. Besides, vitamin C (ascorbic acid), vitamin E (tocopherols) and folate, an important B-group vitamin is also widely distributed in pepper fruits (Nathania and Castello, 2020). A wide range of flavonoids is reported in pepper fruits, which include flavonoid aglycones and glycosides, O-glycosylated flavonols and flavones and C-glycosylated flavones. The levels of flavonoids vary widely among cultivars and maturation stages of the fruit. *p*-Coumaric, caffeic, vanillic, p-hydroxybenzoic, sinapic and ferulic glycosides are characteristic phenolic acid derivatives in pepper fruits (Nathania and Castello, 2020). Aroma also plays a greater role than how "hot" the product is in many pepper varieties. Thus, the aroma so familiar and typical for a green bell pepper is due almost entirely to a single substance, 2-methoxy-3-isobutylpyrazine, which is one of the most odour-intense substances known, with a low odour threshold of 0.002 parts per trillion. Apart from pyrazine, a few other compounds that contribute to the overall characteristic aroma of a pepper are (2Z)-nonenal, (2E, 6Z)-nonadienal, (2E)-nonenal, 2-heptanthiol, ethyl-4-methyl-pentanoate and beta-ionone. Significant differences exist in the aroma profiles of various pepper species. Thus, 2-methoxy-3-isobutylpyrazine is not detected in the volatile components of *C. chinense*, resulting in the complete absence of green pepper aroma (Antonio et al., 2018). In its place is β-ionone, as well as various esters that impart a corresponding fruity/floral bouquet. A supplementary light, nutty scent of *C. pubescens* results from 2-heptanthiol, while a mixture of short-chain esters that include ethyl-4-methylpentanoate imparts a strong and characteristic fruity aroma of the tabasco variety of *C. frutescens* (Govindarajan and Salzer, 1986).

1.4 CAPSAICINOIDS AND CAROTENES: BIOSYNTHESIS AND GENETIC REGULATION

Biosynthesis of capsaicinoids, unique to *Capsicum* species, involves the condensation of vanillylamine, derived from the phenylpropanoid pathway, and a series of branched-chain fatty acid moieties, derived from the branched-chain fatty acid pathway (Figure 1.3). Phenylalanine the primary precursor for the biosynthesis of phenylpropanoids, and either valine or leucine, the precursor for branched-chain fatty acid synthesis, is the initial substrate for capsaicinoid biosynthesis. Acyltransferase 3 (AT3) has been proposed to be capsaicinoid synthase (CS), responsible for the condensation between vanillylamine and branched-chain fatty acids to produce different capsaicinoids. AT3 has been demonstrated to serve as a key regulatory point in pungency (Arce-Rodriguez and Ochoa-Alejo, 2019; Butnariu and Samfira, 2016).

There has been progress in the molecular biology of the capsaicinoid biosynthetic pathway with the characterization of some of the structural and regulatory genes through studies using comparative gene expression in pungent and non-pungent chili pepper fruits (Aza-Gonalez et al., 2011; Arce-Rodríguez and Ochoa-Alejo, 2019; Villa-Rivera et al., 2021). A MYB transcription factor gene, designated *CaMYB31*, acts as a regulator of capsaicinoid biosynthesis. *CaMYB31* silencing causes a diminution of the capsaicin and dihydrocapsaicin content and a decrease in the expression levels of several structural genes, including cinnamate-4-hydroxylase (*C4H*), 4-coumarate CoA

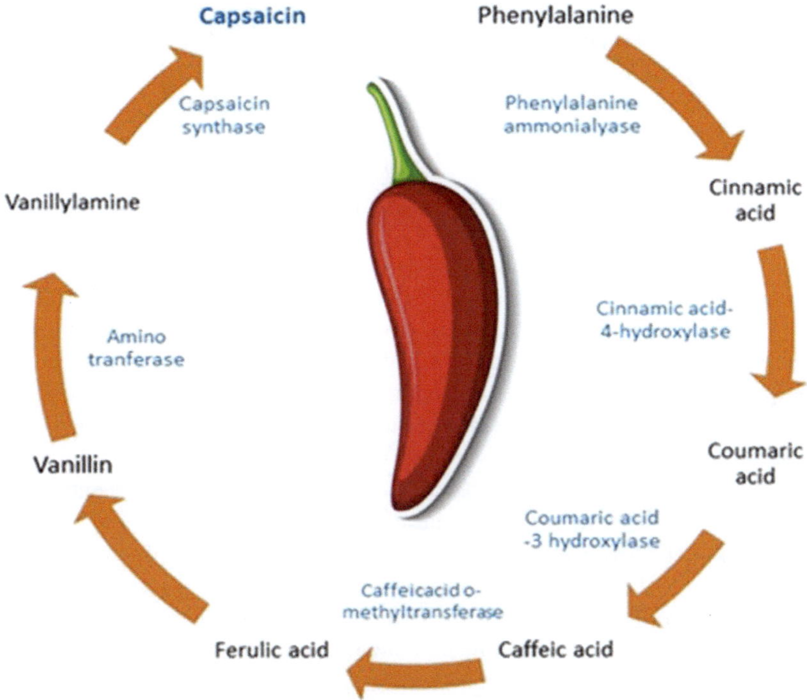

FIGURE 1.3 Biosynthetic pathway of capsaicinoids.

ligase (*4CL*), *p*-coumarate 3-hydroxylase (*C3H*), hydroxycinnamoyl transferase (*HCT*), branched-chain keto acid dehydrogenase (*BCKDH*), in the capsaicinoid biosynthesis pathway, suggesting the participation of this gene in the regulation of the metabolic pathway. Stewart et al. (2005) identified a coenzyme A-dependent acyltransferase encoded by the AT3 gene, which is referred to as *pungent gene 1 (pun1)* as capsaicin synthase (CS).

Ogawa et al. (2015) studied expression levels of *pun1* and *pAMT* genes in pepper cultivars and observed increased expression levels in pungent cultivars, resulting in higher capsaicin content, whereas low expression was noted in the non-pungent peppers. This further validated significance of both the genes for pungency. However, the *pun1* gene has a critical role in the overall capsaicin accumulation since high or low expression levels of *pAMT* will result in vanillylamine (Patrick, 2022). In a recent study of 12 *Capsicum* accessions, RNA-Seq transcriptomic analysis was performed during different stages of growth, development and ripening (Villa-Rivera et al., 2022). This study pioneered in the identification of 83 transcription factors as putative regulators of the carotenogenic pathway, and 54 of these had shown putative binding sites in the promoters of the carotenoid-biosynthesis-related structural genes. Zhang et al. (2022) analyzed the metabolomics and transcriptomics profiles of two peppers (a mildly pungent cultivated pepper, BB3, and its hot progenitor, chiltepin) to gain insights into pungency variation during domestication of the plant species. Accumulation of specific metabolites in BB3 were minimal compared to the wild species, and this correlated well with the gene expression profiles of key regulators, suggesting that there is a loss of pungency during *Capsicum* domestication.

The chili pepper fruits produce different capsaicinoid analogs in various ratios depending on the chili pepper type and maturation stage. Availability or abundance of the substrates vanillylamine and branched-chain fatty acids and variable substrate affinity of capsaicinoid synthase towards these substrates has been shown to be responsible for the variation in the content of these analogs (Arce-Rodriguez and Ochoa-Alejo, 2019). A positive correlation between the fatty acid profile and the capsaicinoid patterns has been observed. Thus, availability of the branched-chain fatty acid moieties has a great impact on the composition and ratios of capsaicinoids. Environmental factors have also been shown to influence the biosynthesis and accumulation of capsaicinoids (Arce-Rodriguez and Ochoa-Alejo, 2019).

A wide range of carotenoids is present within the *Capsicum,* genus which results in variation in colour observed among the different pepper varieties. The majority of the carotenoids are oxygenated derivatives. The carotenoid profile can vary as a function of the maturity stage, species and cultivars, involving a complex biosynthetic pathway. Mevalonic acid, the initial precursor of carotenes, undergoes several reactions to produce the C5 precursors isopentenyl diphosphate and dimethylallyl pyrophosphate, which further condense to produce geranylgeranyl phosphate to form the colourless phytoene. Phytoene is subjected to a series of desaturation and isomerization reactions to produce the red-coloured lycopene. Cyclization of lycopene results in a-carotene and b-carotene (orange-coloured carotenoids). In addition, hydroxylation of carotene can produce the yellow-coloured xanthophylls lutein and zeaxanthin. carotenoids in the *Capsicum* species usually contain nine conjugated double bonds (Hassan et al., 2019).

Qin et al. (2014) sequenced Zunla-1 (*C. annuum* L.) and its wild progenitor Chiltepin (*C. annuum* var. glabriusculum) and found that the genome had ~81% repetitive sequences and 79% of 3.48-Gb scaffolds containing 34,476 protein-coding genes. In a recent study, Song et al. (2022) performed transcriptomics of chili pepper (*Capsicum annuum* var. conoides) at fruit ripening stages and found an inverse relationship between the levels of carotenoid species, with β,β-structures (e.g. β-carotene, zeaxanthin and capsanthin) and of lutein with the β,ϵ-structure. The authors also used the reference genome of chili pepper Zunla-1 and identified 38 homologous genes in the carotenoid biosynthetic pathway and transcription factors involved in the coordinated regulation of synthesis of capsanthin/capsorubin with reduced lutein (Song et al., 2022).

1.5 QUALITY EVALUATION OF PEPPERS

Properties such as colour, pungency and aroma are important criteria for consumer acceptance of peppers in the market (Govindarajan et al., 1987). These parameters are dependent on maturity stage, agricultural practices and various pre- and post-harvest treatments. Proper controlled processing operations are thus needed to prevent alteration in quality during value addition or to avoid quality loss due to activity of enzymes, microorganisms or pest infections. Appearance and colour are the attributes that are perceived first by the consumer in deciding the appropriateness of the product in terms of freshness, maturity and defects (Govindarajan et al., 1987). Thus, colour measurement ensures product quality and commercial acceptance. The most frequently used method is the colorimetric method proposed by the American Spice Trade Association (ASTA) based on measurement of absorption of pigment extract at a determined wavelength, usually at 460 nm (ASTA, 1986). The method provides a rapid evaluation of colour quality but is ineffective in discriminating between paprika and chilli powders that have similar colour despite different pigment composition (Minguez-Mosquera and Perez-Galvez, 1998). The methods also cannot detect possible adulteration with synthetic pigments and oxidative alteration of natural pigments. Various chromatographic techniques such as thin-layer chromatography (TLC), reverse phase TLC (RP-TLC), adsorption and RP-HPLC using both UV/Vis and photodiode array detectors have been successfully advocated for separation, identification and quantification of pigment fractions of paprika, chilli and other *Capsicum* varieties (Cserhati et al., 2000).

A pleasant flavour and aroma of pepper are often associated with high nutritional benefits. The volatile compounds responsible for the aroma of *Capsicum* fruits are reported to have over 200 compounds of varying volatility classified as terpenes, hydrocarbons, alcohols, aldehydes, ketones, acids, esters, lactones, sulphur- and nitrogen-containing compounds and phenolics. Aroma compounds are generally isolated either by steam distillation or by vacuum distillation/continuous solvent extraction depending on their heat sensitivity. These isolates are then analyzed by GC/GC-MS for qualitative and quantitative estimation of individual aroma compounds in the volatile extracts. The GC-olfactory (GC-O) method has aided in odour evaluation by a trained panelist and in understanding the odour profile and, thus, the aroma quality of different *Capsicum* varieties.

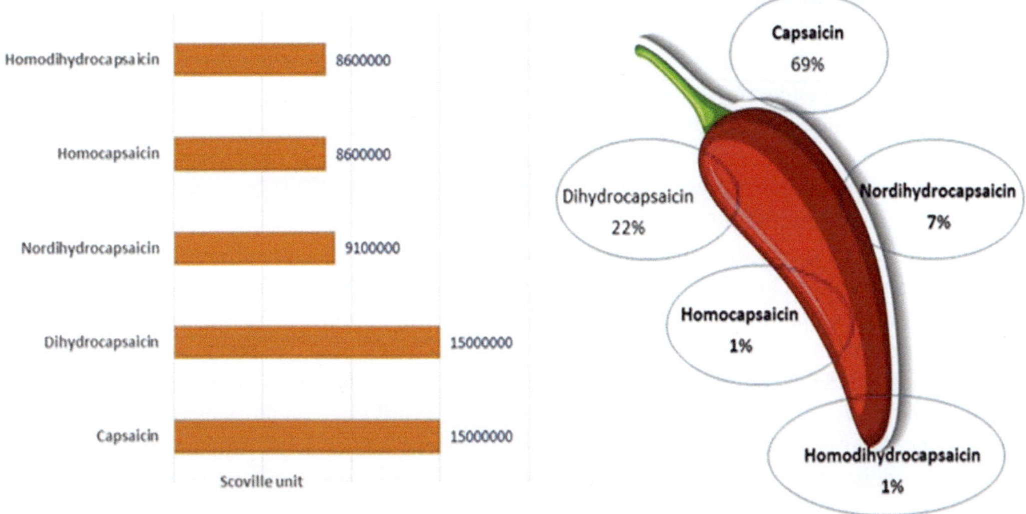

FIGURE 1.4 Pungency levels of capsaicinoids and their derivatives in chilli peppers.

The most important sensory quality of chillies and other fruits of the *Capsicum* species is their pungency, which is attributed to capsaicin and other related constituents (Figure 1.4). Instrumental methods have been employed for analysis and quantification of these components. However, this objective value acquires meaning only when correlated with the measure of sensorily perceived pungency in terms of a recognition threshold. Pungency measured in Scoville heat units (SHU) corresponds to this recognition threshold (Scoville, 1912; Collins et al., 1995). This conventional subjective method is based on the highest dilution of a pepper extract at which heat can be detected by a taste panel. Recently, the American Spice Trade Association has proposed a universal scale for pungency in a given sample based on the concentration of capsaicinoids in parts per million (ppm) (ASTA, 2018). Solvent extraction (using hexane, acetone or methanol), microwave-assisted extraction (MAE) and ultrasound-assisted extraction (UAE), as well as supercritical fluid extraction using liquid CO_2, have been widely employed for obtaining pepper extracts. Some of the methods currently used for estimation of capsaicinoids include TLC, RP-TLC, RP-HPLC using both UV/Vis and photodiode array detectors (Stoica et al., 2016). The result of High Pressure Liquid Chromatography (HPLC) is often converted to Scoville heat units by multiplying the parts per million by 16 in order to correlate sensory and instrumental data (Collins et al., 1995). Direct spectrophotometric determination of capsaicinoid content as a possible alternative to HPLC, with absorbance of the samples monitored at 280 nm, can be also routinely used with a correlation of 0.91 for determining total capsaicinoid content and quality control in pharmaceutical analysis (González-Zamora et al., 2015).

1.6 EFFECT OF PRE- AND POST-HARVEST FACTORS ON THE QUALITY OF PEPPER FRUITS

Climate is the primary determinant that affects the production and productivity of crops as well as the composition of the major active principles of *Capsicum* species. The major environmental factors that influence frequency and severity of events like drought and floods are temperature and precipitation. Generation and accumulation of greenhouse gases (carbon dioxide, methane, nitrous oxide) in the atmosphere also result in the global warming accelerating climate change impacting the chilli yield and quality (Bhutia et al., 2018). Plant pathogen inhibition and animal feeding are also observed due to capsaicin in plants, whereas in the case of human beings, capsaicin is shown to exert heat sensation due to its binding to a membrane channel protein, and in microbes, capsaicin restricts energy production by binding respiratory enzymes (Adams et al., 2020). Further, Adams et al. (2020) have also observed tolerance in 16 pathogenic fungi of seeds of *C. chacoense* and suggested that the fungi could breakdown the capsaicinoids.

Drought stress has been shown to negatively influence pepper yield, dry matter accumulation and nutrient uptake as well as physiological, biochemical and molecular changes. Increased activity of phenylalanine ammonia-lyase (PAL), a crucial enzyme involved in capsaicinoid synthesis, was noted under drought stress, consequently enhancing capsaicinoid content and thus pungency levels of hot pepper cultivars (Mahmood et al., 2021). Drought also significantly affected the synthesis of the colouring pigment capsanthin in pepper fruits. Maximum impact on capsanthin levels was observed during moderate to severe drought, with the levels being restored when water was supplied to the drought-stressed plants at both the fruit development and ripening stages (Shi-Lin et al., 2014).

Worldwide, heat, often in combination with drought or other types of stresses, has been demonstrated to result in huge damage to agricultural crops (Lamaoui et al., 2018). Therefore, high temperatures can significantly affect capsaicin content depending on the stages of maturity of the pepper fruit (Oh and Koh, 2019). High temperatures show negative influence on the accumulation of capsaicinoids in certain varieties of chili peppers, such as jalapeno and de arbol peppers, with a decrease in total capsaicinoids by 61.5% and 32.5%, respectively. In contrast, guajillo and serrano varieties showed a threefold increase, while in puya and ancho, varieties a positive increase of 21% and 8.6%, respectively, in the total capsaicinoids was reported. Tiwari et al. also noted a decrease in capsaicin and dihydrocapsaicin in the naga chili by about 50% under thermal stress. These data demonstrate that thermal effect on capsaicinoid content of peppers does not show a consistent pattern and could be influenced by a combination of age, size and development stage of the fruit as well as genetic regulation and environmental interaction. Higher temperatures also resulted in a rapid change in fruit colour (González-Zamora et al., 2013). A stimulatory effect of elevated CO_2 at higher temperature in maintaining the quality of *Capsicum* fruits was demonstrated by Das et al. (2016), wherein a higher capsaicin content was observed as a result of interaction effect of the two treatments.

High soil salinity poses a serious challenge to horticultural crops, causing a 30% to 50% reduction in productivity. A study by Urrea-Lopez et al. (2014) in habanero peppers (*Capsicum chinense*, Jacq.) showed no significant difference in the capsaicin and dihydrocapsaicin content nor in the content of carotenoids and phenolics compared to the controls at high salinity conditions (179.9 mg/100 g at 7 dS·m–1). Pepper plants have, however, been shown to vary in salt tolerance from highly sensitive to tolerant depending on their genetic makeup (Bojórquez-Quintal et al., 2014). There is a great scope for the application of biotechnological and genetic engineering approaches to produce crop plants tolerant to salt stress (Parmar et al., 2017).

Fresh peppers cannot be stored for extended periods due to high losses in the production areas and consumption centres and during storage and transportation (Sanatombi and Sanatombi, 2019). Thus, several post-harvest preservation processes have been applied to chili peppers. These processes can, however, alter the content of major components, thus impacting quality (Sanatombi and Sanatombi, 2019). Temperature and water activity greatly influence change in the colour of red pepper powder. As temperature and water activity increase, the red colour of pepper powder gradually diminished to brown and dull black, largely due to the breakdown of carotenoid pigments and formation of brown compounds (Rhim and Hong, 2011). One of the oldest techniques used for preservation of peppers is sun drying (Akintunde and Afolabi, 2010). Other pre-treatment techniques include blanching, chemical treatments and cooking procedures such as boiling, steaming and roasting. Dehydration, heating, canning and cooling are the main industrial techniques for processing of peppers (Sanatombi and Sanatombi, 2019). Macronutrients are preserved during hot drying, while micronutrients, polyphenols, flavonoids, antioxidants and the spicy hotness are lowered. Freezing preserves sensorial attributes, vitamin C, chlorophyl and carotenoids, but organic acids are degraded. A decrease in capsaicinoid content, phenols and flavonoids was noted during boiling, pressure cooking and pasteurization. Dried chili has satisfactory levels of vitamin A, known to have good stability under varying conditions of heating and drying. Blanching and drying treatments produced minor changes in essential oil composition but had a major effect on pepper colour (Sanatombi and Sanatombi, 2019). Domestic cooking methods, like boiling, baking, stir-frying, deep-frying, steaming, roasting and microwaving, had a high retention of α-tocopherol, β-carotene and α-tocotrienol in unripe hot red pepper samples, with deep-frying showing the highest retention (Jia et al., 2017). During drying and storage, colour stability, a very important quality parameter, was maintained in paprika powder. Processing of sweet peppers (red, yellow and green) to puree samples had minimal effect on colour, with carotenoids, phenols and flavonoids being retained, while ascorbic acid decreased (Kaur and Kaur, 2021). Capsaicin and dihydrocapsaicin and ascorbic acid in different samples of *C. annum* and *C. chinense* pepper powders was shown to decrease after roasting, while phenolics and flavonoids increased in most samples (Hamed et al., 2019). Freezing preserved capsaicinoids, vitamin C, chlorophyll, carotenoid levels and sensory properties and enhanced bioaccessibility of capsanthin, zeaxanthin, anteraxanthin and violaxanthin. Capsaicin losses in red pepper to an extent of 18% to 36% by boiling (10 to 20 min.) was noted with the effect depending on the ripening stage of pepper and the type of capsaicinoid. Increase in intensity of heat treatments including boiling also led to lowered bioaccessibility of capsaicinoids, capsanthin and zeaxanthin (Sanatombi and Sanatombi, 2019).

1.7 NUTRITIONAL AND FUNCTIONAL PROPERTIES

Green chili fruits are used as vegetables, while the dried ripe fruits find use as spice due to their pungency and striking flavours (Chakrabarty et al., 2017). Safety and effectiveness of capsaicin as a topical analgesic agent in the management of arthritis pain, herpes zoster–related pain, diabetic neuropathy, mastectomy pain and headaches have been demonstrated (Salehia et al., 2018). Capsaicin reduces blood sugar and intestinal problems and aids in improving heart health and in preventing stroke. Together with the presence of other bioactive compounds (violaxanthin, lutein, ferulic acid, sinapic acid etc.), chillis can be considered functional foods with beneficial effects when present in the diet (Chakrabarty et al., 2017). Fresh green and red chilies have a high vitamin C content. They also contain appreciable amounts of other antioxidants such as vitamin A, niacin, pyridoxine (vitamin B6), riboflavin and thiamin (vitamin B1), carotenoids such as β-carotene, α-carotene, lutein, zeaxanthin and cryptoxanthin, and flavonoids. Different minerals such as potassium, manganese, iron and magnesium are also present in chillis. Thus, chillis provide nutritional value to foods in which they are added (Chakrabarty et al., 2017).

1.8 MEDICINAL AND HEALTH BENEFITS

Pepper is known to have diverse preventive and therapeutic properties and finds use in the treatment of various cancers, rheumatism, stiff joints, bronchitis, cough, headache, arthritis and heart arrhythmias and is used as stomachic (Saleh et al., 2018). Chili preparations for disease treatment include standard capsaicin, pharmaceutically prepared gels, creams and plasters, essential oils, powdered pods and water or ethanolic extracts (Saleh et al., 2018).

1.8.1 ANTI-INFLAMMATORY EFFECTS

Capsaicin has the ability to inhibit substance P, a neuropeptide involved in inflammatory processes. It lessens inflammation by increasing blood flow to the site of injury. Capsaicin is effectively used in the treatment of pain associated with inflammatory arthritis, psoriasis and diabetic neuropathy. The antioxidant and anti-inflammatory activities have also been correlated with flavonoids and total phenolic content in chilli. Topical application of capsaicin, either as a cream or plaster, had a short-term effect in reducing nonspecific lower-back pain (Chakrabarty et al., 2017).

1.8.2 CARDIOVASCULAR HEALTH

Red chili such as cayenne has the potential to reduce blood cholesterol, triglyceride levels and platelet aggregation and to enhance the ability of the body to dissolve fibrin, which is primarily responsible for the formation of blood clots. The spice also lowers blood pressure, heart attack frequency, stroke, and pulmonary embolism and prevents fat deposition in the walls of blood vessels triggered by free radicals, thereby reducing the incidence of atherosclerosis (Chakrabarty et al., 2017).

1.8.3 IMMUNITY BOOSTER

Chillis contain high content of β-carotene or provitamin A and a high amount of vitamin C, which are known to play a role in strengthening the body's immune system. Vitamin A is important for maintaining a healthy mucous membrane, which lines the nasal passages, lungs, intestinal tract and urinary tract, and is the first line of defence against invading pathogens in the body. Capsaicin also plays an important role in boosting the defence mechanism of the body (Chakrabarty et al., 2017).

1.8.4 ALLEVIATES PROSTATE CANCER

Capsaicin has been shown to prevent the proliferation of prostate cancer cells. The compound initiates death in primary types of prostate cancer cell lines by mechanisms that are stimulated with or without male hormones. In addition, capsaicin reduces the expression of prostate-specific antigen (PSA) by inhibiting the ability of dihydrotestosterone to activate PSA, thereby directly inhibiting PSA transcription and causing a consequent reduction in PSA levels (Chakrabarty et al., 2017).

1.8.5 PREVENTION OF STOMACH ULCERS

Capsaicin prevents stomach infection caused by *Helicobacter pylori* that cause gastric ulcers by promoting alkaline conditions, mucus secretions and gastric mucosal blood flow, aiding in prevention and healing of ulcers. It also provides protection against injury-causing agents in the stomach by stimulating afferent neuronal signals (Chakrabarty et al., 2017).

1.8.6 REDUCES RISK OF TYPE 2 DIABETES AND OBESITY

Chili pepper reduces the risk of high insulin levels in blood (hyperinsulinemia), a disorder linked to type 2 diabetes. When taken as a regular part of the diet, a positive effect was observed as body mass index increases with a lowering in insulin requirement. A chili-containing diet also resulted in a lower ratio of C-peptide to insulin, indicating an increased clearance of insulin in the liver. Besides capsaicin, other antioxidants present including vitamin C and carotenoids also assist in improving insulin regulation (Chakrabarty et al., 2017).

1.8.7 ANTIRHINITIS AGENT

Capsaicin reduces nasal congestion by clearing mucus from the nose. It can also act against chronic sinus infections due to its antibacterial properties. It releases allergens from the nose due to its anti-inflammatory activity and thus opens up sinus passages (Chakrabarty et al., 2017).

1.9 NOVEL AND DIVERSE PRODUCTS OF PEPPER

Peppers are mainly traded internationally as fresh produce, as spices or as colorants. A significant portion (5%–30 %) of the total fruit produced, which includes peels, stalks, seeds and unused flesh generated during various stages of industrial processing, is discarded as non-marketable products. These by-products, sources of nutrients and secondary metabolites, can be used as alternatives to synthetic and chemical additives for industrial application, thereby contributing to low-waste technology and economic benefits to producers (Baenas et al., 2019). While bioactive compounds present in *Capsicum* peppers and their by-products have various health benefits, these compounds are susceptible to oxidation, have low bio-accessibility and possess a spicy sensory profile (in the case of hot peppers). In order to circumvent these drawbacks, techniques such as encapsulation have been applied to develop food-grade formulations enriched in *Capsicum* bioactive compounds that are safe and have improved bioaccessibility. Encapsulates obtained have high water solubility, low or high pH resistance, protection of volatiles and enhanced bioavailability and are appealing. In combination with suitable wall materials, spray drying produces particles resistant to digestion phases and with control release of specific bioactives (Aguiar et al., 2022).

Capsicum oleoresin emulsions have also been formulated with various techniques to allow better bioaccessibility of compounds and less spicy taste (Aguiar et al., 2022). Phenolic acid derivatives and flavonoids are a major group of phenolic compounds in pepper varieties that show health-promoting properties mainly due to their antioxidant effects. Bioavailability of phenolic compounds is a major concern for beneficial effect of phenolics in human health. In this regard, development of nano-formulations enriched with phenolic compounds administered either orally or topically can increase absorption and synergistic effects of phenolics (Baenas et al., 2019). In combination with certain drugs, they would represent a new line of research and innovation in the formulation of pepper-derived products. With respect to the carotenoids present in peppers such as β-carotene, α-carotene, β-cryptoxanthin, lycopene, lutein and zeaxanthin, incorporation into micelles, mixed with lipoproteins, bile salts and enzymes, improves bioavailability for delivery to gut enterocytes (Molteni et al., 2022).

Green technologies are being sought for preparation of oleoresins from spices. In this direction, use of supercritical carbon dioxide as an extraction process has gained prominence in enabling separation of it into enriched fractions of capsaicinoids, carotenoids and vitamins. Many pepper products available, such as paprika powders, green dehydrated pepper and jalapeño granular dehydrated pepper, are recently used as colour enhancers in the food industry (Baenas et al., 2019). Nanoparticles of paprika oleoresin, when added to meat using water-based carrier systems, resulted in deeper colour penetration and improved colour quality of cooked marinated chicken, thus improving efficiency of marination (Baenas et al., 2019).

Pepper-derived products, such as pepper extracts and seeds that are discarded as waste, can contribute to the development of new and safe antimicrobial and antioxidant agents for food preservation and as replacements to synthetic preservatives that occasionally produce allergic reactions and potential formation of nitrosamines (Baenas et al., 2019). Red bell pepper powders available in diced or granulated formats, when added to foods as antioxidants, substantially reduce cholesterol and docosahexaenoic (DHA) acid degradation during cooking, thereby preventing formation of toxic oxidation products and off-flavours (Baenas et al., 2019). Paprika oleoresin has recently been used as a cosmetic colourant in bath oils, shampoo, soaps, shower gels and several beauty products, including eye makeup and lipstick. The bioactive compounds obtained from pepper by-products can be also used as flavouring or fragrance agents in cosmetic products (Baenas et al., 2019). Some black or violet pepper varieties contain water-soluble anthocyanin pigments, mainly delphinidin, in the free and glycosidic precursors. These pigments find application as colourants and as therapeutic compounds (UV protectors and antioxidants) in the cosmetic pharma industry. Application of natural extracts of red pepper by-products for dyeing woolen fabrics to with acceptable antimicrobial properties has also been reported (Baenas et al., 2019).

1.10 POST-HARVEST PROCESSING FOR IMPROVED PRODUCTION OF BIOACTIVE COMPOUNDS

Colour and pungency are the principal quality attributes of *Capsicum* varieties. Temperature, packaging and storage period can affect the content of capsaicinoids, total carotenoids, ascorbic acid and total phenolics in pepper (Sanatombi and Sanatombi, 2019). The different processing methods generally applied for processing of peppers include physical (conventional sun drying, solar drying, smoke drying, refractance window drying and other thermal treatments) and chemical treatments (sodium chloride, calcium chloride, sucrose, acidification) that can alter its quality. A reduction in the content of ascorbic acid, capsaicinoids, phenolic compounds, protein and other important metabolites has been noted during sun drying, freeze drying, oven drying and other thermal treatments of processed peppers. Shade and room drying of peppers also resulted in a decrease in capsaicin, protein, ascorbic acid, iron and carbohydrate content. On the other hand, methods such as freeze drying, convective drying, osmotic dehydration, drying at 37°C and thermal treatment at low temperature for a short period reduced the degradation of antioxidants, ascorbic acid, protein, carbohydrate, iron and colour compounds of peppers. Refractance window drying is a recent technology and is considered more advantageous than sun drying and mechanical drying methods because it preserves the nutritional components of peppers (Sanatombi and Sanatombi, 2019).

Non-thermal processing technologies available for food processing, such as high-pressure processing, pulsed electric field, ultrasound, pulsed light, ultraviolet light, irradiation and oscillating magnetic field, are cold processes that provide maximum retention of colour, taste, appearance and nutrition content in food (Jadhav et al., 2021). Effects of ultrasound and pulsed light have been investigated for preservation of red bell peppers. Application of ultrasound technologies (US) facilitates enzyme activation or inactivation, surface decontamination, enhanced extraction of bioactive compounds, osmotic dehydration, freezing, drying etc. (Rybak et al., 2020). A pulsed electric field (PEF) causes reversible or irreversible changes in cell membrane due to the electroporation thereby reducing microbial load. When US and PEF treatment and their combinations were used, comparable or slightly lower contents of bioactive compounds (carotenes, vitamin C and phenolics) were obtained in comparison to fresh pepper. The contents were, however, significantly higher than the blanching treatment, which involves thermal processing (Rybak et al., 2020). Exposure to ultraviolet light, blue light–emitting diode and atmospheric plasma was demonstrated to maintain vitamin C, capsaicin and dihydrocapsaicin content in dried red chilli (Kalathil et al., 2022). The effect of gamma-irradiation on colour, pungency, and volatiles of Korean red pepper powder (*Capsicum annuum* L.) was investigated by Lee et al. (2004). A radiation dose of 7 kGy effectively reduced the population of mesophilic bacteria and fungi without affecting major quality factors such as pungency (estimated by capsaicinoid content), colour and sensory quality. Gamma irradiation prevented recontamination and extended the shelf life of pre-packed spice powder. Thus, non-thermal methods of preservation can prevent microbial contamination without affecting quality parameters, thereby ensuring safety and facilitating greater availability.

1.11 CONCLUSIONS

The genus Capsicum consists of approximately five domesticated species and 22 wild species are cultivated in warm-climate regions of the world. Peppers contain chemical constituents with nutritional value, taste, colour and aroma, among these carotenoids and capsaicinoids, contribute to the colour and pungency, respectively. Capsaicinoid biosynthetic pathway is encoded and regulated by key structural and regulatory genes suggesting that omics studies can shed light on knowledge of metabolic pathways. Peppers, and its products showed bioactivity against cancers, rheumatism, stiff joints, bronchitis, cough, headache, arthritis and heart arrhythmias. Post-harvest methods help in preservation of chili peppers which include among others, the non-thermal processing technologies to confer maximum retention of colour, taste, appearance and nutrition profile. Sustainable production and processing of peppers can contribute to various products and byproducts of significance to food, health and other industrial sectors.

REFERENCES

Adams, C. A., Zimmerman, K., Fenstermacher, K., Thompson, M. G., Skyrud, W., Behie, S., & Pringle, A. (2020). Fungal seed pathogens of wild chili peppers possess multiple mechanisms to tolerate capsaicinoids. *Applied and Environmental Microbiology*, *86*(3), e01697–19. https://doi.org/10.1128/AEM. 01697-19

Aguiar, A. C. D., Vigano, J., Anthero, A. G. D. S., Dias, A. L. B., Hubinger, M. D., & Martínez, J. (2022). Supercritical fluids and fluid mixtures to obtain high-value compounds from *Capsicum* peppers. *Food Chemistry: X*, *13*, 100228–100240. https://doi.org/10.1016/j.fochx.2022.100228

Akintunde, T. Y., & Afolabi, T. J. (2010). Drying of chili pepper (*Capsicum frutescens*). *Journal of Food Process Engineering*, *33*, 649–660. https://doi.org/10.1111/j.1745-4530.2008.00294.x

Antonio, A. S., Wiedemanna, L. S. M., & Veiga Junior, V. F. (2018). The genus Capsicum: A phytochemical review of bioactive secondary metabolites. *RSC Advances*, *8*, 25767–25784. https://doi.org/10.1039/c8ra02067a

Arce-Rodriguez, M. L., & Ochoa-Alejo, N. (2019). Biochemistry and molecular biology of capsaicinoid biosynthesis: Recent advances and perspectives. *Plant Cell Reports*, *38*(9), 1017–1030. https://doi.org/10.1007/s00299-019-02406-0

ASTA. (1986). *Official Analytical Methods of the American Spice Trade Association*, 2nd ed. ASTA, Washington, DC.

ASTA. (2018). http://www.astaspice.org/food-safety/astasanalytical-methods-manual/

Aza-González, C., Núñez-Palenius, H. G., & Ochoa-Alejo, N. (2011). Molecular biology of capsaicinoid biosynthesis in chili pepper (*Capsicum* spp.). *Plant Cell Reports*, *30*(5), 695–706.

Baenas, N., Belović, M., Ilic, N., Moreno, D. A., & García-Viguera, C. (2019). Industrial use of pepper (*Capsicum annum* L.) derived products: Technological benefits and biological advantages. *Food Chemistry*, *274*, 872–885. https://doi.org/10.1016/j.foodchem.2018.09.047

Bhutia, K. L., Khanna, V. K., Meetei, T. N. G., & Bhutia, N. D. (2018). Effects of climate change on growth and development of chilli. *Agrotechnology*, *7*, 180. https://doi.org/10.4172/2168-9881.1000180

Bojórquez-Quintal, E., Velarde-Buendía, A., Ku-González, A., Carillo-Pech, M., Ortega-Camacho, D., Echevarría-Machado, I., & Martínez-Estévez, M. (2014). Mechanisms of salt tolerance in habanero pepper plants (*Capsicumchinense*Jacq.): Proline accumulation, ions dynamics and sodium root-shoot partition and compartmentation. *Frontiers in Plant Science*, *5*, 605–618. https://doi.org/10.3389/fpls.2014.00605

Bosland, P., & Votava, E. (2000). *Peppers: Vegetable and Spice Capsicums*. Crop Production Science in Horticulture Series 22, 2nd ed. CABI Publishing, Wallingford, UK.

Buckenhüskes, H. J. (2003). Current requirements on paprika powder for food industry. In: *Capsicum: The Genus Capsicum* (pp. 223–230). Ed. A. Krishna. Taylor and Francis Ltd., London.

Butnariu, M., & Samfira, I. (2016). Aspects relating to mechanism of the capsaicinoids biosynthesis. *Journal of Bioequivalence & Bioavailability*, *5*(5), 10000e32. https://doi.org/10.4172/jbb.10000e32

Chakrabarty, S., Islam, A. K. M. M., & Islam, A. K. M. A. (2017). Nutritional benefits and pharmaceutical potentialities of chili: A review. *Fundamental and Applied Agriculture*, *2*(2), 227–232.

Collins, M. D., Wasmund, L. M., & Bosland, P. W. (1995). Improved method for quantifying capsaicinoids in capsicum using high-performance liquid chromatography. *Horticultural Science*, *30*, 137e139.

Cserhati, T., Forgacs, E., Morais, M. H., Mota, T., & Ramos, A. (2000). Separation and quantitation of colour pigments of chili powder (Capsicum frutescens) by high-performance liquid chromatography–diode array detection. *Journal of Chromatography A*, *896*, 69–73.

Das, S., Das, R., Choudhury, H., & Saikia, A. (2016). Interactive effect of elevated carbon dioxide and high temperature on quality of hot chilli (*Capsicum chinense* Jacq.). *International Journal of Tropical Agriculture*, *34*(7), 1977–1981.

Fortune Business Insight. (2021). Spices and seasonings market size, share & COVID-19 impact analysis, by type (pepper, chili, ginger, cinnamon, cumin, turmeric, nutmeg and mace, cardamom, cloves, and others), by application (meat and poultry, bakery and confectionery, frozen food, snacks & convenience food, and others), and regional forecast, 2022–2029. In: *Market Research Report*, Report ID: FBI101694. https://www.fortunebusinessinsights.com/industry-reports/spices-and-seasonings-market-101694, access date May 1, 2023

González-Zamora, A., Sierra-Campos, E., Luna-Ortega, J. G., Pérez-Morales, R., Rodríguez Ortiz, J. C., & García-Hernández, J. L. (2013). Characterization of different capsicum varieties by evaluation of their capsaicinoids content by high performance liquid chromatography, Determination of pungency and effect of high temperature. *Molecules*, *18*, 13471–13486. https://doi.org/10.3390/molecules181113471

González-Zamora, A., Sierra-Campos, E., Pérez-Morales, R., Vázquez-Vázquez, C., Gallegos-Robles, M. A., López-Martínez, J. D., & García-Hernández, J. L. (2015). Measurement of capsaicinoids in Chiltepin hot pepper: A comparison study between spectrophotometric method and high performance liquid chromatography analysis. *Journal of Chemistry*, *2*, 10–20. https://doi.org/10.1155/2015/709150

Govindarajan, V. S., Rajalakshmi, D., Chand, N., & Salzer, U. J. (1987). Capsicum—production, technology, chemistry, and quality. Part IV. Evaluation of quality. *C R C Critical Reviews in Food Science and Nutrition*, *25*(3), 185–282. https://doi.org/10.1080/10408398709527453

Govindarajan, V. S., & Salzer, U. J. (1986). Capsicum-production, technology, chemistry and quality Part III. Chemistry of color, aroma and pungency stimuli. *C R C Critical Reviews in Food Science and Nutrition*, *24*(3), 245–355. https://doi.org/10.1080/10408398609527437

Hamed, M., Kalita, D., Bartolo, M. E., & Jayanty, S. S. (2019). Capsaicinoids, polyphenols and antioxidant activities of *Capsicum annuum*: Comparative study of the effect of ripening stage and cooking methods. *Antioxidants*, *8*, 364. https://doi.org/10.3390/antiox8090364.

Hassan, N. M., Yusof, N. A., Yahaya, A. F., Rozali, N. N. M., & Othman, R. (2019). Carotenoids of capsicum fruits: Pigment profile and health-promoting functional attributes. *Antioxidants*, *8*, 469–494. https://doi.org/10.3390/antiox8100469

Jadhav, H. B., Annapure, U. S., & Deshmukh, R. R. (2021). Non-thermal technologies for Food processing. *Frontiers in Nutrition*, *8*, 657090–657103. https://doi.org/10.3389/fnut.2021.657090

Jia, C. H., Shin, J. A., Kim, Y. M., & Lee, K. T. (2017). Effect of processing on composition changes of selected spices. *PLoS ONE*, *12*(5), e0176037. https://doi.org/10.1371/journal.pone.0176037

Kalathil, N., Thirunavookarasu, N., & Chidanand, D. V. (2022). Effect of nonthermal techniques (UV, LED and atmospheric plasma) on proximate composition, vitamin C and functional group of dried red chilli (*Capsicum annuum*). *The Pharma Innovation Journal*, *SP-11*(7), 4705–4709.

Kaur, R., & Kaur, K. (2021). Preservation of sweet pepper purees: Effect on chemical, bioactive and microbial quality. *Journal of Food Science and Technology*, *58*(9), 3655–3660. https://doi.org/10.1007/s13197-021-05075-8

Lamaoui, M., Jemo, M., Datla, R., & Bekkaoui, F. (2018). Heat and drought stresses in crops and approaches for their mitigation. *Frontiers in Chemistry*, *6*, 26–39. https://doi.org/10.3389/fchem.2018.00026

Lee, J. H., Sung, T. H., Lee, K. T., & Kim, M. R. (2004). Effect of gamma-irradiation on color, pungency, and volatiles of Korean red pepper powder. *Journal of Food Science*, *69*(8), C585–C592.

Mahmood, T., Rana, R. M., Ahmar, S., Saeed, S., Gulzar, A., Khan, M. A., & Du, X. (2021). Drought stress on capsaicin and antioxidant contents in pepper genotypes at reproductive stage. *Plants*, *10*, 1286–1298. https://doi.org/10.3390/plants10071286

Minguez-Mosquera, M. I., & Perez-Galvez, A. (1998). Color quality of paprika oleoresins. *Journal of Agricultural and Food Chemistry*, *46*, 5124–5127.

Molteni, C., La Motta, C., & Valoppi, F. (2022). Improving the bioaccessibility and bioavailability of carotenoids by means of nanostructured delivery systems: A comprehensive review. *Antioxidants*, *11*, 1931–1961. https://doi.org/10.3390/antiox11101931

Nathania, D. M., & Castello, E. B. A. G. (2020). The role of bioactive components found in peppers. *Trends in Food Science & Technology*, *99*, 229–243. https://doi.org/10.1016/j.tifs.2020.02.032.

Ogawa, K., Murota, K., Shimura, H., Furuya, M., Togawa, Y., Matsumura, T., & Masuta, C. (2015). Evidence of capsaicin synthase activity of the *Pun1*-encoded protein and its role as a determinant of capsaicinoid accumulation in pepper. *BMC Plant Biology*, *15*, 93–102. https://doi.org/10.1186/s12870-015-0476-7

Oh, S. Y., & Koh, S. C. (2019). Fruit development and quality of hot pepper (*Capsicum annuum* L.) under various temperature regimes. *Horticultural Science and Technology*, *37*(3), 313–321. https://doi.org/10.7235/HORT.20190032

Parmar, N., Singh, K. H., Sharma, D., Singh, L., Kumar, P., Nanjundan, J., Khan, Y. J., Chauhan, D. K., & Thakur, A. K. (2017). Genetic engineering strategies for biotic and abiotic stress tolerance and quality enhancement in horticultural crops: A comprehensive review. *Biotechnology*, *7*, 239–273. https://doi.org/10.1007/s13205-017-0870-y

Patrick, C. R. (2022). *Investigation into the Genetic Basis of Capsaicin Production in Peppers Using Next Generation RNA Sequencing and Synthetic Biology Approaches*. Theses and Dissertations, 3008. https://rdw.rowan.edu/etd/3008

Qin, C., Yu, C., Shen, Y., Fang, X., Chen, L., Min, J., & Luo, Y. (2014). Whole-genome sequencing of cultivated and wild peppers provides insights into Capsicum domestication and specialization. *Proceeding of the National Academy of Sciences, USA*, *111*, 5135–5140.

Rhim, J. W., & Hong, S. I. (2011). Effect of water activity and temperature on the color change of red pepper (*Capsicum annuum* L.) powder. *Food Science and Biotechnology*, *20*(1), 215–222. https://doi.org/10.1007/s10068-011-0029-2

Rybak, K., Wiktor, A., Witrowa-Rajchert, D., Parniakov, O., & Nowacka, M. (2020). The effect of traditional and non-thermal treatments on the bioactive compounds and sugars content of red bell pepper. *Molecules*, *25*, 4287–4303. https://doi.org/10.3390/molecules25184287

Saleh, B. K., Omer, A., & Teweldemedhin, B. (2018). Medicinal uses and health benefits of chili pepper (Capsicum spp.): A review. *MOJ Food Processing & Technology*, *6*(4), 325–328.

Salehia, B., Hernández-Álvarezc, A. J., Contreras, M. D. M., Martorell, M., Ramírez-Alarcóne, K., Melgar-Lalanne, G., & Sharifi-Rad, J. (2018). Potential phytopharmacy and food applications of *Capsicum* spp.: A comprehensive review. *Natural Product Communications*, *13*(11), 1543–1556.

Samarth, R. M., Meenakshi Samarth, M., & Matsumoto, Y. (2017). Medicinally important aromatic plants with radioprotective activity. *Future Science OA*, *3*(4), FSO247. https://doi.org/10.4155/fsoa-2017-0061

Sanatombi, K., & Sanatombi, R. (2019). Effect of processing on quality of pepper: A review. *Food Reviews International*, *36*(1), 1–18. https://doi.org/10.1080/87559129.2019.1669161

Scoville, W. L. (1912). Note on Capsicum. *Journal of the American Pharmacists Association*, *1*, 453e454.

Shi-Lin, T., Bo-Ya, L., Zhen-Hui, G., & Shaha, S. N. M. (2014). Effects of drought stress on capsanthin during fruit development and ripening in pepper (*Capsicum annuum* L.). *Agricultural Water Management*, *137*, 46–51.

Singh, S., Jarret, R., Russo, V., Majetich, G., Shimkus, J., Bushway, R., & Perkins, B. (2009). Determination of capsinoids by HPLC-DAD in *Capsicum* species. *Journal of Agricultural and Food Chemistry*, *57*, 3452–3457.

Song, S., Song, S. Y., Nian, P., Lv, D., Jing, Y., Lu, S., & Zhou, F. (2022). Transcriptomic analysis suggests a coordinated regulation of carotenoid metabolism in ripening chili pepper (*Capsicum annuum* var. conoides) fruits. *Antioxidants (Basel)*, *11*(11), 2245. https://doi.org/10.3390/antiox11112245

Spices Board of India (2023) .http://indianspices.com, access date May 1, 2023.

Stewart, C., Kang, B. C., Liu, K., Mazourek, M., Moore, S. L., Yoo, E. Y., Kim, B. D., Paran, I., Molly, M., & Jahn, M. M. (2005). The *Pun1* gene for pungency in pepper encodes a putative acyltransferase. *The Plant Journal*, *42*, 675–688. https://doi.org/10.1111/j.1365-313X.2005.02410.x

Stoica, R., Moscovici, M., Tomulescu, C., & Babeanu, N. (2016). Extraction and analytical methods of capsaicinoids: A review. *Scientific Bulletin, Series F. Biotechnologies*, *XX*, 93–98.

Sung, B., Prasad, S., Yadav, V. R., & Aggarwal, B. B. (2012). Cancer cell signaling pathways targeted by spice-derived nutraceuticals. *Nutrition and Cancer*, *64*, 173–197. https://doi.org/10.1080/01635581.2012.630551

Szydelko, J., Szydelko, M., & Boguszewska-Czubara, A. (2017). Health-promoting properties of compounds derived from *Capsicum* sp. A review. *Herba Polonica*, *63*, 67–87.

Tajkarimi, M. M., Ibrahim, S. A., & Cliver, D. O. (2010). Antimicrobial herb and spice compounds in food. *Food Control*, *21*(9), 1199–1218.

Thampi, P. S. S. (2004). A glimpse of the world trade in Capsicum. Chapter 2. In: *Capsicum, the Genus Capsicum* (pp. 16–24). Ed. A. K. De. CRC Press Inc., Taylor & Francis Group, London.

Tiwari, A., Kaushil, M. P., Pandey, K. S., & Dangi, R. S. (2005). Adaptability and production of hottest chilli variety under Gwalior agro-climatic conditions. *Current Science*, *88*, 1545–1546.

Tradologie.com. *Spice Trade Globally.* https://www.tradologie.com/lp/spice-trade.html, access date May 1, 2023.

Urrea-Lopez, R., Diaz de la Garza, R. I., & Valiente-Banuet, J. I. (2014). Effects of substrate salinity and nutrient levels on physiological response, yield, and fruit quality of habanero pepper. *Horticultural Science*, *49*(6), 812–818.

Villa-Rivera, M. G., and Ochoa-Alejo, N. (2021). Transcriptional regulation of ripening in chili pepper fruits (*Capsicum* spp.). *International Journal of Molecular Sciences*. *22*(22). 12151. https://doi.org/10.3390/ijms222212151

Villa-Rivera, M. G., Martínez, O., & Ochoa-Alejo, N. (2022). Putative transcription factor genes associated with regulation of carotenoid biosynthesis in chili pepper fruits revealed by RNA-Seq co-expression analysis. *International Journal of Molecular Sciences*, *23*(19), 11774. https://doi.org/10.3390/ijms231911774

Zhang, B., Hu, F., Cai, X., Cheng, J., Zhang, Y., Lin, H., & Wu, Z. (2022). Integrative analysis of the metabolome and transcriptome of a cultivated pepper and its wild progenitor Chiltepin (*Capsicum annuum* L. var. glabriusculum) revealed the loss of pungency during capsicum domestication. *Frontiers in Plant Science*, *5*(12), 783496–783509. https://doi.org/10.3389/fpls.2021.783496

2 Genetic Diversity for Fruit Quality Traits in Chilli (*Capsicum* spp.)

Pavani, N., Satyaprakash Barik,
Naresh Ponnam and Madhavi Reddy K

CONTENTS

2.1 INTRODUCTION

Chilli or hot pepper belongs to the genus *Capsicum*. The different species and traditional varieties within this genus offer a wide range of flavours and heat levels, making this crop a versatile spice. Chillis have a rich history dating back thousands of years. They have been integral to the cuisines of regions as diverse as India, Mexico, Thailand and Hungary, shaping culinary traditions and cultural identities. They add flavour, heat and complexity to dishes and are staple in numerous recipes. Chilli is believed to have originated around 7000 B.C. in Mexico and Central America (Dhamodharan et al., 2022). There are about 35 species within the *Capsicum* genus, with five of them being domesticated, namely *C. annuum, C. chinense, C. pubescens, C. frutescens, C. pubescens* and *C. baccatum* (Carrizo García et al., 2016). The domesticated species of assam has been documented in the northeastern region of India as reported by Purkayastha et al. (2012). The wild ancestor of chilli is *C. annuum var. glabriusculum* (Aguilar-Melendez et al., 2009). Chilli plants are primarily cultivated for their fruits, which are rich in capsaicinoids and carotenoids. These fruits have various industrial applications in the food, spice, phytofeed, pharmaceutical and cosmeceutical industries. Capsaicin and other capsaicinoids such as dihydrocapsaicin and nor-dihydro capsaicin have significant pharmaceutical applications for treating neurological complications, acting as musculoskeletal oxidative and to treat inflammatory diseases (Reviewed in, Barik et al., 2022).

India is the largest producer and exporter of chilli (hot pepper) globally, contributing 37.35% of the world's total chilli production (FAOSTAT 2020–21). The fruits of chilli are abundant in beneficial bioactive substances, which include carotenoids like capsanthin, capsaicin, capsorubin, zeaxanthin, lutein and violaxanthin along with Vitamin C, Vitamin E and capsaicinoids. A significant number of these components function as antioxidants and play a role in neutralizing harmful free radicals (Padayatty et al., 2003). One of the most important pepper-related industries is paprika, with 70% of its use as a spice in soups, sauces and meat products. Paprika is typically derived from mild or non-pungent pepper varieties, although completely non-pungent fruits are preferred in the international market. Its distinct red colour enhances its commercial value, and it has been recognized as a food additive (natural dye) (EFSA, 2015). It also serves as a nitrite replacer and inhibitor

DOI: 10.1201/9781003378259-2

of lipid oxidation in pork products. The by-product of pepper oleoresin production known as chilli spent residue (CHSR) holds significant value owing to its rich dietary fibre content (44.4% total dietary fibre, with 34.9% insoluble dietary fibre and 9.5% soluble dietary fibre) and protein content (19.7%). CHSR can be incorporated into the bread manufacturing industry to enhance the physical, sensory and nutritional attributes of bread (Sowbhagya et al., 2015). Chillis are rich sources of carotenoids like α- and β-carotene and β-cryptoxanthin essential for provitamin-A activity (IOM, 2001). α-carotene is particularly abundant in red peppers, while β-cryptoxanthin prevails in yellow and orange varieties. The antioxidant properties in chilli fruits, especially Vitamin C, can exceed the recommended daily intake by up to fivefold (Mohd Hassan et al., 2019; de Sá Mendes and de Andrade Gonçalves, 2020). Moreover, bioactive compounds derived from pepper by-products serve as flavouring and fragrance agents in cosmetic products. Chilli seeds containing 16% to 25% oil, predominantly rich in linoleic acid (68%–72%) and palmitic acid (13%–15%) serve various applications in healthcare and antiaging. Notably, purple-colored peppers contain anthocyanin delphinidin, a significant source of UV protection and antioxidants (Review; Barik et al., 2022).

2.2 PEPPER DIVERSITY, DOMESTICATION AND CONSERVATION

Chilli peppers are native to the Americas, and their genetic diversity is particularly rich in South America. The Andean region, which encompasses parts of present-day Peru, Bolivia, Ecuador and Colombia, is considered one of the primary centres of chilli pepper diversity (D'Arcy et al., 1974). The diverse landscapes, climates and altitudes in the Andes have contributed to the evolution of various *Capsicum* species and landraces. South America is home to multiple *Capsicum* species including *C. annuum*, *C. chinense*, *C. baccatum*, *C. pubescens* and *C. frutescens*. Each species exhibits unique flavours, heat levels and growth habits, contributing to the overall diversity of chilli peppers in the region. Among the domesticated species, *C. annuum var. annuum* is extensively cultivated. The genus also encompasses approximately 37 to 40 wild varieties/species (Barboza et al., 2022; Baral and Bosland, 2002; Khoury et al., 2019; van Zonneveld et al., 2018). The genus C. *annuum* includes mild bell peppers, moderately spicy jalapenos and the flavourful poblano peppers, among others. This highly diverse annual herb or shrub is tall and erect growing up to 1–1.5 m in height and is branched. The fruit type is a berry varying in shape, size, colour (green/yellow/purplish in immature stage and red/brown/orange in mature stage). *Capsicum frutescens* includes tabasco peppers, known for their fiery heat and tiny fruits but potent nature. Thai bird's eye chillis are prominent members of this species. As per recent taxonomical classification, it is considered a variety of *Capsicum annuum*, namely *C. annuum var. glabriusculum* Heiser and Pickersgill (TSN -30493; ITIS, 2022). *Capsicum chinense* is renowned for their intense spiciness including habanero peppers and scotch bonnet peppers. Leaves of *C. chinense* are unique crinkled leaves; they are medium-green to light-green ovate leaves with acute leaf apex that differentiates *C. chinense* from other chilli species (Lim and Tim, 2013). *C. annuum*, *C. frutescens* and *C. chinense* were identified to have elevated levels of L-ascorbic acid and carotenoids at maturity, providing 124% to 338% of the RDA for vitamin C and 0.33–336 retinol equivalents (RE) per 100 grams of provitamin A activity, respectively (Howard et al., 2000). *Capsicum baccatum* includes aji peppers, which offer a unique fruity flavour, while lemon drop peppers provide a citrusy zing. It is a perennial herb/shrub, erect and often branched, growing up to 4.5 m. Plants are characterised by slightly striate and densely pubescent branches. Aji peppers contain coniferyl esters, specifically coniferyl (E)-8-methyl-6-nonenoate (known as capsiconiate) and coniferyl 8-methyl nonanoate. Capsanthin 5,6-epoxide in aji pepper made a notable contribution to the total carotenoid content, ranging from 11% to 27%, which was considerably higher compared to both *C. pubescens* and red *C. annuum* varieties (Kobata et al., 2008). The fifth species, *Capsicum pubescens*, includes rocoto peppers, known for their distinct hairy leaves and robust, earthy flavour. It is a perennial shrub that grows up to 4 m in woody plants and lives up to 15 years, which subsequently gives the plant a treelike appearance. The flowers are characterized by blue-violet-coloured petals with a bright centre along with purple and white anthers. Seeds are dark and rugose. The rocoto pepper exhibited the most significant contributions of provitamin A and carotenoids to the carotenoid fraction (Lim, 2012).

The *Capsicum annuum* peppers are the most widely cultivated in all the continents. *C. frutescens* and *C. chinense* are cultivated in Central America, West and Central Africa and Oceania. The cultivation of *C. baccatum* and *C. pubescens* is currently limited to Central and South America and the Caribbean. Recently, a distinctive domesticated species in Assam has been identified from the northeastern part of India. There are several landraces evolved in India, which have wider applications in crop improvement programmes. The northeastern part of India has a rich natural heterogeneity because of its unique geographic location, where diverse landraces of crops, including *Capsicum* spp., are traditionally produced. Mizoram and Manipur, two northeastern Indian states, exhibit the highest diversity of *Capsicum frutescens* (Dutta et al., 2017). The preservation of *Capsicum* diversity through seed conservation appears viable on a global scale using existing conservation centres, such as the Svalbard Global Seed Vault (SGSV), World Veg, Centro Agronómico Tropical de Investigación y Enseñanza (CATIE) and various national gene banks. According to global genetic resources databases, approximately 18% to 37% of *Capsicum* varieties worldwide are presently safeguarded through duplication (Tripodi et al., 2021). Markedly, the SGSV alone potentially accounts for more than 6,000 *Capsicum* accessions, potentially constituting 13% of the global total. The Genesys plant genetic resources portal encompasses information on about approximately 4 million gene bank accessions, which is nearly half of the projected global total in approximately 450 institutions. Numerous collections are reported through genetic resource networks that represent diverse organizations. Almost 2 million accessions in Genesys originate from the European Cooperative Programme for Plant Genetic Resources (ECPGR) through its online database, EURISCO. Furthermore, national gene banks and many international institutions (WorldVeg, CGIAR gene banks and CATIE) also provide data to the same portal. The Genesys portal reports 32,304 active *Capsicum* accessions, encompassing 17 species and distributed across 73 institutes in 40 countries. WorldVeg, Centre for Genetic Resources (Netherlands), USDA Plant Genetic Resources Conservation Unit, Leibniz Institute of Plant Genetics, Crop Plant Research (IPK) in Germany, two Embrapa institutes in Brazil and the Institute for Agrobotany (RCA) in Hungary are among the institutions that are conserving seven or more *Capsicum* species. *Ex-situ* conservation of *Capsicum* spp. involves preserving these plants outside their natural habitats and is crucial for safeguarding their genetic diversity and ensuring their availability for future research, breeding and cultivation. Out of the 40 institutions surveyed, they report the *ex-situ* conservation of a total of 50,132 *Capsicum* accessions across 27 different countries (Global Crop Diversity Trust, 2022).

Worldwide, the networks and initiatives relevant to *Capsicum* genetic resources are G2P-SOL Network, European Cooperative Programme for Plant Genetic Resources (ECPGR), Simposio de Recursos Genéticos para América Latina y el Caribe (SIRGEALC) and Plant Genetic Resources Management Working Group of the African Union. *In-situ* conservation of *Capsicum* spp. is essential for preserving the natural diversity of these plants and their ecosystems. It complements *ex-situ* conservation efforts and plays a critical role in maintaining the resilience and adaptability of *Capsicum* spp. in the face of environmental challenges. The IUCN (International Union for Conservation of Nature) Red List of Threatened Species presently classifies eight *Capsicum* taxa in which *C. annuum* is categorized as least concern, but *C. caatingae, C. eximium, C. frutescens* and *C. rhomboideum* are also classified as least concern despite declining populations of *C. frutescens* due to various factors. Conversely, *C. lanceolatum, C. eshbaughii* and *C. tovarii* are listed as endangered.

Biotechnological interventions have facilitated the establishment and maintenance of chilli gene banks, preserving a diverse range of chilli germplasm. These resources serve as a genetic treasure trove for future breeding efforts. Through markers, potential identical diversity in Indian and Taiwan chilli gene pools was reported (Krishnamurthy et al., 2015). Diversity of chilli landraces from the northeastern part of India was evaluated using SSRs and genetic diversity determinants (Yumnam et al., 2012). Recently, the genetic makeup of 10,038 pepper varieties from gene banks worldwide has been characterized through genotyping by sequencing (GBS). The genomic data revealed that there were as many as 1,618 duplicate accessions within and between gene banks. This duplication often occurred due to challenges in classifying peppers that result from interspecific hybridization, making their morphological identification difficult. Their results depicted pepper as a highly exclusive cultural commodity that spread rapidly across the globe through major trade routes, both maritime and terrestrial (Tripodi et al., 2021).

In India, several landraces have been recognized with GI tags. These GI tags are crucial in safeguarding the unique identities of chilli varieties associated with specific regions and promoting their recognition and protection (https://search.ipindia.gov.in/GIRPublic/). Naga mircha (chilli) secured the first GI tag with application number 109 (December 2, 2008) from Kohima, Nagaland. Guntur sannam chilli, renowned for its spiciness, acquired a GI tag with application number 143 (May 28, 2010), representing Guntur in Andhra Pradesh. Byadagi chilli, a famous variety from Karnataka, was granted a GI tag under application number 129 (January 27, 2011). Mizoram's mizo chilli secured a GI tag with application number 377 (March 23, 2015). Bhiwapur chilli from Nagpur, Maharashtra, was recognized with a GI tag, application number 473 (November 30, 2016). Khola chilli, originating from Goa, was awarded a GI tag under application number 618 (August 28, 2019). Hathei chilli from Manipur, dalle khursani from Sikkim, harmal chilli from Goa and edayur chilli from Malappuram, Kerala, all obtained their GI tags (September 14, 2021), with application numbers 592, 636, 642 and 662, respectively.

2.3 HYBRIDIZATION AND CROSSABILITY

Numerous intraspecific and interspecific crossings have been conducted to date (Pradeepkumar et al., 2003; Sunil and Rasheed, 1998; Naresh et al., 2016). It has been observed that viable hybrids can be produced through interspecific crosses within the *Capsicum annuum* complex, with varying success rates (Bosland and Baral, 2007; Padayatty et al., 2003; Jarret and Dang, 2004). However, no viable hybrids were obtained from crosses between *Capsicum annuum* and *Capsicum pubescens*. Variable rates of successful fertile hybridization have been achieved through crosses between wild and semi-domesticated *C. annuum var. glabriusculum* and *C. annuum var. annuum* (Hernández-Verdugo et al., 2012; Jarret and Dang, 2004; Guzmán et al., 2005). Although extensive crossability studies encompassing all species within the genus are still pending (Barchenger and Bosland, 2019), there have been documented instances of successful hybridizations between different species, including those from various complexes (Scaldaferro, 2019; Parry et al., 2021). Also, preliminary clades of *Capsicum* spp. have been identified through sequence-based markers within phylogenetic trees based on their position (García et al., 2016).

2.4 GENETIC RESOURCES AND BREEDING OF PEPPERS FOR INDUSTRIAL USE

Due to emerging trends in extensive use of bioactive compounds in various industries, there is extensive consideration for exploring germplasm and breeding peppers with high bioactive compounds. Naga jolokia, originally a local landrace, was identified as the hottest pepper in India (Mathur et al., 2000). This discovery sparked significant interest, and in the following decades, highly pungent genotypes often referred to as super-hot peppers were found and developed. Islam et al. (2018) found high variability of capsaicinoids in 92 accessions of *Capsicum* collected from northeastern states of India, *viz.* Arunachal Pradesh, Assam, Manipur and Sikkim, which ranged from 191,135 to 1,152,832 SHU. Continued exploration for potent sources of high pungency led to the discovery of Trinidad scorpion, a local variety originating from Trinidad and Tobago, which surpasses even bhut jolokia in heat. It is speculated that the exceptionally pungent strains of bhut jolokia, widely grown in northeastern India, might have sympatric origin (Rai et al., 2013). The natural diversity within and among these pungent landraces is conserved locally and should be systematically preserved *ex-situ* to ensure its long-term protection. There are numerous conventional uses for these landraces of chillies, including usage in various ethnic cuisines and application in ethnopharmacology (Bhutia et al., 2019). Butch T scorpion from Australia is another among the hottest known varieties, followed by Trinidad moruga scorpion. Trinidad and Tobago also grow various strains of Trinidad scorpion including Douglah Trinidad chocolate and Trinidad 7-pot Jonah, which are genetically diverse from bhut jolokia (Bosland et al., 2015). These super-hot peppers produce capsaicinoids located on the fruit's placenta. Interestingly, pungent landraces have additional vesicles on the fruit's outer layer (pericarp) in addition to the placenta, resulting in Scoville heat units (SHU) exceeding 1 million (Bosland et al., 2015). A Carolina pepper with SHU of 2.2 million SHUs was bred using "really nastily hot" cultivar and naga viper (Buchanan, 2020). Very recently, new cultivar Pepper X, with 3.18 million SHU, which was bred by Ed Currie, gained a Guinness record (Barik et al., 2022). These genetic resources will help in breeding peppers for industrial uses, mainly the pharmaceutical industry.

Apart from pungent capsaicinoids, peppers are the richest source of non-pungent analogs capsinoids (Kobata et al., 2008). Secondary metabolites, namely capsaicinoids, were identified in chilli fruits, including capsaicin, dihydrocapsaicin, homocapsaicin-I, homocapsaicin-II, homodihydrocapsaicin-I, homodihydrocapsaicin-II, nonivamide and nordihydrocapsaicin. Their concentrations varied from 0.5 to 3,600 micrograms of capsaicin equivalent per gram of product (Kozukue et al., 2005). Several sources rich in capsinoids such as CH-19 sweet, belice sweet (*C. annuum*), S3212 (*C. frutescens*) and SNU11–001 (*C. chinense*) have been reported (Yazawa et al., 1990; Tanaka et al., 2017; Park et al., 2015; Jang et al., 2022).

Pungency compounds are monogenically dominantly *(Pun1)* inherited (Deshpande, 1935; Blum et al., 2002). For pungency, Catf-1 was isolated and reported as the candidate gene (Lang et al., 2006), *C* locus for pungency was mapped on chromosome 2 (Blum et al., 2002), and several QTLs associated with capsaicinoids were identified (Han et al., 2018; Wyatt et al., 2012; Chakradhar et al., 2013). Recent studies showed that level of pungency is quantitively inherited, and genome-wide association analysis identified several SNPs associated with capsaicinoid components (Padma Nimmakayala et al., 2016). Candidate genes such as *C4H, AT Acyl-CoA synthetase (ACS), PAL, BCKDH* and *NADH-GOGAT* and five genes for pungency components have been identified (Zhang et al., 2018).

For carotenoid content, several landraces of chilli evolved, and one such is Byadgi Dabbi, which got the geographical indication certificate in India. Byadgi Dabbi is a native landrace originating from the Byadagi village in Haveri district, Karnataka. It has gained international renown for its deep, dark red color (200 ASTA) and its low or negligible level of pungency (Jhansi et al., 2021). For carotenoid content, several sources rich in carotenoids such Kt-Pl-19, SSP-1999 and ICBD-8/10 (Shiva et al., 2015), Tomato chilli-1 (Prasath et al., 2007), Byadagi hubli (Prasath and Ponnuswami, 2008) and California wonder (Arpacı et al., 2020) have been reported. These sources can be exploited for breeding high-color-value peppers. Sarpras et al. (2018) evaluated 136 different genotypes procured from the northeastern part of India for 65 fruit metabolites and found significant variations for fruit metabolites among the genotypes. Metabolite diversity was analyzed through GC-MS, and researchers found that the metabolites comprised of carboxylic acids, fatty acid and esters, hydrocarbons, aldehydes, terpenoids, alcohol and vitamin E. Specific genotypes rich in these metabolites were identified. Seventeen *C. chinense* genotypes were identified as highly pungent accessions with more than 0.9 million SHU, surpassing the previous reports of pungent peppers. This study indicated the existence of wide variability in metabolites, which can be explored further in breeding programmes for nutritionally rich chillies. Sarpras et al. (2018) studied phytochemical (nutrient and non-nutrient), antioxidant activities and mineral-element diversity in three *Capsicum* species at various fruit developmental stages and reported that higher levels of metabolites were observed at the mature stage of fruit development, and some health beneficial metabolites were exclusively found in *Capsicum chinense* accessions. Six candidate gene markers related to flavonoids and their corresponding gene expression data (expression QTLs) were identified by Wahyuni et al. (2014). They identified Ca-MYB12 transcription factor gene on chromosome 1 and the FS-2 gene on chromosome 6 as potential causal genes responsible for variation in naringenin chalcone and flavone C-glycosides in a population derived from *C. annum* × *C. chinense.*

Six candidate genes play a role in the carotenoid biosynthesis pathway, with Psy and Ccs being two of these genes that influence the color of ripe fruit (Hyo-Bong Jeong et al., 2020). All these investigations consistently indicate the absence or deletion of the Ccs gene in peppers with yellow and orange colouring. For instance, Bouvier et al. (1994) observed the lack of the Ccs gene in two yellow cultivars, golden summer and jaune de pignerolle. Lang et al. (2004); Popovsky and Paran (2000) identified the deletion of the Ccs gene in same cultivars having yellow fruits at positions 211 bp and 220 bp from the 3' end. But Ha et al. (2007) proposed that in yellow peppers, two structural mutations, such as premature stop codons or frame shifts, led to nonsense-mediated transcriptional gene silencing of the Ccs gene rather than complete deletion, resulting in the yellow ripening of *Capsicums*. In orange cultivars, Guzman et al. (2010) observed the presence of the Ccs gene coding region and identified a novel variant, Ccs-3, in the Fogo cultivar, which had a high beta carotene content. Ccs-3 had deletion of nucleotide base cytosine at the position 1283 bp, leading to a frame shift in the coding sequence and yielding a 423aa sequence. Kim et al. (2010) analyzed the genomic regions of PSY from TF68 and habanero, revealing 12 single nucleotide polymorphisms (SNPs) between them, including Asp13Glu and Gly62Arg. SNPs located in introns, i.e. 12th SNP (position 2683), occurred at a predicted splice acceptor site at the 3' end of the fifth intron and led to the use of a cryptic splice acceptor site present 29 bp downstream of the original dinucleotide in habanero PSY, resulting in a 29 bp deletion, a premature stop codon and, consequently, the production of a truncated PSY that led to the orange fruit colour. Jeong et al. (2019) employed SMRT to detect allelic differences in these six candidate genes in 94 accessions. They found a total of 12, 3, 7, 9, 16 and 10 novel alleles with mutated coding regions in the PSY1, PSY2, Lcyb, CrtZ-2, ZEP and CCS genes, which could serve as valuable resources to identify DNA markers for fruit colours. They emphasized the efficiency of SMRT sequencing because the targeted genes are relatively short, making it possible to obtain sequences in a single run (Rhoads and Au, 2015). Furthermore, GWAS was conducted for capsaicinoids in 94 *Capsicum annuum* accessions, revealing 30 and 56 SNPs associated with capsaicin and dihydrocapsaicin, respectively (Nimmakayala et al., 2016).

2.5 CONCLUSIONS

In conclusion, the remarkable diversity in *Capsicum* species is a valuable and often untapped resource for agricultural and culinary innovation. The benefits of exploring this diversity are numerous and far-reaching. Pepper diversity offers a palette of flavours, colours and heat levels that can cater to different culinary preferences and cultural traditions. This diversity in flavour and heat allows for the creation of unique and exciting dishes, enriching culinary experiences around the world. The cultivation of rare or specialty chilli varieties can open niche markets and opportunities for small-scale farmers and local economies. Additionally, the development of novel chilli varieties with desirable traits can enhance the competitiveness of the chilli industry in global markets. As the global demand for chilli continues to rise, biotechnology will likely play an increasingly vital role in ensuring the sustainability and diversity of this essential crop. Plant breeders have the opportunity to utilize the genetic diversity found in chilli to create novel and superior varieties that exhibit improved characteristics.

REFERENCES

Aguilar-Meléndez, A., Morrell, P. L., Roose, M. L., & Kim, S. C. (2009). Genetic diversity and structure in semiwild and domesticated chiles (Capsicum annuum; Solanaceae) from Mexico. *American Journal of Botany*, *96*(6), 1190–1202.

Arpacı, B. B., Baktemur, G., Keleş, D., Kara, E., Erol, Ü. H., & Taşkın, H. (2020). Determination of color and heat level of some resistance sources and improved pepper genotypes. *Crop Breeding, Genetics and Genomics*, *2*(1).

Baral, J. B., & Bosland, P. W. (2002). An updated synthesis of the *Capsicum* genus. *Capsicum Eggplant Newsletter*, *21*, 11–21.

Barboza, G. E., Carrizo García, C., de Bem Bianchetti, L., Romero, M. V., & Scaldaferro, M. (2022). Monograph of wild and cultivated chilli peppers (*Capsicum* L., Solanaceae). *PhytoKeys*, *200*, 1–423.

Barchenger, D. W., & Bosland, P. W. (2019). Wild chile pepper (*Capsicum* L.) of North America. Edited by Greene, S. L., Williams, K. A., Khoury, C. K., Kantar, M. B., Marek, L. F. *North American crop wild relatives, volume 2: Important species* (pp. 225–242). Cham, Switzerland: Springer.

Barik, S., Ponnam, N., Reddy, A. C., Dc, L. R., Saha, K., Acharya, G. C., & Reddy, M. (2022). Breeding peppers for industrial uses: Progress and prospects. *Industrial Crops and Products*, *178*, 114626.

Bhutia, K. L., Bhutia, N. D., & Khanna, V. K. (2019). Rich genetic diversity of *Capsicum* species in Northeast India, as a potential source for chilli crop improvement. *Journal of Agriculture and Forest Meteorology Research*, 2(2), 77–83.

Blum, E., Liu, K., Mazourek, M., Yoo, E. Y., Jahn, M., & Paran, I. (2002). Molecular mapping of the *C* locus for presence of pungency in *Capsicum*. *Genome*, *45*, 702–705.

Bosland, P. W., & Baral, J. B. (2007). 'Bhut Jolokia'—The world's hottest known chile pepper is a putative naturally occurring interspecific hybrid. *HortScience*, *42*(2), 222–224.

Bosland, P. W., Coon, D., & Cooke, P. H. (2015). Novel formation of ectopic (nonplacental) capsaicinoid secreting vesicles on fruit walls explains the morphological mechanism for super-hot chile peppers. *Journal of the American Society for Horticultural Science*, *140*(3), 253–256.

Bouvier, F., Hugueney, P., d'Halingue, A., Kuntz, M., & Camara, B. (1994). Xantophyll biosynthesis in chromoplasts: Isolation and molecular cloning of an enzyme catalyzing the conversion of 5,6- epoxycarotenoid into ketocarotenoid. *The Plant Journal*, *6*, 45–54.

Buchanan, M. (2020). Some like it hot. *Nature Physics*, *16*(1), 112–112.

Carrizo Garcıa, C., Barfus, M. H. J., Sehr, E. M., Barboza, G. E., Samuel, R., Moscone, E. A., & Ehrendorfer, F. (2016). Phylogenetic relationships, diversification and expansion of chilli peppers (*Capsicum*, Solanaceae). *Annals of Botany*, *118*, 35–51.

Chakradhar, T., Reddy, B. P., Srinivas, N., & Reddy, P. (2013). Identification and validation of an allele specific marker associated with pungency in Capsicum spp. *Current Trends in Biotechnology and Pharmacy*, *7*(1), 544–551.

D'Arcy, W. G., Eshbaugh, W. H. (1974). New World peppers (Capsicum–Solanaceae) north of Colombia: A resume. *Baileya*. 19, 93–105.

de Sá Mendes, N., & de Andrade Gonçalves, É. C. B. (2020). The role of bioactive components found in peppers. *Trends in Food Science & Technology*, *99*, 229–243.

Deshpande, R. B. (1935). Studies in Indian chillies. 4. Inheritance of pungency in *Capsicum annuum* L. *Indian Journal of Agricultural Sciences*, *5*, 513–516.

Dhamodharan, K., Vengaimaran, M., & Sankaran, M. (2022). Pharmacological Properties and Health Benefits of Capsicum Species: A Comprehensive Review. In: (Eds. Orlex B. Baylen Yllano) *Capsicum – New Perspectives*.

Dutta, S. K., Singh, S. B., Saha, S., Akoijam, R. S., Boopathi, T., Banerjee, A., ... & Roy, S. (2017). Diversity in bird's eye chilli (Capsicum frutescens L.) landraces of north-east India in terms of antioxidant activities. *Proceedings of the National Academy of Sciences, India Section B: Biological Sciences*, *87*, 1317–1326.

EFSA. (2015). Annual Report of the European Food Safety Authority for 2015 (pp. 1–54). https://www.efsa.europa.eu/en

FAOSTAT. (2021). https://www.fao.org/faostat//in

García, C. C., Michael, H. J. Barfuss, M. H. J., Eva M. Sehr, E. M., Barboza, G. E., Samuel, R., Moscone, E. A. & Ehrendorfer, F. (2016). *Annals of Botany*, *118*(1), 35–51.

Global Crop Diversity Trust. (2022). *Genesys global portal of plant genetic resources for food and agriculture.* Accessed February 9, 2022. https://www.croptrust.org/pgrfa-hub/data-and-information-systems/genesys/

Guzmán, F. A., Ayala, H., Azurdia, C., Duque, M. C., & De Vicente, M. C. (2005). AFLP assessment of genetic diversity of Capsicum genetic resources in Guatemala: Home gardens as an option for conservation. *Crop Science*, *45*(1), 363–370.

Guzman, I., Hamby, S., Romero, J., Bosland, P. W., & O'Connell, M. A. (2010). Variability of carotenoid biosynthesis in orange colored Capsicum spp. *Plant Science*, *179*(1–2), 49–59.

Ha, S. H., Kim, J. B., Park, J. S., Lee, S. W., & Cho, K. J. (2007). A comparison of the carotenoid accumulation in Capsicum varieties that show different ripening colours: Deletion of the capsanthin-capsorubin synthase gene is not a prerequisite for the formation of a yellow pepper. *Journal of Experimental Botany*, *58*(12), 3135–3144.

Han, K., Lee, H. Y., Ro, N. Y., Hur, O. S., Lee, J. H., Kwon, J. K., & Kang, B. C. (2018). QTL mapping and GWAS reveal candidate genes controlling capsaicinoid content in *Capsicum*. *Plant Biotechnology Journal*, *16*, 1546–1558.

Hernández-Verdugo, S., Porras, F., Pacheco-Olvera, A., López-España, R. G., Villarreal-Romero, M., Parra-Terraza, S., & Osuna Enciso, T. (2012). Caracterización y variación ecogeográfica de poblaciones de chile (Capsicum annuum var. glabriusculum) silvestre del noroeste de México. *Polibotánica*, *33*, 175–191.

Howard, L. R., Talcott, S. T., Brenes, C. H., & Villalon, B. (2000). Changes in phytochemical and antioxidant activity of selected pepper cultivars (*Capsicum spp*) as influenced by maturity. *Journal of Agricultural and Food Chemistry, 48*(5), 1713–1720.

Institute of Medicine (IOM). (2001). *Crossing the quality chasm: A new health care system for the 21st century*. National Academy Press, Washington, DC.

Islam, M. R., Sultana, T., Haque, M. A., Hossain, M. I., Sabrin, N., & Islam, R. (2018). Growth and yield of chilli influenced by nitrogen and phosphorus. *Journal of Agriculture and Veterinary Science, 11*(5), 54–68.

Jang, S., Kim, G. W., Han, K., Kim, Y. M., Jo, J., Lee, S. Y., & Kang, B. C. (2022). Investigation of genetic factors regulating chlorophyll and carotenoid biosynthesis in red pepper fruit. *Frontiers in Plant Science, 13*, 922963.

Jarret, R. L., & Dang, P. (2004). Revisiting the waxy locus and the Capsicum annuum L. complex. *Georgia Journal of Science, 62*(3), 118.

Jeong, H. B., Jang, S. J., Kang, M. Y., Kim, S., Kwon, J. K., & Kang, B. C. (2020). Candidate gene analysis reveals that the fruit color locus C1 corresponds to PRR2 in pepper (Capsicum frutescens). *Frontiers in Plant Science, 11*, 399.

Jeong, H. B., Kang, M. Y., Jung, A., Han, K., Lee, J. H., Jo, J., Lee, H. Y., An, J. W., Kim, S., Kang, B. C. (2019). Single-molecule real-time sequencing reveals diverse allelic variations in carotenoid biosynthetic genes in pepper (*Capsicum* spp.). *Plant Biotechnology Journal, 17*(6), 1081–1093.

Jhansi, B., Kalal, A. N., Nagnur, S., Nayak, M. R., & Babitha, B. (2021). Involvement of APMC women labourers in post-harvest activities of dry chilli-a comparative study in guntur and byadgi APMCs. *Current Topics in Agricultural Sciences, 4*, 131–138.

Khoury, C. K., Carver, D., Barchenger, D. W., Barboza, G., van Zonneweld, M., Jarret, R., Bohs, L., Kantar, M. B., Uchanski, M., Mercer, K., Nabhan, G. P., Bosland, P. W., & Greene, S. L. (2019). Modeled distributions and conservation status of the wild relatives of chile peppers (*Capsicum* L). *Diversity and Distributions, 26*(2), 209–225.

Kim, O. R., Cho, M. C., Kim, B. D., & Huh, J. H. (2010). A splicing mutation in the gene encoding phytoene synthase causes orange coloration in Habanero pepper fruits. *Molecules and Cells, 30*, 569–574.

Kobata, K., Tate, H., Iwasaki, Y., Tanaka, Y., Ohtsu, K., Yazawa, S., & Watanabe, T. (2008). Isolation of coniferyl esters from Capsicum baccatum L., and their enzymatic preparation and agonist activity for TRPV1. *Phytochemistry, 95*(5), 1179–1184.

Kozukue, N., Han, J. S., Kozukue, E., Lee, S. J., Kim, J. A., Lee, K. R., Levin, C. E., & Friedman, M. (2005). Analysis of eight capsaicinoids in peppers and pepper-containing foods by high-performance liquid chromatography and liquid chromatography-mass spectrometry. *Journal of Agricultural and Food Chemistry, 53*(23), 9172–9181.

Krishnamurthy, S. L., Prashanth, Y., Rao, A. M., Reddy, K. M., & Ramachandra, R. (2015). Assessment of AFLP marker based genetic diversity in chilli (*Capsicum annuum* L. & *C. baccatum* L.). *nopr.niscpr.res.in, 14*(1).

Lang, Y. Q., Yanagawa, S., Sasanuma, T., & Sasakuma, T. (2004). Orange fruit color in Capsicum due to deletion of capsanthin-capsorubin synthesis gene. *Breeding Science, 54*(1), 33–39.

Lang, Y. Q, Yanagawa, S., Sasanuma, T., & Sasakuma, T. (2006). A gene encoding a putative acyl-transferase involved in pungency of Capsicum. *Breeding Science, 56*(1), 55–62.

Lim, T. K. (2012). *Edible medicinal and non-medicinal plants* (Vol. 1, pp. 656–687). Springer, Dordrecht, The Netherlands.

Lim, T. K., & Lim, T. K. (2013). Capsicum chinense. Edible medicinal and non-medicinal plants: Volume 6, *Fruits*, 205–212.

Mathur, R., Dangi, R. S., Dass, S. C., & Malhotra, R. C. (2000). The hottest chilli variety in India. *Current Science, 79*(3), 287–288.

Mohd Hassan, N., Yusof, N. A., Yahaya, A. F., Mohd Rozali, N. N., & Othman, R. (2019). Carotenoids of capsicum fruits: Pigment profile and health-promoting functional attributes. *Antioxidants, 8*(10), 469.

Naresh, P., Reddy, M. K., Reddy, P. H. C., & Reddy, K. M. (2016). Screening chilli (*Capsicum spp.*) germplasm against Cucumber mosaic virus and Chilli veinal mottle virus and inheritance of resistance. *European Journal of Plant Pathology, 146*, 451–464.

Nimmakayala, P., Abburi, V. L., Saminathan, T., Alaparthi, S. B., Almeida, A., Davenport, B., & Reddy, U. K. (2016). Genome-wide diversity and association mapping for capsaicinoids and fruit weight in Capsicum annuum L. *Scientific Reports, 6*(1), 38081.

Padayatty, S. J., Katz, A., Wang, Y., Eck, P., Kwon, O., Lee, J.-H., Chen, S., Corpe, C., Dutta, A., Dutta, S. K., & Levine, M. (2003). Vitamin C as an antioxidant: Evaluation of its role in disease prevention. *Journal of the American College of Nutrition, 22*(1), 18–35.

Park, Y. J., Nishikawa, T., Minami, M., Nemoto, K., Iwasaki, T., & Matsushima, K. (2015). A low-pungency S3212 genotype of Capsicum frutescens caused by a mutation in the putative aminotransferase (p-AMT) gene. *Molecular Genetics and Genomics, 290*, 2217–2224.

Parry, C., Wang, Y.-W., Linand, S.-W., & Barchenger, D. W. (2021, March 2). Reproductive compatibility in capsicum is not necessarily reflected in genetic or phenotypic similarity between species complexes. Edited by Dengcai Liu. *PLoS ONE, 16*(3), e0243689.

Popovsky, S., & Paran, I. (2000). Molecular genetics of the y locus in pepper: Its relation to capsanthin-capsorubin synthase and to fruit color. *Theoretical and Applied Genetics, 101*, 86–89.

Pradeepkumar, T., Karihaloo, J. L., Archak, S., & Baldev, A. (2003). Analysis of genetic diversity in Piper nigrum L. using RAPD markers. *Genetic Resources and Crop Evolution, 50*, 469–475.

Prasath, D., & Ponnuswami, V. (2008). Breeding for extractable colour and pungency in Capsicum: A review. *Indian Academy of Sciences, 35*(1), 1–9.

Prasath, D., Ponnuswami, V., & Muralidharan, V. (2007). Evaluation of chilli (Capsicum spp.) germplasm for extractable colour and pungency. *Indian Journal of Genetics and Plant Breeding, 67*(1), 97–98.

Purkayastha, J., Alam, S. I., Gogoi, H. K., Singh, L., & Veer, V. (2012). Molecular characterization of 'Bhut Jolokia' the hottest chilli. *Journal of Biosciences, 37*, 757–768.

Rai, V. P., Kumar, R., Kumar, S., Rai, A., Kumar, S., Singh, M., . . . & Paliwal, R. (2013). Genetic diversity in Capsicum germplasm based on microsatellite and random amplified microsatellite polymorphism markers. *Physiology and Molecular Biology of Plants, 19*, 575–586.

Rhoads, A., & Au, K. F. (2015). PacBio sequencing and its applications. *Genomics, Proteomics & Bioinformatics, 13*(5), 278–289.

Sarpras, M., Chhapekar, S. S., Ahmad, I., Abraham, S. K., & Ramchiary, N. (2018). Analysis of bioactive components in Ghost chilli (*Capsicum chinense*) for antioxidant, genotoxic and apoptotic effects in mice. *Drug and Chemical Toxicology, 43*(2), 182–191.

Scaldaferro, M. A. (2019). Molecular cytogenetic evidence of hybridization in the "purple corolla clade of the genus *Capsicum* (*C. eximium* × *C. cardenasii*). *Plant Biosystems*, 1–7.

Shiva, K. N., Prasath, D., Gobinath, P., Leela, N. K., & Zachariah, J. T. (2015). Variability and character association in paprika and paprika alike chillies. *Journal of Spices and Aromatic Crops, 24*(1), 61–65.

Sowbhagya, H. B., Soumya, C., Indrani, D., & Srinivas, P. (2015). Physico-chemical characteristics of chilli spent residue and its effect on the rheological, microstructural and nutritional qualities of bread. *Journal of Food Science and Technology, 52*, 7218–7226.

Sunil, K. P., & Rasheed, A. (1998). Cross compatibility in five species of Capsicum. *Journal of Spices and Aromatic Crops, 7*(1), 35–38.

Tanaka, Y., Nakashima, F., Kirii, E., Goto, T., Yoshida, Y., & Yasuba, K. I. (2017). Difference in capsaicinoid biosynthesis gene expression in the pericarp reveals elevation of capsaicinoid contents in chilli peppers (Capsicum chinense). *Plant Cell Reports, 36*, 267–279.

Tripodi, P., Rabanus-Wallace, M. T., Barchi, L., Kale, S., Esposito, S., Acquadro, A., et al. (2021). Global range expansion history of pepper (*Capsicum* spp.) revealed by over 10,000 genebank accessions. *Proceedings of the National Academy of Sciences of the USA, 118*(34), e2104315118.

van Zonneveld, M., Larranaga, N., Blonder, B., Coradin, L., Hormaza, J. I., & Hunter, D. (2018). Human diets drive range expansion of megafauna-dispersed fruit species. *Proceedings of the National Academy of Sciences, 115*, 3326–3331.

Wahyuni, Y., Stahl-Hermes, V., Ballester, A. R., de Vos, R. C., Voorrips, R. E., Maharijaya, A., . . . & Bovy, A. G. (2014). Genetic mapping of semi-polar metabolites in pepper fruits (Capsicum sp.): Towards unravelling the molecular regulation of flavonoid quantitative trait loci. *Molecular Breeding, 33*, 503–518.

Wyatt, L. E., Eannetta, N. T., Stellari, G. M., & Mazourek, M. (2012). Development and application of a suite of non-pungency markers for the Pun1 gene in pepper (Capsicum spp.). *Molecular Breeding, 30*, 1525–1529.

Yazawa, S., Sato, T., & Namiki, T. (1990). Dwarfism and a virus-like syndrome in interspecific hybrids of genus *Capsicum. The Japanese Society for Horticultural Science, 58*, 935–943.

Yumnam, J. S., Tyagi, W., Pandey, A., Meetei, N. T., & Rai, M. (2012). Evaluation of genetic diversity of chilli landraces from North Eastern India based on morphology, SSR markers and the Pun1 locus. *Plant Molecular Biology Reporter, 30*, 1470–1479.

Zhong, Y., Cheng, Y., Ruan, M., Ye, Q., Wang, R., Yao, Z., Zhou, G., Liu, J., Yu, J., & Wan, H. (2021). High-throughput SSR marker development and the analysis of genetic diversity in *Capsicum frutescens. Horticulturae, 7*(7), 187.

3 Contribution of Interspecific Hybridization for Improvement in Quality and Productivity in Pepper

Kazım Mavi

CONTENTS

3.1 INTRODUCTION

Interspecies hybridization is used to improve properties by transferring certain traits, such as resistance to biotic and abiotic stress, from their wild relatives to cultivated plants. When applicable, this approach is a very effective method of gene transfer. In nature, 30% to 35% of flowering plant species are formed by interspecies hybridization followed by doubling of chromosome numbers (Stebbins, 1971). Based on interspecies hybridization, allopolyploids such as allotetraploids can also be used to double the chromosome number of F1 hybrids. Successful allopolyploid acquisition can result in the formation of a new genome combination or the creation of a previously nonexistent species. However, a great deal of effort may be required to hybridize cultivars and wild species. It is reported that the first interspecies hybridization with human hands was made between carnation (*Dianthus caryophyllus* L.) and *Dianthus barbatus* L. in 1717 (Stalker, 1980). Since then, thousands of interspecies hybridizations have been attempted, with very limited success. Chromosomal, genetic, cytoplasmic or mechanical isolation barriers between species can prevent successful interspecies crossing. It took nearly 100 years for plant breeders to breed a new plant species, triticale, from the hybridization of wheat (*Triticum aestivum* L.) and rye (*Secale cereale* L.) (Zillinsky, 1985). Interspecific hybridization has become an important goal for geneticists and plant breeders with its benefits and difficulties in success.

Interspecific hybridization in the *Solanaceae* family has been studied in some genera such as Solanum (Pandey, 1962), Nicotiana (Lammerts, 1930), Datura (Schieder, 1980), Petunia (Wijsman and De Jong, 1985), Lycium (Bernardello et al., 1995) and Physalis (Azeez and Faluyi, 2018). Amphidiploid (2n=36) F1 individuals were obtained from crossing *Nicotiana tabacum* (2n=48) and *Nicotiana glauca* (2n=24) species within the genus *Nicotiana* (Trojak-Goluch and Berbec, 2003).

DOI: 10.1201/9781003378259-3

However, interspecific hybridization in the Solanum genus in the *Solanaceae* family has been used successfully for some species (Fatokun, 1989).

The peppers cultivated today are in the genus *Capsicum*, which is one of the 98 genera of the *Solanaceae* family (Eshbaugh, 2012). The number of species in the genus *Capsicum* was revised to 47 with newly discovered species (Eshbaugh, 2012; Barboza et al., 2019; Barboza et al., 2020; Mavi, 2020; Barboza et al., 2022) (Table 3.1).

TABLE 3.1

Species in the Genus Capsicum, Culture Conditions, Chromosome Numbers, Distribution Areas

********	**Species**	*****	**CN****	**Distribution area*****	**References**
1a	*C. annuum* var. *annuum*	D	12	All around The World	Eshbaugh, 2012
1b	*C. annuum* var. *glabriusculum*	D	12	USA, North/Northeast Brazil	Carrizo Garcia et al., 2016
2a	*C. baccatum* var. *baccatum*	D	12	Colombia, Northern Argentina, Southeast Brazil	Carrizo Garcia et al., 2016
2b	*C. baccatum* var. *pendulum*	D	12	USA, Central and South America, India	Carrizo Garcia et al., 2016
2c	C. baccatum var. *praetermissum*	W	12	Paraguay, Southeast and Central Brazil	Carrizo Garcia et al., 2016
3	*C. benoistii*	W	?	Equator (endemic)	Barboza et al., 2019
4	*C. buforum*	W	13	Brazil: SP ve RJ	Eshbaugh, 2012
5	*C. caatingae*	W	12	Brazil: BA, PE, and North MG (endemic)	Carrizo Garcia et al., 2016
6	*C. caballeroi*	W	?	Bolivia (endemic)	Eshbaugh, 2012
7	*C. campylopodium*	W	13	Brazil: ES, MG, RJ (endemic)	Eshbaugh, 2012
8	*C. carrassense*	W	?	Southeast Brazil (endemic)	Barboza et al., 2020
9	*C. cardenasii*	W	12	Bolivia (endemic)	Eshbaugh, 2012
10	*C. ceratocalyx*	W	?	Bolivia (endemic)	Eshbaugh, 2012
11	*C. chacoense*	W	12	South Bolivia, Paraguay, North and Central Argentina	Eshbaugh, 2012
12	*C. chinense*	D	12	India, Central and South America, China and Japon	Eshbaugh, 2012
13	*C. coccineum*	W	?	Peru and Bolivia	Eshbaugh, 2012
14	*C. cornutum*	W	13	Brazil: RJ, SP (endemic)	Eshbaugh, 2012
15	*C. dimorphum*	W	?	Colombia and Equator	Eshbaugh, 2012
16	*C. dusenii*	W	?	Southeast Brazil	Eshbaugh, 2012
17	*C. eshbaughii*	W	12	South-Central Bolivia	Eshbaugh, 2012
18	*C. eximium*	W	12	South Bolivia and North Argentina	Eshbaugh, 2012
19	*C. flexuosum*	W	12	Paraguay, South and Southeast Brazil, Northeast Argentina	Eshbaugh, 2012
20	*C. friburgense*	W	13	Brazil: RJ (endemic)	Eshbaugh, 2012
21	*C. frutescens*	D	12	America, Africa, Asia	Eshbaugh, 2012
22	*C. galapagoense*	W	12	Galapagos Island (endemic)	Eshbaugh, 2012
23	*C. geminifolium*	W	?	Colombia, Equator, Peru	Eshbaugh, 2012
24	*C. hookerianum*	W	?	South Equator, Peru (endemic)	Eshbaugh, 2012
25	*C. hunzikerianum*	W	13	Brazil: SP (endemic)	Eshbaugh, 2012
26	*C. lanceolatum*	W	13	Mexico, Guatemala	Eshbaugh, 2012
27	*C. leptopodum*	W	?	Brazil	Eshbaugh, 2012
28	*C. longidentatum*	W	12	Brazil: BA, PE (endemic)	Carrizo Garcia et al., 2016
29	*C. longifolium*	W	13	Andean	Barboza et al., 2019
30	*C. lycianthoides*	W	13	Andean	Khoury et al., 2020
31	*C. minutiflorum*	W	?	Bolivia	Khoury et al., 2020
32	*C. mirabile*	W	13	Brazil: MG, RJ, SP (endemic)	Eshbaugh, 2012

(Continued)

TABLE 3.1 *(Continued)*

Species in the Genus Capsicum, Culture Conditions, Chromosome Numbers, Distribution Areas

****	**Species**	*	**CN****	**Distribution area****	**References**
33	*C. mirum*	W	?	Southeast Brazil (endemic)	Barboza et al., 2022
34	*C. muticum*	W	?	Central-eastern Brazil: RJ	Barboza et al., 2022
35	*C. neei*	W	?	Bolivia (endemic)	Barboza et al., 2019
36	*C. parvifolium*	W	12	Colombia, Venezuela, Northeast Brazil	Eshbaugh, 2012
37	*C. pereirae*	W	13	Brazil: ES, RJ, SP (endemic)	Eshbaugh, 2012
38	*C. piuranum*	W	13	Peru	Barboza et al., 2019
39	*C. pubescens*	D	12	Mexico, Central and South America	Eshbaugh, 2012
40	*C. rabenii*	W	12	Brazil, Paraguay	Barboza et al., 2022
41	*C. recurvatum*	W	13	Brazil: RJ, SP, PR, SC (endemic)	Eshbaugh, 2012
42	*C. regale*	W			Barboza et al., 2022
43	*C. rhomboideum*	W	13	Brazil: MG, RJ, SP (endemic)	Eshbaugh, 2012
44	*C. schottianum*	W	13	Brazil: MG, RJ, SP (endemic)	Eshbaugh, 2012
45	*C. scolnikianum*	W	?	Peru, Equator (endemic)	Eshbaugh, 2012
46	*C. tovarii*	W	12	Peru (endemic)	Eshbaugh, 2012
47a	*C. villosum* var. *villosum*	W	13	Brazil: MG, RJ, SP (endemic)	Carrizo Garcia et al., 2016
47b	*C. villosum* var. *muticatum*	W	13	Southeast Brazil	Carrizo Garcia et al., 2016

* D: Domesticated, W: Wild

** CN Chromosome number, ? Species with an unknown number of chromosomes

*** RJ Rio de Janeiro, SP São Paulo, MG Minas Gerais, ES Espirito Santo, BA Bahia, PE Pernambuco

**** The letters after the sequence numbers represent the subspecies within that species.

Five (*C. annuum* L., *C. frutescens* L., *C. baccatum* L., *C. chinense* Jacq. and *C. pubescens* Ruiz & Pav.) of them are domesticated. It was determined that the chromosome numbers of the species in the genus *Capsicum* changed as 2n=24 and 2n=26. There are 13 *Capsicum* species whose chromosome numbers have not been determined yet (Table 3.1). Although the genetic variation is wide within the *Capsicum annuum* species, this diversity is insufficient in some cases, especially in diseases such as root rot (*Phytophthora capsici*) and anthracnose (*Colletotrichum acutatum, C. capsici* etc.). The genetic diversity in pepper species has always been in the field of interest of breeders due to the differences in their resistance to biotic and abiotic stress conditions. For this reason, interspecies hybridization gains importance and is used by breeders to solve problems. Interspecies hybridization is the process of pollinating the female organ of one species with pollen of another species (Figure 3.1).

The first known cross-breeding study in peppers was carried out between *Capsicum annuum* and *C. frutescens* species by Halstead in 1912. Historically, various approaches (traditional and biotechnological) to interspecies crossbreeding in peppers have been used to overcome pollination and fertilization barriers between different species.

3.2 SYSTEMATIC ADVANCES IN INTERSPECIFIC HYBRIDIZATION

The first successful interspecific hybrids were obtained by crossing *Capsicum pubescens* and *C. cardenasii*. F1 hybrids in this study were found to have high fertility (Yakub and Smith, 1971). In recent studies, the species of the genus have been divided into three different clade in terms of their ability to interbreed. The first of these is *C. annuum* L. clade that includes *C. annuum* [var.

FIGURE 3.1 Crossing stages: (A) balloon flower shape suitable for emasculation, (B) after emasculation, (C) pollination with a different species, (D) flower isolation after pollination, (E) fertilization and labelling. (Original: Kazım Mavi)

annuum and *glabriusculum*], *C. frutescens* L., *C. chinense* Jacq., *C. chacoense* Hunz. and *C. galapagoense* Hunz. The second is *C. baccatum* clade that includes *C. baccatum* L. [var. *baccatum*, *praetermissum* and *pendulum* (Willd.) Eshbaugh], and *C. tovarii* Eshbaugh, Smith & Nickerent. The third is *C. pubescens* clade that includes *C. cardenasii* Heiser & Smith, *C. eximium* Hunz., and *C. pubescens* Ruiz & Pav. species (Pickersgill, 1991). In the most recent studies, the relatedness of species was determined by molecular techniques (Plastid DNA markers matK and psbA-trnH), and the species in the genus were classified as 11 different groups (Carrizo Garcia et al., 2016).

The development and breeding of cultivars are closely related to the successful use of wild species in interspecies crossing studies. Morphological comparisons between species (Baral and Bosland, 2004), pollen tube development (Zijlstra et al., 1991), mechanism of incompatibility (Onus and Pickersgill, 2004), interbreeding (Martins et al., 2015), chromosome pairing (Lippert et al., 1966), isoenzyme variability (McLeod et al., 1983) and DNA variation are needed to understand interspecies phylogenetic relationships within the genus *Capsicum* (Kochieva et al., 2004). Although the group numbers of the species in each study varied, the basic phylogenetic trees obtained from different studies were found to be similar. For example, while *Capsicum annuum* and other cultivated species were found closer together, *C. longidentatum* alone formed a separate group, and other wild species were in different groups from the cultivated species. The group of *C. dimorphum* was the most distantly related group in the genus (Carrizo Garcia et al., 2016).

Capsicum tovarii was found and described on the banks of the Mantaro River in Peru by Eshbaugh et al. in 1983 (Figure 3.2). Studies show that *C. tovarii* is genetically more closely related to the *C. baccatum* group than to the *C. annuum* and *C. pubescens* groups. Researchers have reciprocal hybridizations with a wide variety of species (*C. annuum, C. chinense, C. frutescens, C. chacoense, C. galapogense, C. baccatum, C. praetermissum, C. cardenasii, C. eximium and C. pubescens*). They concluded that F1, F2 and backcross were obtained from the crossing of *C. baccatum* × *C. tovarii* (Tong and Bosland, 1999). However, Ibiza et al. (2012) and Carrizo Garcia et al. (2016) classified the species in a different and distant group from the Baccatum group according to their groupings (Carrizo Garcia et al., 2016).

F1 hybrids obtained as a result of crosses between *C. tovarii* and other *Capsicum* species were analyzed cytologically, and researchers tried to better understand the genetic relationship of the

FIGURE 3.2   General appearance and morphological traits of *Capsicum tovarii* (Eshbaugh et al., 1983; Tong and Bosland, 1999).

species with other species. As a result, when the meiotic cells of *C. tovarii* and *C. baccatum* at the diakinesis and metaphase-I stages were analyzed, it was determined that 100% of the cells had 12 pairs of chromosomes, and no chromosome bridge or chromosomal aberration (lagging) was observed in these species. When cells in diakinesis and metaphase-I stages of F1 (*C. baccatum* × *C. tovarii*) plants obtained from this crossing were examined, quite complex chromosome relationships were detected. While most chromosomes are paired as bivalent, univalent, tetravalent (quadrivalent) and one quinquevalent pairing have been detected. Despite this situation, some of the F1s were obtained by selfing and crossing with F2 plants and backcross plants by crossing with the mother (Tong and Bosland, 1999).

Heterochromatin content and distribution in *C. tovarii* suggested that this species may be more closely related to the purple flower species group (Pickersgill, 1991). Moscone et al. (2007) reports that the species is closer to the species of the *C. baccatum* group. It has been reported that both genotypes used in the studies are morphologically similar. Thus, differences in heterochromatin content and distribution can be explained by natural variability in genotypes within *C. tovarii*. Ince et al. (2010) determined that the *C. tovarii* species was close to *C. baccatum* with their molecular study using RAPD markers. Scaldaferro et al. (2013) also support its inclusion in the *C. baccatum* complex with their study. Unlike this case, Ibiza et al. (2012) showed in their analysis using AFLP and SSR primers that although *C. tovarii* appeared a little closer to the Annuum group, it was classified in a different group. It is thought that this situation may be due to the use of different genotypes by the researchers.

In pepper species with 2n=26 chromosome number, cross-breed combinations with fertile individuals have not been reported. For this purpose, wild species *C. lanceolatum* (2n=26) and *C. buforum* (2n=26) were studied. Although Tong and Bosland (2003) reported *C. buforum* as having 2n=24 chromosomes, there is no other study confirming this. *C. buforum* is a species reported as self-incompatible with *C. cardenasii*. This strain was more closely related to *C. pubescens*, *C. cardenasii* and *C. eximium* than other *Capsicum* species and showed different levels of compatibility (Batista, 2016). These studies form the basis for species combinations to be used for interspecific hybridization.

3.3 PROBLEMS AND SOLUTIONS IN INTERSPECIFIC HYBRIDIZATION

3.3.1 Hybridization Barriers

Many studies have shown that *Capsicum* has a strong barrier to interspecific crossing. The reasons for the incompatibility of crosses between cultivated *Capsicum* species and their wild relatives are not fully understood. Incompatibility is characterized as delayed pollen development, failure of pollen to germinate at the stigma, cessation of elongation of the pollen tube within the stigma, failure of the pollen tube to enter the ovary, inhibition of cell division of the zygote and abnormalities in the endosperm part (Onus, 2001, 2002; Onus and Pickersgill, 2004; Martins et al., 2015).

Some traditional approaches have been used to overcome crossbreeding barriers between species within the genus *Capsicum*. For this purpose, genetic bridge use (Kannangara et al., 2017), growth regulators (Yoon et al., 2006), pollen irradiation (Onus and Pickersgill, 2004) and *in vitro* pollen tube development studies have been performed (Zijlstra et al., 1991). It has been suggested that biotechnological techniques such as somatic hybridization can also be used to overcome these interspecies barriers in the genus *Capsicum* (Morrison et al., 1986). However, these techniques are not sufficient to eliminate all the interspecies hybridization barriers in pepper species. For example, due to prezygotic barriers that prevent the pollen tube from growing in the stigma and postzygotic barriers that can be seen even if the pollen tube develops, it is not possible to hybridize with the other four domesticated species (Zijlstra et al., 1991).

Capsicum baccatum is one of the most studied species in interspecies hybridization. It has shown unsuccessful crossing results with *C. annuum* in this species (Yoon et al., 2004), but the fact that this species is an important source of resistance to diseases causes it to be used as a genetic resource in breeding studies. Persistent efforts in crossing *C. annuum* with *C. baccatum* var. *pendulum* species have resulted in the development of anthracnose-resistant varieties from the resulting backcross (Yoon et al., 2017).

Zijlstra et al. (1991) determined that the interbreeding barrier between *C. chinense* × *C. annuum* and *C. baccatum* × *C. annuum* species occurred because the pollen tube elongation level could not reach the egg cell. The same authors also found that barriers to crossbreeding between *C. chinense* × *C. pubescens* and *C. baccatum* × *C. pubescens* were due to the cessation of pollen tube growth within the style. In another study in which *C. chinense* and *C. baccatum* species were hybridized reciprocally, it was determined that the pollen tubes continued to develop in the stigma and could enter the ovary and reach the ovules 24 hours after pollination. Especially, they obtained healthy seeds with high germination percentage from *C. baccatum* × *C. chinense* hybridization and healthy growing seedlings from these seeds (Kamvorn et al., 2014). For this reason, interspecific hybridization problems in peppers are not only post-zygotic, but also different crossing barriers can be seen in different species.

3.3.2 Post-Fertilization Abnormalities and Embryo Rescue

In advanced plants, the postzygotic failure of hybrid embryos is usually due to incompatibility problems in the endosperm, not to incompatibility between parental chromosomes. In such cases, early embryos need to be rescued from fruits that develop after interspecific hybridization; otherwise, it will result in failure of germinating seeds due to embryo underdevelopment and/or endosperm degeneration. Embryos can sometimes be used to produce new plants even if they are immature or lack endosperm. Embryo rescue technique, which is used integrated with tissue culture, is an important technique that will allow further progress in interspecific hybridization.

Interspecific crosses within the genus *Capsicum*, the conversion rate of hybrid embryos to plant remains very low due to some postzygotic incompatibility barriers (Hossain et al., 2003). Hossain et al. (2003) found that embryos could develop successfully in tissue culture medium when the basic MS medium was supplemented with casein hydrolyzate, yeast extract, coconut water, gibberellin (GA) and naphthaleneacetic acid (NAA) 28 to 33 days after pollination. Very limited attention has

FIGURE 3.3 (A, B) Samples of torpedo and cotyledon embryos that can be used in embryo rescue technique, (C) cotyledon type embryo (30 days after pollination in tissue culture medium), (D) development in tissue culture 60 days after pollination (Manzur, 2013), (E) plant development with embryo rescue technique in *C. chinense* × *C. annuum* (Debbarama et al., 2013), (F) embryo rescue 32–34 days after pollination in hybrids of *C. annuum* × *C. chinense* (Sui and Hui, 2015).

been paid to this technique for pepper, but recent studies have shown it to have significant potential for interspecific hybridization (Suprunova et al., 2010; Debbarama et al., 2013; Manzur et al., 2015; Sui and Hui, 2015; Cremona et al., 2018). As a result of these studies, it was reported that plants were successfully obtained from some interspecies hybrids by embryo rescue technique (Figure 3.3).

3.3.3 Infertility (Sterility) in Interspecific Hybrids

When the studies on interspecific hybridization in pepper are examined, it is stated that the majority of hybrids obtained after hybridization show sterility problems. It was concluded that infertility problems in the studies were related to meiotic abnormalities and hindered the usability of interspecific crossbreeding. The ability to interbreed is seen as a beacon of hope for transferring desired traits from wild species to cultivated species. However, the rate of obtaining fertile individuals in interspecies crossing within the genus *Capsicum* was very low in some combinations and completely unsuccessful in the others. The main reason for this is the incompatibilities between species, which are shown as one-sided conflicts, resulting in the failure of the embryo to develop after fertilization and obtaining male sterile individuals. The hybridizability statuses compiled from interspecific hybridization studies are summarized in Figure 3.4. The more distantly related species are genetically related, their hybrids are the more sterile. Monteiro et al. (2011) used different genotypes of six species and did not even achieve fruit set by crossing *C. pubescens*, which is the most distant relative of other species.

Easy separation of the fruit from the calyx is a desirable feature in Maraş peppers. Studies are continuing to develop varieties with easy breaking properties. In one of the studies carried out for this purpose, it was aimed to transfer this trait to Maraş-type peppers with *C. annuum* varieties (fruit detachment force between 3.86 and 7.34 N) by using a cultivar of *C. chacoense* with weak

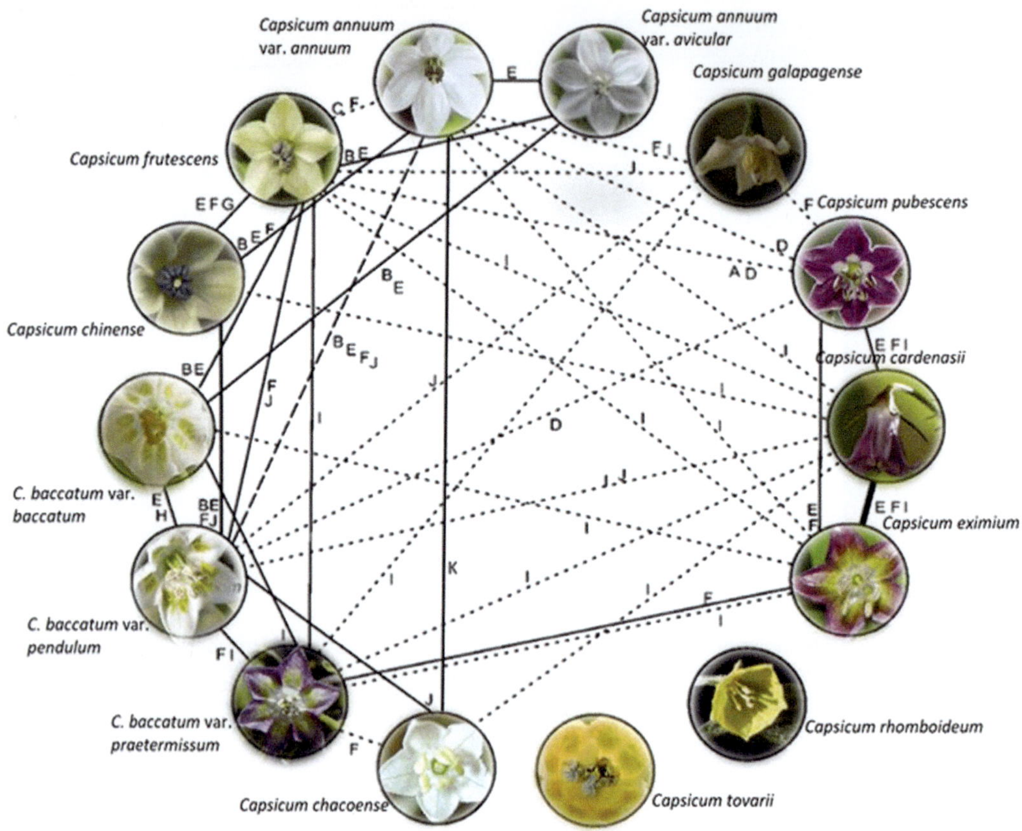

FIGURE 3.4   Some species and success cases used in interspecific hybridization studies of *Capsicum*; ━ ━ highly fertile hybrids, ── viable hybrids and seeds,──F1 hybrids sterile, - - - . Completely sterile, no line non-pollination and fertilization successes, A – (Heiser and Smith, 1948); B – (Pickersgill, 1971); C – (Smith and Heiser, 1951); D – (Heiser and Smith, 1953); E – (Eshbaugh, 1975); F – (Lippert et al., 1966); G – (Smith and Heiser, 1957); H – (Smith and Heiser, 1957); I – (Heiser and Smith, 1958); J – (Emboden, 1961); K – (Arpacı and Yaralı Karakan, 2018)

fruit detachment force (0.199 N) by crossbreeding. F1 hybrids were successfully obtained by crossing Dila, Sena, K7 and K8 Maraş pepper genotypes with *C. chacoense*. It was determined that all of these F1s had a fruit detachment force between 0.22 and 0.26 N and was quite low when compared to the Maraş genotypes (Figure 3.4) (Arpacı and Yaralı Karakan, 2018). However, in addition to these successful results, it has been reported in different studies that crossbreeding between species creates sterile individuals. Cremona et al. (2018) obtained F1 hybrids in their study used as a female parent of the *C. annuum* (cv. Friariello) species, and male parent of the *C. baccatum* var. *pendulum* (cv. P04), but they reported that the pollen viability of the hybrids were very low and male sterile (Figure 3.5). Spots on flowers, which are a characteristic of the male parent, are also found in hybrids, although smaller. F1 fruits appear to be prone to parthenocarpy (Figure 3.5).

Generally, there are studies reporting that hybridization combinations between *C. frutescens* × *C. chinense* species are possible. The bhut jolokia cultivar is reported to be a hybrid of *C. frutescens* × *C. chinense*. It is among the hottest varieties in the world with its 1,001,304 Scoville heat units (Bosland and Baral, 2007) (Figure 3.5). Pickersgill (1991) and Zijlstra et al. (1991) reported that *C. frutescens* × *C. chinense* species could produce fertile F1 individuals as a result of crossing them because they are in the same genetic complex. Examining the same species in terms of

FIGURE 3.5   Fruits of bhut jolokia on plants grown in the field (Bosland and Baral, 2007).

chromosomal homology, Moreira et al. (2017) also determined that hybrids obtained from interspecific hybridization showed normal meiosis and had a bivalent chromosome number of 12, and the viability of seeds obtained from hybrids was 73%. On the other hand, it is much more difficult to produce efficient interspecific hybrids than crosses between species belonging to different genetic complexes (Figure 3.4).

3.4   DETECTION OF INTERSPECIFIC HYBRIDS BY MOLECULAR TECHNIQUES

Cytogenetic and molecular analyses help to elucidate the level of relationship, evolutionary genetics and karyotypic stability of many species. Interspecies crosses have also been used successfully to generate genetic mapping populations, to label desired loci and to obtain marker-based linkage maps (Table 3.2). In the first study on the mapping of hybrid individuals with molecular markers, the identification of hybrids of *C. annuum* × *C. chinense* species was performed using RFLP markers (Tanksley et al., 1988).

As a result of molecular identifications made to determine the relationship of interspecific hybrids with parents, compatibility and fertility status between species and F1 hybrids can also be revealed. Molecular marker techniques such as AFLP (Kang et al., 2001), RFLP (Kang et al., 2001; Rao et al., 2003; Lee et al., 2004; Yi et al., 2006), COS II (Wu et al., 2009) and SSR (Lee et al., 2004; Yi et al., 2006; Tan et al., 2015; Arjun et al., 2018) are also successful used in mapping interspecific hybrids, besides the RAPD (Livingstone et al., 1999) technique.

3.5   QUALITY DEVELOPMENT THROUGH INTERSPECIFIC HYBRIDIZATION

It has been reported that the ripe fruit color of peppers can be yellow, orange, brown and red tones. Stommel and Griesbach (2008) crossbred parents with orange and red ripe fruit color in their study to explain the inheritance of different pepper traits. They reported that mature fruit colors of F1, F2 and backward hybrids were orange and red. In a study using *C. baccatum* genotypes, a much wider color range (white, yellow, light yellow, dark yellow, orange, dark orange and red) was observed as a result of crossing of parents with yellow and red ripe fruit color (Figure 3.6) (Mavi et al., 2021). Interspecific hybridization may be an alternative to transfer this wider color variety to varieties within the Annuum complex. Ripe fruit color occurs as a result of the decrease in chlorophyll and

TABLE 3.2

Molecular Studies and Molecular Marker Maps of Pepper Interspecific Hybrids

Parents	Population	Marker type	Marker numbers	References
C. annuum cv. Doux des landes × *C. chinense* cv. PI 159234	BC	RFLP, isozyme	85	Tanksley et al. (1988)
C. annuum cv. NuMex RNaky × *C. chinense* cv. PI 159234	F_2	RFLP	192	Prince et al. (1993)
C. annuum cv. NuMex RNaky × *C. chinense* cv. PI 159234	F_2	AFLP, RAPD, RFLP	1007	Livingstone et al. (1999)
C. annuum cv. TF68 × *C. chinense* cv. Habanero	F_2	AFLP, RFLP	580	Kang et al. (2001)
C. annuum cv. Maor × *C. frutescens* cv. BG2816	BC_2	RFLP	92	Rao et al. (2003)
C. annuum cv. TF68 × *C. chinense* cv. Habanero	F_2	SSR, RFLP	333	Lee et al. (2004)
C. annuum cv. TF68 × *C. chinense* cv. Habanero	F_2	SSR, EST-SSR, RFLP	243	Yi et al. (2006)
C. annuum cv. NuMex RNaky × *C. frutescens* cv. BG 2814-6	F_2	COS II	263	Wu et al. (2009)
C. annuum cv. BA3 × *C. frutescens* cv. YNXML	F_2	SSR	64	Tan et al. (2015)
C. annuum cv. FL 201 × *C. galapagoense* cv. TC 07245	F_2	SSR	400	Arjun et al. (2018)

FIGURE 3.6   Phenotypic diversity for fruit shape and color in F3 plants of *Capsicum baccatum* var. pendulum (Mavi et al., 2021).

anthocyanin pigments and the accumulation of carotenoid pigments. Numerous genes influence the color formation in ripe fruits, affecting the accumulation of more than 30 different carotenoids (Matus et al., 1991). The dominant *B* gene controls the high β-carotene content in ripe fruits. The *t* gene interacting with the *B* gene causes differences in the β-carotene level (Brauer, 1962). It has been determined that brown fruits are formed by the combination of the *y+* (red fruit color) gene and the *cl* (chlorophyll-retaining) gene (Smith, 1948). It has been determined that fruit shape is transferred to subsequent individuals in relation to immature fruit color and independent of mature fruit color (Lippert et al., 1965).

The variegation of leaf color and the accumulation of anthocyanins in peppers are controlled by many genes. The variegated leaf color of ornamental peppers is a desirable feature in varieties to be used as ornamental plants. Varieties with variegated leaf color have been developed. Disease and pest resistance in these varieties may be weak. For this reason, their resistance can be increased by crossing them with different species. In peppers, the *pi* gene has been found to cause green and white coloration on the leaves by causing irregularity in the plastids (Lippert et al., 1965). Expression of the *A* and *MoA* genes causes violet mottling on variegated and non-variegated green leaves (Peterson, 1959; Lippert et al., 1965; Wang and Bosland, 2006).

The inheritance of bitterness in peppers is one of the most studied subjects. Sweet and hot varieties and species were hybridized for this purpose, and it was determined that bitterness was dominant over sweetness. It is estimated that the first sweet variety in peppers emerged from hot peppers by mutation. In other words, all wild species were considered hot (Mavi, 2013; Güveloğlu et al., 2021). However, in recent studies, it has been reported that some wild species have sweet genotypes. *C. longidentatum* species and some species in the Andean complex were found to have sweet peppers (Carrizo-Garcia, 2016). Varieties such as bhut jolokia, Trinidad moruga scorpion (over 1,019,687 Scoville heat units), which are the hottest varieties breeding today, belong to the *C. chinense* species (Bosland et al., 2012). As a result of the molecular analysis, interestingly, it was determined that the genetic structure of bhut jolokia cultivar contains *C. frutescens* blood. Based on this, it was stated that the variety is a naturally occurring interspecific hybrid (Bosland and Baral, 2007). This makes *C. chinense* an important genetic resource to increase capsaicin content.

Depending on the cultures and demands, the preferences of hot and sweet varieties and the level of bitterness may vary. Bitterness may increase slightly depending on the stress conditions, but it is not possible for a sweet variety to be hot or a hot variety to be sweet. Especially, as mentioned before, since bitterness is a dominant character, two sweet varieties must be used to obtain sweet varieties, and even in this case, hot hybrids are likely to occur in the variation to be obtained. Interspecific hybridizations can be used to increase the bitterness ratio and slight bitterness of sweet genotypes as well as to transfer the disease resistance of the species. In a study conducted to examine the inheritance of bitterness, the bitterness status of hybrids was determined by using varieties belonging to *C. annuum* (cv. Takanotsume medium hot) and *C. frutescens* (cv. Ac 1443 very hot) species. F1 individuals obtained from crossbreeding were inbred, and bitterness values of all F2 individuals obtained were determined. Eighteen of the 22 F2 hybrids obtained were determined to be hotter than the *C. annuum* cv. Takanotsume cultivar used as a female parent. They found that nine of the F2 hybrids were also hotter than the male parent species *C. frutescens* (Ohta, 1962). Successful results were obtained from *C. chinense* × *C. annuum* hybridizations in order to develop small fruited sweet pickled peppers (Fırat et al., 2021).

The possibilities of utilizing interspecies hybrids for the production of high-quality peppers were investigated in the production of ornamental peppers. For this purpose, *C. annuum*, *C. chinense*, *C. baccatum*, and *C. frutescens* species were crossed reciprocally. Successful interspecific hybrids were obtained between *C. annuum* × *C. frutescens*, *C. chinense* × *C. baccatum* and *C. annuum* × *C. chinense* species. No fertile individuals occurred from the other combinations. In successful combinations, an increase in flower size, fruit number, yield and plant height was obtained (Nascimento et al., 2019).

3.6 CONCLUSION AND FUTURE PERSPECTIVES

Interspecific hybridization in peppers has been studied more intensively between domesticated species. There are still many unknown issues regarding the hybridizability of wild species. In this review, information is presented on some of the previous studies on some subjects. In particular, the chance of success is higher in hybridizing species in closely related groups. It is necessary to make crosses between species in a programming framework for purposes and targets. Species that can be used for resistance to diseases and pests in the species, resistance to abiotic stress conditions and fruit quality should be determined, and studies should be continued to transfer superior characteristics to the desired species by crossing them with domesticated species. For successful hybridization, it is necessary to examine and reveal the prezygotic and postzygotic barriers between the species separately. As a result, new information obtained by investigating these relationships will enable us to be successful in hybridizing peppers between domesticated and wild species. In addition, it will pave the way for breeders to take part in breeding programs of any successful species as a source of variation.

The resistance of species to biotic and abiotic stress conditions is one of the primary targets in cross-breeding studies. In terms of some secondary traits, there is a unique genetic potential in different pepper species. Cross-breeding studies should be continued in order to transfer the characteristics of different species to other species, especially in terms of heat levels, aroma compounds, color and some ornamental plant traits (variegated leaf color, short internodes etc.).

The traits that contributed the most to the variability between parents and crosses analyzed in interspecific crosses were fruit yield and leaf width per plant. It is not easy to obtain peppers with an ideal phenotype for producers and consumers by cross-breeding. Studies should continue in the future to determine the most successful combinations by conducting new studies on the success of hybridization between species with desired characteristics.

REFERENCES

Arjun, K., M. S. Dhaliwal, S. K. Jindal, and B. Fakrudin. 2018. Mapping of fruit length related QTLs in interspecific cross (*Capsicum annuum* L. × *Capsicum galapagoense* Hunz.) of chilli. *Breeding Science* 68:219–226.

Arpacı, B. B., and F. Yaralı Karakan. 2018. Inter-specific (*Capsicum chacoense* Hunz. and *Capsicum annuum* L.) inheritance of fruit detachment force trait in hot pepper. *Scientific Papers. Series B, Horticulture* 57:391–394.

Azeez, S. O., and J. O. Faluyi. 2018. Hybridization in four Nigerian *Physalis* (Linn.) species. *Notulae Scientia Biologicae* 10(2):205–210.

Baral, J. B., and P. W. Bosland. 2004. Unraveling the species dilemma in *Capsicum frutescens* and *C. chinense* (*Solanaceae*): A multiple evidence approach using morphology, molecular analysis, and sexual compatibility. *Journal of the American Society for Horticultural Science* 129(6):826–832.

Barboza, G. E., B. L. Bianchetti, and J. R. Stehmann. 2020. *Capsicum carassense* (*Solanaceae*), a new species from the Brazilian Atlantic Forest. *PhytoKeys* 140:125–138.

Barboza, G. E., C. Carrizo García, S. Leiva González, M. Scaldaferro, and X. Reyes. 2019. Four new species of *Capsicum (Solanaceae)* from the tropical Andes and an update on the phylogeny of the genus. *PLoS ONE* 14(1):e0209792.

Barboza, G. E., C. C. García, L. de Bem Bianchetti, M. V. Romero, and M. Scaldaferro. 2022. Monograph of wild and cultivated chili peppers (Capsicum L., Solanaceae). *PhytoKeys* 200:1–423.

Batista, F. R. C. 2016. Cytogenetics in *Capsicum* L. In *Production and Breeding of Chilli Pepper (Capsicum spp.)*, ed. E. R. do Rêgo et al., 41–56. Springer.

Bernardello, L., I. Rodriguez, L. Stiefkens, and L. Galetto. 1995. The hybrid nature of *Lycium ciliatum* × *cestroides* (Solanaceae): Experimental, anatomical and cytological evidence. *Canadian Journal of Botany* 73:1995–2005.

Bosland, P. W., and J. B. Baral. 2007. 'Bhut Jolokia'-The world's hottest known chile pepper is a putative naturally occurring interspecific hybrid. *HortScience* 42(2):222–224.

Bosland, P. W., D. Coon, and G. Reeves. 2012. Trinidad Moruga Scorpion pepper is the world's hottest measured chile pepper at more than two million Scoville heat units. *Horttechnology* 22:534–538.

Brauer, O.1962. Untersuchungen ueber qualitatseigenschaften in F_1 hybriden von paprika, *Capsicum annuum* L. Z. *fuer Pflanzenzuecht* 48: 259–276.

Carrizo Garcia, C., M. H. J. Barfuss, E. M. Sehr, G. E. Barboza, R. Samuel, E. A. Mascone, and F. Ehrendorfer. 2016. Phylogenetic relationships, diversification and expansion of chili peppers (*Capsicum, Solanaceae*). *Annals of Botany* 118:35–51.

Cremona, G., M. Iovene, G. Festa, C. Conicella, and M. Parisi. 2018. Production of embryo rescued hybrids between the landrace "Friariello" (*Capsicum annuum* var. *annuum*) and *C. baccatum* var. *pendulum*: Phenotypic and cytological characterization. *Euphytica* 214(8):129.

Debbarama, C., V. K. Khanna, W. Tyagi, M. Rai, and N. T. Meetei. 2013. Wide hybridization and embryo-rescue for crop improvement in *Capsicum.Agrotechnology* 11:2–6.

Emboden, W. A. Jr. 1961. A preliminary study of the crossing relationships of *Capsicum baccatum.Butler University Botanical Studies* 14:Article 2.

Eshbaugh, W. H. 1975. Genetic and biochemical systematic studies of chili peppers (*Capsicum—Solanaceae*). *Bulletin of the Torrey Botanical Club* 102:396–403.

Eshbaugh, W. H. 2012. The taxonomy of the genus *Capsicum*. In *Peppers Botany, Production and Uses*, ed. Vincent M. Russo, 14–28. CAB International.

Eshbaugh, W. H., P. G. Smith, and D. L. Nickrent. 1983. *Capsicum tovarii* (*Solanaceae*), a new species of pepper from Peru. *Brittonia* 35(1):55–60.

Fatokun, C. A.1989. Cytological studies of the F_1 interspecific hybrid between *Solanum aethiopicum* and *S. gilo* Raddi. *Cytologia* 54:425–428.

Fırat, C., K. Karataş, B. B. Arpacı, and K. Mavi. 2021. Crossbreeding studies of sweet ornamental pepper suitable for pickle industry. *Mustafa Kemal Üniversitesi Tarım Bilimleri Dergisi* 26(3):679–691.

Güveloğlu, Y., F. Uzunoğlu, and K. Mavi. 2021. The morphological characterization of some ornamental pepper lines in the genetics collection of Mustafa Kemal University. In *Overview on Horticulture*, ed. Arzu Çığ, Chapter 10:247–271. Iksad Publications.

Halstead, H. D.1912. Experiments with peppers. *New Jersey Agricultural Experimental StationAnnual Report* 33:365–368.

Heiser, C. B. Jr, and P. G. Smith. 1948. Observations on another species of cultivated pepper, *Capsicum pubescens* R & P. *Proceedings of the American Society of Horticultural Science* 52:331–335.

Heiser, C. B. Jr, and P. G. Smith. 1953. The cultivated *Capsicum* species. *Economic Botany* 7:214–227.

Heiser, C. B. Jr, and P. G. Smith. 1958. New species of *Capsicum* from South America. *Brittonia*10:194–201.

Hossain, M. A., M. Minami, and K. Nemoto. 2003. Immature embryo culture and interspecific hybridization between *Capsicum annuum* L. and *C. frutescens* L. via embryo rescue. *Japanese Journal of Tropical Agriculture* 47:9–16.

Ibiza, V. P., J. Blanca, J. Canizares, and F. Nuez. 2012. Taxonomy and genetic diversity of domesticated *Capsicum* species in the Andean region. *Genetic Resources and Crop Evolution* 59:1077–1088.

Ince, A. G., M. Karaca, and N. Onus. 2010. Genetic relationships within and between *Capsicum* species. *Biochem Genetics* 48:83–95.

Kamvorn, W., S. Techawongstien, S. Techawongstien, and P. Theerakulpisut. 2014. Compatibility of interspecific crosses between *Capsicum chinense* Jacq. and *Capsicum baccatum* L. at different fertilization stages. *Scientia Horticulturea* 179:9–15.

Kang, B. C., S. H. Nahm, J. H. Huh, H. S. Yoo, J. W. Yu, M. H. Lee, and B. D. Kim. 2001. An interspecific (*Capsicum annuum* × *C. chinense*) F, linkage map in pepper using RFLP and AFLP markers. *Theoretical and Applied Genetics* 102:531–539.

Kannangara, K. N., D. De Costa, D. K. N. G. Pushpakumara, and I. P. Wickramasinghe. 2017. Tri-species bridge crosses (*C. annuum* L. × *C. chinense* Jacq.) × (*C. chinense* Jacq. × *C. frutescens* L.) as an alternative approach for introgression of Cucumber Mosaic Virus (CMV) and Chilli Veinal Mosaic Virus (CVMV) resistance from *C. frutescens* L. into *C. annuum* L. *Tropical Agricultural Research* 28(4):472–489.

Khoury, C. K., D. Carver, D. W. Barchenger, G. E. Barboza, M. van Zonneveld, R. Jarret, L. Bohs, M. Kantar, M. Uchanski, K. Mercer, G. P. Nabhan, P. W. Bosland, and S. L. Greene. 2020. Modeled distributions and conservation status of the wild relatives of chile peppers (*Capsicum* L). *Diversity and Distributions* 26:209–225.

Kochieva, E. Z., N. N. Ryzhova, W. Van Dooijeweert, I. W. Boukema, P. Arens, and R. E. Voorrips. 2004. Assessment of genetic relationships in the genus *Capsicum* using different DNA marker systems. In *XIIth Meeting on Genetics and Breeding of Capsicum and Eggplant*, 44–50. Noordwijkerhout, The Netherlands.

Lammerts, W. E.1930. Interspecific hybridization in *Nicotiana*. XII. The amphidiploid *rustica-paniculata* hybrid; its origin and cyto-genetic behavior. *Genetics* 16:191–211.

Lee, J. M., S. H. Nahm, Y. M. Kim, and B. D. Kim. 2004. Characterization and molecular genetic mapping of microsatellite loci in pepper. *Theoretical and Applied Genetics* 108:619–627.

Lippert, L. F., B. O. Bergh, and P. G. Smith. 1965. Gene list for the pepper. *Journal of Heredity* 56:30–34.

Lippert, L. F., P. G. Smith, and B. O. Bergh. 1966. Cytogenetics of the vegetable crops. Garden pepper, *Capsicum* sp. *The Botanical Review* 32:24–55.

Livingstone, K. D., V. K. Lackney, J. R. Blauth, R. I. K. Van Wijk, and M. K. Jahn. 1999. Genome mapping in *Capsicum* and the evolution of genome structure in the Solanaceae. *Genetics* 152:1183–1202.

Manzur, J. P. 2013. *Técnicas y estrategias de mejora para facilitar la hibridación interespecífica y el acortamiento del ciclo generacional en el género Capsicum.* Para optar al título de Doctor Ingeniero Agrónomo por la Universidad Politécnica de Valencia.

Manzur, J. P., A. Fita, J. Prohens, and A. Rodriguez-Burruezo. 2015. Successful wide hybridization and introgression breeding in a diverse set of common peppers (*Capsicum annuum*) using different cultivated Ají' (*C. baccatum*) accessions as donor parents. *PLoS ONE* 10(12):e0144142.

Martins, K. C., T. N. S. Pereira, S. A. M. Souza, R. Rodrigues, and A. T. D. Amaral Junior. 2015. Crossability and evaluation of incompatibility barriers in crosses between *Capsicum* species. *Crop Breeding and Applied Biotechnology* 15:139–145.

Matus, Z., J. Deli, and J. J. Szaaboles, 1991. Carotenoid composition of yellow pepper during ripening–isolation of b-cryptoxanthin 5,6-epoxide. *Journal of Agricultural and Food Chemistry* 39:1907–1914.

Mavi, K. 2013. It is a small, but huge hot taste: Ornamental pepper. *Agrokop* August: 24–28.

Mavi, K. 2020. Interspecific hybridization in pepper species. *International Journal of Life Sciences and Biotechnology* 3(3):386–706.

Mavi, K., H. Hacbekir, F. Uzunoğlu, and M. Türkmen. 2021. The use of volatile compounds as an alternative method in pepper breeding (*Capsicum baccatum* var. *pendulum*). *Ciência Rural* 51(12):e20201066.

McLeod, M. J., S. I. Guttman, W. H. Eshbaugh, and R. E. Rayle. 1983. An electrophoretic study of the evolution in *Capsicum (Solanaceae). Evolution* 37:562–574.

Monteiro, C. E. S., T. N. P. Pereira, and K. P. Campos. 2011. Reproductive characterization of interspecific hybrids among *Capsicum* species. *Crop Breeding and Applied Biotechnology* 11:241–249.

Moreira, N. F., T. N. S. Pereira, and K. C. Martins. 2017. Meiotic analysis of interspecific hybrids between *Capsicum frutescens* and *Capsicum chinense. Crop Breeding and Applied Biotechnology* 17:159–163.

Morrison, R. A., R. E. Koning, and D. A. Evans. 1986. Anther culture of an interspecific hybrid of *Capsicum. Journal of Plant Physiology* 126(1):1–9.

Moscone, E. A., M. A. Scaldaferro, M. Grabiele, N. M. Cecchini, Y. S. Garcia, R. Jarret, J. R. Davina, D. A. Ducasse, G. E. Barboza, and F. Ehrendorfer. 2007. The evolution of chili peppers (*Capsicum*–Solanaceae): A cytogenetic perspective. *Acta Horticulturae* 745:137–170.

Nascimento, N. F. F., E. R. Rego, M. F. Nascimento, C. H. Bruckner, F. L. Finger, and M. M. Rego. 2019. Evaluation of production and quality traits in interspecific hybrids of ornamental pepper. *Horticultura Brasileira* 37:315–323.

Ohta, Y.1962. Physiological and genetic studies on the pungency of *Capsicum*, V. Inheritance of pungency. *The Japanese Journal of Genetics* 37:169–175.

Onus, A. N. 2001. A study on crossability relationships between some of the white and purple flowered *Capsicum* species. *Akdeniz Üniversitesi Ziraat Fakültesi Dergisi* 14(1):101–106.

Onus, A. N. 2002. A study on unilateral incompatibility in *Capsicum.Anadolu* 12(2):75–86.

Onus, A. N., and B. Pickersgill. 2004. Unilateral incompatibility in *Capsicum* (Solanaceae): Occurrence and taxonomic distribution. *Annals of Botany* 94:289–295.

Pandey, K. K. 1962. Interspecific incompatibility in Solanum species. *American Journal of Botany* 49(8): 874–882.

Peterson, P. A.1959. Linkage of fruit shape and color genes in *Capsicum. Genetics* 44:407–419.

Pickersgill, B. 1971. Relationships between weedy and cultivated form in some species of chili peppers (genus *Capsicum).Evolution* 25:683–691.

Pickersgill, B. 1991. Cytogenetics and evolution of *Capsicum* L. In *Chromosome Engineering in Plants: Genetics, Breeding, Evolution, Part B*, ed. T. Tsuchiya and P. K. Gupta, 139–160. Elsevier.

Prince, J. P., E. Pochard, and S. D. Tanksley. 1993. Construction of a molecular linkage map of pepper and a comparison of synteny with tomato. *Genome* 36:404–417.

Rao, U. G., A. Ben Chaim, Y. Borosky, and I. Paran. 2003. Mapping of yield related QTLs in pepper in an interspecific cross of *Capsicum annuum* and *Capsicum frutescens. Theoretical and Applied Genetics* 106:1457–1466.

Scaldaferro, M. A., M. Grabiele, and E. A. Moscone. 2013. Heterochromatin type, amount and distribution in wild species of chili peppers (*Capsicum-Solanaceae). Genetic Resources and Crop Evolution* 60:693–709.

Schieder, O.1980. Somatic Hybrids of *Datura innoxia* Mill. + *Datura discolor* Bernh. and of *Datura innoxia* Mill. + *Datura stramonium* L. var. tatula L. II. Analysis of progenies of three sexual generations. *Molecular Genetics and Genomics* 179:387–390.

Smith, P. G. 1948. Brown, mature fruit color in pepper (*Capsicum frutescens). Science* 107:345–346.

Smith, P. G., and C. B. Jr. Heiser. 1951. Taxonomic and genetic studies on the cultivated peppers *Capsicum annuum* L. and *C. frutescens* L. *American Journal of Botany* 38:362–368.

Smith, P. G., and C. B. Jr. Heiser. 1957. Taxonomy of *Capsicum chinense* Jacq. and the geographic distribution of the cultivated *Capsicum* species. *Bulletin of the Torrey Botanical Club* 84:413–420.

Stalker, H. T.1980. Utilization of wild species for crop improvement. *Advances in Agronomy* 33:111–147.

Stebbins, G. L. 1971. *Chromosomal evolution in higher plants.* Addison-Wesley, 216.

Stommel, J. R., and R. J. Griesbach. 2008. Inheritance of fruit, foliar and plant habit attributes in *Capsicum* L. *Journal of the American Society for Horticultural Science* 133:396–407.

Sui, Y. H., and N. B. Hui. 2015. Acquisition, identification and analysis of an inter-specific *Capsicum* hybrid (*C. annuum* × *C. chinense*). *The Journal of Horticultural Science and Biotechnology* 90:31–38.

Suprunova, T. P., E. A. Dzhos, O. N. Pishnaya, N. A. Shmikova, and M. I. Mamedov. 2010. Production and analysis of interspecific hybrids among four species of the genus *Capsicum*. In *Proceedings of the XIVth EUCARPIA meeting on genetics and breeding of capsicum & eggplant*. Universitat Politècnica de València.

Tan, S., J. W. Cheng, L. Zhang, C. Qin, D. G. Nong, W. P. Li, and K. L. Hu. 2015. Construction of an inter-specific genetic map based on indel and SSR for mapping the QTLs affecting the initiation of flower primordia in pepper (*Capsicum* spp.). *PLoS ONE* 10(3):e0119389.

Tanksley, S. D., R. Bernatzky, N. L. Lapitan, and J. P. Prince. 1988. Conservation of gene repertoire but not gene order in pepper and tomato. *Proceedings of the National Academy of Sciences* 85(17):6419–6423.

Tong, N., and P. W. Bosland. 1999. *Capsicum tovarii*, a new member of the *Capsicum baccatum* complex. *Euphytica* 109:71–77.

Tong, N., and P. W. Bosland. 2003. Observations on interspecific compatibility and meiotic chromosome behavior of *Capsicum buforum* and *C. lanceolatum*. *Genetic Resources and Crop Evolution* 50:193–199.

Trojak-Goluch, A., and A. Berbec. 2003. Cytological investigations of the interspecific hybrids of *Nicotiana tabacum* L. × *N. glauca* Grah. *Journal of Applied Genetics* 44(1):45–54.

Wang, D., and P. W. Bosland. 2006. The genes of *Capsicum*. *HortScience* 41(5):1169–1187.

Wijsman, H. J. W., and J. H. De Jong. 1985. On the interrelationships of certain species of Petunia IV. Hybridization between *P. linearis* and *P. calycina* and nomenclatorial consequences in the Petunia group. *Acta Botanica Neerlandica* 34(3):337–349.

Wu, F., N. T. Eannetta, Y. Xu, R. Durrett, M. Mazourek, M. M. Jahn, and S. D. Tanksley. 2009. A COSII genetic map of the pepper genome provides a detailed picture of synteny with tomato and new insights into recent chromosome evolution in the genus *Capsicum*. *Theoretical and Applied Genetics* 118:1279–1293.

Yakub, C. M., and P. Smith. 1971. Nature and inheritance of self-incompatibility in *Capsicum pubescens* and *C. cardenasii*. *Hilgardia* 40(12):459–470.

Yi, G., J. M. Lee, S. Lee, D. Choi, and B. D. Kim. 2006. Exploitation of pepper EST-SSRs and an SSR-based linkage map. *Theoretical and Applied Genetics* 114:113–130.

Yoon, J. B., J. W. Do, D. C. Yang, and H. G. Park. 2004. Interspecific cross compatibility among five domesticated species of *Capsicum* genus. *Journal of Korean Society of Horticultural Science* 45:324–329.

Yoon, J. B., J. Lee, and J. W. Do. 2017. *Breeding anthracnose resistance in chili pepper: From genetic resources to commercialization.* APSA Seed Congress. https://apsaseed.org/wp-content/uploads/2017/02/1.-Breeding-Anthracnose-Resistance-in-Chilli-from-genetic-resources-to-commercialization-J.B.-Yoon.pdf (accessed: March 03, 2023).

Yoon, J. B., D. C. Yang, J. W. Do, and H. G. Park. 2006. Overcoming two post-fertilization genetic barriers in interspecific hybridization between *Capsicum annuum* and *C. baccatum* for introgression of anthracnose resistance. *Breeding Science* 56:31–38.

Zijlstra, S., A. C. Purimahu, and P. Lindhout. 1991. Pollen tube growth in interspecific crosses between *Capsicum* species. *HortScience* 26:585–586.

Zillinsky, F. J. 1985. Triticale: An update on yield, adaptation, and world production. In *Triticale*, ed. R. A. Forsberg, 1–7. Crop Science Society America Special Publication No. 9. American Society of Agronomy.

4 Chemical Diversity and Functionality of Capsaicinoids

Inder Pal Singh, Nobuyuki Mase, Ankur Kumar Tanwar, Neha Sengar and Olivia Chatterjee

CONTENTS

ABBREVIATIONS:

CAPS: Capsaicinoids, TRPV1: Transient receptor potential vanilloid 1, SAR: Structure activity relationship, Ki: Inhibitory constant, RNA: Ribonucleic acid, CB1: Cannabinoid receptor Type 1, LDL: Low-density lipoprotein, HDL: High- density lipoprotein, VLDL: Very low-density lipoprotein, HMG-COA: 3-hydroxy-3-methylglutaryl coenzyme A, ROS: Reactive oxygen species, RNS: Reactive nitrogen species, BHA: Butylhydroxy anisole, TBARS: Thiobarbituric acid reactive substance, DNA: Deoxyribonucleic acid, TNF-α: Tumor necrosis factor-α, IL-1β: Interleukin-1 β, ASK1: Apoptosis signal regulating kinase 1, ATP: Adenosine triphosphate, ADP: Adenosine diphosphate, AMPK: Activated protein kinase, EMT: Epithelial mesenchymal Transition, VEGFA: Vascular endothelial growth factor-A, TIMP 1: Tissue inhibitors of metalloproteinases-1, ALCAM: Leukocyte cell adhesion molecule, GLP 1: Glucagon-like peptide-1, NOD: Non-obese diabetic, CVS: Cardiovascular system, 5-HT: 5-hydroxytryptamine, CNS: Central nervous system, MPTP: 1-methyl-4-phenyl-1,2,3,6-tetrahydropyridine, GABA: Gamma-aminobutyric acid, NSAID: Non-steroidal anti-inflammatory drug, NF-κB: Nuclear factor kappa

DOI: 10.1201/9781003378259-4

B, cAMP: Cyclic adenosine monophosphate, CAM: Computer-aided manufacturing, N-AVAMs: N-acyl-vanillylamides, UN-AVAMs: Unsaturated N-acylvanillamides, PAL: phenylalanine ammonia lyase, C4H: cinnamate4-hydroxylase, 4CL: 4-coumaroyl-CoAligase, HCT: hydroxycinnamoyl transferase, C3H: coumaroyl shikimate/quinate3-hydroxylase, CCoAOMT: caffeoyl-CoA 3-*O*-methyltransferase, COMT: caffeicacid-*O*-methyltransferase, HCHL: 4-hydroxycinnamoylCo-Ahydratase/lyase, pAMT: putative aminotransferase, BCAT: branched-chain amino acid transferase, KAS: ketoacyl-ACP synthase, ACL: acyl carrier protein, FAT: acyl-ACP thioesterase, ACS: acyl-CoA synthetase, CS: capsaicinoid synthase, N-acyl vanillylamides (N-AVAM).

4.1 INTRODUCTION

Peppers are mostly used as food additives throughout the world because of their characteristic aroma and pungency. The chemicals that give many peppers their spiciness and heat are known as capsaicinoids. Capsaicin (*trans*-8-methyl-*N*-vanillyl-6-nonenamide) and dihydrocapsaicin (8-methyl-*N*-vanillylnonanamide) are the two main capsaicinoids found in the majority of hot pepper species (Laskaridou-Monnerville, 1999). Along with these two principal capsaicinoids, hot peppers also contain smaller amounts of other minor capsaicinoids, including homocapsaicin, nordihydrocapsaicin, nornorcapsaicin, norcapsaicin, nonivamide, and homodihydrocapsaicin (Figure 4.1) (Constant et al., 1996; Constant et al., 1995). A vanillyl ring and an acyl chain linked by an amide bond is the structural feature of capsaicinoids that defines their spicy characteristics. The amount of capsaicinoids in each type of pepper varies greatly from one another. After drying, the capsaicinoid content in the less spicy pepper ranges from 0.003% to 0.01%. This range is between 0.01% and 0.3% in mildly spicy pepper, while in the extremely hot spicy pepper, the capsaicinoid content is above 0.3% of the total weight (Perucka and Oleszek, 2000). Numerous methods, such as Soxhlet, magnetic stirring, maceration, supercritical fluids extraction, microwave-assisted extraction, and enzymatic extraction have been used for the extraction of capsaicinoids from peppers (Contreras-Padilla and Yahia, 1998; Kirschbaum-Titze et al., 2002; Korel et al., 2002; Santamaria et al., 2000; Sato et al., 1999; Williams et al., 2004).

Traditionally, capsaicin was used for improving circulation, headaches, for gastrointestinal protective activity, and for muscular pain. Also, capsaicin has a synergistic action with other herbs, because of which it is added to herbal formulations to increase their absorption. Additionally, there is overwhelming evidence that capsaicin and other capsaicinoids may be effective in the treatment of a variety of cancers. During the last decade, plenty of research has been done to understand the mechanism of action of capsaicin. Capsaicin has been shown to bind to the transient receptor potential vanilloid 1 receptor (TRPV 1), which is mostly produced by sensory neurons. Additionally, numerous capsaicin analogues have been isolated, synthesised, and tested for various bioactivities (Uarrota et al., 2021).

4.1.1 Capsaicinoids

Capsaicinoids (CAPS) are amphiphilic alkaloids, produced only in the *Capsicum* genus, and these are responsible for the pungency of peppers (Antonio et al., 2018). The *Capsicum* genus contains many species of peppers, amongst which only five have been explored scientifically, these include *C. pubescens*, *C. annuum*, *C. baccatum*, *C. frutescens*, and *C. chinense* (de Sá Mendes and de Andrade Gonçalves, 2020). A vanillyl group is attached to an amide and alkyl chain to form the chemical structure of capsaicinoids (Fattorusso and Taglialatela-Scafati, 2007). Among the four major CAPS present in larger amounts in plants, the most prevalent are capsaicin and dihydrocapsaicin, which constitute more than 90%, while homocapsaicin and nordihydrocapsaicin are present in minor quantities (Conforti et al., 2007). To evaluate the pungency (hotness) of chilli peppers, the Scoville scale is used; according to the scale, the *Capsicum*'s range lies between 0 and 100 million Scoville units, with 16 million of those units belonging to capsaicin (González-Zamora et al., 2013). The molecular diversity is due to the carbon chain length, saturation, and unsaturation in the acyl group, as shown in Figure 4.1 (Alberti et al., 2008). The CAPS content varies with genotype, fruit maturity, cultivation, and environmental conditions (Duelund and Mouritsen, 2017). The placenta of the fruit contains the highest quantity (more than 85%), and the seeds are the second-largest part containing CAPS (Fattorusso and Taglialatela-Scafati, 2007).

CAPS show a broad range of pharmacological activities such as antiobesity, analgesic, antioxidant, anticancer, anti-inflammatory, etc. (Knotkova et al., 2008; E.-J. Lee et al., 2010; Macho et al., 2003; Rosa et al., 2002; Shimoda et al., 2007). When exposed to temperatures above 80°C, CAPS quickly decompose to about 50% of their initial concentration due to the disruption of the carbon–nitrogen bond resulting in the formation of the vanillyl group and an acyl chain (Shimoda et al., 2007). The biological activity is mostly attributed to the vanillyl group because it functions as a proton donor, stabilising the radical species, and interacting with cellular membranes, enzymes, and neurological receptors (Loizzo et al., 2015; Mohammad et al., 2013). It often exerts its effects through receptor-dependent mechanisms via transient receptor potential vanilloid subfamily member 1 (TRPV1), which is a non-selective cation channel receptor that not only produces the pungent sensation but also alleviates pain by interacting with proton sites and vanilloid functional groups of different molecules. Mechanistically, inflammatory hyperalgesia is caused by the up-regulation of TRPV1 in inflamed tissues, while desensitisation of the receptor by CAPS over an extended period of time would diminish inflammatory hyperalgesia (Hayman and Kam, 2008; Luo et al., 2011).

4.1.2 CAPSIATES (NATURAL CAPSAICIN ANALOGUES)

Capsinoids are the sweet and non-pungent analogues of capsaicinoids, which were first isolated from the sweet pepper. Capsinoids are present in low amounts in the pericarp as well as in the placenta of fruit. Major capsinoids include capsiate, dihydrocapsiate, and nordihydrocapsiate. All these analogues are distinct from capsaicin in the middle region where the vanillyl group is linked to the carbon chain through an ester group rather than an amide group, as shown in Figure 4.2. Capsinoids

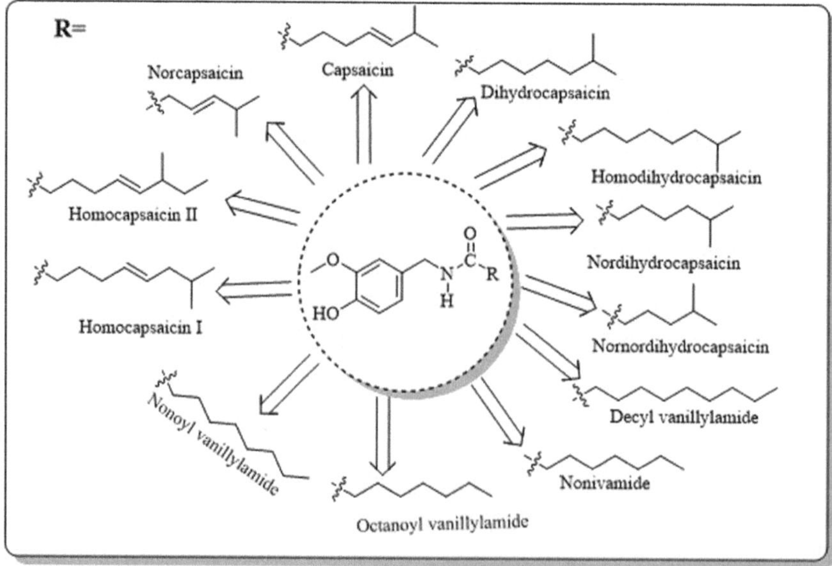

FIGURE 4.1 Chemical structures of different capsaicinoids (pungent) and their analogues.

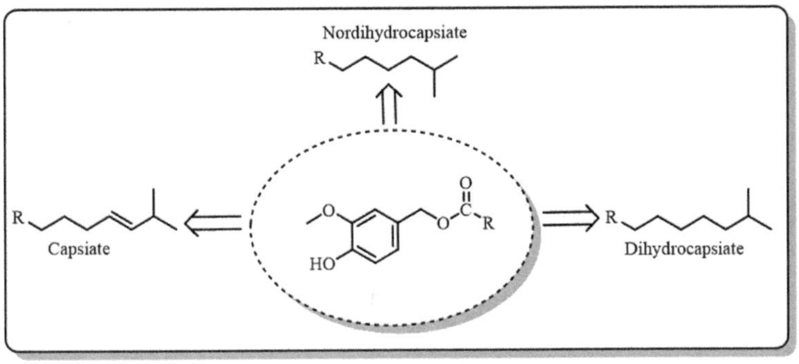

FIGURE 4.2 Chemical structure of different capsiates (natural capsaicin analogues).

show similar pharmacological properties to capsaicin in alleviating pain, as anticancer drugs, and in lowering the risk of cardiovascular and gastrointestinal disorders. Due to their high toxicity and limited selectivity, capsaicinoids (such as capsaicin) have relatively fewer clinical applications. On the other hand, capsinoids (such as capsiate) have a modified chemical structure and lower toxicity. Therefore, in therapeutic applications including cancer, cardiovascular diseases, etc., capsinoids may be more advantageous than capsaicinoids (Luo et al., 2011; Macho et al., 2003).

4.1.3 BIOSYNTHESIS OF CAPSAICINOIDS

Biosynthesis of capsaicin follows two distinct pathways: one route runs through the synthesis of vanillylamine via phenylpropanoid shikimic acid, and the second route is synthesis of branched chain fatty acid from L-valine, as shown in Figure 4.3 (Ananthan et al., 2018; Garcés-Claver et al., 2006). The biosynthesis of vanillylamine starts from L-phenylalanine, which is sequentially converted to cinnamic, p-coumaric, caffeic, ferulic acid, and vanillin and then finally vanillylamine in the presence of a pyrophosphate-dependent aminotransferase enzyme, which is present in larger amounts in the placenta of the fruit of *Capsicum*. The biosynthesis of acyl fatty acid precursor occurs via a sequence of enzyme-dependent reactions starting from the amino acid valine, followed by transamination, decarboxylation, elongation, and reduction. The formation of capsaicin by acid amine coupling or condensation occurs via enzymatic association of vanillylamine and an acyl acid precursor (Fattorusso and Taglialatela-Scafati, 2007). The concentration of vanillylamine or the acyl precursor may be one of the constraints in the synthesis of CAPS. Its biosynthesis occurs mainly within the placenta, where vanillylamine is abundant. An alkaloid that resembles capsaicin in pungency is called nonivamide (N-(4-hydroxy-3-methoxybenzyl)-nonanoic acid amide). It is biosynthesised in lesser amounts in plants and comprises about 3% of the total quantity of the CAPS (Constant et al., 1996). It is frequently employed as an adulterant in pepper-based essentials due to its pungency similar to capsaicin. It differs from capsaicin in biological function and dietary benefits. Therefore, it produces products with low efficiency for the pharmaceutical and food industries (Constant et al., 1996; Walker et al., 2017).

4.1.4 CAPSAICIN

Capsaicin (trans-8-methyl-N-vanillyl-6-nonenamide), a lipophilic alkaloid with the molecular formula $C_{18}H_{27}NO_3$ and a molecular weight of 305.40 g/mol, is a chemically inert, colourless, and crystalline substance. It is insoluble in water and highly soluble in alcohol, lipids, and organic solvents. Capsaicin was originally isolated in 1846, crystallised in 1876 by Tresh, and its molecular structure elucidated in 1919 by Nelson and Dawson (Nelson, 1919; Reyes-Escogido et al., 2011; Sawynok,

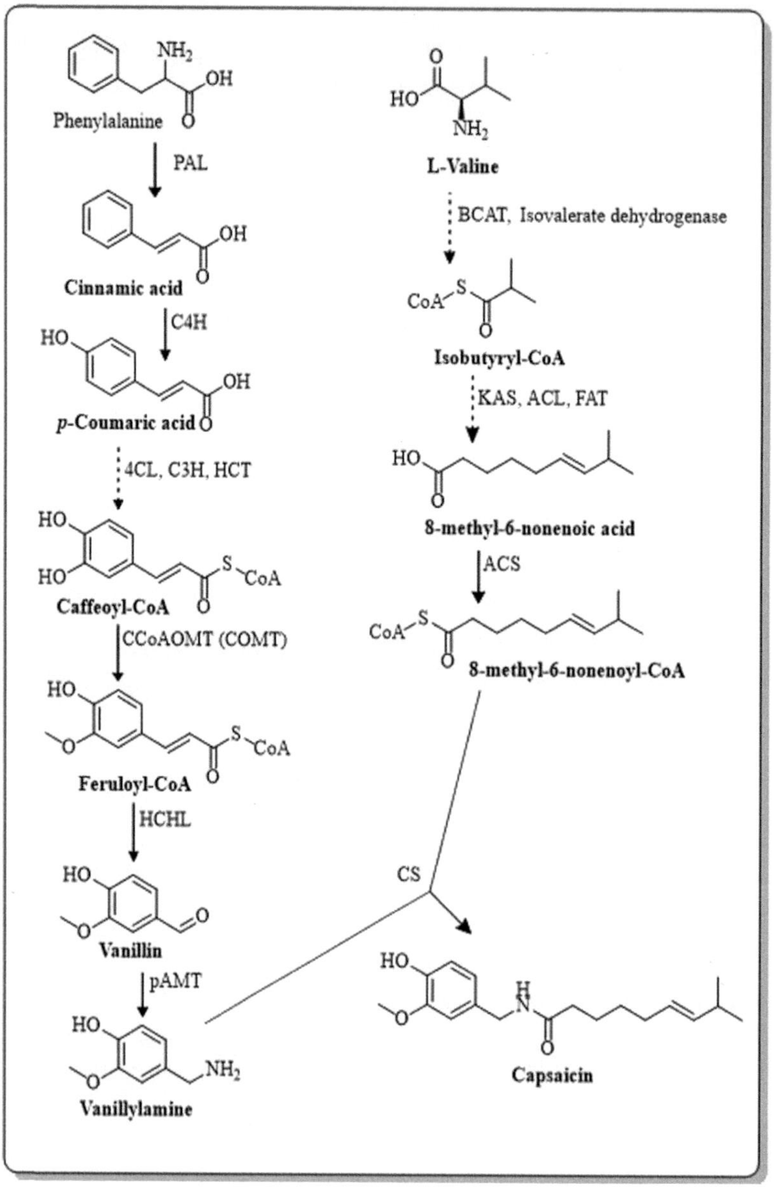

FIGURE 4.3 Biosynthesis of capsaicin by the phenylpropanoid pathway.

2005). The TRPV1 receptor, which resides in small sensory afferent nerve endings, is activated by the capsaicin, which causes a release of inflammatory neuropeptides and an influx of calcium that are responsible for receptor desensitisation to painful stimuli, leading to analgesic action (Caterina et al., 1997; Cortright and Szallasi, 2004).

4.1.5 Metabolism of Capsaicin

Biosynthesis of CAPS produces capsaicin in higher concentration within 1 to 2 months of flowering; the pungency of the entire fruit is higher at this time; after that, pungency subsequently decreases due to the plant metabolism. The primary oxidative domain in capsaicinoids is the vanillyl group, where the peroxidase enzyme or hydrogen peroxide acts to facilitate the metabolic reaction. Peroxidase facilitates the oxidation of phenol, which generates a phenoxy or aryl radical, that undergoes a radical coupling reaction to form a dimer (5,5'-dicapsaicin) and 4'-O-dicapsaicinether, respectively, as shown in Figure 4.4 (Caterina et al., 1997).

4.2 PHARMACOLOGICAL ACTIVITY OF CAPSAICINOIDS

Capsaicin, which is the major constituent in *Capsicum*, majorly contributes to the burning sensation and is directly involved for activating the transient receptor potential vanilloid type 1 channel (TRPV1) receptor. It exerts its action by attaching itself to an intracellular site. TRPV1 acts as a sensor for capsaicin and is permeable to Ca^{2+}, Na^+, etc. Consequently, its stimulation ion causes influx of cations, resulting in cell depolarisation and sometimes calcium ion overload Figure 4.5. Prolonged capsaicin-induced activation of TRPV1 can lead to desensitisation of the channels (Munjuluri et al., 2022).

4.2.1 ANTI-INFLAMMATORY ACTIVITY

The stimulation of vanilloid receptor type 1 is considered the most probable way to fight inflammation and hyperalgesia. The neuropeptide substance P is responsible for causing pain and

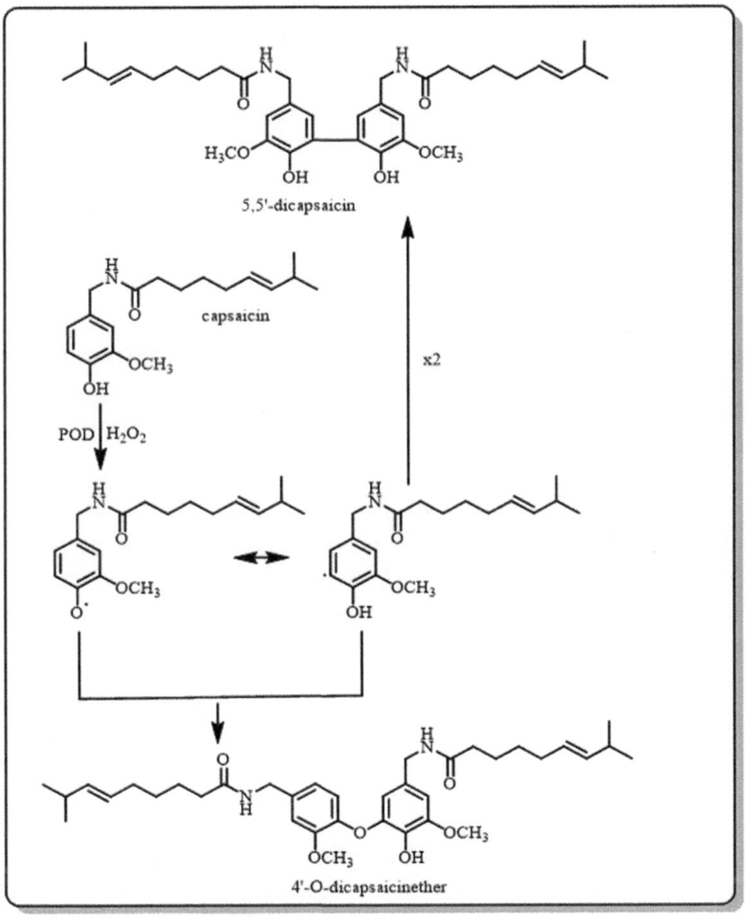

FIGURE 4.4 Metabolic pathway of capsaicin.

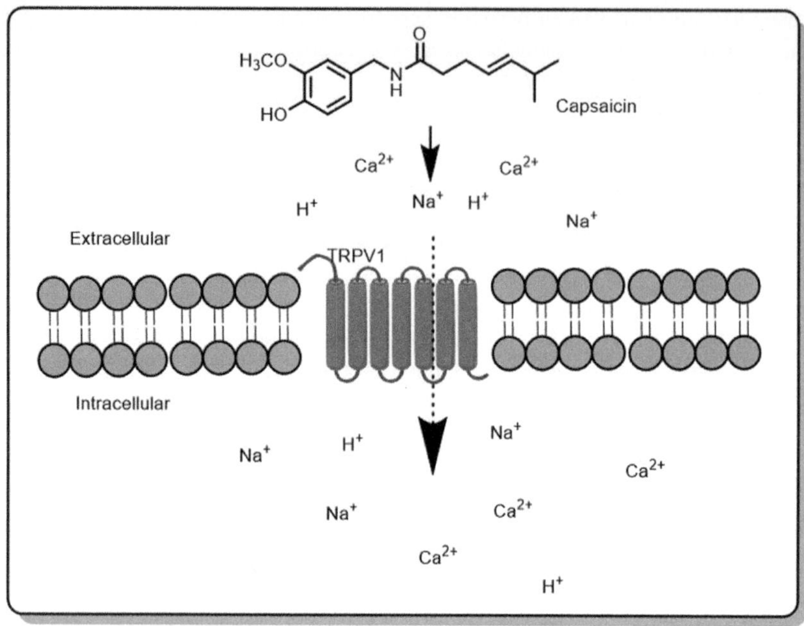

FIGURE 4.5 Sensitisation of TRPV1.

inflammation in arthritis, and it was found that treatment with capsaicin can cause inactivation of sensory neurons of substance P. The structural activity experiments showed that the capsaicin analogues with different side chain length exhibit enhanced analgesic activity with increased binding affinity to TRPV1 receptor (Batiha et al., 2020).

Studies have also shown that pre-treatment of rats with *C. baccatum* fruit juice helped in reducing the concentration of pro-inflammatory cytokines TNF-α and IL-1β. Capsaicin was able to exert its action by inhibiting the migration of white blood cells towards inflammation. Capsaicin induced activation of TRPV1, also responsible for decreasing TNF-α levels in an animal model of contact dermatitis. Two coniferyl esters, which were isolated from *C. baccatum* fruit, namely coniferyl-*E*-8-methyl-6-nonenoate (capsiconiate) and coniferyl-8-methylnonanoate (dihydrocapsiconiate), were found to exhibit stimulatory activity for TRPV1 in HEK293 cells (Kobata et al., 2008). It was found that capsaicin provides neuropathic pain relief in adults. An 8% capsaicin patch provides effective relief from hyperalgesia in post-herpetic neuralgia pain occurring after reactivation of the dormant herpes zoster virus (Group, 1992).

4.2.2 ANTIDIABETIC ACTIVITY

C. annuum exerts its potency as a hypoglycaemic agent through the inactivation of different enzymes like α-glucosidase, α-amylase, etc. *C. annuum* can exert its action as a hypoglycemic agent by increasing insulin sensitivity in peripheral tissues, increasing glucose tolerance, preventing apoptosis of β cells, stimulating the secretion of GLP1, and decreasing fasting glucose level in plasma. It was found that insulin resistance and β-cell stress in pre-diabetic NOD mice can be resisted by removing TRPV1$^+$ neurons (Srinivasan, 2016).

Different reports have suggested that regular consumption of hot *Capsicum* may cause an improvement in the postprandial glucose levels. There have been reports suggesting a better efficacy of *C. chinense* against α-amylase than α-glucosidase enzymes (Tundis et al., 2011). *C. frutescens* presents antioxidant and antiglycation properties (Khan et al., 2014). In humans, the effect of *C. frutescens* on plasma glucose level showed a decrease in plasma glucose levels along with increased plasma insulin levels. *C. frutescens* possesses potency in controlling type 2 diabetes mellitus by dipping plasma glucose level and keeping a steady level of insulin (Weerapan Khovidhunkit, 2009). It is also found that 5 mg/dl of capsaicin in a diet for a month can decrease postprandial hyperglycemia and fasting lipid metabolic disorders in case of gestational diabetes mellitus (Salehi et al., 2018). Moreover, it was found that dietary capsaicin can cause decrease in blood glucose concentration in diabetic rats, suggesting its therapeutic efficacy as an antidiabetic agent (Xiong et al., 2016).

4.2.3 Antioxidant Activity

The antioxidant activity of *Capsicum* is mainly because it contains polyphenolic compounds like hydroxycinnamic acids and flavonoid glycosides. A study showed that, red California pepper, Fino lemon, and red onion showed the highest antioxidant capacity in the TEAC, FRAP, and ORAC assays, respectively. The antioxidant activity can be connected with the amount of polyphenolic class of compounds found in extracts of *C. baccatum* (Kappel et al., 2008).

Predominantly, in *C. annuum* and *C. chinense*, with increase in the amount of phytoconstituents like flavonoids, carotenoids, and phenolic acids due to ripening of fruit, there was an increase in the antioxidant activity, observed via *in vitro* studies. Capsaicin possessed an antioxidant potency similar to that of butylhydroxy anisole and has the potential to prevent the oxidation of LDL and copper ion–induced lipid peroxidation. A different research suggested that the main antioxidant phytoconstituents from *C. annuum* were feruloyl and sinapoyl glycosides, and that from green pepper was quercetin-3-O-*L*-rhamnoside, apart from dihydrocapsaicin and capsaicin (Materska and Perucka, 2005). It was found that capsaicin resists peroxidation of lipid induced by ascorbic acid and ferrous sulphate in the membranes of red blood cells, suggesting its action against free radicals to protect erythrocyte integrity. It was found that methanolic extract of *Capsicum* inhibits HNE-induced and H_2O_2-induced DNA damage on human leucocytes and human colorectal adenocarcinoma cells (HT-29) (J.-H. Park et al., 2012). *In vivo* study of antioxidant activity of *C. annuum* was evaluated, and it was observed that the extracts were well tolerated, independently from the ripening stage (Salehi et al., 2018). Capsaicin, because of its antioxidant property, is able to protect cardiac and skeletal muscles from ROS, which are also responsible for ulcers, etc. (Srinivasan, 2016).

4.2.4 CNS Activity

A correlation between a capsaicin-rich diet in healthy human beings with decreased levels of serum β-amyloid and improved cognition, indicating a decreased risk of Alzheimer's disease, was observed. It was shown that dietary capsaicin can reduce its risk in rats with type 2 diabetes by reduction in tau protein phosphorylation in the hippocampus of diabetic rats, a risk factor for Alzheimer's disease.

Capsaicin was found to reduce neurodegeneration, decrease impairment of motor neurons in animal models of Parkinson's disease due to decreased production of ROS and pro inflammatory cytokines, lessen the microglial activation, and reduce neuroinflammation, thus resulting in beneficial effects on Parkinson's disease. An MPTP mouse model of Parkinson's disease revealed that capsaicin, when administered intraperitoneally, resulted in an increase in the number of dopaminergic (tyrosine hydroxylase positive) neurons in the substantia nigra (Wang et al., 2020).

It is commonly believed that Ca^{2+} ions have a significant role in pathogenesis of epilepsy, and thus, the TRPV1 channel, which is highly permeable to calcium ions, is an important target for drug action. The TRPV1 channel, once activated, can lead to membrane depolarisation and result in enhanced glutaminergic activity. On top of that, some studies state that there is a decrease in GABA release upon TRPV1 channel activation. Investigating antiepileptic effects of capsaicin through invitro studies, induction of epileptiform events in cortical neurons were observed, and the antiepileptic effect was most likely because of TRPV1-independent mechanism of action. In contrast, the influence of capsaicin on TRPV1 channels expressed in hippocampal neurons is responsible for its pro-epileptic effects (Chung et al., 2017). It is also proposed that its anti-epileptic action predominates in the cerebral cortex, whereas its pro-epileptic action prevails in some specific portion of brain such as the hippocampus (Carletti et al., 2016).

4.2.5 Cardiovascular Activity

Capsaicin exhibits a triphasic blood pressure response, which includes decrease followed by a subtle increase and a later slow decrease in blood pressure without affecting respiration and heart rate. TRPV1 expression on endothelial cells indicates substantial effect of capsaicin on endothelial TRPV1 channels to stimulate dilation of blood vessels. There have been reports that endothelial TRPV1, when activated by capsaicin, promotes dilation of blood vessels and reduces blood pressure in genetically hypertensive rats. It was found that the aggregation of the TRPV1 channel, expressed in human platelets, can cause myocardial infarction in patients already suffering from coronary atherosclerosis, inferring the role of TRPV1 in the release of 5-HT and ADP (Mankowski et al., 2017). It is also probable that TRPV1 can activate the production of human platelets connected to inflammatory mediators like 12- and 15-hydroxyeicosatetraenoic acid produced in atherosclerotic plaques and platelets (Mankowski et al., 2017). Consequently, capsaicin can result in the increase of intracellular Ca^{2+} concentration of platelets necessary for their degranulation. A study involving ApoE−/− mice confirms the role of capsaicin in reduction of endothelial dysfunction, which is an indispensable factor for atherosclerosis. It was found that capsaicin stimulated the increase in UCP2 expression in TRPV1 and protein kinase in a dependent manner in the endothelium and a decrease in generation of mitochondrial ROS (Harper et al., 2009). It is suggested, though not confirmed in clinical studies, that addition of capsaicin in diet helps in resisting coronary heart disease (Munjuluri et al., 2022). Angiogenesis can be defined as development of new blood vessels from the previously present vasculature via a series of steps such as activation of endothelial cells followed by proliferation, invasion, and migration of chemotactic factors, finally resulting in formation of new vasculature. The vital reason behind suppressing angiogenesis being extremely effective for treating of various cancers is related to neovascularisation, which is often observed when a local carcinoma transforms into an invasive cancer. The remarkable antiangiogenic activity shown by capsaicinoids in mice models as wells as cell cultures suggests it may have relevant therapeutic efficacy. Capsaicin inhibits proliferation of VEGF-induced endothelial cell and invasion, which has been confirmed using Matrigel model systems, chicken chorioallantoic membrane (CAM) models, *ex vivo* rat aortic rings models, and *in vivo* Matrigel plug experiments. Non-pungent capsinoids, dihydrocapsiate, and capsiate have also shown similar results (Park et al., 2000).

4.2.6 Chemopreventive Activity

Capsaicin has also shown potential to cause apoptosis in prostate, lung, colon, oesophageal, bladder, liver, skin, blood, pancreatic, and endothelial cell lines keeping other cells unaffected. The anticancer activity is exerted by targeting signalling pathways and cancer-associated genes at various stages of carcinoma development. A study showed that proliferation of basal carcinoma (BC) cells lines can be resisted by stimulating the autophagic process, altering ADP/ATP ratio together with AMPK pathway when treated with capsaicin (Amantini et al., 2016). The role of peppers against stomach cancer was also revealed in another study, where it was observed in human gastric cell lines that capsaicin destroys cancerous cells via bcl-2-sensitive apoptotic pathway, whereas in human colorectal cancer cells *in vitro*, capsaicin has shown to suppress cell proliferation involving the suppression of transcriptional activity of β-catenin (S.-H. Lee et al., 2012). In addition to that, capsaicin provokes death of colon cancer cells by activating caspase-3, -8, and -9 enzymes, up-regulating pro-apoptotic bax proteins, and down-regulating antiapoptotic bcl-2 and also by inducing cell cycle G_0/G_1 phase arrest. Moreover, carotenoids found in the fruits of *C. annuum* shows chemo-preventive activity by inhibiting the activation of the Epstein-Barr virus (Maoka et al., 2001). Capsaicin was responsible for killing prostate cancer cells and inhibiting the spread of leukemic cells while blocking breast cancer cells (Baba et al., 2006). Compounds similar to capsaicin such as capsanthin, capsanthindiester, capsanthin-3-*O*-ester, capsorubin, capsanthin-3,6-epoxide, capsorubindiester, cucurbitaxanthin A-30 ester, and *β*-carotene isolated from *C. annuum* have also shown *in vitro* anti–tumor-promoting activity (Maoka et al., 2001). Potent chemopreventiveness has been exhibited by nordihydrocapsiate in an *in vivo* mouse skin carcinogenesis model. These capsiates, expressing better proapoptotic results in Jurkat cells in comparison to capsaicin, along with positive results against skin cancer, generally inhibit the transcription factor nuclear factor κB, increase ROS, and cause the loss of mitochondrial membrane potential. N-AVAMs, which are synthetic analogues of non-pungent capsaicin, are extensively studied because they possess anticancer activity (Min et al., 2004). The research area involving the study of anticancer effect of capsaicin analogues when administered along with traditional chemotherapy or radiation has been of keen interest (Batiha et al., 2020).

4.2.7 Hypocholesterolemic and Hypolipidemic Activity

Multiple studies have shown that peppers demonstrate an antiobesity effect by production of heat, increase of energy expenditure, oxidation of fat, lesser energy intake, prevention of production of adipocyte and increased lipolysis in adipose tissues, inhibition of lipase enzymes, inhibition of the differentiation of adipocytes, and modulation of adipokine release from adipose tissues (Sanati et al., 2018). It has been found that dietary capsaicin helps in the formation of bile acids from cholesterol in liver by increasing the activity of cholesterol-7-α-hydroxylase enzyme. A decrease in liver and plasma cholesterol levels was observed when 5% red pepper or similar amount of capsaicin was added to a choline-free hydrogenated diet (M. Srinivasan et al., 1980). Similar results were obtained when only 0.2 mg% of capsaicin (equivalent to 0.06% of pepper) was added to the diet of fat-fed rats. In a sub-chronic toxicity study, rats were treated with capsaicin by gavage or crude extract of *Capsicum* fruit for 60 days, and it was found that after 30, 40, 50, and 60 days, total cholesterol levels together with triglycerides and phospholipids in blood plasma were strikingly reduced (Monsereenusorn, 1983). It was reported that total cholesterol and HDL cholesterol levels in blood of rats get increased when 0.15, 1.5, and 15 mg% of capsaicin is added to the diet for 7 days. The effect of capsaicin as a hypo-cholesterolemic agent was confirmed by observing that serum cholesterol levels in rats on a 1% cholesterol and 5% red pepper diet were lower than the untreated ones. Cholesterol level was also found to be decreased in the groups treated with red pepper and capsaicin. This therapeutic effect of capsaicin has been found in rats having a high-cholesterol diet, but similar results were not observed in Streptozotocin-induced diabetic rats. Capsaicin also seemed to have a positive effect on the levels of the plasma and total cholesterol as well as triglycerides in rabbits on a 0.5% diet. A serum cholesterol–lowering effect was observed in turkeys when kept on a 2 mg to 3 mg capsaicin/kg diet in addition to 0.5% cholesterol. Birds treated with dihydrocapsaicin showed a dip in LDL and total cholesterol. Both the capsaicin and dihydrocapsaicin exerted effect by reducing VLDL cholesterol and increasing HDL cholesterol in the group that had cholesterol in its diet. The hypo-cholesterolemic potential of pressure-cooked and uncooked red pepper was evaluated in rats, and it was confirmed that heat causes significant loss of capsaicin, but its hypolipidemic potency was not affected (Srinivasan, 2016). The hypolipidemic effect of capsaicin is due to its antilithogenic activity. Capsaicin can stimulate bile secretion that can result in decrease in the development of gall bladder stones. A major reduction in number of gallstones was found when capsaicin was added to the diet of hamsters and mice. On top of that, capsaicin also caused a significant reduction of previously formed gall bladder stones in mice. Cholesterol-to-bile acid ratio in the bile, elevated cholesterol saturation index, and ratio of cholesterol to phospholipid caused by lithogenic diet were decreased by capsaicin (Srinivasan, 2016).

4.2.8 Antiulcer Property

C. frutescens, when administered in the form of an extract, helps in treating ulcers caused by aspirin by improving the histology alterations. A study stated the adverse effect of capsaicin in causation of duodenal and peptic ulcers due to inhibition of *in vitro* development of *Helicobacter pylori* at 10 µg/ml. Capsaicin is responsible for stimulation of prostaglandin formation by sensitising the nerves that ultimately results in reduction of gastric lesions generally observed in NSAIDs usage (Batiha et al., 2020).

Stomach infection caused due to *H. pylori* results in excess secretion of acid due to disrupted inhibitory control for acid secretion that destroys the mucosal barrier, leading to gastric ulcer. The phospholipase, often associated with the colonised *H. pylori* in the stomach, damages the protective layer of the stomach. Red pepper (chilli) is often avoided by people suffering from ulcers due to its acid-secreting and irritant nature, but it has been shown that capsaicin present there has no role in ulcer formation but instead helps in protecting gastro-intestinal mucosa (Pasierski and Szulczyk, 2022). It has been shown that capsaicin causes stimulation of alkali and mucus secretions and enhances flow of gastric mucosal blood, helping in disposing of acid from the stomach, thus resisting as well as curing ulcers. Capsaicin, being a phosphodiesterase inhibitor, causes powerful vasodilation, and enhanced cAMP levels exert its antiulcer effect (Satyanarayana, 2006).

4.3 STRUCTURE ACTIVITY RELATIONSHIP (SAR) OF CAPSAICINOIDS

An alkyl chain and a vanillyl group joined through an amide bond form the basic chemical structure of capsaicinoids (Fattorusso and Taglialatela-Scafati, 2007). Because of the variations in the carbon chain such as (i) saturated and ramified, like in dihydrocapsaicin, (ii) unsaturated and ramified, like in capsaicin, or (iii) saturated and linear, like in nonivamide, capsaicinoids exhibit a vast chemical and biological diversity. This chemical structure gives the CAPS an amphiphilic nature devoid of basicity as compared to what is generally seen in alkaloids (Alberti et al., 2008; Cheok et al., 2017). Generally, the SAR of capsaicinoids is investigated by dividing the structure of capsaicin into three parts: (i) the "A" region, which is called the aromatic region, (ii) the "B" region, which is called the amide bond region, and (iii) the "C" region, which is called the hydrophobic side chain (Figure 4.6) (Jordt and Julius, 2002).

4.3.1 Substitution in A Region

The substitution in the aromatic ring (A region) of the capsaicinoids was initially studied by Walpole et al. using an *in vitro* experiment that assessed $^{45}Ca^{2+}$ absorption into neonatal rat dorsal root ganglia neurons (Walpole et al., 1993c). In this study, the analogues with substitution

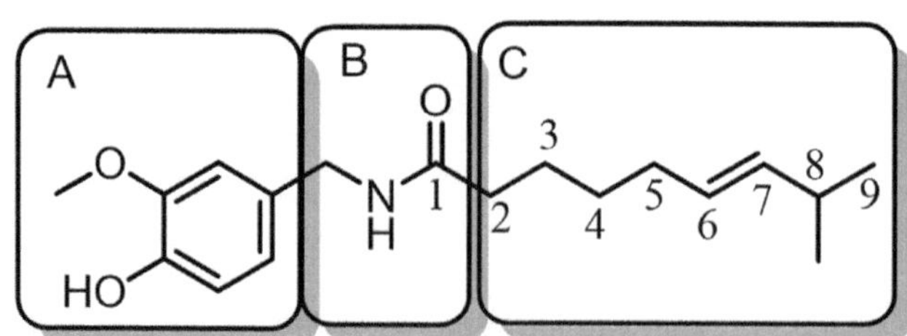

FIGURE 4.6   A, B, and C regions of the capsaicin.

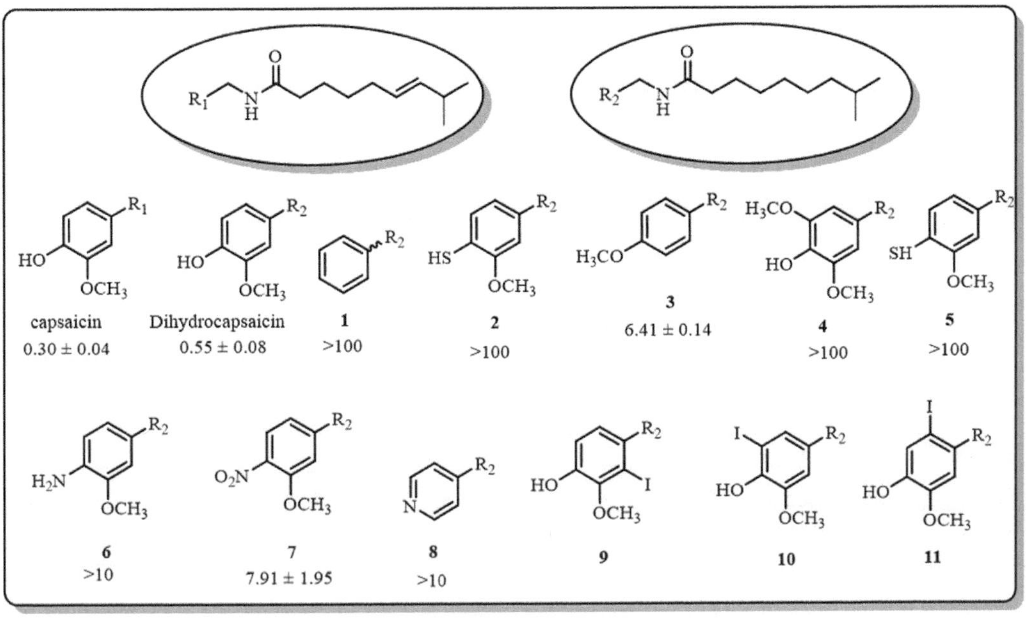

FIGURE 4.7 SAR of capsaicin in A region, their modified analogues, and their 45Ca^{2+} influx.

at the aromatic ring of the capsaicin showed the antinociceptive and analgesic property. The most effective analogues were those having the "parent" 3-methoxy-4-hydroxy benzyl moiety (Figure 4.7). Either singly or in combination, the substitution at positions 2, 5, and 6 on the aromatic ring resulted in either poorly active or inactive molecules (**4**). Activity was simply decreased or eliminated by simple alkylation of the 4-OH or the removal of the 4-OH group. It was notable that residual activity was kept with the nitro substituent (**7**) in contrast to other modifications of the 4-substituent, such as SH or NH$_2$, which eliminated or decreased the activity (**5, 6,** and **8**).

Iodination in the A region (**9, 10,** and **11**) and its impact on antioxidant activity were also studied (Rosa et al., 2009). The antioxidant activity of nonivamide generally decreases when an iodine atom is added to the vanillyl moiety, and this effect is dependent on the location of the iodine atom attached to the aromatic ring.

Capsaicinoids are poorly absorbed when taken orally and are hardly soluble in water and therefore difficult to use as medications or food additives. Capsaicinoids may be glycosylated to give glycoconjugates, which are more water soluble and less pungent, potentially resulting in pro-drugs and weight-loss formulations. Shimoda et al. cultivated the cells of *C. roseus* to glycolyze capsaicin and 8-nordihydrocapsaicin in order to decrease their pungency and increase their solubility in water. Capsaicinoid glycosides are also of physiological interest, since saponins and other glycosides seem to be the active ingredients in many commonly used traditional medicines. More water-soluble capsaicin analogues have been produced through biotransformations (Figures 4.8 & 4.9). However, till now, there have been few studies on the physiological actions of glycosylated analogues (Shimoda et al., 2007; Sivakumar and Divakar, 2007). In addition to synthetic analogues, a novel capsaicin derivative 6,7-dihydro-5,5-dicapsaicin was isolated from the fruits of *Capsicum annuum*. The natural compound possessed nearly the same antioxidant activity to capsaicin, but it lacked the pungent taste (Ochi et al., 2003).

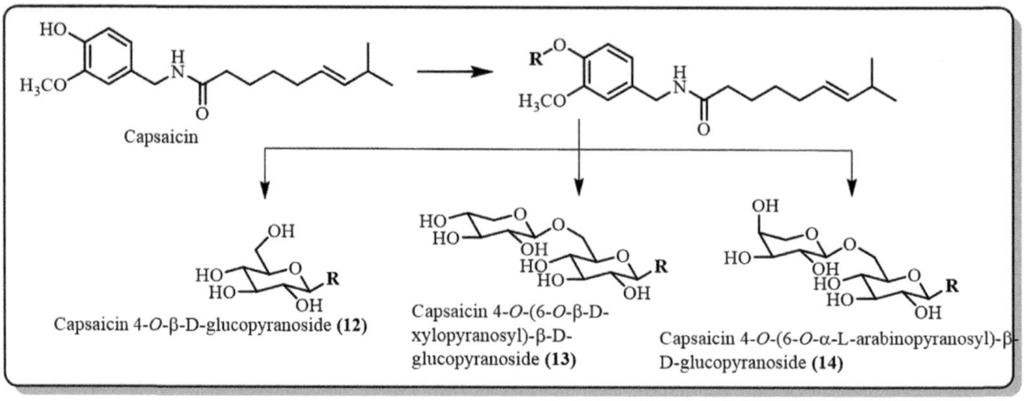

FIGURE 4.8   Glycosylated analogues of capsaicin.

The structures are as follows. Structure of capsaicin as a benzene ring single bonded to an OH group on top left vertex, single bonded to a CH₃O group on bottom left vertex, and single bonded to a carbon chain containing an NH group and an O atom on the top right vertex leading to a structure as a benzene ring single bonded to an RO group on top left vertex, single bonded to a CH₃O group on bottom left vertex, and single bonded to a carbon chain containing an NH group and an O atom on the top right vertex. This compound leads to three different compounds.

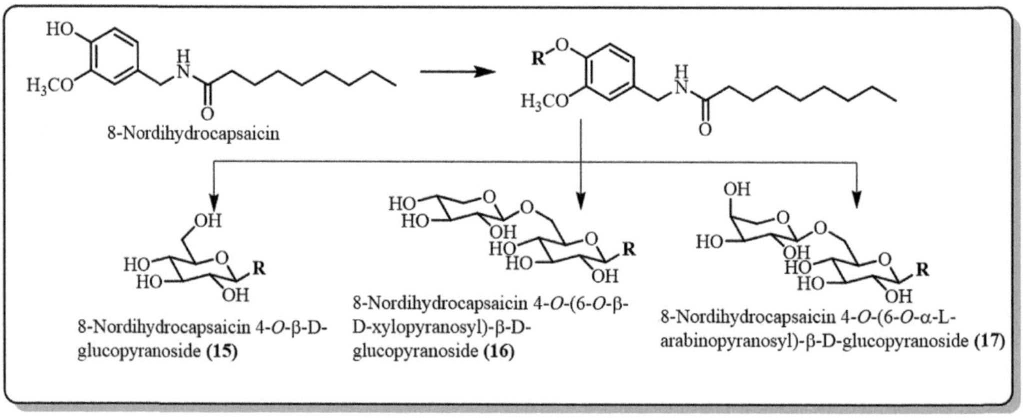

FIGURE 4.9   Glycosylated analogues of 8-nordihydrocapsaicin.

The structures are as follows. Structure of 8-nordihydrocapsaicin as a benzene ring single bonded to an OH group on top left vertex, single bonded to a CH₃O group on bottom left vertex, and single bonded to a carbon chain containing an NH group and an O atom on the top right vertex leading to a structure as a benzene ring single bonded to an RO group on top left vertex, single bonded to a CH₃O group on bottom left vertex, and single bonded to a carbon chain containing an NH group and an O atom on the top right vertex. This compound leads to three different compounds.

Thomas et al. recently synthesised a number of capsaicin analogues with modified A-ring pharmacophores (Figure 4.10). Additionally, molecular modelling was employed to assess the TRPV1 binding of the various analogues that offered a theoretical justification for the variations in agonist potency and toxicity (Thomas et al., 2011).

Analogu Name	R_1	R_2
Nonivamide (18)	OCH_3	OH
N-Benzylnonanamide (19)	H	H
N-(3-Methoxybenzyl)nonanamide (20)	OCH_3	H
N-(3,4-Dimethoxybenzyl)nonanamide (21)	OCH_3	OCH_3
N-(3-Hydroxy-4-methoxybenzyl)nonanamide (22)	OH	OCH_3
N-(3,4-Dihydroxybenzyl)nonanamide (23)	OH	OH
3-Methoxy-4-(nonamidomethyl)phenyl sulfate (24)	OCH_3	SO_4
N-(4-Trifluoromethylbenzyl)nonanamide (25)	H	CF_3
N-(4-Hydroxybenzyl)nonanamide (26)	H	OH

FIGURE 4.10 Nonivamide and their structure analog in the A region.

The structure is a benzene ring single bonded to an R_2 group at the bottom vertex, single bonded to an R_1 group on bottom right vertex, and single bonded to a carbon chain containing an NH group and an O atom on the top vertex. The analogu names with R_1 and R_2 are as follows.

1. Nonivamide (18), with R_1 as OCH_3 and R_2 as OH.
2. N-B enzylnonanamide (19), with R_1 as H and R_2 as H.
3. N-(3-Methoxybenzyl) nonanamide (20), with R_1 as OCH_3 and R_2 as OH.
4. N-(3,4-Dimethoxybenzyl) nonanamide (21), with R_1 as OCH_3 and R 2 as OCH_3.
5. N-(3-Hydroxy-4-methoxybenzyl) nonanamide (22), with R_1 as OH and R_2 as OCH_3.
6. N-(3,4-Dihydroxybenzyl) nonanamide (23), with R_1 as OH and R_2 as OH.
7. 3-Methoxy-4-(nonamidomethyl)phenyl sulfate (24), with R_1 as OCH_3 and R_2 as SO_4.
8. N-(4-Trifluoromethylbenzyl)nonanamide (25), with R_1 as H and R_2 as CF_3.
9. N-(4-Hydroxybenzyl)nonanamide (26), with R_1 as H and R_2 as OH.

4.3.2 Substitution in B Region

Capsiates have an ester bond between the fatty acid side chain and the vanillyl moiety instead of amide bond and have been isolated from the red pepper (Figure 4.11) (Iida et al., 2003). The non-pungent capsiate is a TRPV1 agonist and has the potential to activate peripheral nociceptors. On the other hand, when applied to the oral cavity, eyes, and skin surfaces of mice, capsiate failed to show the significant responses. It is reported that at C1 position substitution of ester for amide and at C4 position substitution of alkoxy for hydroxy provides the correlation among the analgesic activity and molecular structure of capsinoid analogues (He et al., 2009). Compared to indomethacin and vanillyl decanoate (capsinoid precursor), the capsinoid derivatives showed stronger antinociceptive activity, and this property improved by increasing the alkoxyl chain at C4 position. Walpole et al. demonstrated that the capsaicin shows strong agonism action (influx of Ca^{2+} into dorsal root ganglia neurones) when the amide bond of B region is replaced (Walpole et al., 1993a).

One carbon atom seems to be the ideal length requirement for connecting the A region and the dipolar B region moiety; for example, **34** as opposed to **35** and **36** was more effective and potent (Figure 4.12). The amide derivative (**34**) as well as the reverse amide derivative (**37**) showed the same potency. Other analogous pairs include the N-methylamides **40** and **41**, and the higher

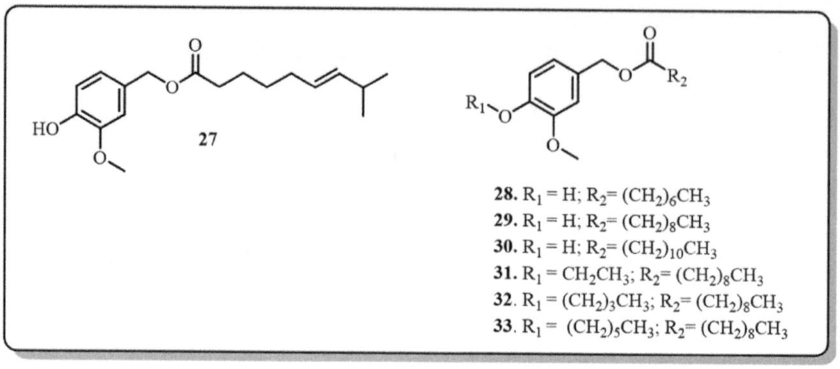

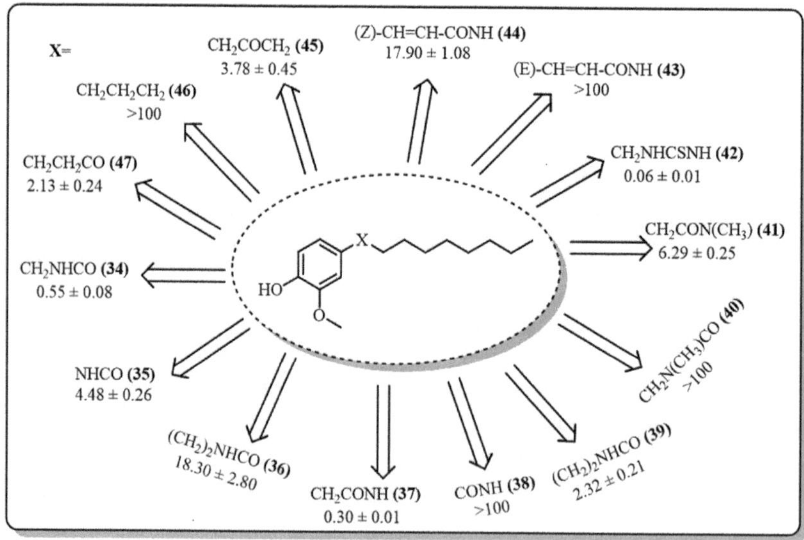

FIGURE 4.11 Capsaicin analogues containing an ester bond.

The structure on the left is a benzene ring single bonded to an OH group on the bottom left vertex, single bonded to an O atom at bottom vertex, and single bonded to a carbon chain containing two O atoms on the top right vertex. The structure on the right is a benzene ring single bonded to an R_1O group on the bottom left vertex, single bonded to an O atom at bottom vertex, and single bonded to a carbon chain containing two O atoms and an R_2 group on the top right vertex. The analogues are listed as follows.

28. $R_1 = H$, $R_2 = (CH_2)_6 CH_3$.
29. $R_1 = H$, $R_2 = (CH_2)_8 CH_3$.
30. $R_1 = H$, $R_2 = (CH_2)_{10} CH_3$.
31. $R_1 = CH_2CH_3$, $R_2 = (CH_2)_8 CH_3$.
32. $R_1 = (CH_2)_3 CH_3$, $R_2 = (CH_2)_8 CH_3$.
33. $R_1 = (CH_2)_5 CH_3$, $R_2 = (CH_2)_8 CH_3$.

FIGURE 4.12 SAR of capsaicin in B region, their modified analogues, and their $45Ca^{2+}$ influx Activity [EC_{50} (μM)].

homologues **36** and **39**, in each of these pairs, the "reverse" analogue is clearly more effective; however, all were considerably less active than the parent amides. Alternatively, sp^2 hybridisation can be necessary at the bridging group. For instance, all weakly active or inactive compounds were those in which the atom adjacent to the aromatic ring is sp^2 hybridised (**38, 43,** and **44**). In this series of analogues, the thiourea moiety **42** showed the greatest efficacy (EC_{50} = 0.06 μM). Numerous other attempts to decrease or eliminate the dipolar functions of the B region resulted in diminution or loss of efficacy (seen in compounds **45** to **47**). The effects of conformational constraints of the A region in regard to the B region on the biological activity were studied (Walpole et al., 2002). Saturated ring systems of different sizes were introduced to create conformational constraint. Other than capsaicinoids, some analogs of capsaicin were synthesised chemically (Figure 4.13). In addition, a moderately potent antagonist, capsazepine **(55)** (having high affinity against TRPV1), a scaffold of capsaicin, has been used to make the synthetic analogs in B region. The C-29, CIDD-24, CIDD-111, CIDD-99 analogs possessed higher anticancer activity (Figure 4.14) (Brown et al., 2023).

4.3.3 SUBSTITUTION IN C REGION

A number of capsaicin analogues **60–68** and vanillylthioureas **69–78** (Figure 4.15) have been synthesised to include the hydrophobic side chain (C region) based on the analgesic effect of capsaicin analogues with modified A region (aromatic ring) and B region (amide bond) (Walpole et al., 1993b). The potencies of dihydro-capsaicin, capsaicin, and octylamide **(60)** were almost equal, suggesting that the total size or the hydrophobicity tends to be significant compared to the double bond and the branched side chain. Except for two compounds **63** and **64**, the substitute with shorter side chains (compounds **62** and **65**) and compound **(61)** with longer polar side chains exhibited minimal or no effect. Numerous studies have been conducted on Olvanil **(64)** as a potential local analgesic. Compounds having a wide range of alkyl chain structures were active in antinociceptive and anti-inflammatory studies (Janusz et al., 1993).

In the aforementioned experiments, short-chain compounds were effective when administered systemically, but they still possessed strong pungency and acute toxicity similar to capsaicin. In comparison to capsaicin, the long-chain substituted analogues (compounds **79** and **80**) were less pungent, showed minimal acute toxicity, and were orally active. On human breast cancer MCF-7

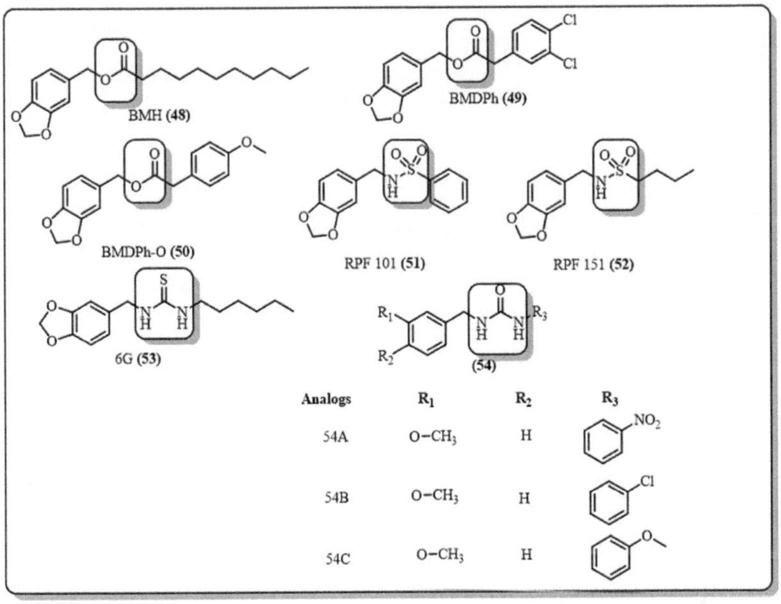

FIGURE 4.13 Synthetic analogs of capsaicin in B region.

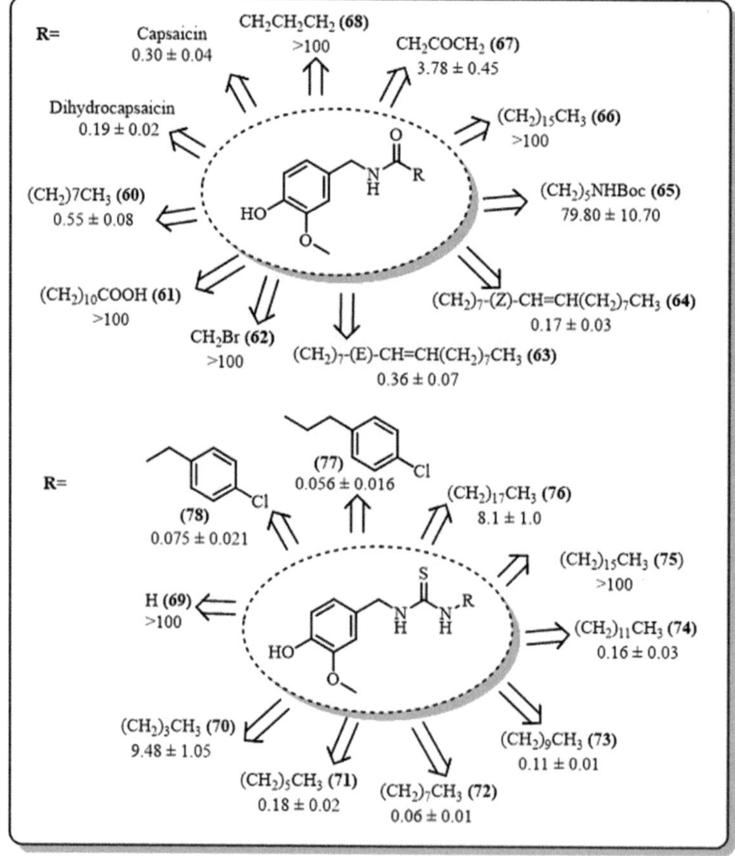

FIGURE 4.14 Capsaicin modified region B in capsazepine and its analog.

FIGURE 4.15 SAR of capsaicin in C region, their modified analogues, and their 45Ca²⁺ influx activity [EC$_{50}$ (µM)].

cells, which lack the expression of caspase-3, Tuoya et al. investigated the effects of capsaicin and its analogue **81**, having a longer and more unsaturated fatty acyl chain. The findings suggested that both vanilloids have the effects that inhibit growth and fragmentation of DNA. However, the potency of capsaicin was less than that of **81** (Tuoya et al., 2006). It is evident that for the series **69–74**, there is an increase in the potency with increase in chain length up to 8 to 12 carbon atoms. The compounds having long chain lengths are inactive or poorly active (Figure 4.16).

The most potent analogue from these compounds, *p*-chloro phenethyl thiourea (**77**), was chosen for in vivo testing to determine its potential as an analgesic drug. Compound **77** had short duration of action, excitatory side effects, and poor oral activity, and its further development was precluded. Initial research on metabolism had revealed that phenol moiety of **77** was quickly glucuronidated *in vivo*. The phenol group was subsequently subjected to modification, producing compounds **82–87** (Figure 4.17) that retained *in vitro* efficacy. Two of these compounds, **85** and **87** had significantly better *in vivo* profiles than the "parent" phenol series (Wrigglesworth et al., 1996). Another analogue, **88**, had antihyperalgesic effects in rats and guinea pigs in addition to producing analgesia in mice. In dose range of the antihyperalgesic/analgesic, **88** did not cause undesirable adverse reactions such as alterations in blood pressure and bronchoconstriction, unlike capsaicin (Urban et al., 2000).

Labelled [³H]RTX studies have shown that RTX binds specifically to the capsaicin binding site at dorsal root ganglia. A putative pharmacophore model for interaction with the capsaicin binding region of VR1 has been proposed, which states that the main pharmacophores are represented by the C20-ester, 4-hydroxy-3-methoxyphenyl, orthophenyl, and C3-keto functional groups. In a [³H]RTX binding experiment on DRG neurons, Lee et al. synthesised two ultrapotent VR1 agonists, **89** and **90**, with K_i values of 19 nM and 11 nM, respectively; the compounds were consequently around 280- and 480-fold more potent than capsaicin, respectively (Lee et al., 2001). A number of powerful antagonists with high binding affinities were produced as a result of the isosteric replacement of the phenolic OH group with the alkylsulfonamido group (Lee et al., 2003). A $^{45}Ca^{2+}$ uptake assay was used to evaluate the agonistic and antagonistic actions of these analogues *in vitro*.

The effects of alterations to A and C regions of 4-methylsulfonamide ligands on VR1 agonism and antagonism were studied using structure activity analysis (**91–96**). Compound **91** showed a potent and full VR1 antagonist activity, and by substituting 3-fluoro-4-methylsulfonylamino pattern in the A region, it also had a remarkable analgesic activity *in vivo*. It was found that the antagonism is shown by the 3-fluoro substitution in the A region and a 4-tert-butylbenzyl moiety in the C region, while the agonism is shown by the substitution of 3-methoxy group in the A region and 3-acyloxy-2-benzylpropyl moieties in the C region (compound **92**) (Figure 4.18) (Chung et al., 2007).

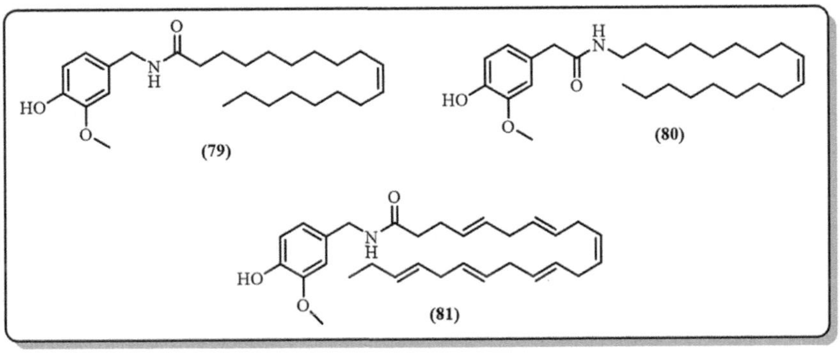

FIGURE 4.16 Capsaicin analogues containing long chain in C region.

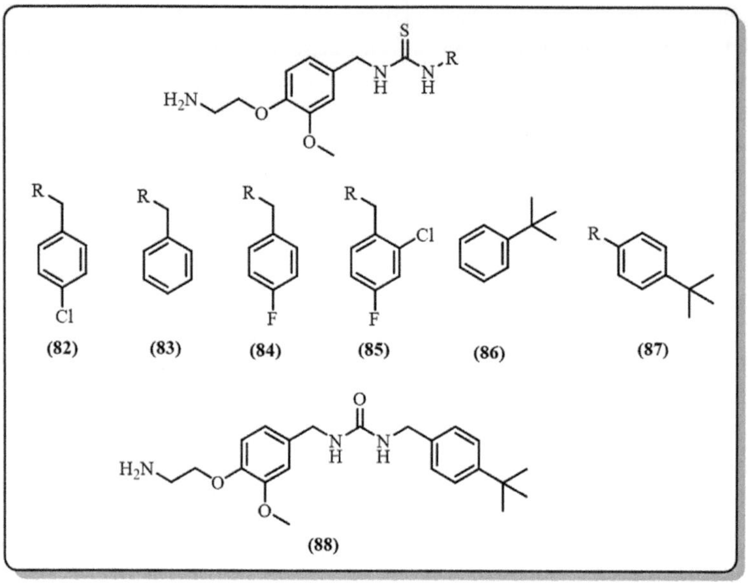

FIGURE 4.17 Capsaicin analogues containing 2-aminoethyl group.

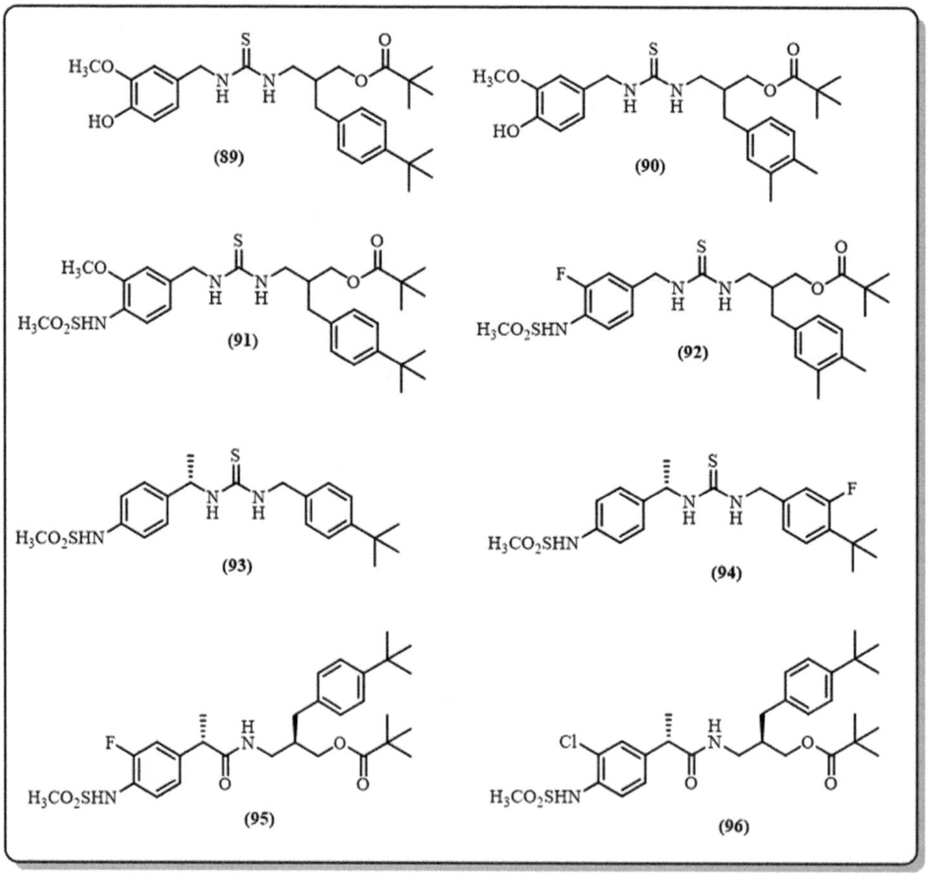

FIGURE 4.18 Analogues of capsaicin containing alkylsulfonamido group.

Adding a methyl group with a (*R*)-configuration increased selective binding to the receptor. The *R*-configuration-containing compounds **93** and **94** showed excellent potencies (K_i, respectively, of 41 and 39.2 nM and K_i (ant), respectively, of 4.5 and 37 nM). The lead compounds were further optimised using a series of their amide and *R*-substituted amide surrogates, and new chiral N-(2-benzyl-3-pivaloyloxypropyl) 2-[4-(methylsulfonylamino) phenyl]-propionamide analogues were identified as powerful and stereospecific rTRPV1 antagonists. The compounds **95** and **96** demonstrated strong antagonistic activity in rTRPV1/CHO cells, with K_i values of 0.58 and 5.2 nM, respectively, and significant binding affinities with K_i values of 4.12 and 1.83 nM, respectively (Ryu et al., 2008).

The toxic catalytic and condensing agents, the production of by-products, and protection and deprotection of aromatic hydroxyl processes are some of the reasons for the partial success in the chemical synthesis of capsaicin analogues from vanillylamine and acyl donors. One of the most spectacular outcomes of the interplay between organic chemistry and enzymology is the use of enzymes for innovative and effective synthetic transformations of organic molecules. As a result, the usual chemical synthesis of capsaicin has been replaced by the intriguing enzyme-catalysed production of its analogues. Castillo et al. synthesised the capsaicin analogs by using a lipase catalysed reaction under favourable thermodynamic conditions, utilising different substrates of acyl chain up to 18 carbon and amine donors (Figure 4.19) (Castillo et al., 2007).

The N-AVAM-containing short acyl chain, acyl groups with short-branched chain, and polar compounds with short acyl groups in the C region showed no activity or weak pain-relieving activity. A saturated long-chain alkyl group in region C yielded molecules having analgesic properties. The addition of unsaturated long-chain fatty acyl side groups resulted in non-toxic, non-spicy, and orally active analogs of capsaicin. By measuring the absorption of radiolabelled calcium (^{45}Ca) into cultured dorsal root ganglia neurons, the pharmacological activity of Compounds (**99–108**) was determined. The antinociceptive and anti-inflammatory efficacy of these substances was evaluated using mouse hot tail flick and mouse ear inflammation models induced by croton oil (Figure 4.20) (Kobata et al., 2010; Richbart et al., 2021).

4.4 CONCLUSION

Capsaicin shows great promise as a potential treatment for various conditions, including chronic cough, migraine, diabetes, cardiovascular conditions, central nervous system, and overactive bladder. Additionally, it exhibits potent anticancer and analgesic properties. However, its intense pungency and nociceptive effects hinder its widespread use in food and medicine. Future research

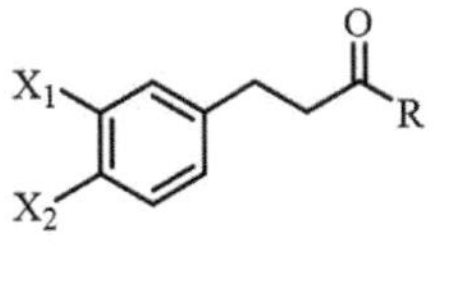

Compound	X_1	X_2	Compound	R
97	MeO-	OH-	a	-C_3H_7
98	H-	H-	b	-C_5H_{11}
98	MeO-	H-	c	-C_7H_{15}
			d	-C_9H_{19}
			e	-$(CH_2)_7CH{=}CH(CH_2)_7CH_3$

FIGURE 4.19 Capsaicin analogues synthesised via lipase catalysed reaction using different acyl and amine substrate.

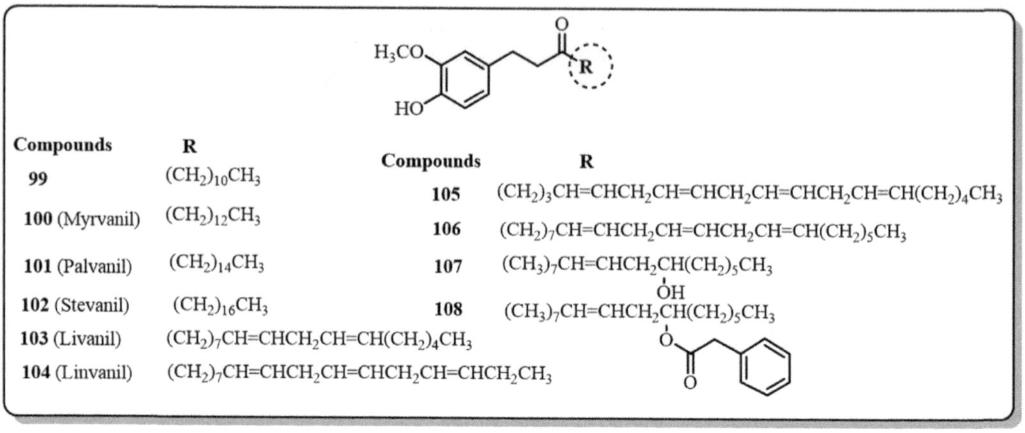

FIGURE 4.20 SAR in C region of the capsaicin with long-chain acyl groups and its analogues having growth-inhibitory activity.

focused on comprehending the mechanisms of action of TRPV1 receptor agonists and antagonists, as well as understanding the structural activity relationships between them, holds the potential to lead to the development of more efficient and tolerable therapeutic agents.

REFERENCES

Alberti, A., Galasso, V., Kovac, B., Modelli, A., & Pichierri, F. (2008). Probing the molecular and electronic structure of capsaicin: A spectroscopic and quantum mechanical study. *Journal of Physical Chemistry, A, 112*(25), 5700–5711. https://doi.org/10.1021/jp801890g

Amantini, C., Morelli, M. B., Nabissi, M., Cardinali, C., Santoni, M., Gismondi, A., & Santoni, G. (2016). Capsaicin triggers autophagic cell survival which drives epithelial mesenchymal transition and chemoresistance in bladder cancer cells in an Hedgehog-dependent manner. *Oncotarget, 7*(31), 50180. https://doi.org/10.18632%2Foncotarget.10326

Ananthan, R., Subhash, K., & Longvah, T. (2018). Capsaicinoids, amino acid and fatty acid profiles in different fruit components of the world hottest Naga king chilli (Capsicum chinense Jacq). *Food Chemistry, 238*, 51–57.

Antonio, A., Wiedemann, L., & Junior, V. V. (2018). The genus Capsicum: A phytochemical review of bioactive secondary metabolites. *RSC Advances, 8*(45), 25767–25784.

Baba, N., Shimoishi, Y., Murata, Y., Tada, M., Koseki, M., & Takahata, K. (2006). Apoptosis induction by dohevanil, a DHA substitutive analog of capsaicin, in MCF-7 cells. *Life Sciences, 78*(13), 1515–1519.

Batiha, G. E.-S., Alqahtani, A., Ojo, O. A., Shaheen, H. M., Wasef, L., Elzeiny, M., & Zaragoza-Bastida, A. (2020). Biological properties, bioactive constituents, and pharmacokinetics of some Capsicum spp. and capsaicinoids. *International Journal of Molecular Sciences, 21*(15), 5179.

Brown, K. C., Modi, K. J., Light, R. S., Cox, A. J., Long, T. E., Gadepalli, R. S., & Dasgupta, P. (2023). Anticancer activity of region B capsaicin analogs. *Journal of Medicinal Chemistry, 66*(7), 4294–4323. http://doi.org/10.1021/acs.jmedchem.2c01594

Carletti, F., Gambino, G., Rizzo, V., Ferraro, G., & Sardo, P. (2016). Involvement of TRPV1 channels in the activity of the cannabinoid WIN 55,212–2 in an acute rat model of temporal lobe epilepsy. *Epilepsy Research, 122*, 56–65.

Castillo, E., Torres-Gavilán, A., Severiano, P., Arturo, N., & López-Munguía, A. (2007). Lipase-catalyzed synthesis of pungent capsaicin analogues. *Food Chemistry, 100*(3), 1202–1208.

Caterina, M. J., Schumacher, M. A., Tominaga, M., Rosen, T. A., Levine, J. D., & Julius, D. (1997). The capsaicin receptor: A heat-activated ion channel in the pain pathway. *Nature, 389*(6653), 816–824.

Cheok, C. Y., Sobhi, B., Adzahan, N. M., Bakar, J., Rahman, R. A., Ab Karim, M. S., & Ghazali, Z. (2017). Physicochemical properties and volatile profile of chili shrimp paste as affected by irradiation and heat. *Food Chemistry, 216*, 10–18.

Chung, J.-U., Kim, S. Y., Lim, J.-O., Choi, H.-K., Kang, S.-U., Yoon, H.-S., & Kang, B. (2007). α-Substituted N-(4-tert-butylbenzyl)-N'-[4-(methylsulfonylamino) benzyl] thiourea analogues as potent and stereo-specific TRPV1 antagonists. *Bioorganic & Medicinal Chemistry, 15*(18), 6043–6053.

Chung, Y. C., Baek, J. Y., Kim, S. R., Ko, H. W., Bok, E., Shin, W.-H., & Jin, B. K. (2017). Capsaicin prevents degeneration of dopamine neurons by inhibiting glial activation and oxidative stress in the MPTP model of Parkinson's disease. *Experimental & Molecular Medicine, 49*(3), e298–e298.

Conforti, F., Statti, G. A., & Menichini, F. (2007). Chemical and biological variability of hot pepper fruits (*Capsicum annuum* var. acuminatum L.) in relation to maturity stage. *Food Chemistry, 102*(4), 1096–1104.

Constant, H. L., Cordell, G. A., & West, D. P. (1996). Nonivamide, a constituent of Capsicum oleoresin. *Journal of Natural Products, 59*(4), 425–426.

Constant, H. L., Cordell, G. A., West, D. P., & Johnson, J. H. (1995). Separation and quantification of capsa-icinoids using complexation chromatography. *Journal of Natural Products, 58*(12), 1925–1928.

Contreras-Padilla, M., & Yahia, E. M. (1998). Changes in capsaicinoids during development, maturation, and senescence of chile peppers and relation with peroxidase activity. *Journal of Agricultural and Food Chemistry, 46*(6), 2075–2079.

Cortright, D. N., & Szallasi, A. (2004). Biochemical pharmacology of the vanilloid receptor TRPV1: An update. *European Journal of Biochemistry, 271*(10), 1814–1819.

de Sá Mendes, N., & de Andrade Gonçalves, É. C. B. (2020). The role of bioactive components found in pep-pers. *Trends in Food Science & Technology, 99*, 229–243.

Duelund, L., & Mouritsen, O. G. (2017). Contents of capsaicinoids in chillies grown in Denmark. *Food Chemistry, 221*, 913–918.

Fattorusso, E., & Taglialatela-Scafati, O. (2007). *Modern alkaloids: Structure, isolation, synthesis, and biol-ogy.* Wiley-VCH Verlag GmbH & Co. KgaA.

Garcés-Claver, A., Arnedo-Andrés, M. S., Abadía, J., Gil-Ortega, R., & Álvarez-Fernández, A. (2006). Determination of capsaicin and dihydrocapsaicin in Capsicum fruits by liquid chromatography-electrospray/time-of-flight mass spectrometry. *Journal of Agricultural and Food Chemistry, 54*(25), 9303–9311.

González-Zamora, A., Sierra-Campos, E., Luna-Ortega, J. G., Pérez-Morales, R., Ortiz, J. C. R., & García-Hernández, J. L. (2013). Characterization of different capsicum varieties by evaluation of their capsa-icinoids content by high performance liquid chromatography, determination of pungency and effect of high temperature. *Molecules, 18*(11), 13471–13486.

Group, C. S. (1992). Effect of treatment with capsaicin on daily activities of patients with painful diabetic neuropathy. *Diabetes Care, 15*(2), 159–165.

Harper, A., Brownlow, S., & Sage, S. (2009). A role for TRPV1 in agonist-evoked activation of human plate-lets. *Journal of Thrombosis and Haemostasis, 7*(2), 330–338.

Hayman, M., & Kam, P. C. (2008). Capsaicin: A review of its pharmacology and clinical applications. *Current Anaesthesia & Critical Care, 19*(5–6), 338–343.

He, G.-J., Ye, X.-L., Mou, X., Chen, Z., & Li, X.-G. (2009). Synthesis and antinociceptive activity of capsinoid derivatives. *European Journal of Medicinal Chemistry, 44*(8), 3345–3349.

Iida, T., Moriyama, T., Kobata, K., Morita, A., Murayama, N., Hashizume, S., & Tominaga, M. (2003). TRPV1 activation and induction of nociceptive response by a non-pungent capsaicin-like compound, capsiate. *Neuropharmacology, 44*(7), 958–967.

Janusz, J. M., Buckwalter, B. L., Young, P. A., LaHann, T. R., Farmer, R. W., Kasting, G. B., & Maddin, C. S. (1993). Vanilloids. 1. Analogs of capsaicin with antinociceptive and antiinflammatory activity. *Journal of Medicinal Chemistry, 36*(18), 2595–2604.

Jordt, S.-E., & Julius, D. (2002). Molecular basis for species-specific sensitivity to "hot" chili peppers. *Cell, 108*(3), 421–430.

Kappel, V. D., Costa, G. M., Scola, G., Silva, F. A., Landell, M. F., Valente, P., & Moreira, J. C. (2008). Phenolic content and antioxidant and antimicrobial properties of fruits of Capsicum baccatum L. var. pendulum at different maturity stages. *Journal of Medicinal Food, 11*(2), 267–274.

Khan, I., Ahmad, H., & Ahmad, B. (2014). Anti-glycationand Anti-oxidation properties of Capsicum frutes-cens and Curcuma longa fruits: Possible role in prevention of diabetic complication. *Pakistan Journal of Pharmaceutical Sciences, 27*(5), 1359–1362.

Kirschbaum-Titze, P., Hiepler, C., Mueller-Seitz, E., & Petz, M. (2002). Pungency in paprika (Capsicum ann-uum). 1. Decrease of capsaicinoid content following cellular disruption. *Journal of Agricultural and Food Chemistry, 50*(5), 1260–1263.

Knotkova, H., Pappagallo, M., & Szallasi, A. (2008). Capsaicin (TRPV1 Agonist) therapy for pain relief: Farewell or revival? *The Clinical Journal of Pain, 24*(2), 142–154.

Kobata, K., Saito, K., Tate, H., Nashimoto, A., Okuda, H., Takemura, I., & Watanabe, T. (2010). Long-chain N-vanillyl-acylamides from Capsicum oleoresin. *Journal of Agricultural and Food Chemistry, 58*(6), 3627–3631. https://doi.org/10.1021/jf904280z

Kobata, K., Tate, H., Iwasaki, Y., Tanaka, Y., Ohtsu, K., Yazawa, S., & Watanabe, T. (2008). Isolation of coniferyl esters from Capsicum baccatum L., and their enzymatic preparation and agonist activity for TRPV1. *Phytochemistry, 69*(5), 1179–1184.

Korel, F. G., Bağdatlioğlu, N. M., Balaban, M. Ö., & Hişil, Y. (2002). Ground red peppers: Capsaicinoids content, Scoville scores, and discrimination by an electronic nose. *Journal of Agricultural and Food Chemistry, 50*(11), 3257–3261.

Laskaridou-Monnerville, A. (1999). Determination of capsaicin and dihydrocapsaicin by micellar electrokinetic capillary chromatography and its application to various species of Capsicum, Solanaceae. *Journal of Chromatography A, 838*(1–2), 293–302.

Lee, E.-J., Jeon, M.-S., Kim, B.-D., Kim, J.-H., Kwon, Y.-G., Lee, H., & Kim, T.-Y. (2010). Capsiate inhibits ultraviolet B-induced skin inflammation by inhibiting Src family kinases and epidermal growth factor receptor signaling. *Free Radical Biology and Medicine, 48*(9), 1133–1143.

Lee, J., Lee, J., Kang, M., Shin, M., Kim, J.-M., Kang, S.-U., & Park, H.-G. (2003). N-(3-Acyloxy-2-benzylpropyl)-N '-[4-(methylsulfonylamino) benzyl] thiourea analogues: Novel potent and high affinity antagonists and partial antagonists of the vanilloid receptor. *Journal of Medicinal Chemistry, 46*(14), 3116–3126.

Lee, J., Lee, J., Kim, J., Kim, S. Y., Chun, M. W., Cho, H., & Marquez, V. E. (2001). N-(3-Acyloxy-2-benzylpropyl)-N'-(4-hydroxy-3-methoxybenzyl) thiourea derivatives as potent vanilloid receptor agonists and analgesics. *Bioorganic & Medicinal Chemistry, 9*(1), 19–32.

Lee, S.-H., Richardson, R. L., Dashwood, R. H., & Baek, S. J. (2012). Capsaicin represses transcriptional activity of β-catenin in human colorectal cancer cells. *The Journal of Nutritional Biochemistry, 23*(6), 646–655.

Loizzo, M. R., Pugliese, A., Bonesi, M., Menichini, F., & Tundis, R. (2015). Evaluation of chemical profile and antioxidant activity of twenty cultivars from Capsicum annuum, Capsicum baccatum, Capsicum chacoense and Capsicum chinense: A comparison between fresh and processed peppers. *LWT-Food Science and Technology, 64*(2), 623–631.

Luo, X.-J., Peng, J., & Li, Y.-J. (2011). Recent advances in the study on capsaicinoids and capsinoids. *European Journal of Pharmacology, 650*(1), 1–7.

Macho, A., Lucena, C., Sancho, R., Daddario, N., Minassi, A., Muñoz, E., & Appendino, G. (2003). Non-pungent capsaicinoids from sweet pepper: Synthesis and evaluation of the chemopreventive and anticancer potential. *European Journal of Nutrition, 42*, 2–9.

Macho, A., Sancho, R., Minassi, A., Appendino, G., Lawen, A., & Muñoz, E. 2003. Involvement of reactive oxygen species in capsaicinoid-induced apoptosis in transformed cells. *Free Radical Research, 37*(6), 611–619.

Mankowski, C., Poole, C. D., Ernault, E., Thomas, R., Berni, E., Currie, C. J., & Zafeiropoulou, E. (2017). Effectiveness of the capsaicin 8% patch in the management of peripheral neuropathic pain in European clinical practice: The ASCEND study. *BMC Neurology, 17*(1), 1–11.

Maoka, T., Mochida, K., Kozuka, M., Ito, Y., Fujiwara, Y., Hashimoto, K., & Tokuda, H. (2001). Cancer chemopreventive activity of carotenoids in the fruits of red paprika *Capsicum annuum* L. *Cancer Letters, 172*(2), 103–109.

Materska, M., & Perucka, I. (2005). Antioxidant activity of the main phenolic compounds isolated from hot pepper fruit (*Capsicum annuum* L.). *Journal of Agricultural and Food Chemistry, 53*(5), 1750–1756.

Min, J.-K., Han, K.-Y., Kim, E.-C., Kim, Y.-M., Lee, S.-W., Kim, O.-H., & Kwon, Y.-G. (2004). Capsaicin inhibits in vitro and in vivo angiogenesis. *Cancer Research, 64*(2), 644–651.

Mohammad, R., Ahmad, M., & Heng, L. Y. (2013). An amperometric biosensor utilizing a ferrocene-mediated horseradish peroxidase reaction for the determination of capsaicin (chili hotness). *Sensors, 13*(8), 10014–10026.

Monsereenusorn, Y. (1983). Subchronic toxicity studies of capsaicin and capsicum in rats. *Research Communications in Chemical Pathology and Pharmacology, 41*(1), 95–110.

Munjuluri, S., Wilkerson, D. A., Sooch, G., Chen, X., White, F. A., & Obukhov, A. G. (2022). Capsaicin and TRPV1 channels in the cardiovascular system: The role of inflammation. *Cells, 11*(1), 18.

Nelson, E. (1919). The constitution of capsaicin, the pungent principle of capsicum. *Journal of the American Chemical Society, 41*(7), 1115–1121.

Ochi, T., Takaishi, Y., Kogure, K., & Yamauti, I. (2003). Antioxidant activity of a new capsaicin derivative from *Capsicum annuum*. *Journal of Natural Products, 66*(8), 1094–1096. http://doi.org/10.1021/np020465y

Park, J.-H., Jeon, G.-I., Kim, J.-M., & Park, E. (2012). Antioxidant activity and antiproliferative action of methanol extracts of 4 different colored bell peppers (*Capsicum annuum* L.). *Food Science and Biotechnology, 21*, 543–550.

Park, J.-S., Choi, M.-A., Kim, B.-S., Han, I.-S., Kurata, T., & Yu, R. (2000). Capsaicin protects against ethanol-induced oxidative injury in the gastric mucosa of rats. *Life Sciences, 67*(25), 3087–3093.

Pasierski, M., & Szulczyk, B. (2022). Beneficial effects of capsaicin in disorders of the central nervous system. *Molecules, 27*(8), 2484.

Perucka, I., & Oleszek, W. (2000). Extraction and determination of capsaicinoids in fruit of hot pepper Capsicum annuum L. by spectrophotometry and high-performance liquid chromatography. *Food Chemistry, 71*(2), 287–291.

Reyes-Escogido, M. D. L., Gonzalez-Mondragon, E. G., & Vazquez-Tzompantzi, E. (2011). Chemical and pharmacological aspects of capsaicin. *Molecules, 16*(2), 1253–1270.

Richbart, S. D., Friedman, J. R., Brown, K. C., Gadepalli, R. S., Miles, S. L., Rimoldi, J. M., & Finch, P. T. (2021). Nonpungent N-AVAM capsaicin analogues and cancer therapy. *Journal of Medicinal Chemistry, 64*(3), 1346–1361.

Rosa, A., Appendino, G., Melis, M. P., Deiana, M., Atzeri, A., Alessandra, I., & Dessì, M. A. (2009). Protective effect and relation structure-activity of nonivamide and iododerivatives in several models of lipid oxidation. *Chemico-Biological Interactions, 180*(2), 183–192. https://doi.org/10.1016/j.cbi.2009.01.002

Rosa, A., Deiana, M., Casu, V., Paccagnini, S., Appendino, G., Ballero, M., & Dessí, M. A. (2002). Antioxidant activity of capsinoids. *Journal of Agricultural and Food Chemistry, 50*(25), 7396–7401.

Ryu, H., Jin, M.-K., Kim, S. Y., Choi, H.-K., Kang, S.-U., Kang, D. W., & Morgan, M. A. (2008). Stereospecific high-affinity TRPV1 antagonists: Chiral N-(2-benzyl-3-pivaloyloxypropyl) 2-[4-(methylsulfonylamino) phenyl] propionamide analogues. *Journal of Medicinal Chemistry, 51*(1), 57–67.

Salehi, B., Hernández-Álvarez, A. J., del Mar Contreras, M., Martorell, M., Ramírez-Alarcón, K., Melgar-Lalanne, G., & Nadeem, M. (2018). Potential phytopharmacy and food applications of Capsicum spp.: A comprehensive review. *Natural Product Communications, 13*(11), 1934578X1801301133.

Sanati, S., Razavi, B. M., & Hosseinzadeh, H. (2018). A review of the effects of Capsicum annuum L. and its constituent, capsaicin, in metabolic syndrome. *Iranian Journal of Basic Medical Sciences, 21*(5), 439.

Santamaria, R., Reyes-Duarte, M., Barzana, E., Fernando, D., Gama, F., Mota, M., & Lopez-Munguia, A. (2000). Selective enzyme-mediated extraction of capsaicinoids and carotenoids from chili guajillo puya (*Capsicum annuum* L.) using ethanol as solvent. *Journal of Agricultural and Food Chemistry, 48*(7), 3063–3067.

Sato, K., Sasaki, S. S., Goda, Y., Yamada, T., Nunomura, O., Ishikawa, K., & Maitani, T. (1999). Direct connection of supercritical fluid extraction and supercritical fluid chromatography as a rapid quantitative method for capsaicinoids in placentas of Capsicum. *Journal of Agricultural and Food Chemistry, 47*(11), 4665–4668.

Satyanarayana, M. (2006). Capsaicin and gastric ulcers. *Critical Reviews in Food Science and Nutrition, 46*(4), 275–328.

Sawynok, J. (2005). Topical analgesics in neuropathic pain. *Current Pharmaceutical Design, 11*(23), 2995–3004.

Shimoda, K., Kwon, S., Utsuki, A., Ohiwa, S., Katsuragi, H., Yonemoto, N., & Hamada, H. (2007). Glycosylation of capsaicin and 8-nordihydrocapsaicin by cultured cells of *Catharanthus roseus*. *Phytochemistry, 68*(10), 1391–1396. https://doi.org/10.1016/j.phytochem.2007.03.005

Sivakumar, R., & Divakar, S. (2007). Syntheses of N-vanillyl-nonanamide glycosides using amyloglucosidase from Rhizopus and β-glucosidase from sweet almond. *Biotechnology Letters, 29*(10), 1537–1548. https://doi.org/10.1007/s10529-007-9424-4

Srinivasan, K. (2016). Biological activities of red pepper (*Capsicum annuum*) and its pungent principle capsaicin: A review. *Critical Reviews in Food Science and Nutrition, 56*(9), 1488–1500.

Srinivasan, M., Sambaiah, K., Satyanarayana, M., & Rao, M. (1980). Influence of red pepper and capsaicin on growth, blood constituents and nitrogen balance in rats. *Nutrition Reports International, 21*(3), 455–467.

Thomas, K. C., Ethirajan, M., Shahrokh, K., Sun, H., Lee, J., Cheatham, T. E., & Reilly, C. A. (2011). Structure-activity relationship of capsaicin analogs and transient receptor potential vanilloid 1-mediated human lung epithelial cell toxicity. *Journal of Pharmacology and Experimental Therapeutics, 337*(2), 400–410. http://doi.org/10.1124/jpet.110.178491

Tundis, R., Loizzo, M. R., Menichini, F., Bonesi, M., Conforti, F., Statti, G., & Menichini, F. (2011). Comparative study on the chemical composition, antioxidant properties and hypoglycaemic activities of two *Capsicum annuum* L. cultivars (Acuminatum small and Cerasiferum). *Plant Foods for Human Nutrition, 66*, 261–269.

Tuoya, Baba, N., Shimoishi, Y., Murata, Y., Tada, M., Koseki, M., & Takahata, K. (2006). Apoptosis induction by dohevanil, a DHA substitutive analog of capsaicin, in MCF-7 cells. *Life Sciences, 78*(13), 1515–1519. https://doi.org/10.1016/j.lfs.2005.07.019

Uarrota, V. G., Maraschin, M., de Bairros, Â. d. F. M., & Pedreschi, R. (2021). Factors affecting the capsaicinoid profile of hot peppers and biological activity of their non-pungent analogs (Capsinoids) present in sweet peppers. *Critical Reviews in Food Science and Nutrition, 61*(4), 649–665. https://doi.org/10.1080/10408398.2020.1743642

Urban, L., Campbell, E. A., Panesar, M., Patel, S., Chaudhry, N., Kane, S., & James, I. F. (2000). In vivo pharmacology of SDZ 249–665, a novel, non-pungent capsaicin analogue. *Pain, 89*(1), 65–74.

Walker, J., Ley, J. P., Schwerzler, J., Lieder, B., Beltran, L., Ziemba, P. M., & Krammer, G. E. (2017). Nonivamide, a capsaicin analogue, exhibits anti-inflammatory properties in peripheral blood mononuclear cells and U-937 macrophages. *Molecular Nutrition & Food Research, 61*(2), 1600474.

Walpole, C. S., Bevan, S., Bovermann, G., Boelsterli, J. J., Breckenridge, R., Davies, J. W., & Winter, J. (2002). The discovery of capsazepine, the first competitive antagonist of the sensory neuron excitants capsaicin and resiniferatoxin. *Journal of Medicinal Chemistry, 37*(13), 1942–1954.

Walpole, C. S., Wrigglesworth, R., Bevan, S., Campbell, E. A., Dray, A., James, I. F., & Winter, J. (1993a). Analogs of capsaicin with agonist activity as novel analgesic agents; structure-activity studies. 2. The amide bond "B-region". *Journal of Medicinal Chemistry, 36*(16), 2373–2380.

Walpole, C. S., Wrigglesworth, R., Bevan, S., Campbell, E. A., Dray, A., James, I. F., & Winter, J. (1993b). Analogs of capsaicin with agonist activity as novel analgesic agents; structure-activity studies. 3. The hydrophobic side-chain "C-region". *Journal of Medicinal Chemistry, 36*(16), 2381–2389.

Walpole, C. S., Wrigglesworth, R., Bevan, S., Campbell, E. A., Dray, A., James, I. F., & Winter, J. (1993c). Analogs of capsaicin with agonist activity as novel analgesic agents; structure-activity studies. 1. The aromatic "A-region". *Journal of Medicinal Chemistry, 36*(16), 2362–2372.

Wang, J., Sun, B.-L., Xiang, Y., Tian, D.-Y., Zhu, C., Li, W.-W., & Jin, W.-S. (2020). Capsaicin consumption reduces brain amyloid-beta generation and attenuates Alzheimer's disease-type pathology and cognitive deficits in APP/PS1 mice. *Translational Psychiatry, 10*(1), 230.

Weerapan Khovidhunkit, M. (2009). Pharmacokinetic and the effect of capsaicin in Capsicum frutescens on decreasing plasma glucose level. *Journal of Medical Association of Thailand, 92*(1), 108–113.

Williams, O. J., Raghavan, G. V., Orsat, V., & Dai, J. (2004). Microwave-assisted extraction of capsaicinoids from capsicum fruit. *Journal of Food Biochemistry, 28*(2), 113–122.

Wrigglesworth, R., Walpole, C. S., Bevan, S., Campbell, E. A., Dray, A., Hughes, G. A., & Winter, J. (1996). Analogues of capsaicin with agonist activity as novel analgesic agents: Structure–activity studies. 4. Potent, orally active analgesics. *Journal of Medicinal Chemistry, 39*(25), 4942–4951.

Xiong, S., Wang, P., Ma, L., Gao, P., Gong, L., Li, L., & He, H. (2016). Ameliorating endothelial mitochondrial dysfunction restores coronary function via transient receptor potential vanilloid 1–mediated protein kinase A/uncoupling protein 2 pathway. *Hypertension, 67*(2), 451–460.

5 Analytical Methods for Capsaicinoids and Other Bioactive Metabolites

Inder Pal Singh, Dharmistha Rajput and Debanjan Chatterjee

CONTENTS

ABBREVIATIONS

ATPS: Aqueous two-phase system, BJO: Bhut jolokia oleoresin, CARS: Competitive adaptive reweighted sampling, CAP: Capsaicin, CC: Capsaicin content, CCD: Central composite design, CTAB: Cetrimonium bromide, DHC: Dihydrocapsaicin, GO–Fe_3O_4: Graphene oxide–magnetite, GY: Global yield, HC: Homocapsaicin, HDHC: Homodihydrocapsaicin, HPLC: High-performance liquid chromatography, HPTLC: High-performance thin-layer chromatography, IAC: Immune-affinity chromatography, MC-UVE: Monte Carlo uninformative variable elimination, MISER: Multiple injections in a single experimental run, MSPE: Magnetic solid phase extraction, MS; Mass spectrometry, NDHC: Nordihydrocapsaicin, NVM: N-vanillylnonanamide, PLS: Partial least squares, RMSEPv: Root mean square error of prediction for the validation set, RPDv: Residual prediction deviation for the validation set, RSM: Response surface methodology, SDS: Sodium dodecylsulfate, SFE: Supercritical fluid extraction, SHU: Scoville heat units, SPME: Solid phase micro-extraction, TPC: Total phenolic content, VABPLS: Variable adaptive boosting partial least squares, VIP: Variable importance in the projection.

5.1 INTRODUCTION

Capsaicinoids, the non-toxic alkaloids found in chilli peppers, are responsible for the peppery taste of chillies. They are the amides of branched-chain fatty acids (C_9-C_{11}) and vanillylamine. Capsaicin (CAP) and dihydrocapsaicin (DHC), which have concentrations of 0.1% to 1.0% and a ratio of 1:1–2:1, respectively, comprise 80% to 90% of the capsaicinoids. CAP (8-methyl-N-vanillyl-6-nonenamide), a fat-soluble solid, is the key capsaicinoid, having melting point of 62°C to 65°C and a molecular weight of 305.4 kDa. It has pain-relieving, anti-inflammatory, anticancer, antioxidant, anticholesterolemic, antiobesity, antimicrobial, and insect-repellent properties (Kaiser et al., 2018).

Tresh identified and crystallized capsaicin in 1876, whereas Nelson and Dawson determined its molecular structure in 1919 (Asnin and Park, 2013). Plants produce capsaicin by two biosynthetic pathways, the phenylpropanoid pathway and the fatty acid pathway for the phenolic structure and

DOI: 10.1201/9781003378259-5

fatty acids, respectively (Liu and Zou, 2020). After reaching its peak between 40 and 50 days during fruit development, capsaicin tends to degrade into secondary chemicals as a result of peroxidase activity (Aranha et al., 2017).

The CAP molecule consists of an aromatic ring, an amide bond, and a hydrophobic side chain (Chen et al., 2019). The major capsinoids of sweet peppers are esters of capsaicin, namely capsiate and its dihydro-derivatives. The vanillyl moiety attached to the hydrophobic chain is connected by an amide bond (capsaicin-type compounds) and an ester (capsiate-type compounds) bond, which is the characteristic difference between capsaicinoids and capsiates (Sancho and Minassi, 2003). Due to the steric barrier of the *cis* form, capsaicin is only detected as the *trans* isomer (Naves et al., 2019).

Wilbur L. Scoville created the earliest technique for gathering quantitative data in 1912 (Siudem et al., 2020). The Scoville method, which measures the quantity of sugar needed to balance a pepper's pungency, is an organoleptic method. A Scoville heat unit (SHU) is used to measure the outcome of such sugar neutralization (Wang et al., 2018). The irritation brought on by the pungency of capsaicin and other capsaicinoids frequently restricts the existing and potential innovative applications of these compounds. The Scoville scale, which rates pungency based on taste, determines "hotness" of chilli peppers. A habanero pepper rates between 2,000 and 3,000 Scoville units, a *Capsicum* (or bell pepper) scores between 0 and 100 Scoville units, and pure CAP rates between 2,000 and 16 million Scoville units. Due to this, researchers have been looking for analogous substances or methods to lessen the pungency, but not at the price of the existing or prospective future pharmacological effects of capsaicinoids (Contreras-padilla and Yahia, 1998).

The potential of capsaicinoids for application in food, nutritional supplements, pharmaceuticals, and self-defense goods has recently heightened interest in them. It is imperative to establish an effective qualitative and quantitative approach of capsaicinoids for consumer rights and to uphold the regulation regarding capsaicinoids concentration in commercial products. Thin-layer chromatography (TLC), gas chromatography (GC), high-performance liquid chromatography (HPLC), mass spectrometry (MS), liquid chromatography (LC), solid-phase micro-extraction (SPME), and nuclear magnetic resonance (NMR) have been used to analyze these substances (Woodman and Negoescu, 2019; Gomez-Calvario et al., 2016). This chapter discusses appropriate extraction and analytical techniques for determining capsaicinoids in a range of samples.

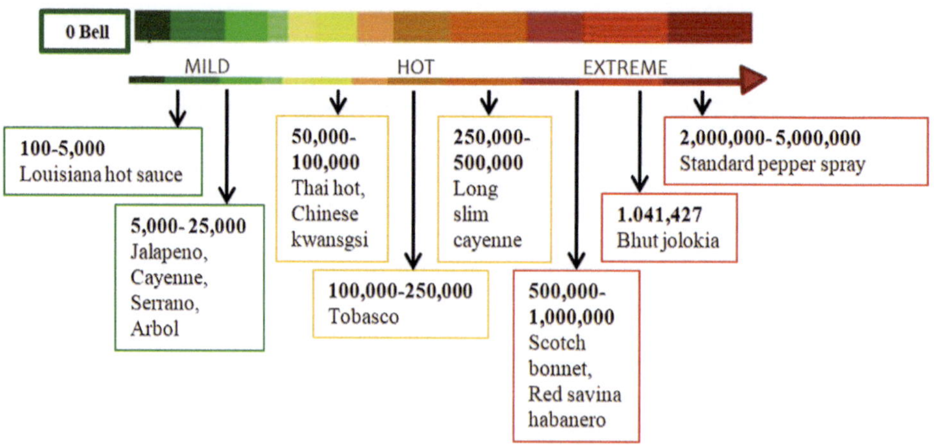

FIGURE 5.1 SHU of various peppers.

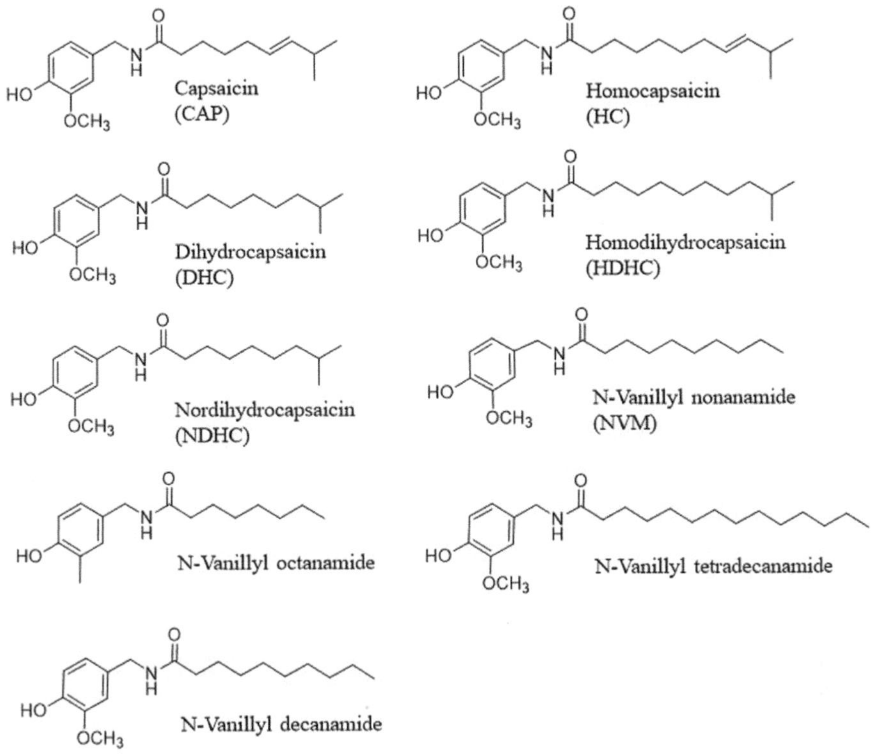

FIGURE 5.2 Structures of major capsaicinoids.

5.2 EXTRACTION TECHNIQUES

Extraction of capsaicin (CAP) from a commercial oleoresin that was acquired from Shandong Datang Phytochemicals Co., Ltd. in Qingdao, China using the aqueous two-phase system (ATPS) technique was studied (Zhao et al., 2015). A number of compositions consisting of alcohols (1-propanol, ethanol, and 2-propanol) and salts such as ammonium sulphate, potassium carbonate, sodium citrate, and dipotassium phosphate were tried. The maximum yield of capsaicin was obtained in the ATPS consisting of ethanol-potassium carbonate and 1-propanol-sodium citrate systems. The ethanol-potassium carbonate system was considered better ATPS because 1-propanol is known to be toxic, while sodium citrate leads to crystallization in ATPS. The ethanol-potassium carbonate system, having 20% w/w ethanol and 22.3% w/w potassium carbonate (pH of buffer solution = 2.74), gave a yield of 85.4% of CAP in the upper ethanol-rich phase due to its less polar nature. The lower phase contained the oil and capsanthin ester. The extraction followed by two-step chromatography on macro-porous resin and reverse-phase resin gave 85% pure CAP. The method was later optimized for large-scale isolation of capsaicin. Two chromatography steps gave CAP recoveries of 93% and 80%, respectively, with the productivity of 1.86 and 4.2 g CAP/L oleoresin. The developed method effectively separated and purified capsaicin on a large scale by combining ATPS and chromatography with advantages in terms of selectivity, recovery, cost-effectiveness, and scalability.

The yield (%) of CAP in ATPS was calculated using the formula

$$Y(\%) = C_{Product} \times V_{Product} / M_{Added} \times C_{Added}$$

Where

$C_{product}$ = concentration of CAP in the product phase (upper alcohol-rich phase) (CAP concentration was determined using HPLC)

$V_{product}$ = volume of product phase (upper alcohol-rich phase)

M_{added} = mass of CAP oleoresin added

C_{added} = content of CAP in CAP oleoresin (CAP content was determined by the U.S. Pharmacopoeia method) (Zhao et al., 2015).

The extraction of capsaicinoids, primarily CAP, DHC, and nordihydrocapsaicin (NDHC), was studied using two ultrasonic-assisted extraction (UAE) methods: one utilizing a homogenizer and an ultrasonic probe (UPR) with a frequency of 20 kHz for 30 minutes and the second utilizing an ultrasonic bath (UBA) and ultrasonic cleaner with a frequency of 35 kHz and 50W for 30 minutes (Bajer et al., 2016). Both the methods were compared with one another and with conventional Soxhlet (SOX) method (extraction carried out for 2 hours). Sample and solvent ratio for UAE taken was 100 mg:45 ml methanol and 100 mg:100 ml methanol for SOX. Cayenne pepper was utilized as a sample for optimization of extraction methods. Seven chilli fruit samples taken for the analysis from three different species of peppers included Trinidad scorpion morunga, yellow habanero and Jamaican red hot pepper from *Capsicum chinense*; yellow bedder, ring of fire and chiltepin from *Capsicum annum,* and tabasco from *Capsicum frutescens*. The content of capsaicinoids was determined using HPLC-MS.

The SHU was calculated from the total content of capsaicinoids using the formula

$$SHU = 16.1 \times [CAP(PPM) + DHC(PPM)] + 9.3 \times NDHC(PPM)$$

Where the coefficient of the heat value is 9.3 for NDHC and 16.1 for both CAP + DHC.

The highest content of CAP, DHC, and NDHC was observed in Trinidad scorpion morunga (CAP & DHC) and tabasco chilli fruits. It was also noted that the yield of capsaicinoids elevated with the rise in power supplied to the system. The result showed the UPR method to be the better for highest content yield. This study showed that the UPR method gave higher yields of capsaicinoids when compared to SOX extraction. The benefits include fast extraction, ease of use, and minimal effort; higher ultrasound power and lower frequencies yielded the best results (Bajer et al., 2016).

The method for extraction of principal capsaicinoids (CAP & DHC) from habanero pepper was developed by Martínez et al., 2019, who compared four extraction techniques: near-infrared (NIR), microwave (MW), ultrasonic (US), and SOX. The non-conventional methods were used: (i) irradiation without solvent, (ii) irradiation with solvent, and (iii) two-step process in which the sample was first irradiated in the absence of solvent and then in the presence of solvent. The study also focused

TABLE 5.1

Total Content of Capsaicinoids (mg/g) Obtained in Different Chilli Fruits Using UPR, UBA & SOX

Variety of chilli fruits	UPR		UBA		SOX	
	Total content	SHU (ppm)	Total content	SHU (ppm)	Total content	SHU (ppm)
Trinidad scorpion morunga	838.11	1,345,146	817.06	1,313,213	613.84	985,400
Yellow habanero	1.16	18,625	1.15	18,480	0.98	15,596
Jamaican red hot	4.51	70,917	4.41	69,287	3.45	54,162
Yellow bedder	9.15	143,563	9.25	145,265	5.3	83,503
Ring of fire	4.99	76,010	4.42	67,295	3.97	60,549
Chiltepin	0.62	9,444	0.57	8,734	0.576	8,788
Tabasco	7.97	121,034	7.85	118,856	6.63	100,386

on increasing the ratio of CAP:DHC apart from developing a green approach of extraction. The capsaicinoids were determined by GC-MS without need of any prior derivatization.

The concentrations of CAP and DHC were calculated using

$$Weight\ of\ extract(without\ solvent) \times \frac{GC\ peak\ area\%}{100} \times Weight\ of\ chilli\ sample$$

For the solvent-free condition, 1 g sample of pepper was kept in a flask without solvent and then subjected to the three different energy sources: NIR for 15 min at 80°C, ultrasonication bath for 35 min at 65°C, or MW for 6 min at 65°C. The hold time was 5 min at 65°C, 100 W powers. After cooling, the samples were washed with 30 ml of ethanol, filtered, and analyzed by GC-MS. The results showed that MW irradiation without solvent gave the highest yield of capsaicinoids. These methods enhanced the extraction of CAP and DHC, surpassing the performance of Soxhlet extraction. Additionally, the processes are simple, safe, highly reproducible, and environment friendly (Martínez et al., 2019).

Another study examined the effect of various surfactants on extraction ability of various solvents for the extraction of CAP from *capsicum annum L.* fruits (Abbas et al., 2022). The solvents dichloromethane, acetonitrile, methanol, toluene, acetic acid, ethyl acetate, acetone, glycerol, and surfactants cetrimonium bromide or CTAB (cationic), sodium dodecylsulfate or SDS (anionic), tween-80 (non-ionic) were used. Dry chilli powder was macerated with the solvents in the presence and absence of the surfactants. The yield of extracts was calculated from the absorbance at 280 nm.

Ethyl acetate was a highly efficacious solvent for extracting capsaicin from chilli and provided the highest yield compared with other solvents, while toluene was the least efficient. The extraction efficiency order for other solvents was, dichloromethane > acetone > glycerol > acetonitrile > methanol > acetic acid. Among surfactants used, tween-80 efficaciously increased the extracting capability of acetone and dichloromethane. The results suggested the possibility of developing techniques that are optimized to produce a significant amount of CAP from chilli peppers (Abbas et al., 2022).

The extraction of capsaicinoids from bhut jolokia fruits with the non-hybrid supercritical fluid extraction (SFE) method was optimized using a central composite design (CCD) of the response surface methodology (RSM) (Ksh et al., 2023). RSM helps to explore the relationships between several explanatory variables and one or more response variables. The experimental layout was prepared by

TABLE 5.2

Comparison of Four Extraction Methods for CAP and DHC Extraction

Extraction method	Extraction time (min)	Ethanol (ml)	Concentration (mg/g)		Yield (%)	
			CAP	DHC	CAP	DHC
NIR[a]	15	0	337.20	237.19	33.72	23.71
NIR[b]	15	30	266.73	158.49	26.67	15.85
NIR[c]	30	30	453.57	307.03	45.3	30.70
US[a]	35	0	362.73	253.37	36.27	25.34
US[b]	35	30	367.95	255.60	36.80	25.56
US[c]	70	30	438.77	294.35	43.88	29.44
MW[a]	6	0	398.13	268.96	39.81	26.90
MW[b]	6	30	480.97	308.68	48.90	30.87
MW[c]	12	30	377.48	266.34	37.75	26.63
SOX	180	250	235.05	221.84	23.51	22.19

[a]Irradiation without solvent, [b]Irradiation with solvent, [c]Two-step process

TABLE 5.3

Extraction Yield Obtained in Different Solvents and Effect of Surfactants

Solvent used for extraction	Yield (mg/10 g)	Yield (mg/10 g) (solvent + surfactants)		
		Tween 80	CTAB	SDS
Toluene	2.77	—	—	—
Acetic acid	4.08	—	—	—
Methanol	7.37	—	—	—
Acetonitrile	7.87	—	—	—
Glycerol	17.78	—	—	—
Acetone	28.8	61.6	46.7	43.7
Dichloromethane	35.98	71.20	35.57	46.53
Ethyl acetate	73.97	72.74	72.62	72.29

making use of design expert software. The optimized bhut jolokia oleoresin (BJO) was analyzed for functional quality with regards to its total phenolic content. The oleoresin or the global yield (GY) of bhut jolokia powder was examined using the SOX method. Bhut jolokia powder (3 g) in 100 ml hexane was kept for extraction for 6 h. Then the extract was concentrated, and capsaicin content (CC) was determined. The supercritical carbon dioxide (SC-CO$_2$) extraction parameters such as temperature (40, 50, 60, 70, and 80°C), time (30, 60, 90, 120, and 150 min), and pressure (150, 225, 300, 375, and 450 bar) were investigated using RSM so that CC and GY of BJO could be maximized. Flow rate of SC-CO$_2$ was set at 35 g/min. Further, emphasizing the extraction capability, 5% (w/w) ethanol was employed as a co-solvent. CAP was quantified by ultra-fast liquid chromatography (UFLC) on a shim-pack solar reverse phase C-18 column (4.6 × 250 mm; 5 μm particle size) with an isocratic elution of acetonitrile and water (70:30), flow rate 0.8 ml/min, run time 15 min, detection 280 nm, and column temperature 40 °C. Injection volume was 20 μl. The samples to be analyzed were prepared from a 1000 ppm solution of the CAP in acetonitrile as a stock solution. For analysis, the concentrations of the standard prepared were 50, 100, 200, 300, and 500 ppm. The total phenolic content (TPC) was determined for the optimized BJO. The BJO solution of 1000 ppm was prepared initially in acetonitrile; 100 μl of 650 μg ml^{-1} solution of the optimized BJO was added to 3 ml of distilled water, Folin-Ciocalteau reagent 500 μl was added, and it was kept at incubation for 2 min. Later on, 2 ml of 20 % sodium carbonate solution was added, and absorbance was observed at 750 nm against a blank reagent following incubation for 30 min. The results were calculated as

$$\text{Gallic acid equivalents} \frac{(GAE)(mg)}{100(g)(DW)}$$

BJ powder consisted of carbohydrates (64.79%), proteins (11.29%), crude fats (4.98%), ash (7.88%), and moisture (11.05 %). The conventional Soxhlet extraction carried out for 6 h had yield of 11.73 ± 1.15 % of oleoresin (GY) and CC of 485.83 ± 45.79 mg/g. The optimized extraction parameters were 68.31°C/347.98 bars/102.50 min, of SFE, where ethanol was used as a co-solvent and SC-CO$_2$ was used as the main solvent yielded 7.23 ± 2.15 % of oleoresin and CAP content of 367.14 ± 1.12 mg/g. The optimized method of SFE resulted in good yield in less time compared to conventional SOX. It also provides advantages such as non-toxicity, affordability, non-inflammability, chemical inertness, greater availability, and when kept to room conditions (25°C, 1 atm) removal from the extract becomes easy (Ksh et al., 2023).

The study showed that combining ATPE and chromatography efficiently separated (1.86 and 4.2 g CAP/L oleoresin) and purified (93% and 80%) CAP at a large scale, highlighting benefits in selectivity, recovery, cost-effectiveness, and scalability. The UPR method surpassed SOX extraction, offering higher capsaicinoid yields in a shorter processing time. Ultrasound-assisted extraction, using higher power and lower frequencies, proved superior, ensuring simplicity, safety, reproducibility, and eco-friendliness. The optimized SFE method provided a quick, non-toxic, and affordable means of CAP extraction from chili peppers.

5.3 ANALYTICAL TECHNIQUES

5.3.1 CHROMATOGRAPHIC TECHNIQUES

Capsaicin and ascorbic acid were quantified in 24 genotypes of Indian chilli (*Capsicum annuum L.*) by using HPTLC on pre-coated silica gel 60 F_{254} plates for the experiment and chloroform: methanol: acetic acid (9.5: 0.5: 0.1, v/v/v) was used as the mobile phase for developing the TLC plate. The TLC plate was then derivatized by Gibb's reagent and exposed to ammonia vapors in a twin trough chamber. Due to the spontaneous reaction, blue violet zones were observed. Densitometric scanning was performed at 254 nm. The CAP content in each sample was quantified using the calibration graph. The regression co-efficient, relative standard deviation, slope, and intercept on the Y-axis were determined. The limit of detection (LOD) and the limit of quantification (LOQ) were calculated from the relative standard deviation. CAP content was turned to SHU by multiplying content of pepper (g) by heat value co-efficient for CAP (1.6×10^7). The ascorbic acid content was estimated volumetrically and calculated by the formula

$$\text{Ascorbic acid}\left(\frac{mg}{g}\right) = \left(\frac{0.5mg}{V_{1ml}}\right) \times \left(\frac{V_2}{5_{ml}}\right) \times \frac{100_{ml}}{Weight\ of\ sample} \times 100$$

Where

V_1 = 5 ml standard ascorbic acid (0.1 mg/ml) with 4% oxalic acid titrated against 2, 6-dichlorophenol indophenol dye till the visibility of persistent pink colour, which was considered an end point. The quantity of dye consumed (V_1 ml) is equal to the quantity of ascorbic acid.

V_2 = 5 ml of sample (2.5 g of fruit in 100 ml of 4% oxalic acid) having 10 ml of 4% oxalic acid was titrated against dye, and the quantity of dye consumed was considered V_2.

CC was found to be in the range of 500 to 51,000 mg/100 g DW, where 2011/CHIVAR-8 had possessed the minimum CAP content of 500 mg/100 g DW, and 2011/CHIVAR-6 had the highest content of 50,890 mg/100 g DW. SHU as measured for the mentioned pepper genotypes ranged from 80,000 to 8,200,000 SHU; where 2011/CHIVAR-8 had the lowest 80,000 SHU and 2011/CHIVAR-6 had highest 8,142,400 SHU as compared to other genotypes. Ascorbic acid content in green chilli ranged from 26 to 112 mg/100 g fresh weight (FW), and in red chilli, it ranged from 45 to 318 mg/100 g FW. This study demonstrated that genetic variability in pepper germplasm can be used to create chili varieties with higher CC and higher vitamin C content, meeting a range of dietary requirements. The variation in pepper germplasm that has been observed in terms of CAP and ascorbic acid content presents good chances for breeding new varieties with improved nutritional characteristics (Pradhan et al., 2018).

Eight distinct varieties of Capsicum annuum L. from the Comarca Lagunera region in North Mexico were evaluated for their capsaicinoids content (Gonzalez-Zamora et al., 2013). Puya, serrano, guajillo, ancho, bell pepper, chiltepín, de arbol, and jalapeno were chosen for the study, where bell pepper was considered a negative reference for other types because it had no pungency. The main goal was to study the effects of temperature on capsaicinoid content and their pungency levels on eight types of Capsicum annuum L. HPLC-DAD (high-performance liquid chromatography—diode

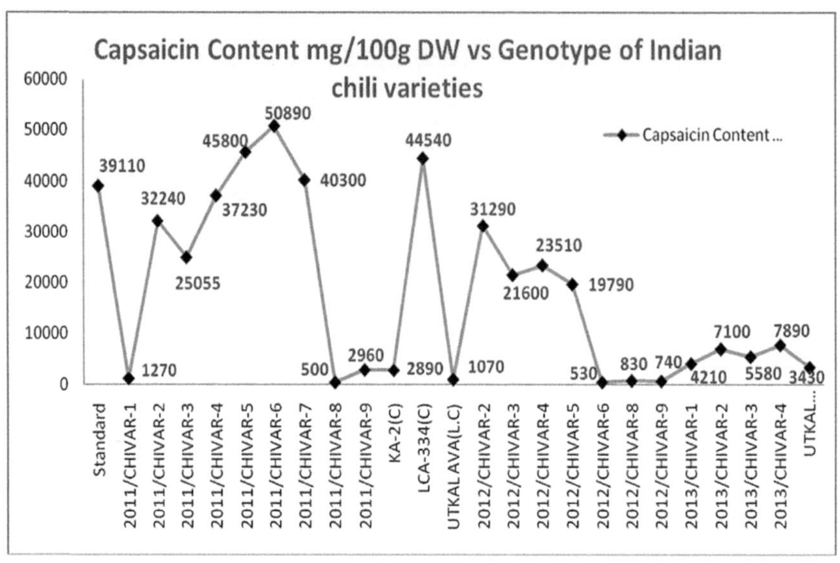

FIGURE 5.3 Capsaicin content of chilli varieties.

array) was used with reversed-phase column Kromasil Eternity-5-C18 (4.6 × 150 mm), isocratic elution using a mixture of water:acetonitrile (50:50) with run time of 15 min, detection at 222 and 280 nm. The main capsaicinoids observed in the samples were CAP, DHC, homocapsaicin (HC), nordihydrocapsaicin (NDHC), and homodihydrocapsaicin (HDHC). Retention times of CAP and DHC were 7 and 10 mins, respectively. Calibration curve was prepared using standards CAP and DHC. NDHC and HDHC were quantified using DHC, whereas HC was quantified using CAP because they were considered their equivalent.

Results showed that chiltepin under thermal stress possessed the highest amount of capsaicinoids (31.84 mg/g DW), and in conditions without thermal stress, serrano had the highest amount of capsaicinoids (18.05 mg/g DW). The pungency of serrano was found to be 262,916 SHU which makes it most pungent among other chilli type under thermal stress condition and Chiltepin was found to be the most pungent among other chilli type with 483,089 SHU under without thermal stress condition. The outcome of the study highlights the effect of high temperatures on the accumulation of capsaicinoids in particular pepper cultivars, such as guajillo, serrano, jalapeno, and de arbol (Gonzalez-Zamora et al., 2013).

The HPLC method using buffer-free eluent was developed by Daood et al., 2015 to measure the capsaicinoids in Hungarian paprika. Three types of columns, namely conventional ODS-3, bifunctional C-18 Sphinx, and cross-linked ISIS columns, were utilized. The fluorescent detector was kept at 280 nm (extinction wavelength) and 320 nm (emission wavelength). The isocratic elution was carried out using three mobile phases: (A) acetonitrile—phosphate buffer (pH 4), (B) acetonitrile—water, and (C) acetonitrile—0.1% formic acid in water at different proportions, 0.8 ml/min flow rate, at room temperature. The cross-linked ISIS column showed better separation of capsaicinoids using ACN: water as mobile phase in the ratio of 1:1. The non-pungent ground spice red pepper samples were spiked with the known amounts of capsaicinoids, which provided 94% to 96% recovery of the capsaicinoids. Capsaicinoids content was analyzed in three samples: pungent Hungarian paprika, chili samples, and non-pungent paprika. The capsaicinoid content observed for the organic chilli-3 in chili samples was (2.81 mg/g), Rubin blend -1 in pungent Hungarian paprika (1.66 mg/g), and Rubin blend -1 in non-pungent paprika was (0.05 mg/g). To effectively separate all capsaicinoids in pepper products, a new HPLC method with short run time was developed. The method is also

TABLE 5.4

Effect of Temperature on Capsaicinoid Content and Its Pungency Degree (SHU)

Capsicum annuum L. type	Serrano	Puya	Ancho	Guajilo	De arbol	Chiltepin	Jalapeno	Bell pepper
Without thermal stress (mg/g DW)								
CAP	1.52	1.18	0.29	0.17	5.22	15.36	8.03	ND
HC	ND	0.08	ND	ND	0.13	0.22	0.16	ND
DHC	3.54	2.32	0.77	0.61	6.25	13.39	9.39	ND
NDHC	0.53	0.55	0.25	0.12	1.07	2.17	2.48	ND
HDHC	0.64	0.68	0.58	0.56	1.29	0.69	0.97	ND
Total Capsaicinoids	6.25	4.8	1.88	1.46	13.96	31.84	21.03	ND
SHU (ppm)	86,427	61,526	19,335	13,704	194,591	483,089	303,602	ND
Thermal stress (mg/g DW)								
CAP	4.76	1.54	1.32	1.32	4.20	—	2.35	ND
HC	0.36	—	—	—	0.04	—	—	ND
DHC	10.14	3.37	0.63	2.83	4.19	—	4.18	ND
NDHC	2.47	0.90	0.44	0.27	0.64	—	1.19	ND
HDHC	0.31	ND	0.60	0.33	0.33	—	0.40	0.41
Total Capsaicinoids	18.05	5.81	2.05	4.74	9.4	—	8.11	0.41
SHU (ppm)	262,916	87,440	20,275	69,210	141,072	—	116,158	ND

(*ND = Not detected)

useful for generating precise paprika product fingerprints to support authenticity, traceability, and the evaluation of pungency variations across various pepper products (Daood et al., 2015).

A UHPLC method on BEH column with UV and fluorescence detector was developed for better separation as well as to study the content CAP and DHC in different parts of chilli fruits of 19 chilli samples of 10 varieties from two species that are *Capsicum annum L.* and *Capsicum chinense* collected from Austria. A mobile phase 0.1% formic acid in both water (eluent-A) and acetonitrile (eluent-B) was used in gradient elution, run time (0–4 mins) at 60°C, and flow rate at 1 ml/min. LOD and LOQ observed for CAP was 0.040 µg ml^{-1} (UV detector), 0.001 µg ml^{-1} (fluorescence detector), and 0.136 µg ml^{-1} (UV detector), 0.003 µg ml^{-1} (fluorescence detector), respectively, whereas for DHC, it was 0.046 µg ml^{-1} (UV detector), 0.001 µg ml^{-1} (fluorescence detector), and 0.145 µg ml^{-1} (UV detector), 0.003 µg ml^{-1} (fluorescence detector). The lowest capsaicin content was about 0.13 mg/g DW for fruit flesh of a jalapeno variety from *Capsicum annum L.* and highest at 21.9 mg/g DW in whole fruit of bhut jolokia variety from *Capsicum chinense.* Similarly, DHC content was observed to be lowest 0.061 mg/g DW in fruit flesh of the jalapeno variety from *Capsicum annum L.* and highest content 7.91 mg/g DW in seeds of the bhut jolokia variety from *Capsicum chinense.* Pericarp of chilli fruit showed the lowest concentration of 4120 ± 271 µg/g and 1070 ± 56.1 µg/g DW of CAP and DHC, whereas placenta and seeds had 12,400 ± 40.9 µg/g of CAP and 3720 ± 8.88 µg/g of DHC, respectively. In this study, a rapid and efficient UHPLC method has been proposed for quantifying CAP and DHC, achieving quick separation and utilizing fluorescence detection. The method had superior sensitivity compared to UV detection, allowing for precise analysis of capsaicinoid content variations among chili species, varieties, and different fruit parts (Schmidt et al., 2017).

In conclusion, various analytical techniques have been employed for the analysis of capsaicinoids in peppers. The HPTLC method was developed to explore the genetic diversity in pepper germplasm in chilli varieties with higher CAP and vitamin C content, addressing diverse dietary needs. HPLC was used to elucidate the intricate relationship between high temperatures and the accumulation of capsaicinoids, a pivotal aspect for enhancing the nutritional profile of chilli varieties. Additionally, other HPLC and UPLC methods were developed for separation and quantitation of CAP and DHC, offering efficient tools for precise analysis of chilli variations in terms of pungency and nutritional content.

5.3.2 Spectroscopic Techniques

A single-scan non-destructive near-infrared spectroscopy (NIR) method was designed for predicting the hotness of fresh red pepper (*Capsicum frutescens L.*). One hundred sixty pepper samples were collected for study, and 40 fresh peppers were gathered after a month to do the external validation study. The near-stalk, equator, and near-navel regions of chilli fruit single-scan spectra were recorded between 898 and 1720 nm. Different models such as competitive adaptive re-weighted sampling (CARS), Monte Carlo uninformative variable elimination (MC-UVE), variable adaptive boosting partial least squares (VABPLS), and boosting partial least squares (PLS) was optimized and evaluated using root mean square error of prediction for the validation set (RMSEPv) and residual prediction deviation for the validation set (RPDv) for analysing CAP, DHC, and pungency degree. The content of capsaicin at the near-stalk region was the highest, whereas the amounts at the equator and near-navel regions were considerably lower. The cause might be that the CAP and DHC predominantly exist at the near-stalk region of fresh peppers. This research presented a non-destructive method using portable near-infrared spectroscopy (NIRS) and chemometric techniques to predict the hotness of fresh peppers. The study found that NIRS models near the stalk area were the most accurate for hotness prediction, offering a simple and cost-effective quality assessment method for the pepper industry (Chen et al., 2022).

A rapid and convenient method for quantifying capsaicinoids using ^{1}H NMR spectroscopy was developed by Valim et al., 2019. This method was applied to analyze various pepper samples, including malagueta, bhut jolokia, habanero, Trinidad scorpion, and Carolina reaper. The analysis focused on determining the levels of capsaicinoids, specifically CAP and DHC, as well as their SHU. For qNMR analysis, 80 µL solution of maleic acid in deuterated water at a concentration of 6.7 mg/ml was used as an internal standard. Various dilutions of CAP and DHC were prepared using a 1:1 mixture of triple-distilled water and methanol, in the scale of 0.5 mg/ml to 5 mg/ml. The validation of this process was carried out by evaluating its linearity, selectivity, precision, detection limit, accuracy, quantification limit, and robustness. Distinct non-overlapping signals were identified; the

FIGURE 5.4 Chilli fruit regions.

signal at δ 5.4 ppm for the unsaturation of CAP and the signal at δ 1.5 ppm for DHC were used for quantification. Total capsaicinoids were quantified from the signal δ 6.7 ppm assigned to the aromatic moiety (H-2 and H-3) for this class of compounds.

The following equation was used for quantitation:

$$Wx = \left(\frac{Ix}{Istd}\right) \times \left(\frac{Nstd}{Nx}\right) \times \left(\frac{Mx}{Mstd}\right) \times Wstd$$

Where

I_x = integral of the analyte; I_{std} = integral of the standard; N_x = no. of nuclei (proton) of analyte; N_{std} = no. of nuclei (proton) of standard; M_{std} = molecular weight of standard; M_x = molecular weight of analyte; W_{std} = weight of internal standard; W_x = amount of analyte

The LOD values were found to be 0.639 mg/ml at δ_H 5.4 ppm, 0.960 mg/ml at δ_H 6.7 ppm for CAP, and 1.111 mg/ml at δ_H 6.7 ppm for DHC. On the other hand, the LOQ values were determined to be 1.076 mg/ml at δ_H 5.4 ppm, 2.246 mg/ml at δ_H 6.7 ppm for CAP, and 2.633 mg/ml at δ_H 6.7 ppm for DHC. Additionally, HPLC analysis was conducted to validate the method. The study introduced a method for quantifying capsaicinoids in commercial peppers using ^{1}H NMR analysis without deuterated solvents. The method was validated by linearity, selectivity, accuracy, precision, and robustness studies. The use of non-deuterated solvents did not interfere with the method, demonstrating superior precision compared to previous methods (Valim et al., 2019).

For quantifying the total capsaicinoids and CAP in dried chilli and chilli oleoresin from 15 distinct landraces of chilli from four different species—*Capsicum pubescens*, *Capsicum chinense*, *Capsicum annuum*, and *Capsicum frutescens*—found in northeastern India, Bora et al., 2021 developed a qNMR method. Benzene was used as an internal standard (IS). Approximately 5.0 to 36.0 mg of the capsaicinoid-enriched fraction was acquired from 100 mg of oleoresin of these species. The whole obtained amount was mixed with 0.5 mg of IS in CDCl$_3$ to make a total volume of 0.6 ml. The peak area ratio (integrals) of specific signals for total capsaicinoids (δ_H 4.35 ppm) or 1 (δ_H 0.94 ppm) relative to IS (δ_H 7.36 ppm) was calculated from discrete spectra, and quantities of CAP and total capsaicinoids were determined through a linear regression equation.

$$\frac{M_{Analyte}}{M_{IS}} = \left[Slope \times \left(\frac{A_{Analyte}}{A_{IS}}\right) + Intercept \right]$$

Where $M_{Analyte}/M_{IS}$ and $A_{Analyte}/A_{IS}$ = mole area ratio and peak area ratio of the analyte versus IS, respectively.

Oleoresin from the Mynsain village landrace-I chilli, a variety of *Capsicum frutescens*, exhibited the highest concentration of capsaicinoids at 11.52%, while the Raja chilli landrace-I, belonging

---

TABLE 5.5

SHU and Capsaicin Content in Different Chilli Fruits by ^{1}H NMR

Types of chili fruits	Integration	SHU (ppm)	Capsaicin amount (mg/g)
Malagueta pepper	1.54	31.481	2.0
Habanero	2.13	121,633	7.6
Bhut jolokia	1.96	190,986	11.9
Trinidad scorpion	2.03	214,359	13.3
Carolina reaper	2.21	254.310	15.8

to the *Capsicum chinense* species, had the highest capsaicinoid content at approximately 2.50%. Bhut jolokia chilli from the *Capsicum chinense* species had the highest capsaicin content at about 1.49%. The developed ^{1}H NMR spectroscopic method quantitatively analyzed CAP and total capsaicinoid content in dry chili and chili oleoresin samples and proved to be sensitive and reliable. The study could aid in selecting suitable chili landraces for intended pungency levels in oleoresin production (Bora et al., 2021).

Thread mass spectrometry, a novel ambient ionization technique developed to ease the direct analysis of capsaicinoids in pepper products, was designed by the Cooks group in 2004; the desorption electro-spray ionization uses charged micro-droplets for *in situ* analyte MS analysis in the open air. In this method, a 35-mm-long thread is pulled from a fabric, and then it is placed in a glass capillary (1.63 mm OD and 30 mm length). A small diameter of 225 to 302 µm of the thread enables the tip to fit easily in the front of a MS spectrometer, where no modification or sharpening is required; it provides good electrical conductivity. A high direct current (DC) voltage was applied to the thread to electrically charge the glass capillary where analyte of interest is placed. Optimization of this method was done by the polyester (35/65), 100% cotton, nylon, and cotton: 100% polyester fabric materials available commercially. Then it was applied for *in situ* analysis of seven distinct pepper fruits and pepper spray remnants on fabrics. Field-induced charged droplets were generated by applying DC voltage and solvent to analyze the pepper remnants present on thread. Some advantages of thread spray ionization are:

- Ion current sustained for longer time as evaporation of solvent was limited
- Large degree of flexibility in experimental design due to availability of various thread types
- Thread substrates have sharp tip that leads to the generation of stable ion signal
- Thread is ductile, allows ease of storage and transport

This study showed the effectiveness of thread spray ionization for non-destructive sampling of the interior of pepper fruits and analyzing pepper spray residues on clothing. It is useful in the food industry for capsaicinoid analysis as well as forensic applications for pepper spray detection, all without sample pretreatment. The approach simplifies mass spectrometric analysis, offering rapid, cost-effective, and on-site testing (Jackson et al., 2018).

In a nutshell, non-destructive methods using portable near-infrared spectroscopy (NIRS) and ^{1}H NMR have been developed for predicted pepper hotness accurately and quantifying capsaicinoids.

5.3.3 Hyphenated Techniques

A GC-MS procedure was established for quantitative assessment of capsaicinoids present in both dried and fresh *Capsicum chinense* chillies using HP-5 MS column. Initially, temperature was kept at 90°C, increased by 5°C in gradient manner until it reached 290°C. This study introduces a simple and rapid method for simultaneous estimation of capsaicinoid, tocopherol, and phytosterols in chili peppers using direct GC-MS analysis without derivatization. This approach is effective for various pepper types and provides accurate quantitation along with the detection of possible interferences and minor constituents (Saha et al., 2015).

An immune-affinity chromatography-liquid chromatography-mass spectrometry (IAC-LC–MS/MS) process was carried out for determining the capsaicinoids, mainly CAP and DHC, in vegetable oil samples and inedible waste oil samples (total nine samples) collected from local restaurant grease traps. IAC-LC-MS/MS is an immune-affinity chromatography (IAC) combined with liquid chromatography–tandem mass spectrometry. In this technique, the antibodies are generated against capsaicinoids and applied to the IAC cleanup. Then the immune-sorbents are obtained via covalent coupling of highly precised capsaicinoids polyclonal antibodies with CNBr-activated Sepharose 4B and then packed into a polyethylene column. The Hypersil GOLD C18 column was used, and gradient elution was carried out using mobile phase having 0.1% aqueous formic acid (solvent A) and acetonitrile (solvent B). The highest and lowest amount of CAP was 45.2 µg/kg in inedible waste oil

TABLE 5.6
Quantity and Retention Time of Capsaicinoids in Chilli Types by GC-MS

Capsaicinoids	R_t (min)	Chili-1	Chili -2	Chili-3	Chili-4
NDHC	32.81	ND	1.05	0.42	2.94
NVM	33.43	0.32	1.51	ND	1.55
NNDHC	33.95	D	ND	ND	ND
CAP	34.19	1.29	0.55	0.14	0.65
DHC	34.55	0.89	0.33	0.23	0.18

D = Detectable, ND = Not Detectable

sample 9 and 2.8 µg/kg in inedible waste oil sample 4. DHC content was highest at 26.7 µg/kg in inedible waste oil sample 3 and lowest at 3.4 µg/kg in inedible waste oil sample 1. CAP and DHC were not detected in edible vegetable oils. This study presented a novel approach in which antibodies were generated against capsaicinoid compounds and applied in an IAC-LC-MS/MS method for detecting CAP and DHC in vegetable oil samples. The method demonstrated high sensitivity, accuracy, and precision, effectively removing potential interference from vegetable oils, revealing that capsaicinoid compounds were absent in market vegetable oils and only present in inedible waste oils, making it suitable for routine monitoring and adulteration detection in vegetable oil samples (Ma et al., 2016).

CAP and DHC were quantified in dermal micro-dialysis samples from rats using LC-MS/MS with a C-18 column. The method was validated with a linear range of 0.5 to 100 ng/mL for CAP and 0.25 to 100 ng/mL for DHC. Recovery studies showed dependence on flow rate but independence from drug concentration. *In vitro* and *in vivo* methods were employed to determine the relative recovery of micro-dialysis probes. Similarities between dialysis and retro-dialysis *in vitro* suggested minimal drug binding to probe tubing and membrane, while retro-dialysis *in vivo* indicated the relative recovery of the drug. For CAP, calibration of the *in-vitro* micro-dialysis probes by dialysis and retro-dialysis resulted in statistically related drug recovery of 68.5% ± 5.9% and 77.8% ± 6.6%, respectively, at a flow rate of 0.5 µl/min. For DHC, the recovery by dialysis was lesser as compared to retro-dialysis, at 51.4% ± 6.6% and 92.6% ± 2.4%, respectively. This differentiation was due to the binding of DHC with plastic tubing as determined experimentally. *In vivo* recoveries observed were 75.7% ± 6.3% for CAP and 81.9% ± 1.5% for DHC at the same flow rate (Lorenzoni et al., 2019).

Ultra-high-performance liquid chromatography-tandem mass spectrometry (UHPLC-MS/MS) was utilized for analyzing CAP, DHC, and N-vanillylnonamide (NVM) extracted in the gutter oil. Gutter oils are the subservient oils that are disposed of as waste but illicitly recycled. Based on the sources, gutter oil can be classified into deep-frying oil recycled from drains, residual oil extracted from animal fat, cooking oil recovered from food waste, and other types of inedible oils. The processed gutter oil showed no dissimilarity in appearance or taste on comparison with edible oil except being detrimental. Generally, capsaicinoids are not eliminated via illegitimate gutter oil processing methods. Therefore, CAP, DHC, and NVM can be used as distinctive markers for distinguishing the gutter oil. For the analysis, the nano-composite of graphene oxide–magnetite (GO–Fe_3O_4) was applied to the magnetic solid phase extraction (MSPE) process. GO-Fe_3O_4 was employed for capsaicinoid extraction in vegetable oil using the MSPE process. The method achieved a low LOD, satisfactory and low RSDs demonstrating the accuracy and sensitivity of capsaicinoid determination in oil through MSPE coupled with UHPLC. This method holds significance in identifying gutter oil and monitoring food safety (Lu et al., 2020).

Multiple injections in a single experimental run (MISER) LC-MS analysis was done to quantify pungent components in chilli papers and hot sauces. The Poroshell SB-C18 column was used, and isocratic elution was carried out using mobile phase 2 mM ammonium formate in water, pH 3.5

(solvent A), and 2 mM ammonium formate in 90/10 acetonitrile/water, pH 3.5 (solvent B), in the ratio of 20:80. The column and samples were maintained at a temperature of 25°C at a flow rate of 0.75 ml/min. Serano pepper sauce was found to be the most pungent, with SHU of 4859, and curly pepper sauce was less pungent with SHU of 252 (Welch et al., 2014).

In conclusion, a rapid, derivatization-free GC-MS method for simultaneous estimation of capsaicinoid, tocopherol, and phytosterols in chili peppers was developed for aiding nutraceutical assessment of pepper products. Additionally, an IAC-LC-MS/MS method with generated antibodies detected capsaicinoids in vegetable oils, demonstrating high sensitivity and suitability for routine monitoring of edible oils.

5.4 SUMMARY

Capsaicinoids, chiefly CAP and DHCAP, being the crucial elements, are responsible for the unique peppery taste of chilli peppers. CAP and DHCAP fascinate the culinary world and also captivates scientific attention due to their numerous beneficial properties like pain-relieving, anti-inflammatory, anticancer, and antiobesity effects. With a history of discovery and structural determination dating back to the 19th century, the intricate chemical nature of capsaicinoids and their biosynthetic pathways have been extensively studied.

Their potential applications in food, pharmaceuticals, and even self-defense products have recently elevated interest in capsaicinoids. This interest comes with a necessity for accurate and efficient methods for quantifying capsaicinoids in various products. Analytical methods like GC, HPLC, and MS have proven essential in this regard. Despite their extensive applications, the sensory impact of capsaicinoids presents a challenge, as their pungency often limits their use. The current chapter has focused on efficient large-scale capsaicinoid extraction, discussing the developed methods such as combining ATPS and chromatography, ultrasonic-assisted extraction (UPR), and optimized supercritical fluid extraction (SFE) to obtain better yield. Techniques such as HPTLC, NIRS, ^{1}H NMR, and thread spray ionization used for quantitative analysis have been reviewed. Additionally, rapid GC-MS for chili peppers, IAC-LC-MS/MS and UHPLC methods for vegetable oils, and adulteration with gutter oil or use of recycled gutter oil are discussed.

Acknowledgement: The authors are thankful to Director NIPER-S.A.S. Nagar for support.

REFERENCES

Abbas, W., Ahmed, D., & Qamar, M.T. (2022). Surfactant-mediated extraction of capsaicin from *Capsicum annuum L.* fruit in various solvents. *Heliyon. 8*(8).

Aranha, B.C., Hoffmann, F., Barbieri, L., Valmor, C., & Clasen, F. (2017). Untargeted metabolomic analysis of capsicum spp. by GC—MS. *Phytochem. Anal. 3*: 25–37.

Asnin, L., & Park, S.W. (2013). Isolation and analysis of bioactive compounds of capsicum peppers. *Crit. Rev. Food Sci. Nutr. 8*: 37–41.

Bajer, T., Bajerova, P., Kremr, D., Eisner, A., Adam, M., & Ventura, K. (2016). Comparison of two ultrasonic systems with different settings for extraction of capsaicinoids from chilli peppers. *Int. J. Food Eng. 12*: 567–576.

Bora, P.K., Kemprai, P., Barman, R., Das, D., Nazir, A., Saikia, S.P., Banik, D., & Haldar, S. (2021). A sensitive 1H NMR spectroscopic method for the quantification of capsaicin and capsaicinoid: Morphochemical characterisation of chilli land races from northeast India. *Phytochem. Anal.32*: 91–103.

Chen, K., Feng, L., Feng, S., Yan, Y., Ge, Z., Li, Z., & Chen, Z. (2019). Multiple quantitative structure—pungency correlations of capsaicinoids. *Food Chem.11*: 245–265.

Chen, M.J., Yin, H.L., Liu, Y., Wang, R.R., Jiang, L.W., & Li, P. (2022). Non-destructive prediction of the hotness of fresh pepper with a single scan using portable near infrared spectroscopy and a variable selection strategy. *Anal. Methods. 14*: 114–124.

Contreras-padilla, M., & Yahia, E.M. (1998). Changes in capsaicinoids during development, maturation, and senescence of chilli peppers and relation with peroxidase activity. *J. Agric. Food Chem. 8561*: 2075–2079.

Daood, H.G., Halasz, G., Palotas, G., Palotas, G., Bodai, Z., & Helyes, L. (2015). HPLC determination of capsaicinoids with cross-linked C18 column and buffer-free eluent. *J. Chromatogr. Sci. 53*: 135–143.

Gomez-Calvario, V., Garduno-Ramírez, M.L., Leon-Rivera, I., & Rios, M.Y. (2016) ^{1}H and ^{13}C NMR data on natural and synthetic capsaicinoids. *Magn. Reson. Chem. 54*: 268–290.

Gonzalez-Zamora, A., Sierra-Campos, E., Luna-Ortega, J.G., Perez-Morales, R., Ortiz, J.C., & García-Hernandez, J.L. (2013). Characterization of different capsicum varieties by evaluation of their capsaicinoids content by high performance liquid chromatography, determination of pungency and effect of high temperature. *Molecules. 18*: 13471–13486.

Jackson, S., Swiner, D.J., Capone, P.C., & Badu-Tawiah, A.K. (2018). Thread spray mass spectrometry for direct analysis of capsaicinoids in pepper products. *Anal. Chim. Acta.1023*: 81–88.

Kaiser, M., Higuera, I., & Goyocoolea, F.M. (2018). Capsaicinoids: Occurrence, chemistry, biosynthesis and biological effects. In E.M. Yahia (Ed.), *Advances in Food Science and Technology* (pp. 499–513). John Wiley and Sons.

Ksh, V., Anand, V., Rana, V.S., Mishra, J., Varghese, E., Upadhyay, N., & Kaur, C. (2023). Extraction of capsaicin from *Capsicum chinense* (cv Bhut Jolokia) using supercritical fluid technology and degradation kinetics. *Chem. Pap. 28*: 1–5.

Liu, F., & Zou, X. (2020). Genome-wide identification and capsaicinoids biosynthesis-related expression analysis of the R2R3-MYB gene family in capsicum. *Front. Genetics. 11*: 1–12.

Lorenzoni, R., Barreto, F., Contri, R.V., de Araujo, B.V., Pohlman, A.R., Dalla Costa, T., & Guterres, S.S. (2019). Rapid and sensitive LC-MS/MS method for simultaneous quantification of capsaicin and dihydrocapsaicin in microdialysis samples following dermal application. *J. Pharm. Biomed. Anal. 173*: 126–133.

Lu, Q., Guo, H., Li, D., & Zhao, Q. (2020). Determination of capsaicinoids by magnetic solid phase extraction coupled with UPLC-MS/MS for screening of gutter oil. *J. Chromatogr. B. 1158*: 122344.

Ma, F., Yang, Q., Matthaus, B., Li, P., Zhang, Q., & Zhang, L. (2016). Simultaneous determination of capsaicin and dihydrocapsaicin for vegetable oil adulteration by immunoaffinity chromatography cleanup coupled with LC–MS/MS. *J. Chromatogr. B. 1021*: 137–144.

Martínez, J., Rosas, J., Perez, J., Saavedra, Z., Carranza, V., & Alonso, P. (2019). Green approach to the extraction of major capsaicinoids from habanero pepper using near-infrared, microwave, ultrasound and Soxhlet methods, a comparative study. *Nat. Prod. Res. 33*: 447–452.

Naves, E.R., Silva, L.D.A., Sulpice, R., Araujo, W.L., Nunes-nesi, A., Peres, L.E.P., & Zsogon, A. (2019). Capsaicinoids: Pungency beyond capsicum. *Trends. Plant. Sci. 24*: 109–120.

Pradhan, K., Nandi, A., Das, A., Sahu, N., Senapati, N., Mishra, S.P., Patnaik, A., & Pandey, G. (2018). Quantification of capsaicin and ascorbic acid content in twenty four Indian genotypes of chilli (*Capsicum annuum L.*) by HPTLC and volumetric method. *Int. J. Pure Appl. Biosci. 6*: 1322–1327.

Saha, S., Walia, S., Kundu, A., Kaur, C., Singh, J., & Sisodia, R. (2015). Capsaicinoids, tocopherol, and sterols content in chilli (Capsicum sp.) by gas chromatographic-mass spectrometric determination. *Int. J. Food Prop. 18*: 1535–1545.

Sancho, R., & Minassi, A. (2003). Non-pungent capsaicinoids from sweet pepper Synthesis and evaluation of the chemopreventive and anticancer potential. *Eur. J. Nutr. 9*: 2–9.

Schmidt, A., Fiechter, G., Fritz, E.M., & Mayer, H.K. (2017). Quantitation of capsaicinoids in different chillies from Austria by a novel UHPLC method. *J. Food Compos. Anal. 60*: 32–37.

Siudem, P., Bukowicki, J., Wawer, I., & Paradowska, K. (2020). Structural studies of two capsaicinoids: Dihydrocapsaicin and nonivamide. 13C and 15N MAS NMR supported by genetic algorithm and GIAO DFT calculations. *RSC. Adv. 10*: 18082–18092.

Valim, T.C., Cunha, D.A., Francisco, C.S., Romao, W., Filgueiras, P.R., dos Santos, R.B., de Souza Borges, W., Conti, R., Lacerda, V., & Neto, A.C. (2019). Quantification of capsaicinoids from chilli peppers using 1H NMR without deuterated solvent. *Anal. Methods. 11*: 1939–1950.

Wang, R., Liu, Y., Sun, S., Si, Y., Liu, X., Liu, X., Zhang, S., & Wang, W. (2018). Capsaicinoids from hot pepper (*Capsicum annuum L.*) and their phytotoxic effect on seedling growth of lettuce (*Lactuca sativa L.*). *Nat. Prod. Res. 8*: 1–5.

Welch, C.J., Regalado, E.L., Welch, E.C., Eckert, I.M., & Kraml, C. (2014). Evaluation of capsaicin in chilli peppers and hot sauces by MISER HPLC-ESIMS. *Anal. Methods. 6*: 857–862.

Woodman, T.J., & Negoescu, E. (2019). A simple ^{1}H NMR based assay of total capsaicinoids levels in Capsicum using signal suppression in non-deuterated solvent. *J. Agric. Food Chem. 99*: 1765–1771.

Zhao, P.P., Lu, Y.M., Tan, C.P., Liang, Y., & Cui, B. (2015). Aqueous two-phase extraction combined with chromatography: New strategies for preparative separation and purification of capsaicin from capsicum oleoresin. *Appl. Biochem. Biotechnol. 175*: 1018–1034.

6 Transcriptomics of Chili Pepper Fruit with Emphasis on the Carotenoid Biosynthetic Pathway

Maria Guadalupe Villa-Rivera, Octavio Martínez and Neftalí Ochoa-Alejo

CONTENTS

6.1 INTRODUCTION

Chili pepper is an important horticultural crop worldwide; during 2021, the gross production value of green chili peppers was 31.419 billion USD, while that of dry chili peppers was 6.147 billion USD (FAOSTAT 2021). All chili peppers belong to the *Capsicum* genus of the Solanaceae family, which also includes some other important species such as tomato (*Solanum lycopersicum*), potato (*Solanum tuberosum*), tobacco (*Nicotiana tabacum*), eggplant (*Solanum melongena*), and petunia (*Petunia hybrida*). *Capsicum* genus includes approximately 37 species, from which five have been domesticated: *Capsicum annuum* L., *Capsicum baccatum* L., *Capsicum chinense* Jacq., *Capsicum frutescens* L., and *Capsicum pubescens* Ruiz & Pav. (Jarret et al. 2019).

Chili pepper fruits are used for diverse purposes: as spice, vegetable, flavoring, aroma, and colorant, among others. They can be used fresh, dried, in conserves, or as chili pepper sauces. Also, pharmaceutical and cosmetic industries use chili oleoresins and other enriched extracts of high pungency and low color as enhancers of physical properties of some products and to improve product stability and preservation due to the reduction of the oxidation of active substances and excipients in the medicinal products. Also, bioactive compounds from chili peppers have been applied as flavoring or fragrance agents in the cosmetic industry (Baenas et al. 2019; Idrees et al. 2020). In addition to their use as flavoring agents, chili pepper fruits have been used in traditional medicine to cure muscle pain and toothache, to improve gastric activity and blood circulation, and to treat neuralgia,

DOI: 10.1201/9781003378259-6

arthritis, rheumatic diseases, migraine, and painful diabetic neuropathy. Moreover, the antimicrobial, antifungal, antiviral, insecticidal, antihelmintic and larvicidal effects have been reported as traditional use of chili peppers (Badia et al. 2017; Batiha et al. 2020). Interestingly, the antimicrobial and antifungal effect of extracts of *Capsicum* spp. fruits have been demonstrated against a wide variety of human pathogens (Omolo 2014).

The nutritional and nutraceutical properties of chili peppers are a consequence of the accumulation of important amounts of bioactive compounds with notable nutraceutical potential. These bioactive compounds comprise capsaicinoids (capsaicin and dihydrocapsaicin mainly), phenolic compounds (phenolic acids, flavonoids, stilbene, and lignans), carotenoids (carotenes and xanthophylls), and vitamins (C and E) (de Sá Mendes and Branco de Andrade Gonçalves 2020; Azlan et al. 2022). Capsaicinoids are secondary metabolites responsible for the pungent sensation of chili pepper fruits. They are biosynthesized in the pericarp of the fruits with an ambiguous boundary in the placental septum and pericarp (Aza-González et al. 2011; Tanaka et al. 2021). Capsaicinoids have become relevant because of their therapeutic uses as anticancer and antiobesity agents and for cardiovascular effect, dermatological potential, gastrointestinal properties, and weight control (Hernández-Pérez et al. 2020).

On the other hand, phenolic compounds are produced from plant secondary metabolism. Particularly, anthocyanins are metabolites responsible for the purple pigmentation of fruits from different species of *Capsicum*, for which important health benefits have been reported, such as prevention of cardiovascular diseases; anticancer, antidiabetes, antiobesity, and antimicrobial effects; and promoting of visual health (Aza-González et al. 2012; Khoo et al. 2017). In addition, chili pepper fruits are rich in vitamin E (tocopherol) and vitamin C (ascorbic acid) for humans. The consumption of ascorbic acid is very important for good heath, since it acts as an antioxidant, analgesic, anticancer agent, COVID-19 preventative, skin health promotor, and preventer of ill effects induced by radiation, and it has been used in the treatment of cancer and psychiatric disorders (Villa-Rivera and Ochoa-Alejo 2023).

Finally, carotenoids are bioactive compounds accumulated in the pericarp of chili pepper fruits; they confer the characteristic colors yellow, orange, and red of mature fruits. Carotenoids in chili pepper fruits comprise lutein, β-carotene, β-cryptoxanthin, violaxanthin, zeaxanthin, capsorubin, and capsanthin; the profile of accumulation of carotenoids depends on the stage of ripening and the cultivar (Maoka 2020; Guo et al. 2021). Notable nutraceutical potential has been reported for carotenoids, which have powerful antioxidant activity and are critical for human health. Several studies have been conducted using carotenoids as cancer-preventive agents and for treatment in different types of cancer; as anti-inflammatory, antiobesity, analgesic, and antinociceptive agents; for promoting skin health; and for preventing cardiovascular illness, atherosclerosis, diabetes, and Alzheimer's disease (Villa-Rivera and Ochoa-Alejo 2020).

Because carotenoids represent important bioactive compounds of chili pepper, the carotenoid biosynthesis pathway in the *Capsicum* genus has been widely studied from different perspectives; nevertheless, the transcriptional regulation of genes encoding carotenoid biosynthetic enzymes remains poorly understood. In these sense, novel approaches have been proposed to identify the molecular mechanisms that regulate this biosynthetic pathway, and transcriptomic analysis has emerged as a good source of information to study it. Here, we describe different transcriptomic results and present an RNA-Seq network co-expression analysis to identify potential transcription factor candidates to regulate the carotenoid biosynthetic pathway in chili pepper fruits.

6.2 APPLICATIONS OF TRANSCRIPTOMIC ANALYSIS IN CHILI PEPPER

Because of the nutritional, nutraceutical, and cultural importance of chili pepper, it has been studied from different perspectives; in recent years, next-generation sequencing techniques have represented a cost-effective and rapid alternative to characterize plant genomes and gene expression profiles and to identify molecular markers in species of interest (Ahn et al. 2014). In fact, the Expressed Sequence Tags (EST) database of chili pepper is available with the aim of providing *in silico* tools to identify genes and to analyze their expression patterns in different developmental tissues and under stress conditions (Kim et al. 2008). Particularly, transcriptomic data have been useful to identify candidate genes as molecular markers such as quantitative trait loci (QTLs), simple sequence repeat (SSR), and specific single nucleotide polymorphism (SNPs) in several species of *Capsicum*. These molecular markers have been applied for genetic studies, linkage mapping, and association mapping research for assisted breeding, characterization of some important agronomic traits, and identification of molecular mechanisms such as phytopathogen resistance, among others (Lu et al. 2011, 2012; Ashrafi et al. 2012). In addition, polymorphism markers were predicted in transcriptomic sequences of *Capsicum annuum*, providing information about flower development (Deng et al. 2020).

Some other applications of transcriptomic analysis of chili pepper include the identification of common regulons encoding signaling components of biotic and abiotic stress (Lee and Choi 2013); comparison of transcriptomes from cultivars of *Capsicum annuum* tolerant and susceptible to heat stress for the identification of differential expression of genes encoding heat shock proteins; heat shock transcription factors (TFs); hormone-related and kinase and calcium signaling pathways related to generate knowledge on the molecular mechanisms of responses to heat stress of different genotypes of chili pepper (Li et al. 2015); and on the function of hormones such as epibrassinolides in response to chilling stress (Li et al. 2016).

Interestingly, the integration of transcriptomic and metabolomic data has permitted scientists to expand the knowledge about different regulatory mechanisms, such as responses to heat stress, among others (Wang et al. 2019). Additionally, the integration of transcriptome data, methylation of DNA, and global metabolite contents has provided insights into the manifestation of heterosis of *Capsicum* (Jaiswal et al. 2022). On the other hand, microRNA (miRNA) transcriptome of hot chili pepper has reported relevant information to understand the function of miRNAs in this species in relation to other plants (Hwang et al. 2013).

Fruit development and ripening constitute complex processes that are highly regulated in angiosperm plants (Seymour et al. 2013), and transcriptomic analyses have provided important information. This kind of analysis has allowed the comprehensive understanding of biological events such as metabolic regulation during climacteric and non-climacteric fruit development (Osorio et al. 2012). Likewise, the analysis of transcriptomic data obtained from the pericarp and placenta of chili pepper during pathogen infection and fruit ripening have permitted the prediction of molecular networks of fruit development and resistance to diseases in this important crop (Kim et al. 2018). In addition, the analysis of transcriptomic data sequenced from hot pepper *C. frutescens* during ripening (fruit large green, initial coloring brown, and full red stages) and the identification of differentially expressed genes (DEGs) related to phytohormone pathways suggested an important role of ethylene and a complex role of abscisic acid (ABA) in the ripening process (Hou et al. 2018). With the aim to elucidate the mechanism of resistance of roots from *C. annuum* to the pathogen *Phytophtora capsici*, the dynamic transcriptome using RNA-Seq data was employed. Different transcriptional responses between resistant and susceptible accessions of chili pepper were detected, with high levels of expression of genes implicated in the biosynthesis of secondary metabolites; particularly, it has been proposed that phenylpropanoid compounds (cinnaldehyde and lignin) play a central role during the defense response of chili pepper roots against *P. capsici* (Li et al. 2020).

Alternatively, analysis of data from RNA-Seq and metabolic profiling obtained from *Capsicum* species during early green, mature green, and braker stages were useful to identify candidate genes with alternative splicing involved in development, ripening, and metabolite production of chili pepper fruits (Rawoof et al. 2022).

We have proposed a novel tool for transcriptomic analysis of chili pepper fruits during development and ripening: Salsa, an auto-contained R package, compiles analyzed and curated data from 168 RNA-Seq libraries obtained from 12 accessions of *Capsicum annuum* (four wild, six domesticated, and two reciprocal crosses between domesticated and wild-type accessions). These data are represented as standardized expression profiles (SEPs) of genes expressed during seven steps of the ripening process of chili pepper fruit [0, 10, 20, 30, 40, 50, and 60 days after anthesis (DAA)]. This invaluable database allows us to perform a variety of analysis with different approaches in chili pepper and also permits comparison with other angiosperms (Escoto-Sandoval et al. 2020, 2021). For instance, an analysis of the expression profiles of genes involved in the cell cycle using the Salsa database showed that changes in the expression levels between domesticated and wild accessions could be derived from selection pressures during the domestication process and could explain phenotypic differences such as fruit size (Martínez et al. 2021). According to that, transcriptomic analysis of two varieties of *Capsicum annuum* (wild and cultivated) at two stages of maturation (immature and mature) exhibited differences in the expression patterns of genes associated with size, shape, and biosynthesis of secondary metabolites and ethylene, confirming that differential gene expression is related to domestication process (Razo-Mendivil et al. 2021). In addition, utilizing an algorithm of the R package Salsa, it was possible to construct gene co-expression networks to understand functional interactions between genes and to find transcription factor candidates (TFC) to regulate genes of interest within a network (Flores-Díaz et al. 2023).

Another interesting application of transcriptomic analysis comprises the generation of information for the improvement of nutritional qualities of *Capsicum* fruits. The integration of transcriptome analysis and amino acid metabolism profiles allowed to perform co-expression and co-expression network analysis of chili pepper at green and red stages of ripening, contributing to the development of amino acid nutritional supplements and chili pepper as a functional food (Fei et al. 2022).

Finally, noncoding and coding transcriptomic analysis, specifically strand-specific RNA sequencing, has allowed the determination of the role of non-coding RNAs (lncRNAs) and potential targets protein-coding (PC) genes during the chili pepper fruit development (Ou et al. 2017).

6.2.1 TRANSCRIPTOMIC ANALYSIS IN THE STUDY OF BIOACTIVE COMPOUNDS

Transcriptomic analyses have also been applied to the study of biosynthetic pathways of bioactive compounds and their transcriptional regulation (Villa-Rivera and Ochoa-Alejo 2021). Using transcriptomic data obtained from the placenta and pericarp of *C. frutescens*, was possible to identify genes involved in the capsaicinoid biosynthetic pathway (Liu et al. 2013). Moreover, RNA-Seq was used to obtain transcript sequences from *Capsicum annuum* cv. serrano Tampiqueño 74 at different stages of development and ripening at 10, 20, 40, and 60 DAA. With this information, analysis of expression profiles of genes and groups of genes associated to biological processes, ontologies, and metabolic pathways were conducted. As a conclusion of these analyses, higher levels of expression of genes encoding capsaicinoid and ascorbic acid biosynthetic pathways were detected at 20 DAA, while genes encoding enzymes related to biosynthesis of carotenoids reached their maximum levels of expression at 60 DAA (Martínez-Lopez et al. 2014). Furthermore, transcriptomic analysis has been applied to determine the expression levels of genes involved in the biosynthetic pathway, recycling and degradation of ascorbic acid, and during development and ripening of the sweet pepper, providing information about differential accumulation of vitamin C in varieties of chili pepper (Chiaiese et al. 2019). On the other hand, using metabolome and transcriptome comparisons, as well as gene co-expression network analyses, modules implicated with flavonoid synthesis and candidate genes to be regulators (TF) of flavonoid biosynthesis and transport were identified in chili pepper (Liu et al. 2020).

As mentioned, transcriptome analysis has been also used to identify TFC to be regulators of the genetic expression during chili pepper fruit development, stress responses, or biosynthetic pathways of bioactive compounds. For instance, RNA-Seq analysis of genes encoding DNA-binding One Zinc Finger (Dof) TFs showed different expression patterns during the first stages of fruit growth and tissue types in response to abiotic stress, in particular, salt stress (Wu et al. 2016). Also, the expression profiles of genes encoding auxin response factors (ARFs) of *C. annuum* in response to hormones and abiotic treatments have been obtained by quantitative real-time RT-PCR (qRT-PCR). As a result from this work, 22 *ARF* genes were identified, and they were differentially expressed in distinct tissues and organs of chili pepper; in addition, the expression of some genes in response to stress and hormones suggests their participation in plant tolerance to abiotic stress (Yu et al. 2017).

On the other hand, transcriptomic data of *C. annuum* genes encoding GRAS TFs, a family which includes GAI, RGA, and SCR members, showed that 21 *CaGRAS* genes were differentially expressed under salt, drought, cold, and gibberellin acid treatments, suggesting that they could be involved in plant responses to abiotic stress (Liu et al. 2018). Regarding the family of *Myeloblastosis* (MYB) TFs, RNA-Seq analyses have revealed that *CaR2R3-MYB* genes are highly expressed during the placenta development process, and they co-expressed with capsaicinoid biosynthetic genes, suggesting them as strong candidates for the transcriptional regulation of the capsaicinoid biosynthetic pathway (Wang et al. 2020). Using transcriptomic data from *C. chinense* and *C. annuum*, it was possible to identify 54 and 81 *MYB* genes, respectively, that presented differential expression profiles during three stages (early green, mature green, and breaker) of fruit development. In addition, co-expression analysis demonstrated that several *MYB* genes were clustered with capsaicinoid, carotenoid, anthocyanin, vitamin C, phenylpropanoid, and flavonoid biosynthetic-related genes and were implicated in the fruit size and shape determination. Moreover, an SSR motif in *MYB* genes was identified (Islam et al. 2021). Similarly, the integration of RNA-Seq expression with other analysis such as phylogenetic relationships, conserved domains, and gene structure organization of MYB transcription factors family led researchers to propose CaMYB as candidates for the regulation of lignin, phenylpropanoid, capsaicinoid, ascorbic acid, and carotenoid biosynthetic pathways, suggesting the role of MYB TFs in the regulation of secondary metabolism (Arce-Rodríguez et al. 2021). On the other hand, correlation analysis between gene expression profiles obtained from transcriptomic data and accumulation of metabolites has been performed in chili pepper. In this sense, it has been reported that expression patterns of genes encoding ethylene responsive factors (ERF) were consistent with the accumulation of carotenoids and capsaicinoids in the pericarp (Song et al. 2020). Likewise, expression patterns obtained from transcriptomic data from genes encoding basic helix-loop-helix (bHLH) TFs demonstrated that they are implicated in plant development and stress responses (Zhang et al. 2020). Additionally, *bHLHs* which present similar gene expression profiles and carotenoids accumulation patterns in the pericarp and capsaicinois in the placenta of *Capsicum annuum* fruits have been proposed as candidates to be regulators of structural biosynthetic genes (Liu et al. 2021). Another type of TF that has been studied in chili pepper is the WRKY family. By using global expression patterns obtained by RT-PCR analysis, 61 *CaWRKYs* were identified, and from them, 60% were expressed during the ripening process. Some other WRKY transcription factors have been associated with stress-related signaling pathways, such as disease and drought, during fruit ripening (Cheng et al. 2016, 2019).

6.3 TRANSCRIPTOMICS AND TRANSCRIPTIONAL REGULATION OF THE CAROTENOID BIOSYNTHETIC PATHWAY

6.3.1 Carotenoid Biosynthesis Pathway in Chili Pepper

Carotenoids are tetraterpenoids composed by eight isoprenoid units and, commonly, 40 atoms of carbon. Their structure is characterized by the presence of conjugated double bonds. Some carotenoids contain atoms of oxygen as part of their structure, and they are denominated xanthophyll (Meléndez-Martínez et al. 2019). This group of pigments is present in bacteria, microalgae, archaea, fungi, plants, and animals (Alcaino et al. 2016; Maoka 2020). In plants, carotenoids have relevant functions; for example, they are responsible for the pigmentation of flowers, seeds, and fruits, are key components of photosystems, and participate in photosynthesis and photoprotection. Additionally, carotenoids are precursors of phytohormones such as abscisic acid and strigolactone. Finally, the apocarotenoid signaling has effects on the development and plant growth (Sun et al. 2022).

The ripening process of chili pepper fruits is characterized by the loss of chlorophyl and anthocyanin and accumulation of high amounts of carotenoids; therefore, chloroplasts from the pericarp are differentiated into chromoplasts, the plastids where the mix of carotenoids occurs. Because of these chemical and biological processes, mature chili pepper fruits turn from green or brown color to yellow, orange, red, or dark red. The variation in fruit color is given by the composition and content of carotenoids and depends on the cultivar, differences in genotypes, climatic conditions, post-harvest storage, processing, and preparation (Mohd Hassan et al. 2019). Particularly, the colors red and dark red correlate with the accumulation of capsanthin and its esters, pigments found exclusively in fruits of the *Capsicum* genus, and its accumulation is associated with the expression of a putative carotenoid acyl transferase and an increase in the content of fibrillin in the chromoplasts (Berry et al. 2019).

The carotenoid biosynthetic pathway has been widely studied in plants. In chili pepper, as in other plants, carotenoids are synthetized via the mevalonate pathway (MVA) from a five-carbon unit, the isoprene, from which the isopentenyl pyrophosphate (IPP) and dimethylallyl pyrophosphate (DMAPP) are formed. According to Alcaino et al. (2016), the carotenoid biosynthetic pathway is divided into three major phases: (i) the synthesis of IPP and the formation of DMAPP, (ii) the synthesis of geranylgeranyl pyrophosphate (GGPP), and (iii) the synthesis of carotenoids *per se*, being the phytoene, the initial step determining the carotenoid production. As shown in Figure 6.1, the carotenoid biosynthesis pathway includes reactions of isomerization, desaturation, cyclization, hydroxylation, and epoxidation among others, catalyzed by around 10 enzymes along the pathway (Gómez-García and Ochoa-Alejo 2013; Alcaino et al. 2016; Rosas-Saavedra and Stange 2016; Sathasivam et al. 2021).

Deferential expression profiles of eight carotenoid biosynthetic-related genes were reported in a transcriptome analysis (Song et al. 2022); for instance, the study of patterns of expression of genes encoding carotenoid biosynthetic enzymes, such as the carotenoid cleavage oxygenases (CCOs), the enzymes which catalyze the formation of conjugated double bonds in carotenoid and apocarotenoids. Analysis of genetic expression of *CCOs* genes, performed from RNA-Seq data, revealed that the *CCO* genes exhibited distinct expression profiles in the roots, stems, leaves, and fruit, suggesting that these genes have important roles in the different vegetative and reproductive processes of chili pepper (Zhang et al. 2016).

The time expression pattern in genes involved in carotenoid biosynthesis during fruit development can vary depending on the specific fruit species and cultivar as well as on the environmental factors. However, in general, the expression of carotenoid biosynthetic genes is typically upregulated during the later stages of fruit development, particularly during the ripening phase. Expression profiles of nine genes encoding structural biosynthetic enzymes of carotenoids showed in Figure 6.2, were determined by RNA-Seq and corroborated by quantitative qRT-PCR techniques, and at least six genes showed similar expression profiles. During the early stages of fruit development (from 10 to 20 DAA), there was typically little or no expression of carotenoid biosynthetic genes; likewise, from 20 to 40 DAA, steady levels of expression were observed. However, as the fruit matures and approaches the ripening stage (from 40 to 60 DAA), the expression of these genes increases dramatically. This caused a significant accumulation of carotenoids in the fruit, giving its characteristic color (Martínez-Lopez et al. 2014).

On the other hand, comparisons of metabolome and transcriptome analysis conducted in varieties of chili pepper with different-color fruit showed differences in the accumulation of carotenoids and flavonoids and in the levels of transcripts of biosynthetic genes at 50 DAA, depending on the color of fruit, suggesting a very fine regulation of the biosynthetic pathways (Liu et al. 2020). For instance, it has been described that differences in the regulation of CCS resulted in the formation of different pericarp colors in chili pepper fruits (Li et al. 2021).

A positive correlation between the amount of capsanthin and its esters and the expression levels of the 1-deoxy-D-xylulose 5-phosphate synthase (*DXS*) and phytoene synthase-1 (*PSY-1*) genes was reported in chili pepper fruits (Berry et al. 2019). Some other structural biosynthetic carotenoid-related genes that showed abundant transcription levels during the accumulation of carotenoids in chili pepper were those of the geranylgeranyl pyrophosphate synthase (*GGPS*), phytoene synthase (*PSY*), β-carotene hydroxylase (*CRTZ-2*), and capsanthin-capsorubin synthase (*CCS*). Moreover, a significant correlation in the expression levels of *CCS* with *CRTZ-2* and *PSY* with *CCS* and *CRTZ-2*

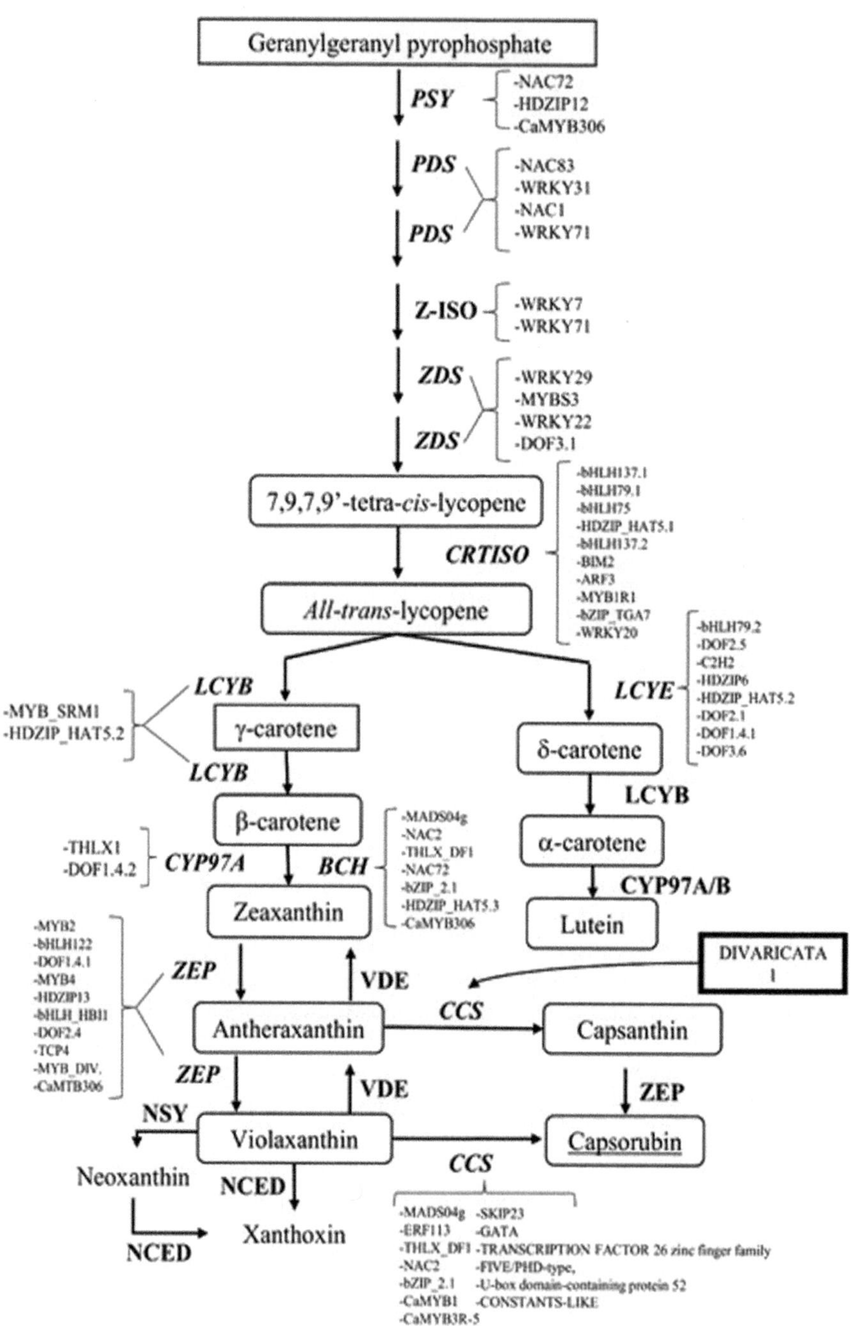

FIGURE 6.1 Chili pepper fruit carotenoid biosynthetic pathway and transcription factor candidates to be regulating this pathway (suggested by (Arce-Rodríguez et al. 2021; Li et al. 2021; Ma et al. 2022; Villa-Rivera et al. 2022; Song et al. 2023). PSY (phytoene synthase), PDS (phytoene desaturase), Z-ISO (z-carotene isomerase), ZDS (z-carotene desaturase), CRTISO (carotene isomerase), LCYB (β-lycopene cyclase), LCYE (ε-lycopene cyclase), BCH (β-carotene hydroxylase), CYP (β-carotene hydroxylase cytochrome 450 types A and B), ZEP (zeaxanthin epoxidase), VDE (violaxanthin epoxidase), and CCS (capsanthin-capsorubin synthase). Modified from (Gómez-García and Ochoa-Alejo 2013).

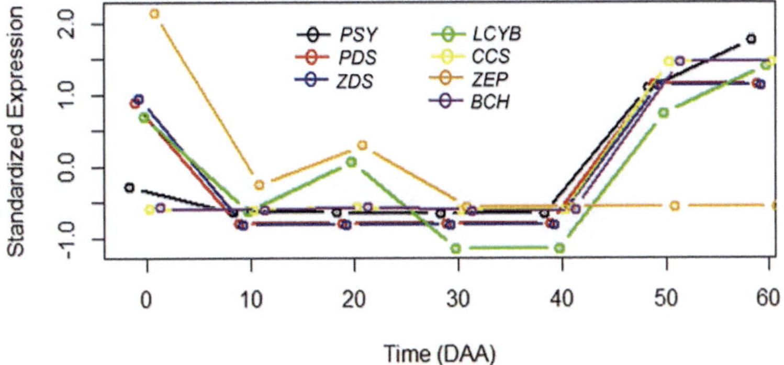

FIGURE 6.2 Expression profiles of seven genes encoding structural biosynthetic enzymes of carotenoids. *PSY* (phytoene synthase), *PDS* (phytoene desaturase), *ZDS* (ζ–carotene desaturase), *LCYB* (β-lycopene cyclase), *BCH* (β-carotene hydroxylase), *ZEP* (zeaxanthin epoxidase), and *CCS* (capsanthin-capsorubin synthase). Modified from (Martínez-Lopez et al. 2014).

was detected, while the expression of lycopene cyclase B (*LYC-B*) gene was correlated with that of the ζ-carotene dehydrogenase (*ZDS*). Finally, the expression of *LYC-B*, phytoene desaturase (*PDS*), and *ZDS* showed strong correlation with that of the *GGPS* gene (Guo et al. 2021).

6.3.2 Transcriptional Regulation of Carotenoid Biosynthetic Pathway

The biosynthesis of carotenoids is regulated at the transcriptional, post-transcriptional, post-translational, and epigenetic levels. At the transcriptional level, developmental program, phytohormone, and environmental signaling (light, temperature, circadian rhythm) activate or inactivate transcription factors to regulate carotenogenic gene expression (Sun et al. 2022). Unlike with the tomato (*Solanum lycopersicum*), a climacteric model, where TFs that regulate the expression of carotenoid biosynthetic genes are well known, there are few reports about TFs that regulate this biosynthetic pathway in chili pepper (a non-climacteric fruit) (Stanley and Yuan 2019).

A comparative transcriptomic analysis conducted in tomato and chili pepper fruits demonstrated that genes implicated in the biosynthesis of ethylene were not induced in chili pepper fruits; however, as it occurs in tomato, genes involved in cell wall metabolism or carotenoid biosynthetic genes were induced in *Capsicum* fruits, suggesting different mechanisms of regulation in both fruits (Osorio et al. 2012). Nevertheless, through genome-wide analysis of *C. annuum*, an ortholog (*LOC10787473*) of tomato *MADS -box RIPENING INHIBITOR* (*MADS-RIN*) gene was identified, and higher levels of expression were detected by qRT-PCR in fruits at the braker and mature stages (Dubey et al. 2019). In *S. lycopersicum*, the MADS-RIN TF has been considered as a master regulator of fruit ripening and is implicated in processes such as cell wall metabolism and biosynthesis of carotenoids (Martel et al. 2011; Shima et al. 2013), suggesting a conserved function in both solanaceous species. In other transcriptome analysis, an ortholog of *ARABIDOPSIS PSEUDO RESPONSE REGULATOR2-LIKE* (*APRR2-like*) gene, which is associated with the accumulation of pigments in fruits, was reported in sweet chili pepper (Pan et al. 2013).

Gene expression profiles obtained from RNA-seq data and further correlation analysis have facilitated the identification of genes encoding TFC, related to the regulation of the carotenoid biosynthetic pathway in chili pepper. Functional characterization of these TFC would help us understand different metabolic networks during the ripening process (Song et al. 2022). In this regard, it was determined that the expression profiles of *CaERF66, CaERF82, CaERF97, CaERF101,* and *CaERF107* TF genes were similar to the patterns of accumulation of β-carotene, zeaxanthin, and capsorubin in the pericarp (Song et al. 2020). Likewise, the expression profiles (obtained by RNA-Seq data) of *CabHLH007, CabHLH009, CabHLH032, CabHLH048, CabHLH095,* and *CabHLH100* were consistent with the patterns of synthesis of zeaxanthin, lutein and capsorubin of *C. annuum* (Liu et al. 2021). Moreover, as part of a wide-genome analysis conducted in the MYB transcription factors family from *Capsicum annuum* serrano Tampiqueño 74, it was reported that the expression patterns of *CaDIV1* and *CaMYB3R-5* genes were similar to the expression profile of *CCS*, suggesting them as potential candidates for transcriptional regulation of *CCS* (Arce-Rodríguez et al. 2021). Similarly, through co-expression network analysis derived from RNA-Seq data, it has been proposed that F-box protein SKIP23, GATA TRANSCRIPTION FACTOR 26, zinc finger family FIVE/PHD-type, U-box domain-containing protein 52, and CONSTANTS-LIKE 9 could be implicated in the regulation of the *CCS* gene (Li et al. 2021). Finally, through RNA-Seq co-expression analysis from 12 accessions of *C. annuum* during the ripening process (0, 10, 20, 30 40, 50, and 60 DAA) and the analysis of putative binding sites in the promoters of carotenoid biosynthetic genes, it was possible to identify 54 putative TFC to be regulators of structural genes of the carotenoid biosynthetic pathway (Villa-Rivera et al. 2022).

Undoubtedly, the co-expression studies mentioned are useful to propose potential candidates to be regulators of biosynthetic genes; nevertheless, experimental evidence is necessary to verify their participation in the transcriptional regulation of the carotenogenic pathway. Recently, the TF CaMYB306 from chili pepper was characterized. The level of expression of *CaMYB306* was significantly higher in red fruits, and knockdown experiments showed a complete inhibition of carotenoid biosynthesis as well as a significant decrease in the expression levels of the carotenoid structural biosynthetic genes *CaPSY, CaBCH,* and *CaZEP,* suggesting its participation as a transcriptional regulator of carotenoid biosynthesis (Ma et al. 2022). Recently, a novel R-R-type TF DIVARICATA1 was characterized. It was reported that DIVARICATA 1 binds to the promoter of *CCS* and positively regulates the expression of this gene encoding the enzyme, which catalyzes the synthesis of capsanthin and capsorubin. Interestingly, it was also demonstrated that *DIVARICATA 1* gene expression was regulated by the TF CaMADS-RIN (Song et al. 2023). Transcription factor candidates proposed to be regulating this via pathway are summarized in Figure 6.1.

6.4 GENE FUNCTIONAL NETWORKS OF CAROTENOID BIOSYNTHETIC PATHWAY IN CHILI PEPPER

6.4.1 Time Expression Profiles of Genes Involved in Carotenoid Biosynthesis in Chili Pepper

As mentioned, RNA-Seq data allows us to perform transcriptomic analysis with different applications. Here, we describe *in silico* analyses using the data available in the Salsa R package (Escoto-Sandoval et al. 2020, 2021) (version 1.0) to investigate the standardized time expression profiles, or SEPs (Martínez et al. 2021; Martínez 2022), of the genes involved in the synthesis of carotenoids. We want to test the *"Gene2Gene"* and *"Gene2TF"* algorithms to investigate whether the genes involved in carotenoids form a robust gene functional network (GFN) and, additionally, to determine a set of potential TFs to be regulating the carotenoid biosynthesis. All methods resulted from the application of the algorithms reported by Flores-Díaz et al. (2023).

The hypothesis of this work is that genes involved in carotenoids biosynthesis will have a relatively congruent time expression profile among genotypes. Also, it appears reasonable to expect that the expression pattern of those genes will be correlated with the change of color of the fruit, from the green color present in the early stages to the colors present in the mature phase.

According to Villa-Rivera et al. (2022), 18 genes putatively involved in the carotenoid biosynthetic pathway were identified in Salsa; their gene zexpression profile averages in 12 accessions (four wild, six domesticated, and two reciprocal crosses between domesticated and wild-type accessions) at 0, 10, 20, 30, 40, 50, and 60 DAA were clustered in four groups with similar expression patterns. Nevertheless, with the aim of calculating all possible correlations between SEPs of the different genes within each genotype, a *"Gene2Gene"* algorithm was implemented. Results showed a large heterogeneity in the SEPs of the 18 genes. The number of possible pairs of SEPs comparisons, 18 $(18 - 1)/2 = 153$ pairs per genotype, is consistently expressed in all 12 genotypes. However, we found genes such as *CRTISO2* in accession ancho San Luis, which showed a null model since it did not present significant changes in expression during fruit development, and therefore, that SEP could not be used in the correlation analysis. In Figure 6.3, we can see a large diversity of expression profiles among the genes of interest in the 12 genotypes. This explains why there is only one pair of genes (red and blue lines) that exhibited a fully concordant expression in all 12 genotypes. The number of SEPs that had the maximum expression at each time point is shown as black figures near -2 in the Y-axis, while the brown figures below give the corresponding percentage.

The percentages of SEPs with the maximum at each time point (brown figures in Figure 6.3) present an interesting pattern, where the largest percentage, 36%, was at 0 DAA, while the second-largest one, 26%, was in the other extreme of fruit development, at 60 DAA. This suggests that there could be genes that are involved in carotenoid biosynthesis in the flower—those with maximum at 0 DAA—while others will be responsible of biosynthesis in fruits.

Evaluation of the relations that were consistent in six or more genotypes resulted in nine pairs of genes with highly correlated SEPs. In total, 7 out of the 18 genes studied were found in one or more of the nine estimated relations, shown in Table 6.1, corroborating the high heterogeneity of expression profiles as shown in Figure 6.3. Therefore, those genotypes showed very specific SEPs for the 11 genes.

It is also interesting to note that two of the seven genes (gi=3 and gi=7) are chromoplastic, i.e. specially expressed in the organelle when it is mature and expressing carotenoids.

On the other hand, using the *"Gene2TF"* algorithm, the seven carotenoid-related genes form two segregated sub-networks, the first with four fully connected genes (gi's 1, 3, 6, and 7 in Table 6.1) labeled "In

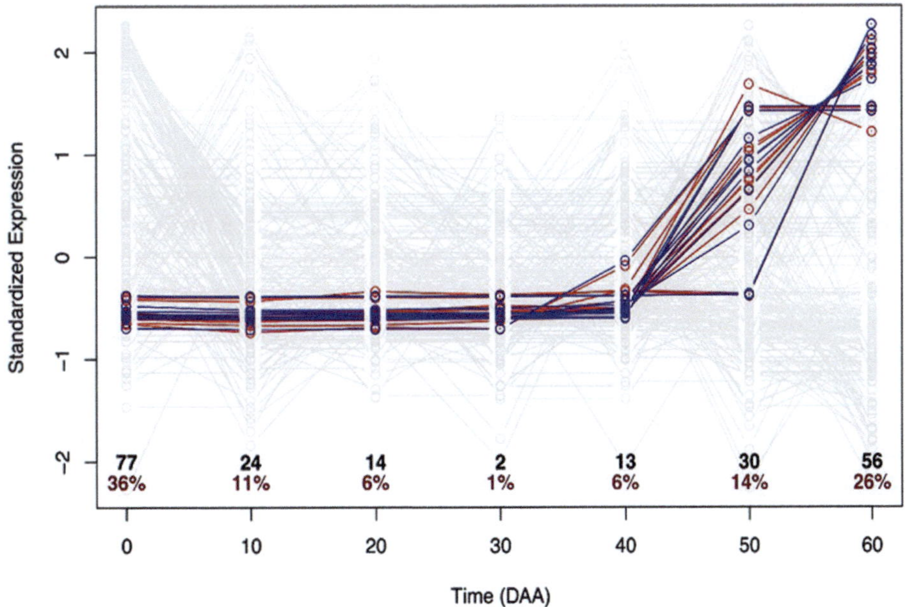

FIGURE 6.3 Plot of 216 individual SEPs showing expression of all the 18-carotenoid biosynthesis-related genes in the 12 genotypes. Red and blue lines correspond to SEPs of β–carotene hydroxylase 1 and capsanthin-capsorubin synthase, respectively, while grey lines correspond to the other 16 carotenoid biosynthesis-related genes. Black figures give the number of SEPs whose maximum expression is at each of the 7 time points, while brown figures give the percentage of those genes.

TABLE 6.1

Genes Present in Significant Relations (consistent in 6 or more genotypes)

gi	Salsa identifier (id)	Protein description	NCBI identifier	Number of relations (n_r)
1	4572	β-carotene hydroxylase 1, chloroplastic	NP_001311784	3
2	10326	Phytoene synthase 2, chloroplastic	XP_016560212.1	2
3	20260	Bifunctional 15-*cis*-phytoene synthase, chromoplastic	XP_016570422	2
4	27689	Lycopene β–cyclase, chloroplastic	XP_016543793	2
5	28396	Zeaxanthin epoxidase, chloroplastic	XP_016561102.1	2
6	28615	Prolycopene isomerase, chloroplastic	XP_016555023.1	3
7	34360	Capsanthin-capsorubin synthase, chromoplastic	NP_001311998	3

fruit" and the second with three fully connected genes (gi's 2, 4, and 5) labeled as "In flower", presented in Figure 6.4. The reason to assign the label "In fruit" to the first sub-network in Figure 6.4 is because genes 1, 3, and 7 had a maximum at the fully mature fruit (60 DAA); moreover, the reason to assign the label "In flower" to the second sub-network in Figure 6.4 is because those genes presented their maximum expression in the mature flower, at 0 DAA, and then in general decremented their expression during fruit development, presenting their minimum expression in the fully mature fruit, at 60 DAA.

To add transcription factor candidates (TFC) to the two sub-networks presented in Figure 6.4, we run the *"Gene2TF"* algorithm. For each one of the seven genes in Table 6.2, the *"Gene2TF"* was run with very high stringency criteria (also see Villa-Rivera et al. 2022), with a minimum value of the determination coefficient, $r^2 \geq 0{:}9$, which implies a correlation $|r| \geq 0{:}9487$, but also, and more importantly, demanding that such relation must be found in all 12 genotypes. This process resulted in the selection of 16 TFC, which are presented in Table 6.2.

In a second step, we run the *Gene2Gene* algorithm, having as input the seven structural genes (SG) (in Table 6.1) plus the 16 TFC, i.e. a total of $7 + 16 = 23$ genes, resulting in a network with 119 relations involving the set of 23 genes, i.e. all genes (the structural 7 SG and the 16 TFC) were included. Furthermore, the minimum number of accessions involved was eight, higher than the minimum threshold of six asked in the input. This means that the estimated network is extremely robust and can be considered to represent relations that are true for a minimum of 67% (8/12) of the genotypes tested. The median of the positive r's was 0.96, which implies a strong linear relation between the pairs of SEPs tested.

As expected, the network estimated consists of two independent subnetworks: one of the genes that we putatively think that is related with carotenoid biosynthesis in the fruit and the second for the genes that are related with carotenoid biosynthesis in the flower. To simplify the interpretation, we ignored all the $119 - (29 + 30) = 60$ connections (edges) that were estimated between pairs of TFC and segregated the two sub-networks in the corresponding organs: fruit and flower, presenting only relations (edges) that link two SG or one SG with a TFC.

In Figure 6.5, we can first note that the topologies between SG, previously presented in Figure 6.4, are again recovered in Figure 6.5, i.e. the fruit network has four fully interconnected SG, while the flower network contains three SG, also fully interconnected. The relations between SG are shown as red lines in both panels in Figure 6.5.

On the other hand, relations between SG and TFC, shown as grey lines in Figure 6.5, indicate that in almost all cases, the TFCs (represented by white squares and labeled tf1 to tf16) were linked to all the corresponding SGs, with only few exceptions. For the fruit network, shown in Figure 6.5A, every TFC could be linked to the four SG, and that indeed was the case for five out of the six TFC, the only exception being tf6, which was not linked to the SG g6. For the flower network, in Figure

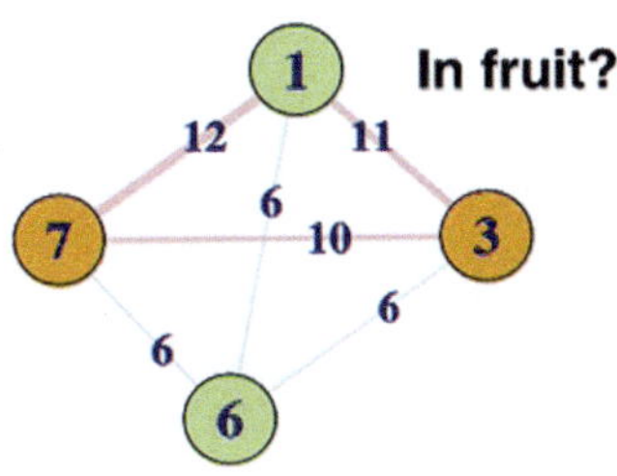

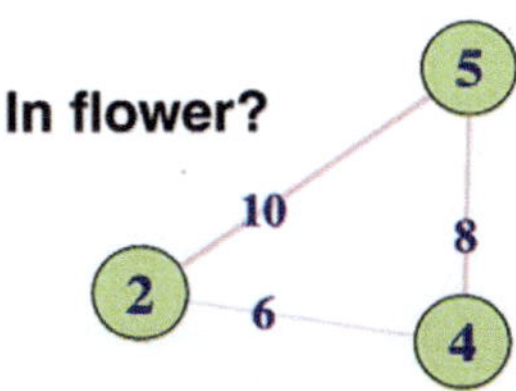

Nodes description
1: beta–carotene hydroxylase 1
2: phytoene synthase 2
3: bifunctional 15–cis–phytoene synthase
4: lycopene beta cyclase
5: zeaxanthin epoxidase
6: prolycopene isomerase
7: capsanthin/capsorubin synthase

FIGURE 6.4 Undirected network with 7 genes (nodes) and 9 relations (edges) in two independent sub-networks, top: gi's: 1, 3, 6, 7 (labeled \In fruit) and bottom: gi's: 2, 4, 5 (labeled \In flower). Color of the nodes is green for chloroplast and orange for chromoplast (see Table 6.1). Color of the edges is pink for positive (r > 0) and light blue for negative (r < 0) relations. Numbers over edges give the number of genotypes included in the relation, and width of the edges is different for the different number of genotypes in the relations (thinner are relations with less genotypes).

TABLE 6.2

Transcription Factor Candidates to Be Regulating Carotenoid Biosynthesis-Related Genes

Tfi	Salsa identifier (id)	Protein description	NCBI identifier	In
1	663	Truncated transcription factor CAULIFLOWER A	XP_016546516.2	Flo.
2	665	MADS-box protein 04g005320	XP_016547230.2	Fru.
3	1393	MYB-related protein 2 isoform X1	XP_016543534.2	Flo.
4	6499	Transcription factor TCP4	XP_016580868.1	Flo.
5	13720	NAC domain-containing protein 72	XP_016541783.1	Fru.
6	15021	Trihelix transcription factor DF1	XP_016551172.2	Fru
7	18678	bZIP transcription factor 11	XP_016560685.2	Flo
8	19849	Dof zinc finger protein DOF3.6	XP_016576870.1	Flo
9	23726	Two-component response regulator ARR12	XP_016579827.2	Flo.
10	24018	NAC domain-containing protein 2-like	XP_016569664.2	Fru.
11	24048	Dof zinc finger protein DOF1.4	XP_016559834.1	Flo.
12	24794	LOB domain-containing protein 1-like	XP_016548470.1	Fru.
13	25505	Transcription factor bHLH122-like	XP_016550196.1	Flo.
14	25737	Zinc finger protein CONSTANTS -LIKE 5	XP_016579622.2	Flo.
15	30278	Truncated transcription factor CAULIFLOWER A	XP_016546516.2	Flo.
16	32729	Ethylene-responsive transcription factor ERF113	XP_016563056.1	Fru

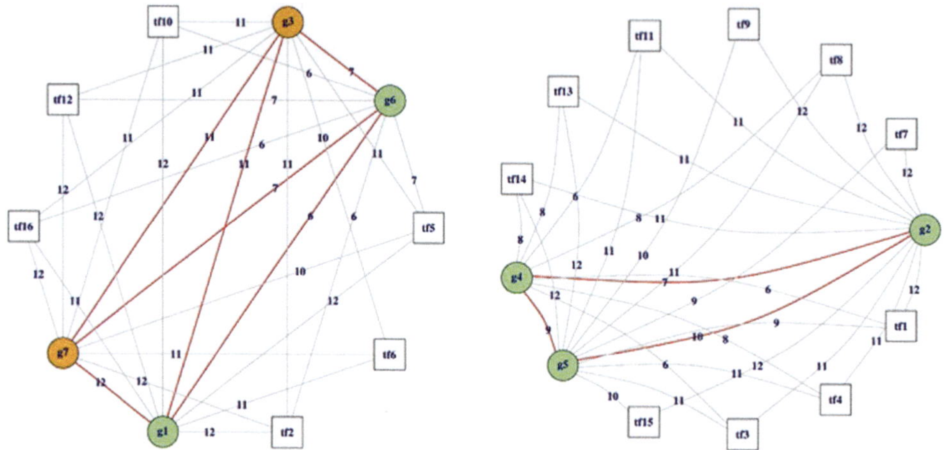

(A) Fruit: 10 genes (nodes); 29 edges (relations).   **(B) Flower**: 13 genes (nodes); 30 edges (relations).

FIGURE 6.5 Networks of carotenoid biosynthesis-related structural genes (SG; circles) and transcription factor candidates (TFC; squares) for Fruit (A) and Flower (B). Chloroplast and chromoplast genes in green and orange, respectively. Relations between SG in red and relations between SG and TFC in grey. Relations between TFC are not shown. Numbers over the edges denote the number of genotypes where the relation is repeated. See Table 6.1 for descriptions of structural genes (g1 to g7) and Table 6.2 for descriptions of TFC (tf1 to tf16).

6.5B, every TFC could be linked to the three SG, and that happened for 7 out of the 10 TFC, the exceptions being tf9, tf7 and tf15, which were only linked to two of the three possible SG.

6.4.2 LOCATION OF PUTATIVE BINDING SITES OF TRANSCRIPTION FACTORS CANDIDATES ON STRUCTURAL GENES

At next step, we analyzed the promoter region of each SG to identify putative binding sequences for each TFC according to the interaction networks shown in Figure 6.5. Some promoter analyses of SG were previously reported by Villa-Rivera et al. (2022), and others were performed according to their reported methodology. A fragment of 2000 nucleotides of promoter region of *BCH, PSY1, PSY2, LCYB2, ZEP, CRTISO2*, and *CCS* genes were analyzed in Table 6.3; putative binding sites for transcription factor candidates located in the promoter region of SGs are shown.

TFCs excepting tf1, tf3, tf7, and tf15 have been previously reported as candidates to be regulators of SG of the carotenoid biosynthetic pathway (Villa-Rivera et al. 2022). Regarding the fruit co-expression network, binding motifs for tf2, tf5, and tf10 were present in all promoters analyzed; similarly, binding sites for tf6 were identified in several promoters excepting in the promoter of *CRTISO2*. In addition, only the promoter of *CCS* contained putative union motifs for tf16 (ethylene-responsive transcription factor ERF113), and no putative binding sites were located for LOB domain-containing protein 1-like (tf12). On the other hand, the analysis of the promoter region of SG grouped in the flower co-expression network, *PSY2* promoter contained putative binding motifs only for tf13, and one binding site for tf11 was in the *LCYB2* promoter region. Interestingly, *cis* sequences for binding of tf1, tf4, tf8, tf9, tf11, tf13, and tf15 were in the *ZEP* promoter region. Finally in Table 6.3, we can see that no binding sites for tf3, tf7, and tf14 were located on the promoter region of SG which interact in the flower GFN.

Here, implementing GFN and promoter region analysis, we propose a reduced list of TFCs for regulation of the carotenoid biosynthetic SG. It has been proposed that if motifs of union are found, it will increase the likelihood of the corresponding TFC to be controlling carotenoid synthesis.

TABLE 6.3

Number of Putative Binding Sites Located in SGs Promoter Region

	g1 BCH	g2 PSY2	g3 PSY1	g4 LCYB2	g5 ZEP	g6 CRTISO2	g7 CCS
tf1	—	0	—	0	3	—	—
tf2	9	—	9	—	—	10	73
tf3	—	0	—	0	0	—	—
tf4	—	0	—	0	1	—	—
tf5	2	—	8	—	—	2	4
tf6	7	—	2	—	—	null	6
tf7	—	0	—	null	0	—	—
tf8	—	0	—	0	42	—	—
tf9	—	0	—	null	16		—
tf10	10	—	13	—	—	2	7
tf11	—	0	—	1	44	—	—
tf12	0	—	0	—		0	0
tf13	—	7	—	0	1	—	—
tf14	—	0-	—	0	0	—	—
tf15	—	0-	—	null	3		—
tf16	0	—	0	—	—	0	4

The next step consists of the experimental corroboration by different techniques or experimental approaches to demonstrate their participation in the regulation of the carotenoid biosynthesis pathway.

6.5 CONCLUSIONS AND FUTURE PERSPECTIVES

Transcriptomic analysis has several applications for the study of different aspects of chili pepper: the search of molecular markers with genetic selection potential and the analysis and increase in knowledge of different metabolic pathways, mechanisms of response to biotic and abiotic stress, and evolutive approaches, among others.

RNA-Seq data have reported important information about the exact timing and pattern of gene expression of the carotenoid biosynthetic genes, influenced by various factors such as temperature, light, and hormonal signals. Overall, the time expression patterns of genes involved in carotenoid biosynthesis during fruit development are complex and can be influenced by a range of internal and external factors. In this sense, co-expression analysis and the integration of transcriptomic and metabolomic data has allowed us to predict putative TFs that could be participating in the transcriptional regulation of carotenoid biosynthesis.

In this chapter, the analysis of time expression profiles of genes and the construction of gene functional networks allowed the identification of seven structural genes (SGs) known to be involved in carotenoid synthesis, which form two separated networks that appear to be involved in the process in fruit (four genes) and flower (three genes). Additionally, we estimated 16 transcription factor candidates (TFCs) to be regulating carotenoid synthesis, 6 in fruit and 10 in flower. Moreover, we noted putative binding sites for TFC in the promoter region of carotenoid SGs, reducing the list of 54 candidates proposed in 2022 and increasing the likelihood of the corresponding TFC to be controlling carotenoid biosynthesis. Of course, experimental evidence is necessary to corroborate the regulatory participation of each TFC; nevertheless, *in silico* analysis could guide future research in this topic.

REFERENCES

Ahn, Y. K., Tripathi, S., Kim, J. H., Cho, Y. I., Lee, H. E., Kim, D. S., et al. (2014). Transcriptome analysis of *Capsicum annuum* varieties Mandarin and Blackcluster: Assembly, annotation and molecular marker discovery. *Gene. 533*(2), 494–499.

Alcaino, J., Baeza, M., and Cifuentes, V. (2016). Carotenoid distribution in nature. *Sub-Cellular Biochemistry. 79*, 3–33.

Arce-Rodríguez, M. L., Martínez, O., and Ochoa-Alejo, N. (2021). Genome-wide identification and analysis of the MYB transcription factor gene family in chili pepper (*Capsicum* spp.). *International Journal of Molecular Sciences. 22*(5).

Ashrafi, H., Hill, T., Stoffel, K., Kozik, A., Yao, J., Chin-Wo, S. R., and Van Deynze, A. (2012). De novo assembly of the pepper transcriptome (*Capsicum annuum*): A benchmark for in silico discovery of SNPs, SSRs and candidate genes. *BMC Genomics. 13*, 571.

Aza-González, C., Nuñez-Palenius, H. G., and Ochoa-Alejo, N. (2011). Molecular biology of capsaicinoid biosynthesis in chili pepper (*Capsicum* spp.). *Plant Cell Reports. 30*(5), 695–706.

Aza-González, C., Nuñez-Palenius, H. G., and Ochoa-Alejo, N. (2012). Molecular biology of chili pepper anthocyanin biosynthesis. *Journal of Mexican Chemical Society 53*(1), 93–98.

Azlan, A., Sultana, S., Huei, C. S., and Razman, M. R. (2022). Antioxidant, anti-obesity, nutritional and other beneficial effects of different chili pepper: A review. *Molecules. 27*(3).

Badia, A. D., Spina, A. A., and Vassalotti, G. (2017). *Capsicum annuum* L.: An overview of biological activities and potential nutraceutical properties in humans and animals. *Journal of Nutritional Ecology and Food Research. 4*(2), 167–177.

Baenas, N., Belovic, M., Ilic, N., Moreno, D. A., and Garcia-Viguera, C. (2019). Industrial use of pepper (*Capsicum annum* L.) derived products: Technological benefits and biological advantages. *Food Chemistry. 274*, 872–885.

Batiha, G. E., Alqahtani, A., Ojo, O. A., Shaheen, H. M., Wasef, L., Elzeiny, M., et al. (2020). Biological properties, bioactive constituents, and pharmacokinetics of some *Capsicum* spp. and capsaicinoids. *International Journal of Molecular Sciences. 21*(15).

Berry, H. M., Rickett, D. V., Baxter, C. J., Enfissi, E. M. A., and Fraser, P. D. (2019). Carotenoid biosynthesis and sequestration in red chilli pepper fruit and its impact on colour intensity traits. *Journal of Experimental Botany. 70*(10), 2637–2650.

Cheng, Y., Ahammed, G. J., Yao, Z., Ye, Q., Ruan, M., Wang, R., et al. (2019). Comparative genomic analysis reveals extensive genetic variations of WRKYs in Solanaceae and functional variations of CaWRKYs in pepper. *Frontiers in Genetics. 10*, 492.

Cheng, Y., Ahammed, G. J., Yu, J., Yao, Z., Ruan, M., Ye, Q., et al. (2016). Putative WRKYs associated with regulation of fruit ripening revealed by detailed expression analysis of the WRKY gene family in pepper. *Scientific Reports. 6*, 39000.

Chiaiese, P., Corrado, G., Minutolo, M., Barone, A., and Errico, A. (2019). Transcriptional regulation of ascorbic acid during fruit ripening in pepper (*Capsicum annuum*) varieties with low and high antioxidants content. *Plants. 8*(7).

Deng, M. H., Zhao, K., Lv, J. H., Huo, J. L., Zhang, Z. Q., Zhu, H. S., et al. (2020). Flower transcriptome dynamics during nectary development in pepper (*Capsicum annuum* L.). *Genetics and Molecular Biology. 43*(2), e20180267.

de Sá Mendes, N., and Branco de Andrade Gonçalves, É. C. (2020). The role of bioactive components found in peppers. *Trends in Food Science & Technology. 99*, 229–243.

Dubey, M., Jaiswal, V., Rawoof, A., Kumar, A., Nitin, M., Chhapekar, S. S., et al. (2019). Identification of genes involved in fruit development/ripening in *Capsicum* and development of functional markers. *Genomics. 111*(6), 1913–1922.

Escoto-Sandoval, C., Flores-Díaz, A., Reyes-Valdés, M. H., Ochoa-Alejo, N., and Martínez, O. (2020). An R package for data mining chili pepper fruit transcriptomes. *Research Square.* https://doi.org/10.21203/rs.3.rs-130806/v1

Escoto-Sandoval, C., Flores-Díaz, A., Reyes-Valdés, M. H., Ochoa-Alejo, N., and Martínez, O. (2021). A method to analyze time expression profiles demonstrated in a database of chili pepper fruit development. *Scientific Reports. 11*(1), 13181.

FAOSTAT. (2021). *Crops.* Italy: FAO.

Fei, X., Hu, H., Luo, Y., Shi, Q., and Wei, A. (2022). Widely targeted metabolomic profiling combined with transcriptome analysis provides new insights into amino acid biosynthesis in green and red pepper fruits. *Food Research International. 160*, 111718.

Flores-Díaz, A., Escoto-Sandoval, C., Cervantes-Hernández, F., Ordaz-Ortiz, J. J., Hayano-Kanashiro, C., Reyes-Valdés, H., et al. (2023). Gene functional networks from time expression profiles: A constructive approach demonstrated in chili pepper (*Capsicum annuum* L.). *Plants. 12*(5).

Gómez-García, M. D. R., and Ochoa-Alejo, N. (2013). Biochemistry and molecular biology of carotenoid biosynthesis in chili peppers (*Capsicum* spp.). *International Journal of Molecular Sciences. 14*(9), 19025–19053.

Guo, Y., Bai, J., Duan, X., and Wang, J. (2021). Accumulation characteristics of carotenoids and adaptive fruit color variation in ornamental pepper. *Scientia Horticulturae. 275*.

Hernández-Pérez, T., Gómez-García, M. D. R., Valverde, M. E., and Paredes-López, O. (2020). *Capsicum annuum* (hot pepper): An ancient Latin-American crop with outstanding bioactive compounds and nutraceutical potential. A review. *Comprehensive Reviews in Food Sciences and Food Safety. 19*(6), 2972–2993.

Hou, B. Z., Li, C. L., Han, Y. Y., and Shen, Y. Y. (2018). Characterization of the hot pepper (*Capsicum frutescens*) fruit ripening regulated by ethylene and ABA. *BMC Plant Biology. 18*(1), 162.

Hwang, D. G., Park, J. H., Lim, J. Y., Kim, D., Choi, Y., Kim, S., et al. (2013). The hot pepper (*Capsicum annuum*) microRNA transcriptome reveals novel and conserved targets: A foundation for understanding MicroRNA functional roles in hot pepper. *PLoS ONE. 8*(5), e64238.

Idrees, S., Hanif, M. A., Ayub, M. A., Hanif, A., and Ansari, T. M. (2020). Chapter 9—Chili pepper. In *Medicinal Plants of South Asia*, ed. Muhammad Asif Hanif, Haq Nawaz, Muhammad Mumtaz Khan and Hugh J. Byrne. Amsterdam, The Netherlands: Elsevier, 113–124.

Islam, K., Rawoof, A., Ahmad, I., Dubey, M., Momo, J., and Ramchiary, N. (2021). *Capsicum chinense* MYB transcription factor genes: Identification, expression analysis, and their conservation and diversification with other Solanaceae genomes. *Frontiers in Plant Science. 12*, 721265.

Jaiswal, V., Rawoof, A., Gahlaut, V., Ahmad, I., Chhapekar, S. S., Dubey, M., and Ramchiary, N. (2022). Integrated analysis of DNA methylation, transcriptome, and global metabolites in interspecific heterotic *Capsicum* F(1) hybrid. *iScience. 25*(11), 105318.

Jarret, R. L., Barboza, G. E., Costa Batista, F. R. D., Berke, T., Chou, Y.-Y., Hulse-Kemp, A., et al. (2019). Capsicum—an abbreviated compendium. *Journal of the American Society for Horticultural Science. 144*(1), 3–22.

Khoo, H. E., Azlan, A., Tang, S. T., and Lim, S. M. (2017). Anthocyanidins and anthocyanins: Colored pigments as food, pharmaceutical ingredients, and the potential health benefits. *Food & Nutrition Research. 61*(1), 1361779.

Kim, H. J., Baek, K. H., Lee, S. W., Kim, J., Lee, B. W., Cho, H. S., et al. (2008). Pepper EST database: Comprehensive in silico tool for analyzing the chili pepper (*Capsicum annuum*) transcriptome. *BMC Plant Biology. 8*, 101.

Kim, M. S., Kim, S., Jeon, J., Kim, K. T., Lee, H. A., Lee, H. Y., et al. (2018). Global gene expression profiling for fruit organs and pathogen infections in the pepper, *Capsicum annuum* L. *Scientific Data. 5*, 180103.

Lee, S., and Choi, D. (2013). Comparative transcriptome analysis of pepper (*Capsicum annuum*) revealed common regulons in multiple stress conditions and hormone treatments. *Plant Cell Reports. 32*(9), 1351–1359.

Li, J., Yang, P., Kang, J., Gan, Y., Yu, J., Calderon-Urrea, A., et al. (2016). Transcriptome analysis of pepper (*Capsicum annuum*) revealed a role of 24-epibrassinolide in response to chilling. *Frontiers in Plant Science. 7*, 1281.

Li, Q.-H., Yang, S.-P., Yu, Y.-N., Khan, A., Feng, P.-L., Ali, M., et al. (2021). Comprehensive transcriptome-based characterization of differentially expressed genes involved in carotenoid biosynthesis of different ripening stages of *Capsicum. Scientia Horticulturae. 288*.

Li, T., Xu, X., Li, Y., Wang, H., Li, Z., and Li, Z. (2015). Comparative transcriptome analysis reveals differential transcription in heat-susceptible and heat-tolerant pepper (*Capsicum annum* L.) cultivars under heat stress. *Journal of Plant Biology. 58*(6), 411–424.

Li, Y., Yu, T., Wu, T., Wang, R., Wang, H., Du, H., et al. (2020). The dynamic transcriptome of pepper (*Capsicum annuum*) whole roots reveals an important role for the phenylpropanoid biosynthesis pathway in root resistance to *Phytophthora capsici. Gene. 728*, 144288.

Liu, B., Sun, Y., Xue, J., Jia, X., and Li, R. (2018). Genome-wide characterization and expression analysis of *GRAS* gene family in pepper (*Capsicum annuum* L.). *PeerJ. 6*, e4796.

Liu, R., Song, J., Liu, S., Chen, C., Zhang, S., Wang, J., et al. (2021). Genome-wide identification of the *Capsicum* bHLH transcription factor family: Discovery of a candidate regulator involved in the regulation of species-specific bioactive metabolites. *BMC Plant Biology. 21*(1), 262.

Liu, S., Li, W., Wu, Y., Chen, C., and Lei, J. (2013). De novo transcriptome assembly in chili pepper (*Capsicum frutescens*) to identify genes involved in the biosynthesis of capsaicinoids. *PLoS ONE. 8*(1), e48156.

Liu, Y., Lv, J., Liu, Z., Wang, J., Yang, B., Chen, W., et al. (2020). Integrative analysis of metabolome and transcriptome reveals the mechanism of color formation in pepper fruit (*Capsicum annuum* L.). *Food Chemistry. 306*, 125629.

Lu, F. H., Cho, M. C., and Park, Y. J. (2012). Transcriptome profiling and molecular marker discovery in red pepper, *Capsicum annuum* L. TF68. *Molecular Biology Reports. 39*(3), 3327–3335.

Lu, F.-H., Yoon, M.-Y., Cho, Y.-I., Chung, J.-W., Kim, K.-T., Cho, M.-C., et al. (2011). Transcriptome analysis and SNP/SSR marker information of red pepper variety YCM334 and Taean. *Scientia Horticulturae. 129*(1), 38–45.

Ma, X., Yu, Y.-N., Jia, J.-H., Li, Q.-H., and Gong, Z.-H. (2022). The pepper MYB transcription factor CaMYB306 accelerates fruit coloration and negatively regulates cold resistance. *Scientia Horticulturae. 295*.

Maoka, T. (2020). Carotenoids as natural functional pigments. *Journal of Natural Medicines. 74*(1), 1–16.

Martel, C., Vrebalov, J., Tafelmeyer, P., and Giovannoni, J. J. (2011). The tomato MADS-box transcription factor RIPENING INHIBITOR interacts with promoters involved in numerous ripening processes in a COLORLESS NONRIPENING-dependent manner. *Plant Physiology. 157*(3), 1568–1579.

Martínez-Lopez, L. A., Ochoa-Alejo, N., and Martínez, O. (2014). Dynamics of the chili pepper transcriptome during fruit development. *BMC Genomics. 15*, 143.

Martínez, O. (2022). Time course gene expression experiments. In *Transcriptome Profiling: Progress and Prospects*, ed. Joongku Lee and Mohammad Ajmal Ali. London: Elsevier, 85–110.

Martínez, O., Arce-Rodríguez, M. L., Hernández-Godínez, F., Escoto-Sandoval, C., Cervantes-Hernández, F., Hayano-Kanashiro, C., et al. (2021). Transcriptome analyses throughout chili pepper fruit development reveal novel insights into the domestication process. *Plants. 10*(3).

Meléndez-Martínez, A. J., Mapelli-Brahm, P., Hornero-Méndez, D., and Vicario, I. M. (2019). CHAPTER 1. Structures, nomenclature and general chemistry of carotenoids and their esters. In *Carotenoid Esters in Foods*, ed. Adriana Z. Mercadante. London: The Royal Society of Chemistry, 1–50.

Mohd Hassan, N., Yusof, N. A., Yahaya, A. F., Mohd Rozali, N. N., and Othman, R. (2019). Carotenoids of *Capsicum* fruits: Pigment and health-promoting functional attributes. *Antioxidants. 8*(10).

Omolo, M. A. (2014). Antimicrobial properties of chili peppers. *Journal of Infectious Diseases and Therapy. 2*(4).

Osorio, S., Alba, R., Nikoloski, Z., Kochevenko, A., Fernie, A. R., and Giovannoni, J. J. (2012). Integrative comparative analyses of transcript and metabolite profiles from pepper and tomato ripening and development stages uncovers species-specific patterns of network regulatory behavior. *Plant Physiology. 159*(4), 1713–1729.

Ou, L., Liu, Z., Zhang, Z., Wei, G., Zhang, Y., Kang, L., et al. (2017). Noncoding and coding transcriptome analysis reveals the regulation roles of long noncoding RNAs in fruit development of hot pepper (*Capsicum annuum* L.). *Plant Growth Regulation. 83*(1), 141–156.

Pan, Y., Bradley, G., Pyke, K., Ball, G., Lu, C., Fray, R., et al. (2013). Network inference analysis identifies an *APRR2*-like gene linked to pigment accumulation in tomato and pepper fruits. *Plant Physiology. 161*(3), 1476–1485.

Rawoof, A., Ahmad, I., Islam, K., Momo, J., Kumar, A., Jaiswal, V., and Ramchiary, N. (2022). Integrated omics analysis identified genes and their splice variants involved in fruit development and metabolites production in *Capsicum* species. *Functional & Integrative Genomics. 22*(6), 1189–1209.

Razo-Mendivil, F. G., Hernández-Godínez, F., Hayano-Kanashiro, C., and Martínez, O. (2021). Transcriptomic analysis of a wild and a cultivated varieties of *Capsicum annuum* over fruit development and ripening. *PLoS ONE. 16*(8), e0256319.

Rosas-Saavedra, C., and Stange, C. (2016). Biosynthesis of carotenoids in plants: Enzymes and color. *Sub-Cellular Biochemistry. 79*, 35–69.

Sathasivam, R., Radhakrishnan, R., Kim, J. K., and Park, S. U. (2021). An update on biosynthesis and regulation of carotenoids in plants. *South African Journal of Botany. 140*, 290–302.

Seymour, G. B., Ostergaard, L., Chapman, N. H., Knapp, S., and Martin, C. (2013). Fruit development and ripening. *Annual Review of Plant Biology. 64*, 219–241.

Shima, Y., Kitagawa, M., Fujisawa, M., Nakano, T., Kato, H., Kimbara, J., et al. (2013). Tomato FRUITFULL homologues act in fruit ripening via forming MADS-box transcription factor complexes with RIN. *Plant Molecular Biology. 82*(4–5), 427–438.

Song, J., Chen, C., Zhang, S., Wang, J., Huang, Z., Chen, M., et al. (2020). Systematic analysis of the *Capsicum* ERF transcription factor family: Identification of regulatory factors involved in the regulation of species-specific metabolites. *BMC Genomics. 21*(1), 573.

Song, J., Sun, B., Chen, C., Ning, Z., Zhang, S., Cai, Y., et al. (2023). An R-R-type MYB transcription factor promotes non-climacteric pepper fruit carotenoid pigment biosynthesis. *Plant Journal. 115*(3), 724–741.

Song, S., Song, S. Y., Nian, P., Lv, D., Jing, Y., Lu, S., et al. (2022). Transcriptomic analysis suggests a coordinated regulation of carotenoid metabolism in ripening chili pepper (*Capsicum annuum* var. conoides) fruits. *Antioxidants. 11*(11).

Stanley, L., and Yuan, Y. W. (2019). Transcriptional regulation of carotenoid biosynthesis in plants: So many regulators, so little consensus. *Frontiers in Plant Science. 10*, 1017.

Sun, T., Rao, S., Zhou, X., and Li, L. (2022). Plant carotenoids: Recent advances and future perspectives. *Molecular Horticulture. 2*(1).

Tanaka, Y., Watachi, M., Nemoto, W., Goto, T., Yoshida, Y., Yasuba, K. I., et al. (2021). Capsaicinoid biosynthesis in the pericarp of chili pepper fruits is associated with a placental septum-like transcriptome profile and tissue structure. *Plant Cell Reports. 40*(10), 1859–1874.

Villa-Rivera, M. G., Martínez, O., and Ochoa-Alejo, N. (2022). Putative transcription factor genes associated with regulation of carotenoid biosynthesis in chili pepper fruits revealed by RNA-Seq coexpression analysis. *International Journal of Molecular Sciences. 23*(19).

Villa-Rivera, M. G., and Ochoa-Alejo, N. (2020). Chili pepper carotenoids: Nutraceutical properties and mechanisms of action. *Molecules. 25*(23).

Villa-Rivera, M. G., and Ochoa-Alejo, N. (2021). Transcriptional regulation of ripening in chili pepper fruits (*Capsicum* spp.). *International Journal of Molecular Sciences. 22*(22).

Villa-Rivera, M. G., and Ochoa-Alejo, N. (2023). Ascorbic acid in chili pepper fruits: Biosynthesis, accumulation, and factors affecting its content. *Journal of the Mexican Chemical Society. 67*(3), 187–199.

Wang, J., Liu, Y., Tang, B., Dai, X., Xie, L., Liu, F., and Zou, X. (2020). Genome-wide identification and capsaicinoid biosynthesis-related expression analysis of the *R2R3-MYB* gene family in *Capsicum annuum* L. *Frontiers in Genetics. 11*, 598183.

Wang, J., Lv, J., Liu, Z., Liu, Y., Song, J., Ma, Y., et al. (2019). Integration of transcriptomics and metabolomics for pepper (*Capsicum annuum* L.) in response to heat stress. *International Journal of Molecular Sciences. 20*(20).

Wu, Z., Cheng, J., Cui, J., Xu, X., Liang, G., Luo, X., et al. (2016). Genome-wide identification and expression profile of *Dof* transcription factor gene family in pepper (*Capsicum annuum* L.). *Frontiers in Plant Science. 7*, 574.

Yu, C., Zhan, Y., Feng, X., Huang, Z. A., and Sun, C. (2017). Identification and expression profiling of the auxin response factors in *Capsicum annuum* L. under abiotic stress and hormone treatments. *International Journal of Molecular Sciences. 18*(12).

Zhang, X. H., Liu, H. Q., Guo, Q. W., Zheng, C. F., Li, C. S., Xiang, X. M., et al. (2016). Genome-wide identification, phylogenetic relationships, and expression analysis of the carotenoid cleavage oxygenase gene family in pepper. *Genetics and Molecular Research. 15*(4).

Zhang, Z., Chen, J., Liang, C., Liu, F., Hou, X., and Zou, X. (2020). Genome-wide identification and characterization of the bHLH transcription factor family in pepper (*Capsicum annuum* L.). *Frontiers in Genetics. 11*, 570156.

7 Chemistry and Post-Harvest Processing Technologies of Peppers (*Capsicum* spp.)

Filipe Kayodè Felisberto dos Santos, Rayssa Ribeiro,
Karine Sayuri Lima Miki, Barbara Elisabeth Teixeira-Costa
and Valdir Florêncio da Veiga Júnior

CONTENTS

7.1 INTRODUCTION

Pepper (*Capsicum* spp.) has a millennial historical use by mankind, being one of the oldest culti-vated plants. It originates from the Americas, more specifically from the region that encompasses Mexico and Central America. Historical data indicates that its domestication occurred about 6,000 years ago. It is believed that the first varieties of pepper were discovered by the Mayan community and the Aztecs who inhabited the region and observed its spicy flavour as well as its medicinal properties. It quickly found its way to other parts of the world through trade routes, being created in different culinary preparations, such as sauces, seasonings and marinades, and as a main ingredient in dishes typical of different cultures (Ali et al., 2016).

In addition to its culinary use, pepper was also used for medicinal purposes. Several studies pointed out that pepper can display analgesic, anti-inflammatory and digestive properties. It was also used as a pain reliever, to treat gastrointestinal problems and to promote general health status. In many other cultures, it was used in traditional medicine to treat headaches, rheumatism, diges-tive problems and colds. These medicinal applications are due to the complex and rich chemical composition of pepper, which can be linked to the content of diverse phenolic compounds, vitamins and carotenoids (Pinar et al., 2023).

Together with capsaicin and its derivatives, which are commonly found in the seeds and responsible for the pungent taste of peppers, other compounds, such as anthocyanins, tocopherol, vitamin C and carotenoids, as well as other phenols and flavonoids have already been identified

DOI: 10.1201/9781003378259-7

(Santos et al., 2022a; Karaman et al., 2023). Such compounds may undergo some type of alteration depending on how the pepper is treated in the post-harvest stages, maximizing or attenuating their amounts and activities. The application of peppers as raw materials for industry depends on the content of these bioactive compounds in their matrix. Peppers are usually applied in industrial processes such as a dry powder (paprika) or an oleoresin (concentrated extract of essential oils, waxes, carotenoids, flavonoids and capsaicinoids) (Antonio et al., 2018). Techniques such as dehydration (removal of moisture), freezing (inhibiting the growth of microorganisms), ultraviolet radiation (inactivation of microorganisms), electric pulse field (cell permeability) and modified atmosphere packaging (bioplastic films or vacuum packaging) are commonly used by the industry for food preservation (Adewoyin et al., 2023).

In this context, this chapter aims to present the main chemical compounds present in peppers and how the post-harvest stages are correlated to promote the extension of their useful life, as well as their advantages and disadvantages.

7.2 PEPPERS AND THEIR CHEMICAL COMPLEXITY

Peppers from different *Capsicum* species are composed of a wide variety of chemical compounds that contribute to their sensorial characteristics as well as to health-related benefits. As the chemical profile of peppers varies with the species, seasonality, environmental conditions and even the life cycle of the plant, it can be difficult to choose the *Capsicum* species and cultivars that are economically profitable for a specific objective. The main chemical constituents found in peppers and the differences observed between the studied varieties are described in detail as follows.

7.2.1 CAPSAICINOIDS

Capsaicinoids are the compounds responsible for the sensation of heat and spiciness in chilli peppers. Capsaicin is the main capsaicinoid present in chilli peppers, while other capsaicinoids such as dihydrocapsaicin, nordihydrocapsaicin and homocapsaicin (Figure 7.1) may also be present. The amount and proportion of capsaicinoids vary among pepper varieties and are responsible for the hotness scale, measured in Scoville heat units (SHU). Milder varieties, such as bell peppers, have lower levels of capsaicinoids, while the spicier ones, such as habaneros and jalapeños, contain higher concentrations.

The capsaicinoids present a high toxicity degree to humans. In mammalian organisms, they act as an inflammatory and carcinogenic agent, causing irritation in mucous membranes in the digestive and respiratory systems and cellular rupture when orally ingested. It is estimated that the capsaicinoids' lethal dose of 50% (LD50) by oral ingestion in humans is about 5.0 g kg-1, with a safe daily intake dose of 2.64 mg of capsaicinoids. For medicinal purposes, cultivars with moderate to low capsaicinoid content are preferable due to the alkaloid toxicity, which can make it difficult to formulate medicines that are safe to be used by humans (Melgar-Lalanne et al., 2017).

Capsicum chinense cultivars are mainly known for their high levels of pungency, generally showing higher concentrations of capsaicin. Therefore, cultivars of *C. chinense* are associated with a high toxicity for human use, which makes their use as a medicine more difficult when compared to other species, such as *Capsicum annuum*, which is popularly applied as a spice or even as a chemical weapon. Additionally, it should be pointed out that *C. annuum* cultivars are more suitable to present species with dihydrocapsaicin as a major capsaicinoid during the maturity stage, although its inversion or even the presence of low levels of capsaicin cannot be directly correlated to low pungency levels, as cultivars such as Toegaard hot banana and guajillo are scaled as have moderate pungency in the Scoville heat scale, with a 15,000 ± 5,300 SHU and 13,704.8 ± 2,275.7 SHU, respectively (Duelund & Mouritsen, 2017).

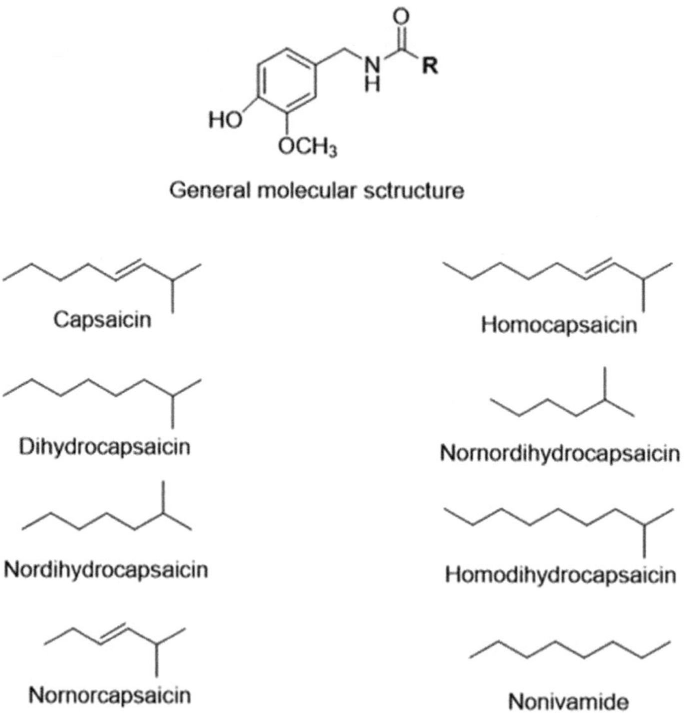

General molecular sctructure

Capsaicin

Homocapsaicin

Dihydrocapsaicin

Nornordihydrocapsaicin

Nordihydrocapsaicin

Homodihydrocapsaicin

Nornorcapsaicin

Nonivamide

FIGURE 7.1 General molecular structure of the capsaicinoids and the main substances described in *Capsicum* species.

7.2.2 CAROTENOIDS

Carotenoids are a group of nearly 750 tetraterpenoid compounds and are mainly responsible for the characteristic colour of peppers, which varies from pale yellow to deep red, depending on the stage of harvest and the variety of pepper, present a low polar molecular structure based on a tetraterpenoid skeleton, in which the most common structures present 40 carbon atoms derived from isoprene units. This class of pigments can be synthesized in all photosynthetic organisms and in some non-photosynthetic organisms playing an important role in photosynthesis, photoprotection and stabilization of free radicals (Huang et al., 2017).

Carotenoids can be divided into two main groups in accordance with their chemical structure: carotenes and xanthophylls. Carotene groups include linear or cyclized hydrocarbons, such as α-carotene and β-carotene, while xanthophylls include carotenoids with oxygenated functions such as hydroxy-, keto-, methoxy-, epoxy- and carbonyl-, such as those that occur in lutein, violaxanthin, neoxanthin and zeaxanthin. These carotenoids contribute to the variety of colours found in different pepper varieties, such as yellow, orange and red. Each of these carotenoids has a unique chemical structure with a conjugated carbon chain that gives its characteristic colour (Usmani et al., 2022). The chemical structures of the main carotenoids present in peppers can be observed in Figure 7.2.

Capsanthin, capsanthin-5,6-epoxide and capsorubin are the main carotenoids linked to the intense red colour found in some varieties of ripe pepper (Kim et al., 2009). Their chemical structures are similar to other carotenoids, with a long and conjugated carbon chain, composed of double bonds and single bonds. Capsanthin has two β-ionone molecules, which are associated with the distinctive red colour in peppers. It is an acyclic carotenoid because it does not contain closed carbon rings (e.g. cyclic carotenoids such as β-carotene) and is classified as a xanthophyll carotenoid due to the presence of oxygenated functional groups in its structure (Rodriguez-Concepcion & Daròs,

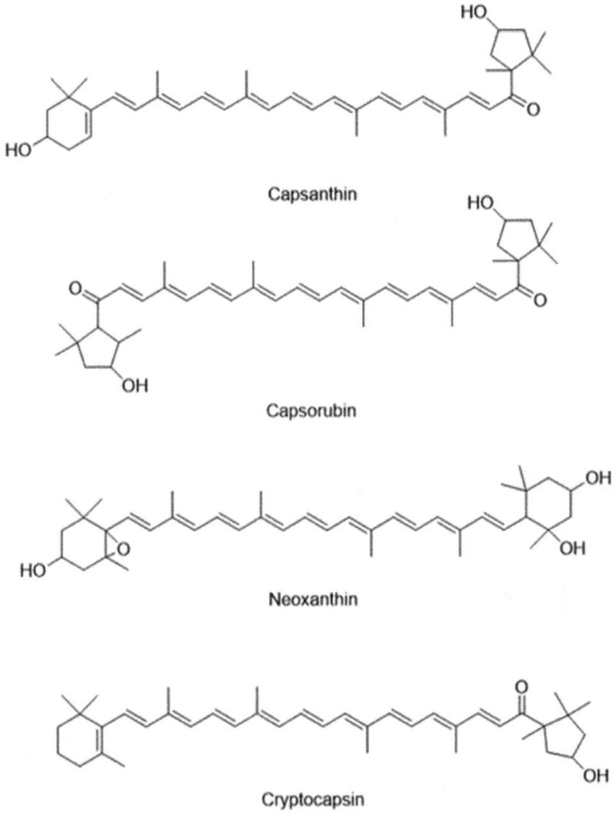

FIGURE 7.2 Main carotenoids observed within the *Capsicum* genus.

2022). Capsanthin, like other xanthophylls, can be considered a functional substance because it can inhibit the photooxidation of polyunsaturated fatty acids and linoleic acid, even greater than β-carotene, lycopene or lutein (Kim et al., 2009).

The intense colour of capsanthin is a result of the conjugation of double bonds and single bonds in its structure. This conjugation allows capsanthin to strongly absorb blue and green light from the electromagnetic spectrum and reflect a red-orange hue (Kim et al., 2009). Another characteristic of capsanthin is that it is soluble in organic solvents such as oils and lipids, but it is insoluble in water. This characteristic influences its distribution in pepper since carotenoids are stored in the lipid structures of plant cells, such as chloroplasts. Exposure to light, heat and oxygen can cause degradation of carotenoids, including capsanthin. Therefore, it is important to protect capsanthin-containing foods such as chilli peppers from prolonged exposure to these factors to preserve their colour and antioxidant properties (Kim et al., 2009).

Despite being used as a natural pigment, capsanthin has also been linked to health-related benefits, such as anti-inflammatory, antioxidant and immunomodulatory effects, contributing to protection against different chronic diseases and strengthening the immune system (Shanmugham & Subban, 2022). Some functional properties of carotenoids, such as antioxidants, are associated with their free radical scavenging and quenching activities (Kim et al., 2009). Narisawa et al. (1996) studied the effect of paprika juice rich in capsanthin (3.54 mg/100 ml) against colon carcinogenesis in F344 rats. These authors found that the paprika juice may affect colon carcinogenesis at the 30th week via the suppression of aberrant crypt foci formation. Despite this relevant result, these authors also found that capsanthin itself failed to inhibit colon tumorigenesis in the short-term assay and that more studies are needed to explain this discrepancy (Narisawa et al., 1996). Capsanthin, capsanthin monoester, capsanthin diester and capsanthin 3, 6-epoxide and others showed great inhibitory tumor-promotion effects when tested in vitro on Epstein-Barr virus early antigen (EBV-EA) activation (Kim et al., 2009).

Capsorubin is another carotenoid that contributes to the deep red colour of ripe peppers. Its chemical structure is similar to that of capsanthin, with a long, conjugated carbon chain containing double bonds and single bonds. Capsorubin also has two β-ionone moieties, giving it a characteristic red colour. Due to this effect, no capsaicin or capsorubin is found in yellow and orange ripe pepper fruits (Wang et al., 2023). Its physicochemical characteristics are very similar to capsanthin, and studies on its activity also point to antioxidant and anti-inflammatory effects (Wang et al., 2023).

7.2.3 FLAVONOIDS

Flavonoids are a class of chemical compounds that consist of more than 7,000 secondary plant metabolites and are based on a structure with 15 carbon atoms, contributing to the antioxidant properties of peppers. They are mainly accumulated in the fruit peel, and although most published papers represent their content as aglycones flavonoids, they can occur in pepper as conjugated O-glycosides and C-glycosides derivates (Castillejo et al., 2022). In the *Capsicum* genus, quercetin and luteolin are described as major flavonoids, representing approximately 41% of the total flavonoid content generally in their hydrolyzed form. Chilli peppers studied by Mi et al. (2022) presented up to 38 flavonoid-related compounds. These authors found that 16 flavonoid-related components were common in Jize and Korean chilli peppers, while this last one solely presented quercetin and isorhamnetin (Mi et al., 2022). These last two substances are very well known for their relevant antioxidant properties. Usually, the content of flavonoids in pepper is expressed as the sum of these cited compounds due to the difficulty of performing acid hydrolysis in several of the other conjugate flavonoids (Mi et al., 2022). The chemical structures from the main flavonoids reported for pepper are showed at Figure 7.3.

7.3 UPDATE ON PEPPERS' MEDICINAL PROPERTIES AND APPLICATIONS

Capsicum fruits have been used throughout the last centuries as both fresh and dried spices. Some species within this genus are considered a rich source of health-related bioactive compounds and have also been applied in traditional medicine for the treatment of cough, toothache, sore throat, parasitic infections, rheumatism and wound healing. Among the reported biological properties, the most notable is their capacity to act as antioxidants to reduce oxidative stress, leading to the prevention of various degenerative diseases. There are several applications for pepper extracts in formulating various medications, such as the use of ointments containing capsaicin for pain relief, migraines, headaches, psoriasis and in the treatment of some viral infections. Additionally, capsaicin is also administered for lack of appetite, flatulence, atherosclerosis, heart diseases and muscle tension (Alonso-Villegas et al., 2023; Khan et al., 2014; Howard et al., 2000). Functional compounds found in *Capsicum* demonstrate remarkable antimicrobial properties, especially against gram-positive pathogenic microorganisms. Furthermore, the nutraceutical potential of phytochemical compounds derived from species of the genus confirms their effectiveness in cancer prevention and heart health. Additionally, *Capsicum* essential oils are being used as antiageing agents in cosmetic products (Akhtar et al., 2021).

Capsicum annuum is the most cultivated spice in the world, primarily linked to a culinary application and, later, to diverse therapeutic uses. Some studies have tested the biological properties of *C. annuum* and found that it can display antimicrobial, anti-inflammatory, antioxidant, antihypertensive, antihyperglycemic and antitumoural activities (Akhtar et al., 2021; Alonso-Villegas et al., 2023).

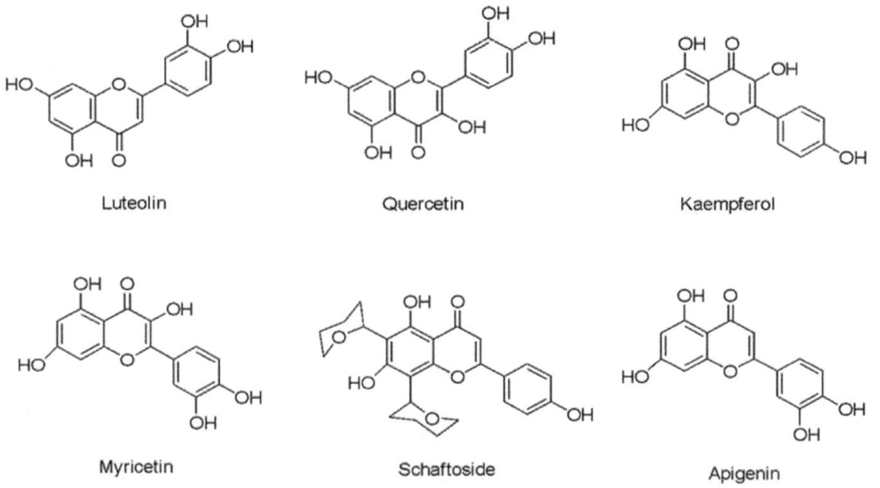

Luteolin

Quercetin

Kaempferol

Myricetin

Schaftoside

Apigenin

FIGURE 7.3 Main flavonoids reported within the *Capsicum* genus.

A recent study conducted by Hamed et al. (2019) analyzed the content of capsaicinoids and other bioactive compounds in pepper cultivars under different maturation stages. These authors reported that the content of capsaicinoids in both green and red/yellow stages was affected by roasting. Regarding the antioxidant assays determined by 2,2-diphenyl-1-picrylhydrazyl—DPPH and 2,2′-azino-bis (3-ethylbenzthiazoline-6-sulphonic) acid—ABTS reagents both raw and roasted peppers possessed great antioxidant activity, 61% to 87% and 73 to 159 µg/g, respectively (Hamed et al., 2019). In another study, the phytochemical and *in vitro* and *in vivo* biological profiles of sweet and hot peppers from the Altino variety (Abruzzo region, Italy) were investigated (Della Valle et al., 2020). This study evaluated the antioxidant properties of *C. annum* extracts by using free radical scavenging assays (DPPH and ABTS) and reducing power assays (cupric-reducing antioxidant capacity—CUPRAC and ferric-reducing antioxidant power—FRAP). The antioxidant assays demonstrated that decoction extracts of hot pepper in methanol exhibited higher values than the hot pepper n-hexane decoction, while the hot pepper microwave extract showed the greatest inhibition value on the in vitro cell growth assay (Della Valle et al., 2020).

The *in vitro* antioxidant and anti-inflammatory effects, as well as the flavonoid contents, in chilli pepper (*Capsicum annuum*), leaves and fruits were studied by Cho et al. (2020). Both extracts, from chilli pepper leaves and fruits, exhibited radical scavenging activities and better lipopolysaccharide-stimulated inflammatory response by reducing the nitric oxide production and interleukin-6 and tumour necrosis factor-alpha levels in RAW 264.7 cells, although the first extract displayed greater response than the second one (Cho et al., 2020). These results indicate that this matrix can be a useful source to prevent oxidative stress and inflammation-related diseases (Cho et al., 2020).

Capsicum chinense is commonly known as habanero pepper. It is popular not only due to its sweet, flavourful, and aromatic fruits but also because it exhibits varying levels of pungency and pharmacological properties. Several studies performed with this species have shown promising clinical results against some pathological microorganisms, unequivocally contributing to the use of natural products in the fight against diseases. Recently, in vitro antiproliferative activity and modulation of cytotoxicity evaluated extracts from this species using the sulforhodamine B method. The extract from leaves of plants grown in black soil, obtained through maceration, exhibited high selective cytotoxicity against colorectal cancer cells (HCT–15), with an IC50 = 16.23 ± 2.89 µg mL-1 (Chel-Guerrero et al., 2022).

The ethyl acetate extract from mature *C. chinense* fruits was evaluated in vitro against the fungi *Sclerotinia sclerotiorum*, *Rhizopus stolonifer* and *Colletotrichum gloeosporioides* using the disc diffusion methodology. At a concentration of 100 µL, the extract was able to inhibit 72.7% of *R. stolonifer* growth, while at 200 µL, it inhibited 98.3% of *C. gloeosporioides* growth and 96.2% of *S. sclerotiorum* fungus (Santos et al., 2022b). The hexane extracts of two Brazilian fruit varieties of *C. chinense* (unripe bode pepper 'HE-UB' and ripe little beak pepper 'HE-RB') were also evaluated against *S. sclerotiorum*, *R. stolonifer* and *C. gloeosporioides* using the same disc diffusion method. 'HE-RB' extract completely inhibited the growth of *S. sclerotiorum*, *R. stolonifer* and *C. gloeosporioides* at doses of 200 µL, 100 µL and 300 µL, respectively. Furthermore, 'HE-UB' extract also achieved 100% fungal growth inhibition at concentrations of 100 µL for *S. sclerotiorum*, 150 µL for *C. gloeosporioides* and 200 µL for *R. stolonifer* (Toigo et al., 2022). Furthermore, the ethanolic extract from *C. chinense* fruits inhibited the growth of the fungus *Aspergillus parasiticus* at a MIC50 concentration of 381 mg/mL (minimum concentration that inhibited 50% of fungal growth, as determined by Probit analysis) (Buitimea-Cantúa et al., 2020).

The investigation of hexane extracts obtained from the fruits of two varieties (unripe bode pepper HE-UB and ripe little beak pepper HE-RB) of the plant *C. chinense* was also evaluated for its anti-acetylcholinesterase, leishmanicidal and antiproliferative activities. Regarding the antiacetylcholinesterase activity by the microplate assay method, the results indicated inhibitory activity against the acetylcholinesterase enzyme, with IC50 values of HE-UB = 41.5 µg/mL and HE-RB = 20.3 µg/mL, suggesting a potential neuroprotective effect. Both varieties (HE-UB and HE-RB) also showed antileishmanial activity against *Leishmania amazonensis* promastigotes, with IC50 values of 67.19 µg/mL and 38.16 µg/mL, respectively. The antiproliferative activity of the extracts was tested in cancer cells to assess their potential in combating the uncontrolled growth of tumour cells. Extracts from both pepper varieties demonstrated cytotoxic activity against various human tumour cell lines with IC50 values ranging from 325.40 to 425.0 µg/mL. These findings suggest that these pepper varieties have the potential to be used as sources of bioactive compounds with possible therapeutic applications (Toigo et al., 2022).

Martínez-Arámburu et al. (2015) used sodium ferulate and sodium caffeate, previously identified in *Capsicum pubescens* extracts, as antibacterial agents against *S. enterica ser. Typhimurium* and *L. monocytogenes*. Sodium ferulate generated a bacteriostatic effect on *S. enterica ser. Typhimurium* at concentrations of 0.3% and 0.6% and *L. monocytogenes* at 0.6%. Meanwhile, the same effect was shown by sodium caffeate in both pathogenic bacteria but at a concentration of 0.15%. Likewise, the effect of hydroxycinnamic and cinnamic acids has been tested on the growth of non-pathogenic bacteria such as LAB (lactic acid bacteria).

Malagueta pepper, the popular name for *Capsicum frutescens*, was evaluated for its antioxidant activity by the DPPH radical scavenging method. The results of the antioxidant activity assay revealed that the ethanol extract of this pepper presents an IC50 (DPPH) value of 270.99 ± 1.50 µg/mL, directly related to capsaicin. Therefore, the ethanolic extract of *Capsicum frutescens* is considered promising for the formulation of products with antioxidant activity (Alarcón et al., 2019). These findings suggest that varieties of pepper species have the potential to be used as sources of bioactive compounds with possible therapeutic applications. However, it is important to emphasize that more research is needed to fully understand the mechanisms of action and effectiveness of these extracts in clinical and human health contexts.

Just as for therapeutic applications, species within the *Capsicum* genus are the subject of research and applications in the food and cosmetic industries. Peppers offer a nutritional content rich in proteins, vitamins and fibres. Recently, the effect of supplementing extra-virgin olive oil with pepper was evaluated. In comparison to un-supplemented oil, higher levels of antioxidant activity were observed when enriched with *Capsicum annuum* pepper. These findings provide a scientific basis for innovation in food technology for new food products (Zellama et al., 2022). Furthermore, the incorporation of various ingredients derived from *Capsicum* has been applied in a wide range of commercial products (Alonso-Villegas et al., 2023). This underscores the importance of incorporating these *Capsicum*-derived ingredients in the cosmetic and food industries.

7.4 PHYSICAL, CHEMICAL AND NON-DESTRUCTIVE POST-HARVEST TECHNOLOGIES

7.4.1 DEHYDRATION

Dehydration is a traditional and widely used technique to extend the shelf life of food, as many foods are prone to deterioration after harvesting. This deterioration is influenced by various internal and external factors, in which water content is one of the primary determinants of food quality over time. Therefore, dehydration plays a crucial role in controlling deterioration by reducing water activity, inhibiting the proliferation of microorganisms and preventing undesirable chemical reactions. In addition to its importance in preserving food quality, dehydration also brings economic benefits, such as reducing costs associated with packaging, handling, storage and transportation, as it decreases the weight and, in some cases, the volume of food products (Wray & Ramaswamy, 2015).

Drying processes can be categorized as natural or artificial. Natural drying involves exposing food to solar energy and harnessing sunlight to remove moisture. While it is a simple approach, this technique can be slow and susceptible to contamination, commonly used in family farming, especially for products like coffee, peppers and legumes. On the other hand, artificial drying employs controlled equipment to precisely adjust temperature, humidity and air velocity. This allows farmers to produce high-quality products more quickly and uniformly, regardless of weather conditions. In this process, water is removed by applying a stream of dry air that extracts moisture from the surface of the product (González-Toxqui et al., 2020). However, it is crucial to understand that the dehydration process can affect various physical and chemical characteristics of foods, including the potential degradation of nutrients and the occurrence of both enzymatic and non-enzymatic reactions. To effectively optimize this procedure, it is essential to consider the mechanisms related to the movement of water inside and outside of foods, which can be influenced by factors such as capillary forces, water diffusion due to concentration gradients, surface diffusion, vapor diffusion in air-filled pores, flow due to pressure gradients, vaporization and water condensation (Ibarz & Barbosa-Cánovas, 2014).

In a specific context, the dehydration of peppers is an ancient practice that plays a significant role in culinary practices, spice blends and the canning industry. Different types of peppers, such as cayenne, ancho, pasilla, mirasol, piquin, and de Arbol, are used in the production of dried pepper. The quality of these products, such as 'red pepper' and paprika, is evaluated based on criteria like spiciness, red colour and flavour, making careful processing and storage essential. These peppers can be marketed in various forms, including whole, in flakes or as a powder (Bosland & Votava, 2012). Historically, the dehydration of peppers was carried out through sun exposure. However, issues related to the presence of birds and rodents led to the adoption of controlled artificial drying methods, which are preferred by most processors in the United States. Preserving the red colour is essential and requires measures to prevent oxidative damage. Peppers undergo dehydration until they reach a moisture content of 4% to 6%, and they are later rehydrated to 8% to 11%, which are ideal levels for storage, typically done at a temperature of 3°C (Bosland & Votava, 2012). However, during the pepper dehydration process, the activation of lipoxygenases occurs, which alters the permeability of cell membranes. As a result of this membrane damage, a series of catabolic reactions is triggered, leading to tissue deterioration. The leakage of electrolytes is an indicator that membrane integrity has been compromised. This behaviour is more pronounced in pepper genotypes that are susceptible to high water loss compared to cultivars less sensitive to dehydration during storage (Kissinger et al., 2005; Lownds et al., 1993).

Maintaining the red colour is crucial for the quality of paprika and 'red pepper' powder. Several factors, such as moisture content, temperature and storage conditions, influence colour retention. Among these factors, pepper variety and storage temperature have the greatest impact, while the initial colour of the fruits is not a reliable indicator of colour loss during storage. Several studies have shown that traditional drying of red peppers at lower temperatures favours enzyme activity and drying, especially for fruits harvested when they are more mature and have a higher content of soluble pectin and calcium pectate. However, higher temperatures have a negative effect on carotenoid content (Gallardo-Guerrero et al., 2010; Zaki et al., 2013). Therefore, dehydration is an effective method for extending the shelf life of peppers by removing the water that promotes microbial growth and deterioration. The choice of dehydration method and proper control of temperature and humidity are crucial for preserving the quality of the final product. Dehydrated peppers are widely used in various products such as seasonings, snacks and culinary dishes, adding flavours and spiciness to many recipes.

7.4.2 Freezing

The freezing method is a widely adopted technique in food preservation, owing to two crucial factors. Firstly, low temperatures inhibit microbial growth, extending the shelf life of food because many microorganisms do not thrive in cold environments. Additionally, during freezing, a portion of the water in food transforms into ice, reducing water activity (Aw) and impacting microbial development, as many of them require higher Aw levels (Ibarz & Barbosa-Cánovas, 2014).

The quality of frozen foods varies with the speed of the process. Rapid freezing results in small ice crystals and less impact on food tissues. Conversely, slower freezing produces larger crystals, affecting quality due to water loss during thawing. In the case of peppers, post-harvest preservation involves specific procedures. New Mexican peppers have their skin removed, while jalapeños and bell peppers undergo blanching before freezing. The latter is verified with a peroxidase test, eliminating the need for preservatives (Wampler & Barringer, 2012).

The 'individual quick freezing' (IQF) method is also applied to diced peppers of various varieties. Furthermore, blanching proves to be an effective process in maintaining the colour of frozen green peppers and extending their shelf life. Unblanched peppers have a shelf life of only 3 to 4 months, whereas blanched ones can be stored for up to 12 months following industry standards. Surprisingly, blanched peppers often maintain their acceptable quality for over 24 months when stored between -32°C and -35°C, although they are labelled with a 12-month shelf life for commercial reasons. However, red peppers retain their colour after freezing, eliminating the need for blanching. Studies indicate that blanching jalapeños before freezing reduces the capsaicinoid content by half compared to fresh peppers (Bosland & Votava, 2012; Rybak et al., 2021).

Another critical factor affecting the quality of frozen peppers is aroma. Inhibiting enzymatic activity, specifically lipoxygenase, can enhance the aroma of processed peppers. Various studies, including Azcarate and Barringer (2010), have highlighted that blanching and frozen storage can significantly impact the volatile compound profile of peppers, with lipoxygenase inhibition contributing to increased concentrations of specific aromatic elements. It's important to note that frozen storage induces notable enzymatic and chemical changes in the volatile compound profile of unblanched peppers. Blanched peppers exhibit greater stability in their aromatic profile during freezing, albeit with a reduction in the overall concentration of volatile compounds (Wampler & Barringer, 2012). Food freezing is influenced by various factors, including process speed, temperature, the concentration of soluble solids and the choice of freezer type, all aimed at meeting specific needs and achieving the desired quality of the end product.

7.4.3 Modified Atmosphere Packaging

Packaging plays a crucial role in preserving the quality and extending the shelf life of fresh fruits and vegetables, which are highly perishable after harvesting. Traditional packaging techniques like controlled atmosphere (CA) and modified atmosphere packaging (MAP) combined with low temperatures have been valuable but face increasing challenges, such as the demand for more innovative and effective packaging solutions. Research in the field of active and intelligent packaging is emerging in response to these needs and challenges (Devgan et al., 2019).

Active packaging emerges as a response to these challenges, incorporating additional components into the packaging or packaging material to enhance its functionality. These systems interact with the internal atmosphere of the packaging to preserve the quality and extend the shelf life of foods. In many cases, the removal of oxygen from the packaging is crucial, as oxygen is one of the main causes of food deterioration (Ibarz & Barbosa-Cánovas, 2014).

Bell peppers (*C. annuum*), known for their low tolerance to cold injuries, face challenges in post-harvest quality preservation. Modified atmosphere packaging has been an approach used to protect these fruits from cold damage by adjusting the concentrations of CO_2 and O_2 in the packaging. However, there is controversy regarding the ideal levels of CO_2, with some studies indicating that higher concentrations are beneficial, while others establish a maximum tolerance limit of 5% CO_2 for bell peppers (Afolabi et al., 2023; Ranjitha et al., 2015). Furthermore, short-term treatment with high CO_2 levels has gained prominence as an effective strategy to reduce firmness loss and delay the senescence of bell peppers. This method inhibits the action of ethylene, slows down fruit ageing and increases resistance to bacterial and fungal attacks. Other crops, such as zucchini and persimmon, have also shown tolerance to high CO_2 levels under MAP conditions. Two recent studies explored methods to extend the shelf life of bell peppers during storage. Devgan et al. (2019) highlighted the success of active modified atmosphere packaging, which extended the shelf life of yellow bell peppers to 28 days compared to the 12 days of passive packaging. Quality, measured by firmness and antioxidant capacity, was also better preserved in active packaging. The study suggested that low-density polyethylene (LDPE) films with oxygen absorbers hold promise, but the number of absorbers should be optimized.

On the other hand, Afolabi et al. (2023) investigated pre-treatment of bell peppers with CO_2 to reduce cold injuries. While different pre-treatment durations did not show significant differences, a 24-hour treatment was commercially recommended. Both 24 and 48-hour treatments outperformed untreated samples in various parameters. MAP with an oxygen transmission rate of 50,000 cc·m−2·day−1·ATM−1 was considered effective. However, the complexity of CO_2 effects is not fully understood. These studies emphasize the importance of innovative preservation strategies for horticultural products like bell peppers, aiming to reduce food waste and meet the demands for high-quality fresh produce.

7.4.4 Application of Edible Coatings

Edible films can be described as a thin layer of material used to coat or separate foods, serving as a safe and consumable barrier. While both are forms of food packaging, they possess significant differences (Teixeira-Costa & Andrade, 2021). Primarily, films are created as solid laminates, undergoing separate preparation, drying, processing and eventual usage to cover food surfaces, separate food portions or act as edible sealed bags. In contrast, coatings are formulated as solutions, directly sprayed or dipped onto the food surface before it is dried (Otoni et al., 2017). As a result, coatings are considered an integral part of the food product, as they are not intended to be removed. Additionally, films can be fabricated as monolayers, bilayers or multilayers. The latter offers enhanced water resistance to food but requires multiple casting and drying steps, making them less commonly employed (Teixeira-Costa & Andrade, 2021).

Recently, researchers have studied the properties of using edible coatings of chitosan, gelatin and chitosan-gelatin applied layer by layer in maintaining the quality of red bell peppers (*C. annum* cv. Kannon (Zeraim Gedera—Syngenta, Revadim, Israel)) during storage (Kehila et al., 2021). These researchers found that a 2% edible chitosan coating was successful in reducing chilling injury and decay incidence and preserving the nutritional quality of the peppers during the storage period. In comparison to the other coatings, the chitosan coating demonstrated higher efficacy (Kehila et al., 2021). Interestingly, all three coating treatments led to a significant decrease in external CO_2

production compared to the uncoated control. Storage temperatures had no significant impact on CO_2 production despite a slightly higher production rate at 1.5°C. Therefore, the chitosan coating displayed favourable CO_2 gas permeability properties and resulted in lower respiration rates of the peppers compared to the other coatings and the uncoated control (Kehila et al., 2021).

In another study, Poverenov et al. (2014) studied the effectiveness of a composite chitosan-gelatin (CH-GL) coating on red bell peppers (*C. annum* cv. Vergasa), along with comparisons to pure chitosan (CH) and gelatin (GL) coatings. In this study, the CH coating demonstrated its ability to prevent microbial spoilage and extend the potential storage period of the peppers. The GL coating contributed to fruit firmness but did not offer prolonged storage capabilities. On the other hand, the composite CH-GL coating resulted in a twofold reduction in microbial decay, significantly improved fruit texture and extended the cold storage period up to 21 days and fruit shelf life up to 14 days. Notably, these positive effects were achieved without impacting the respiratory or nutritional characteristics of the fruit (Poverenov et al., 2014).

7.5 TRENDS AND PERSPECTIVES

The diversified cultivation of *Capsicum*, which includes bell peppers and hot peppers, is influenced by a range of factors such as local culture, cultivation goals, growth conditions, market demand and consumer preferences. In different parts of the world, preferences vary between spicy and sweet varieties, and the diseases affecting these crops are directly related to climate and geographical regions (Selvakumar et al., 2023; Hegena & Tigistu, 2022; Kang, 2018). As consumers become increasingly aware of the health benefits of food and its nutritional value in disease prevention and well-being promotion, the global market for functional foods and nutraceuticals has gained prominence (Daliri & Lee, 2015). *Capsicum* species emerge as a significant source of healthy foods, rich in proteins, trace elements, vitamins, minerals and other beneficial components. In recent years, in line with the growing demand for nutritious foods and increasing health awareness, peppers have stood out as a medicinal resource, a functional food and a cosmetic component, especially in the dietary context. Additionally, their active plant-based compounds play a crucial role in the prevention and treatment of pathological conditions, influencing various cellular pathways (Dhamodharan et al., 2023). Some perspectives on the use of peppers are summarized in Figure 7.4.

In this scenario, genetic improvement plays an essential role in developing *Capsicum* varieties that meet these growing demands. One of the primary goals of pepper genetic improvement is to

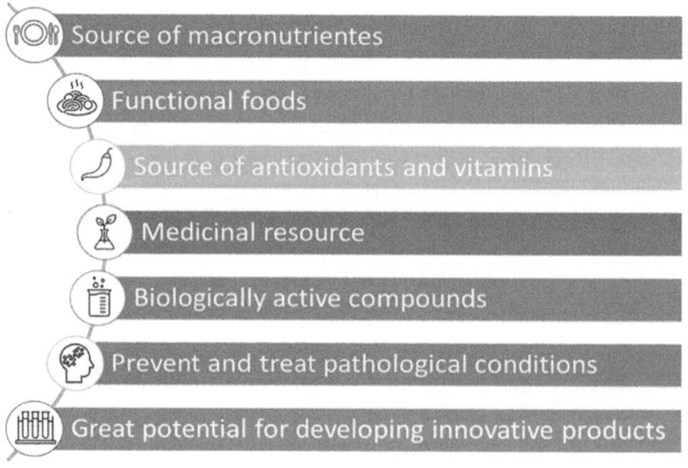

FIGURE 7.4 Some advantages and properties of using peppers.

develop varieties that offer a wide range of colours, from green in the early stage to red, yellow or orange in the mature stage. These varieties are not only desired for their appealing appearance but also for their high nutritional value. Researchers focus their efforts on finding and enhancing varieties rich in antioxidants and vitamins, including flavonoids and carotenoids, which contain vitamin A precursors such as alpha and beta-carotene, as well as β-cryptoxanthin (Dhamodharan et al., 2023). Genetic improvement in peppers aims to enhance productivity and resilience across various climates, addressing challenges like low temperatures, drought and disease resistance. Researchers also prioritize developing varieties suitable for long-term storage while maintaining carotenoid stability.

Efforts are made to combat common pepper diseases such as powdery mildew, anthracnose, Phytophthora fruit rot, bacterial wilt and viruses by creating more robust genotypes. Adapting peppers to tropical conditions expands their availability year-round and in new regions (Selvakumar et al., 2023). In protected environments like greenhouses, the focus is on breeding peppers with indeterminate growth, tolerance to training and pruning and resistance to biotic stressors like nematodes. These genetic advancements contribute to global food security and the provision of nutritious foods. Post-harvest technologies play a vital role in ensuring product quality and global distribution. In the context of developing new products, improved varieties of *Capsicum* offer exciting opportunities. Scientists and producers can harness these varieties to create a wide range of innovative food products, including sauces, seasonings, preserves, baked goods and healthy snacks. The potential is vast for developing tasty yet nutritious foods, capitalizing on their antioxidant benefits and nutritional value.

Furthermore, in the *Capsicum* production cycle, it is essential to emphasize the importance of post-harvest technologies. After harvest, the application of advanced handling, storage and processing techniques plays a crucial role in preserving the freshness, flavour and nutritional value of bell peppers. This includes practices such as proper refrigeration, storage under controlled conditions, the use of specific packaging for preservation and drying or dehydration methods. These technologies not only ensure that bell peppers reach global markets in optimal conditions but also contribute to the availability of these nutritious and flavourful foods worldwide, benefiting food security and public health. Moreover, they open doors for the development of a variety of pepper-derived products, further expanding their potential as functional and nutraceutical foods.

REFERENCES

Adewoyin, O., Famaye, A., Ipinmoroti, R., Ibidapo, A., & Fayose, F. (2023). *Postharvest Handling Methods, Processes and Practices for Pepper. Capsicum—Current Trends and Perspectives*. IntechOpen. https://doi.org/10.5772/intechopen.106592.

Afolabi, A. S., Choi, I. L., Lee, J. H., Kwon, Y. B., Yoon, H. S., & Kang, H. M. (2023). Effect of pre-storage CO_2 treatment and modified atmosphere packaging on sweet pepper chilling injury. *Plants*, 12(3). https://doi.org/10.3390/plants12030671.

Akhtar, A., Asghar, W., & Khalid, N. (2021). Phytochemical constituents and biological properties of domesticated capsicum species: A review. *Bioactive Compounds in Health and Disease*, 4(9), 201–225. https://www.doi.org/10.31989/bchd.v4i9.837.

Alarcón, M. T., Conde, C. G., & Mendez, G. L. (2019). Actividad antioxidante del extracto etanólico de Capsicum frutescens L. *BISTUA Revista de la Facultad de Ciencias Básicas*, 17(2), 102–111.

Ali, A., Bordoh, P. K., Singh, A., Siddiqui, Y., & Droby, S. (2016). Post-harvest development of anthracnose in pepper (Capsicum spp): Etiology and management strategies. *Crop Protection*, 90, 132–141. https://doi.org/10.1016/j.cropro.2016.07.026.

Alonso-Villegas, R., González-Amaro, R. M., Figueroa-Hernández, C. Y., & Rodríguez-Buenfil, I. M. (2023). The genus Capsicum: A review of bioactive properties of its polyphenolic and capsaicinoid composition. *Molecules*, 28(10), 4239. https://doi.org/10.3390/molecules28104239.

Antonio, A. S., Wiedemann, L. S. M., & Veiga-Junior, V. F. (2018). The genus Capsicum: A phytochemical review of bioactive secondary metabolites. *RSC Advances*, 8(45), 25767–25784. https://doi.org/10.1039/c8ra02067a.

Azcarate, C., & Barringer, S. A. (2010). Effect of enzyme activity and frozen storage on jalapeño pepper volatiles by selected ion flow tube-mass spectrometry. *Journal of Food Science*, 75(9), C710–C721. https://doi.org/10.1111/j.1750-3841.2010.01825.x.

Bosland, P. W., & Votava, E. J. (2012). Chapter—9 Postharvest handling. In *Peppers: Vegetable and Spice Capsicums*, 2nd ed. CAB International, 230p.

Buitimea-Cantúa, G. V., Velez-Haro, J. M., Buitimea-Cantúa, N. E., Molina-Torres, J., & Rosas-Burgos, E. C. (2020). GC-EIMS analysis, antifungal and anti-aflatoxigenic activity of Capsicum chinense and Piper nigrum fruits and their bioactive compounds capsaicin and piperine upon Aspergillus parasiticus. *Natural Product Research*, 34(10), 1452–1455.

Castillejo, N., Martínez-Zamora, L., & Artés-Hernández, F. (2022). Postharvest UV radiation enhanced biosynthesis of flavonoids and carotenes in bell peppers. *Postharvest Biology and Technology*, 184, 111774. https://doi.org/10.1016/j.postharvbio.2021.111774.

Chel-Guerrero, L. D., Scampicchio, M., Ferrentino, G., Rodríguez-Buenfil, I. M., & Fragoso-Serrano, M. (2022). In vitro assessment of antiproliferative activity and cytotoxicity modulation of Capsicum chinense by-product extracts. *Applied Sciences*, 12(12), 5818.

Cho, S.-Y., Kim, H.-W., Lee, M.-K., Kim, H.-J., Kim, J.-B., Choe, J. S., Lee, Y.-M., & Jang, H.-H. (2020). Antioxidant and anti-inflammatory activities in relation to the flavonoids composition of pepper (Capsicum annuum L.). *Antioxidants*, 9(10), 986. https://doi.org/10.3390/antiox9100986.

Daliri, E. B.-M., & Lee, B. H. (2015). Current trends and future perspectives on functional foods and nutraceuticals. *Microbiology Monographs*, 221–244. http://doi.org/10.1007/978-3-319-23177-8_10.

Della Valle, A., Dimmito, M. P., Zengin, G., Pieretti, S., Mollica, A., Locatelli, M., Cichelli, A., Novellino, E., Ak, G., Yerlikaya, S., Baloglu, M. C., Altunoglu, Y. C., & Stefanucci, A. (2020). Exploring the nutraceutical potential of dried pepper Capsicum annuum L. on market from Altino in Abruzzo region. *Antioxidants*, 9(5), 400. https://doi.org/10.3390/antiox9050400.

Devgan, K., Kaur, P., Kumar, N., & Kaur, A. (2019). Active modified atmosphere packaging of yellow bell pepper for retention of physico-chemical quality attributes. *Journal of Food Science and Technology*, 56(2), 878–888. https://doi.org/10.1007/s13197-018-3548-5.

Dhamodharan, K., Vengaimaran, M., & Sankaran, M. (2023). *Pharmacological Properties and Health Benefits of Capsicum Species: A Comprehensive Review*. IntechOpen. https://doi.org/10.5772/intechopen.104906.

Duelund, L., & Mouritsen, O. G. (2017). Contents of capsaicinoids in chillies grown in Denmark. *Food Chemistry*, 221, 913–918. https://doi.org/10.1016/j.foodchem.2016.11.074.

Gallardo-Guerrero, L., Pérez-Gálvez, A., Aranda, E., Mínguez-Mosquera, M. I., & Hornero-Méndez, D. (2010). Physicochemical and microbiological characterization of the dehydration processing of red pepper fruits for paprika production. *LWT*, 43(9), 1359–1367. https://doi.org/10.1016/j.lwt.2010.04.015.

González-Toxqui, C., González-Ángeles, Á., López-Avitia, R., & González-Balvaneda, D. (2020). Drying habanero pepper (Capsicum chinense) by modified freeze drying process. *Foods*, 9, 437. https://doi.org/10.3390/foods9040437.

Hamed, M., Kalita, D., Bartolo, M. E., & Jayanty, S. (2019). Capsaicinoids, polyphenols and antioxidant activities of Capsicum annuum: Comparative study of the effect of ripening stage and cooking methods. *Antioxidants*, 8(9), 364. https://doi.org/10.3390/antiox8090364.

Hegena, B., & Tigistu, S. (2022). Push and pull factors of red pepper (capsicum) market supply: Evidence from south western Ethiopia. *Journal of Agriculture and Food Research*, 10. https://doi.org/10.1016/j.jafr.2022.100417.

Howard, L. R., Talcott, S. T., Brenes, C. H., & Villalon, B. (2000). Changes in phytochemical and antioxidant activity of selected pepper cultivars (Capsicum species) as influenced by maturity. *Journal of Agricultural and Food Chemistry*, 48(5), 1713–1720. https://doi.org/10.1021/jf990916t.

Huang, J. J., Lin, S., Xu, W., & Cheung, P. C. K. (2017). Occurrence and biosynthesis of carotenoids in phytoplankton. *Biotechnology Advances*, 35(5), 597–618. https://doi.org/10.1016/j.biotechadv.2017.05.001.

Ibarz, A., & Barbosa-Cánovas, G. V. (2014). Chapter 23—Dehydration in introduction to food process engineering. In *Food Preservation Technology Series*, 1st ed. Taylor & Francis Group—CRC Press, 722p. https://doi.org/10.1201/b14969.

Kang, T. S. (2018). Identification of undeclared ingredients in red pepper products sold on the South Korea commercial market using real-time PCR methods. *Food Control*, 90, 73–80. https://doi.org/10.1016/j.foodcont.2018.02.033.

Karaman, K., Pinar, H., Ciftci, B., & Kaplan, M. (2023). Characterization of phenolics and tocopherol profile, capsaicinoid composition and bioactive properties of fruits in interspecies (Capsicum annuum X Capsicum frutescens) recombinant inbred pepper lines (RIL). *Food Chemistry*, 423, 136173. https://doi.org/10.1016/j.foodchem.2023.136173.

Kehila, S., Alkalai-Tuvia, S., Chalupovicz, D., Poverenov, E., & Fallik, E. (2021). Can edible coatings maintain sweet pepper quality after prolonged storage at sub-optimal temperatures? *Horticulturae, 7*(10). https://doi.org/10.3390/horticulturae7100387.

Khan, F. A., Mahmood, T., Ali, M., Saeed, A., & Maalik, A. (2014). Pharmacological importance of an ethnobotanical plant: Capsicum annuum L. *Natural Product Research, 28*(16), 1267–1274. https://doi.org/10.1080/14786419.2014.895723.

Kim, S., Ha, T. Y., & Hwang, I. K. (2009). Analysis, bioavailability, and potential healthy effects of capsanthin, natural red pigment from Capsicum spp. *Food Reviews International, 25*(3), 198–213. https://doi.org/10.1080/87559120902956141.

Kissinger, M., Tuvia-Alkalai, S., Shalom, Y., Fallik, E., Elkind, Y., Jenks, M. A., & Goodwin, M. S. (2005). Characterization of physiological and biochemical factors associated with postharvest water loss in ripe pepper fruit during storage. *Journal of the American Society for Horticultural Science, 130*(5).

Lownds, N.K., Banaras, M., & Bosland, P.W. (1993). Relationships between postharvest water loss and physical properties of pepper fruit (Capsicum annuum L.). *HortScience HortSci, 28*(12), 1182–1184. https://doi.org/10.21273/HORTSCI.28.12.1182.

Martínez-Arámburu, D., González-Quijano, G. K., Dorantes-Alvarez, L., Aparicio-Ozores, G., & López-Villegas, E. O. (2015). Changes in microstructure of salmonella typhimurium and Listeria monocytogenes exposed to hydroxycinnamic salts. *Revista Mexicana de Ingeniería Química, 14*(2), 347–354.

Melgar-Lalanne, G., Hernández-Álvarez, A. J., Jiménez-Fernández, M., & Azuara, E. (2017). Oleoresins from Capsicum spp.: Extraction methods and bioactivity. *Food Bioprocess Technology, 10*, 51–76. https://doi.org/10.1007/s11947-016-1793-z.

Mi, S., Zhang, X., Wang, Y., Zheng, M., Zhao, J., Gong, H., & Wang, X. (2022). Effect of different genotypes on the fruit volatile profiles, flavonoid composition and antioxidant activities of chilli peppers. *Food Chemistry, 374*, 131751. https://doi.org/10.1016/j.foodchem.2021.131751.

Narisawa, T., Fukaura, Y., Hasebe, M., Ito, M., Aizawa, R., Murakoshi, M., Uemura, S., Khachik, F., & Nishino, H. (1996). Inhibitory effects of natural carotenoids, alpha-carotene, beta-carotene, lycopene and lutein, on colonic aberrant crypt foci formation in rats. *Cancer Letters, 107*(1), 137–142. https://doi.org/10.1016/0304-3835(96)04354-6.

Otoni, C. G., Avena-Bustillos, R. J., Azeredo, H. M. C., Lorevice, M. V., Moura, M. R., Mattoso, L. H. C., & McHugh, T. H. (2017). Recent advances on edible films based on fruits and vegetables-a review. *Comprehensive Reviews in Food Science and Food Safety, 16*, 1151–1169.

Pinar, H., Kaplan, M., Karaman, K., & Ciftci, B. (2023). Assessment of interspecies (Capsicum annuum X Capsicum frutescens) recombinant inbreed lines (RIL) for fruit nutritional traits. *Journal of Food Composition and Analysis, 115*, 104848. https://doi.org/10.1016/j.jfca.2022.104848.

Poverenov, E., Zaitsev, Y., Arnon, H., Granit, R., Alkalai-Tuvia, S., Perzelan, Y., Weinberg, T., & Fallik, E. (2014). Effects of a composite chitosan-gelatin edible coating on postharvest quality and storability of red bell peppers. *Postharvest Biology and Technology, 96*, 106–109. https://doi.org/10.1016/j.postharvbio.2014.05.015.

Ranjitha, K., Sudhakar Rao, D. V., Shivashankara, K. S., & Roy, T. K. (2015). Effect of pretreatments and modified atmosphere packaging on the shelf life and quality of fresh- cut green bell pepper. *Journal of Food Science and Technology, 52*(12), 7872–7882. https://doi.org/10.1007/s13197-015-1928-7.

Rodriguez-Concepcion, M., & Daròs, J. A. (2022). Transient expression systems to rewire plant carotenoid metabolism. *Current Opinion in Plant Biology, 66*. https://doi.org/10.1016/j.pbi.2022.102190.

Rybak, K., Wiktor, A., Witrowa-Rajchert, D., Parniakov, O., & Nowacka, M. (2021). The quality of red bell pepper subjected to freeze-drying preceded by traditional and novel pretreatment. https://doi.org/10.3390/foods.

Santos, F. F. dos, Nascimento, L. D. S., Cavalcante, S. F. D. A., Rezende, C. M., & da Veiga-Junior, V. F. (2022a). TLC-ESI-MS as a QuEChERS approach to detection of capsaicin present in different matrices. *Natural Product Research, 36*(20), 5342–5346. https://doi.org/10.1080/14786419.2021.1922905.

Santos, L. S., Fernandes, C. C., Dias, A. L. B., Souchie, E. L., & Miranda, M. L. D. (2022b). Phenolic compounds and antifungal activity of ethyl acetate extract and methanolic extract from Capsicum chinense Jacq. ripe fruit. *Brazilian Journal of Biology, 84*.

Selvakumar, R., Chandregowda Manjunathagowda, D., & Kumar Singh, P. (2023). *Capsicum: Breeding Prospects and Perspectives for Higher Productivity*. IntechOpen. http://doi.org/10.5772/intechopen.104739.

Shanmugham, V., & Subban, R. (2022). Capsanthin from Capsicum annum fruits exerts anti-glaucoma, antioxidant, anti-inflammatory activity, and corneal pro-inflammatory cytokine gene expression in a benzalkonium chloride-induced rat dry eye model. *Journal of Food Biochemistry, 46*(10), e14352. https://doi.org/10.1111/jfbc.14352.

Teixeira-Costa, B. E., & Andrade, C. T. (2021). Natural polymers used in edible food packaging—history, function and application trends as a sustainable alternative to synthetic plastic. *Polysaccharides*, 3(1), 32–58. https://doi.org/10.3390/polysaccharides3010002.

Toigo, S. E. M., Fernandes, C. C., & Miranda, M. L. D. (2022). Promising antifungal activity of two varieties of Capsicum chinense against Sclerotinia sclerotiorum, Rhizopus stolonifer and Colletotrichum goleosporoides. *Food Science and Technology*, 42.

Usmani, Z., Dar, A. H., Altaf, A., & Jagdale, Y. D. (2022). Pepper oleoresin: Properties and economic importance. In *Handbook of Oleoresins*. CRC Press, pp. 1–9.

Wampler, B., & Barringer, S. A. (2012). Geração de voláteis em pimentão durante armazenamento congelado e descongelamento usando espectrometria de massa com tubo de fluxo de íons selecionado (SIFT-MS). *Journal of Food Science*, 77(6), C677–C683. http://doi.org/10.1111/j.1750-3841.2012.02727.x.

Wang, L., Zhong, Y., Liu, J., Ma, R., Miao, Y., Chen, W., Zheng, J., Pang, X., & Wan, H. (2023). Pigment biosynthesis and molecular genetics of fruit colour in pepper. *Plants*, 12(11), 2156. https://doi.org/10.3390/plants12112156.

Wray, D., & Ramaswamy, H. S. (2015). Development of a microwave–vacuum-based dehydration technique for fresh and microwave–osmotic (MWODS) pretreated whole cranberries (Vaccinium macrocarpon). *Drying Technology*, 33(7), 796–807. https://doi.org/10.1080/07373937.2014.982758.

Zaki, N., Hakmaoui, A., Ouatmane, A., & Fernandez-Trujillo, J. P. (2013). Quality characteristics of Moroccan sweet paprika (Capsicum annuum L.) at different sampling times. *Food Science and Technology*, 33(3), 577–585. https://doi.org/10.1590/S0101-20612013005000072.

Zellama, M. S., Chahdoura, H., Zairi, A., Ziani, B. E. C., Boujbiha, M. A., Snoussi, M., & Chaouachi, M. (2022). Chemical characterization and nutritional quality investigations of healthy extra virgin olive oil flavored with chili pepper. *Environmental Science and Pollution Research*, 1–12.

8 Capsaicin and Capsaicinoids—Sources, Characteristics, Isolation, Food and Medical Uses

Talía Hernández-Pérez and Octavio Paredes-López

CONTENTS

8.1 INTRODUCTION

Chili contains high amounts of outstanding bioactive compounds with broad benefits on health, including mainly capsaicinoids, pigments (*i.e.*, chlorophyll, carotenoids, anthocyanin, and lutein), vitamins, minerals and phenolic compounds. Capsaicin (*trans*-8-methyl-*N*-vanillyl-6-nonenamide) produces the pungency in *Capsicum* spp. fruits and has been related to several positive effects on health (Hernández-Pérez et al. 2020). The capsaicin concentration in peppers depends on their color; green and red samples change from 0.1 to 1%. Chili peppers are used around the world as ingredient of different dishes and beverages, mainly in Latin America, the Caribbean region and South Asia (Chapa-Oliver and Mejia-Teniente 2016; Morales-Soriano and Ugás 2022; Srinivasan 2016). Capsaicin is responsible for the burning and irritant effect of chili peppers. This compound has been used in traditional medicine for centuries because it has the ability to reduce blood lipid, cholesterol and sugar levels, and it also has antioxidant, anti-inflammatory, antiobesity and analgesic properties. Besides, its derivatives have been investigated for the treatment of pruritus, atopic dermatitis, rosacea, non-allergic and allergic rhinitis, chronic idiopathic cough, chronic obstructive pulmonary disease, dry eye syndrome and tumors and cardiovascular and oncological disorders. But its role in nociception has not been studied for long (Basith et al. 2016; Brederson et al. 2013; Kim et al. 2003).

Richards et al. (2012) determined the effect of capsaicin in rheumatoid arthritis; other studies evaluated its benefits on cluster headaches (Matharu 2010) as well as in herpes zoster (Jeon 2015). Besides, it also has been used to prevent and treat neurodegenerative disorders like Alzheimer's disease (Dairam et al. 2008), reducing neurodegeneration and memory impairment. Also, it has been suggested that this compound provides benefits for Parkinson's disease and depression and promotes better neurological outcome in infarction and stroke (Pasierski and Szulczyk 2022). Additionally, capsaicin displays modulatory effects on gut microbiota and may provide positive effects to treat metabolic syndrome and some inflammatory bowel diseases. A possible antimicrobial property of capsaicin, mediated by the beneficial alteration of microbiota, has been elucidated (Rosca et al. 2020). In brief, the aim of this chapter is to provide an overview of the main characteristics and methods of extraction of capsaicin and capsaicinoids as well as their food and medical properties, including antiobesity, analgesic and anticancer potential and help for urinary system disorders.

8.2 MAIN CHARACTERISTICS OF CAPSAICINOIDS

Capsaicinoids are secondary metabolites that produce a strong and hot taste in chili pepper; they are recognized by their pungency. They are alkaloids probably released as deterrents against certain herbivores and fungi. Capsaicin is synthesized in the interlocular septum and placental tissue of chili peppers by addition of a branched-chain fatty acid to vanillylamine; specifically, capsaicin is made from vanillylamine and 8-methyl-6-nonenoyl Coenzyme-A. Their biosynthesis is correlated to the environment conditions as well as to genotype. Capsaicin is the main responsible of the oral sensation of pungency in *Capsicum* fruits, but the other capsaicinoids might produce slight and mild pungency (Uarrota et al. 2021).

Capsaicinoids are synthesized in the fruit placenta through the integration of two different biochemical pathways that involve multiple intracellular compartments. Capsaicinoids consist of over 23 different chemical species (Stellari et al. 2010); both capsaicin (*trans*-8- methyl-*N*-vanillyl-6-nonenamide) and dihydrocapsaicin (8-methyl-*N*-vanillylnonanamide) together account for ~90% of the total fruit capsaicinoid content (Barbero et al. 2014). Additionally, nordihydrocapsaicin, homodihydrocapsaicin, and homocapsaicin are also important natural occurring capsaicinoids, however, they only reach 20% of the total concentration (Table 8.1) (Delgado-Vargas et al. 2000; Giuffrida et al. 2013; Perucka and Oleszek 2000).

As indicated, capsaicinoids consist of mainly two isomers, capsaicin and dihydrocapsaicin, whereas capsinoids also have two major isomers, capsiate and dihydrocapsiate. Capsaicinoids are fatty acid amides linked with vanillylamine, while capsinoids are fatty acid esters linked with vanillyl alcohol (Huang et al. 2014). Several techniques to extract capsaicin have been tested, such

---

TABLE 8.1

Capsaicinoids Concentration in Most *Capsicum* Species (source: Delgado-Vargas et al. 2000; Giuffrida et al. 2013; Perucka and Oleszek 2000)

Capsaicinoid	Molecular formula	Relative abundance (%)
Capsaicin (*trans*-8-methyl-*N*-vanillyl-6-nonenamide)	$C_{18}H_{27}NO_3$	69
Dihydrocapsaicin (8-methyl-*N*-vanillyl-nonanamide)	$Cl_8H_{29}NO_3$	22
Nordihydrocapsaicin (7-methyl-*N*-vanillyl-octamide)	$C_{17}H_{27}NO_3$	7
Homodihydrocapsaicin (9-methyl-*N*-vanillyl-decamide)	$C_{19}H_{31}NO_3$	1
Homocapsaicin (*trans* 9-methyl-*N*-vanillyl-7-decenamide)	$C_{19}H_{29}NO_3$	1

SHU	Capsicum variety
15,000,000-16,000,000	Pure capsaicin
9,100,000	
2,000,000-5,300,000	
855,000-1,041,427	
876,000-970,000	
350,000-577,000	
100,000-350,000	Habanero pepper
100,000-200,000	
50,000-100,000	
30,000-50,000	
10,000-23,000	Serrano pepper
7,000-8,000	
5,000-10,000	
2,500-5,000	Jalapeño pepper
1,500-2,500	
1,000-1,500	
600-800	
500-1,000	
100-500	
0	Pimiento pepper

FIGURE 8.1 Scoville scale units (SHU) of some *Capsicum* varieties (source: Hernández-Pérez et al. 2020).

as organic solvent (Chinn et al. 2011), ultrasound-assisted (Yue et al. 2012), microwave-assisted (Du et al. 2013), supercritical fluid (Kwon et al. 2011), acid and alkali and enzymatic extractions (Salgado-Roman et al. 2008). Capsaicin concentration between chili peppers depends on variety and is quantified as Scoville heat units (SHU) (Scoville 1912) (Figure 8.1).

Hotness is measured in SHU, which indicate the number of times the substance has to be diluted in order that the pungency is not perceived. Capsaicin is present in large quantities in the placental tissue (which holds the seeds), the internal membranes and, to a lesser extent, the other fleshy parts of the fruits of *Capsicum* plants. The seeds themselves do not contain capsaicin, although its highest concentration of capsaicin can be found in the white pith of the inner wall, where the seeds are attached (Govindarajan 1985).

8.3 PUNGENCY

In the 19th century, capsaicinoids were isolated from chili peppers, and the pungent taste or smell was assessed. The most abundant capsaicinoids in *Capsicum* species are capsaicin (69%), followed by dihydrocapsaicin (22%), nordihydrocapsaicin (7%), homocapsaicin (1%), and homodihydrocapsaicin (1%) (relative abundance) (Perry et al. 2007). As indicated before, capsaicin is usually the main capsaicinoid in almost all peppers; however, *C. baccatum*, a scarcely cultivated crop, shows a 51% dihydrocapsaicin concentration (Uarrota et al. 2021).

Since the 19th century, a good number of studies have been conducted on the isolation, purification and uses of capsaicin in order to establish its pharmacological benefits. Therefore, capsaicin is now recognized as a bioactive compound with a great diversity of applications and nutraceutical potential (Papoiu and Yosipovitch 2010).

Capsaicinoid concentration is an important parameter to indicate the hotness of this fruit, which is a result of the capsaicinoid concentration; and pungency level is measured in SHU, whose definition

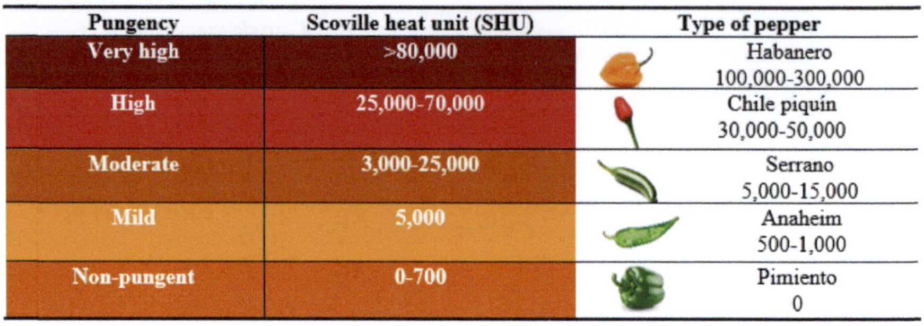

Pungency	Scoville heat unit (SHU)	Type of pepper	
Very high	>80,000		Habanero 100,000-300,000
High	25,000-70,000		Chile piquín 30,000-50,000
Moderate	3,000-25,000		Serrano 5,000-15,000
Mild	5,000		Anaheim 500-1,000
Non-pungent	0-700		Pimiento 0

FIGURE 8.2 Category of pungency of different peppers in Scoville heat units (SHU) (source: Hernández-Pérez et al. 2020).

was described before. There are five levels of pungency in the Scoville scale: non-pungent (0 to 700 SHU), mildly pungent (700 to 3,000 SHU), moderately pungent (3,000 to 25,000 SHU), highly pungent (25,000 to 70,000 SHU) and very highly pungent (> 80,000 SHU) (Baenas et al. 2019; Patowary et al. 2017; Scoville 1912; Sharma et al. 2013) (Figure 8.2). The daily intake of chili pepper in countries where consumption is high is 2.5, 5.0, 15.5 and 20.0 g/person in India, Thailand, Saudi Arabia and Mexico, respectively. The concentration of capsaicin in the species *C. annuum, C. frutescens*, and *C. chinense* ranges between 0.22 and 20 mg total capsaicinoids/g sample (dry weight).

8.4 NON-PUNGENT COMPOUNDS

Capsinoids, the non-pungent capsaicinoid analogues, are more palatable and more preferred than the pungent form, which are obtained from CH-19 sweet (*Capsicum annuum* L.) (Yoneshiro et al. 2012). Capsinoids are produced, in trace amounts, by highly pungent pepper varieties (Uarrota et al. 2021). Some years ago, capsaicin analogues [N-acyl-vanillamide (NAVAM)] were synthesized to obtain non-pungent compounds with analgesic effect. Other NAVAM analogues [N-oleoyl-vanillamide (olvanil) and N-arachidonoy-vanillamide (arvanil)] show antihyperalgesic capacity in chronic and inflammatory pain; they also activate receptors of capsaicin (transient receptor potential of vanilloid type-1 (TRPV1) channel and type-1 cannabinoid receptors). TRPV1 channels are non-selective cation receptors that are gated by a broad array of noxious ligands and are implicated in inflammation, painful stimuli sensation, and mechanotransduction. N-palmitoyl-vanillamide (palvanil), another NAVAM, has high TRPV1 desensitizing effect and thus strong antihyperalgesic capacity (Cui et al. 2016; De Petrocellis et al. 2011).

Several lines of evidence show that NAVAM capsaicin analogues display improved growth-suppressive activity in human cancers relative to capsaicin. An advantage of NAVAM capsaicin analogues is that they suppress the growth of human cancer cells and do not harm normal cells (Han et al. 2020). An exciting finding is that the NAVAM capsaicin analogues enhance the growth-suppressive activity of conventional chemotherapy in both classical and drug-resistant cancers. Certain cancers like small cell lung cancer are known to relapse within a few months, and these relapsed tumors are usually resistant to chemotherapy and radiation (Qin and Kalemkerian 2018).

Capsiate, another non-pungent capsaicin analogue, when administered to mice for 2 weeks, plus exercise, increased energy metabolism and suppressed body fat accumulation during four more weeks of *ad libitum* feeding. It was observed that the body weight as well as abdominal tissue weight in capsiate and exercise groups were significantly lower than in the controls; besides, oxygen consumption increased. Thus, body fat accumulation by capsiate intake is beneficial for maintaining an ideal body weight by exercise (Haramizu et al. 2011).

Capsaicin has shown anticancer capacity, but its undesired secondary effects, *i.e.*, burning sensation, stomach cramps, gut pain and nausea, limit its use in cancer therapies. Hurley et al. (2017) revealed that arvanil and olvanil (non-pungent long chain capsaicin analogues) provide anti-invasive effects associated to capsaicin in small lung cancer cells in humans. These agents may represent the second generation of capsaicin-like compounds, which are more potent than the parent molecule and have a better side effect profile.

Capsaicin and its non-pungent analogues enhance metabolism and energy expenditure and stimulate the browning of white adipose tissue and brown fat activation to counter diet-induced obesity. Capsaicin shows high growth-inhibitory capacity in different types of cancers; however, its viability is hampered by its adverse side effects. Non-pungent capsaicin-like compounds from the NAVAM derivatives of capsaicin are obtained from natural sources or by chemical and enzymatic procedures. Hence, the production of novelty analogues of capsaicin with better stability and growth-suppressive cancer capacity is a great challenge for the medical industry (Richbart et al. 2021). Merritt et al. (2022) found that capsaicin encapsulated in sustained release carriers decreased its undesirable side effects. Hence, it can be used as novel chemotherapeutic and chemosensitization agents.

8.5 NUTRACEUTICAL AND MEDICAL POTENTIAL OF CAPSAICIN

8.5.1 ANTIOBESITY EFFECT

Nowadays, obesity constitutes a real worldwide health pandemic; its prevention and treatment are important public health challenges. Research studies have shown that the onset of obesity has been related to the gut microbiome. Moreover, increased body fat correlates with declining diversity of the gut microbiota. Interestingly, Wang et al. (2020) found that capsaicin improves the diversity of gut microbiota in mice fed capsaicin. It reduced food intake and showed anti-obesity effect regardless of TRPV1 channel activation, which is mediated by changes in the gut microbiota populations and short-chain fatty acids (SCFAs) concentrations. Diversity of gut microbiota in obese people decreases, and many pathogenic bacteria or opportunistic pathogens multiply in large numbers, resulting in dysbacteriosis. Capsaicin has strong antibacterial activity and may reverse gut microbiota dysbiosis by killing many harmful bacteria, resulting in lower lipid levels.

Capsaicin increases energy expenditure as well as thermogenesis induced by diet, which is promoted by the β-adrenergic stimulation and a reduction in the respiratory ratio; this alters substrate oxidation, from carbohydrates to lipids (Ludy and Mattes 2011; Shook et al. 2015; Smeets et al. 2013) (Table 8.2). In animal models, capsaicinoids significantly decreased total serum and low-density lipoprotein cholesterol, as well as triacylglycerols, but high-density lipoprotein cholesterol was not affected. This behavior resulted from the stimulating conversion of cholesterol to bile acids (Zhang et al. 2020).

Saito (2015) found that capsaicin and capsinoids positively affect basal energy expenditure and thus promote a reduction of fat deposits. Patients without active brown adipose tissue (BAT) exposed daily to cold temperatures during 6 weeks showed activation of BAT, and it was associated with a high energy expenditure and reduction of body fat. The daily 6-week intake of capsinoids has a similar effect as the recurrent exposure to low temperatures of BAT.

Regarding carbohydrate metabolism, the activation of TRPV1 receptors by capsaicin promotes an increase of glucagon-like peptide 1 (GLP-1), which improves glucose homeostasis and promotes healthier postprandial insulin concentration. Insulin resistance was increased in TRPV1 knockout mice (Lee et al. 2015). In previous experiments, Diaz-Garcia et al. (2014) determined that insulin sensitivity might be regulated by TRPV1. Besides, capsaicin improves glucose uptake by the

TABLE 8.2

Health Benefits of the Intake of Capsaicin

Potential	Effect	Reference
Antiobesity	Increase energy expenditure, diet-induced thermogenesis	Ludy and Mattes (2011), Shook et al. (2015) and Smeets et al. (2013)
	Boost energy expenditure, decrease fat tissue	Saito (2015)
	Decrease serum total cholesterol, low-density lipoprotein and triacylglycerols	Lee et al. (2015)
	Improve postprandial hyperglycemia, hyperinsulinemia and fasting lipid metabolic disorders	Yuan et al. (2015)
	Improve glucose homeostasis, lower postprandial insulin levels	Zhang et al. (2020)
	Attenuate hunger, decrease fullness, energy expenditure and fat	Tremblay et al. (2016)
	Increase energy expenditure by activating brown adipose tissue	Georgescu et al. (2017) and Narang et al. (2017)
	Browning of white adipose tissue, control obesity in fat-rich diets	Kang et al. (2007)
	Improve diversity of gut microbiota, decrease blood lipids	Wang et al. (2020)
Analgesic	Decrease postsurgical neuropathic pain and itch	Patowary et al. (2017)
	Reduce central sensitization, promote durable pain relief	Gašparini et al. (2020)
	Regeneration and restoration of skin nerve fibers in chemotherapy-induced peripheral neuropathy	Privitera and Anand (2021)
	Lower pain and enhance locomotor function	Issa et al. (2021)
Anticancer	Chemopreventive and antitumor activity in many types of human cancers	Hwang et al. (2011), Lee et al. (2014) and Shin et al. (2008)
	Growth-inhibitory activity, suppress prostate cancer, melanoma, cholangiocarcinoma and fibrosarcoma invasion and migration	Venier et al. (2015)
	Antitumorigenic, cancer-preventive agent	Srinivasan (2016)
	Anticolonic cancer	Bhagwat et al. (2021) and Zhang et al. (2020)
Urinary disorders	Decrease hyperexcitability and neurogenic lower urinary tract dysfunction	Shimizu et al. (2018)
	Larger bladder capacity and lower voiding frequency, greater intravesical pressure threshold for micturition	Ríos-Silva et al. (2019)
	Reduction in bladder capacity, urethral perfusion pressure and bladder compliance	Hokanson et al. (2021)

generation of reactive oxygen species (ROS) and the consequent activation of AMPK and p38 MAPK (Kim et al. 2013).

In women with gestational diabetes, a regular diet supplemented with capsaicin reduced postprandial hyperglycemia and hyperinsulinemia and also fasting lipid metabolic conditions; the occurrence of large-for-gestational-age newborns was also reduced (Yuan et al. 2015). In this context, Tremblay et al. (2016) described that a diet supplemented with capsaicin attenuated hunger and decreased fullness, energy expenditure and fat. Interestingly, the intake of capsaicin 1 hour before low-intensity exercise improved lipolysis in individuals with hyperlipidemia and/or obesity. Otherwise, the intake of capsaicinoids positively affects energy expenditure through brown adipose tissue activation (Georgescu et al. 2017; Narang et al. 2017). In extracts of hot pepper flower, Marrelli et al. (2016) observed the inhibition of pancreatic lipase activity, resulting in a lower absorption of triglycerides in the small intestine.

C. annuum has shown hypoglycemic properties by inhibiting α-amylase activity, which delays carbohydrate uptake and increases its digestion time, promoting lower rates of glucose absorption and better blood glucose postprandial levels (Lee et al. 2013). A healthy carbohydrate metabolism is a key factor to have a healthy weight or even to lose weight. In this context, the intake of chili pepper at different doses, i.e. from one meal to daily consumption for 12 weeks, results in the loss of body fat. The metabolism of lipids, energy expenditure as well as thermogenesis are mainly affected by capsaicinoid ingestion. Capsaicin induces the loss of body fat by activating TRPV1, which augments intracellular calcium concentration, thus activating the sympathetic nervous system. It improves insulin sensitivity, resulting in weight control and aiding in the treatment of obesity, type 2 diabetes and cardiovascular diseases (Janssens et al. 2013; Varghese et al. 2017).

The intake of capsaicin in the regular diet prevents weight gain and promotes the expression of brown fat thermogenes (Baskaran et al. 2018). Saito (2015) also suggested that capsaicin enhances metabolic activity and thermogenesis. Capsaicin stimulates the browning of white fat and controls obesity in fat-rich diets. Mice fed a high-fat diet (HFD) supplemented with 0.01% capsaicin for 12 weeks showed a strengthened microbiome—an increase of short-chain fatty acids (butyrate, propionate, and acetate), produced by *Ruminococcaceae* and *Lachnospiraceae* bacteria, was observed. Capsaicin promoted a "healthy" gut microbiome providing butyrate, which serves as a dominant energy source to colonocytes; hence, capsaicin may act as a therapeutic agent for obesity (Kang et al. 2007).

It has been described that capsaicin activates TRPV1 in high-fat diet (HFD)-induced obesity and promotes the *in vitro* browning of white adipose tissue or *in vivo* inguinal white adipose tissue (iWAT). Baskaran et al. (2018) fed mice with a normal chow diet (NCD) ± capsaicin (0.01% in NCD) or HFD ± capsaicin (0.01% in HFD) for 8 months. Capsaicin over 0.001% reduced obesity in HFD mice. Capsaicin enhanced the expression of Sirtuin-1 and thermogenic uncoupling protein 1 in iWAT; capsaicin did not cause inflammation. Additionally, iWAT hypertrophy and hepatic steatosis related to the high HFD plasma alanine aminotransferase and creatinine was reduced by capsaicin. Thus, capsaicin antagonizes HFD-induced metabolic stress and inflammation, although it does not cause any systemic toxicities and is well tolerated by mice.

Dietary capsaicin is involved in different mechanisms in gastrointestinal (GI) physiology and pathology. It acts on the GI tract via TRPV1-dependent and -independent manners as well as through acute and chronic effects, mostly depending on its consumption concentrations. At high doses, capsaicin is harmful; Xiang et al. (2022) suggested that its proper consumption is preventive or therapeutic to maintain a healthy gut. The daily intake of capsaicin can be considered a safe treatment to manage different GI disorders.

Shanmugham and Subban (2022) observed that mice fed HFD gained body mass and white adipose tissue mass compared to those with normal diet. Oral administration of capsanthin-enriched pellets and capsaicin pellets significantly reduced body mass gain. These pellets have significant impact on obesity biomarkers by increasing adiponectin and decreasing leptin, free fatty acid and insulin concentrations relative to HFD control.

8.5.2 Analgesic Potential

Capsaicin is a recognized drug to alleviate pain and also itch. Capsaicinoids are important agonists of the capsaicin receptor (TRPV1). Vanilloids have several benefits via the receptor-dependent and -independent pathway. Serotonin, neuropeptide substance P and somatostatin are involved in the pharmacological effects of capsaicin (Patowary et al. 2017).

Neuropathic pain associated with chemotherapy-induced peripheral neuropathy (CIPN) was relief for up to three months or longer after a single 30- to 60-min application of a CAP 8% patch. Capsaicin 8% patch is a licensed treatment in the EU/UK for neuropathic pain. This patch also promotes the regeneration and restoration of skin nerve fibers in CIPN and now is often a preferred treatment option for localized neuropathic pain conditions, including feet and hands. Capsaicin 8% patch can be repeated three times monthly, if needed, for a year (Privitera and Anand 2021). Patowary et al. (2017) found that a capsaicin cream reduced postsurgical neuropathic pain; patients preferred this cream to other therapies.

Arthritis is a common chronic joint disease that causes pain, stiffness and disability. Patients diagnosed with either osteoarthritis, rheumatoid arthritis or chronic joint pain started transdermal glucosamine sulfate and capsaicin (TGC-Plus cream) applied twice daily for 12 weeks. It promoted pain relief and joint function improvement (alleviating joint stiffness), and the need for analgesics and the number of doctor's visits decreased. The treatment with the TGC-Plus cream significantly reduces pain and enhances locomotor function in patients with chronic pain who failed to achieve adequate prior pain relief (Issa et al. 2021).

Chronic pain is a painful condition that persists 3 months or more; its prevalence ranges between 11% and 51.3% in the general population. Topical capsaicin acts as a highly selective agonist of TRPV1 of C and Aδ nociceptors, reducing central sensitization. Capsaicin provides effective durable pain relief and reduction of intensity and area of pain in adult patients with chronic pain with a faster onset of analgesia and considerably fewer systemic adverse effects than the conventional treatment. Additionally, it improves sleep, fatigue, depression and quality of life. Topical administration avoids dangerous systemic adverse effects and enables the combination with other drugs and analgesics with limited drug–drug interactions (Gašparini et al. 2020). Capsaicin has a wide application in topical gels and patches to relieve pain due to the anti-inflammatory characteristics (Lu et al. 2020).

8.5.3 Anticancer Activity

Capsaicin plays a "dual role" in tumorigenesis, acting as a carcinogen or as a cancer-preventive agent (Zhang et al. 2020). It displays potent chemopreventive and antitumor activity in many types of human cancers (Srinivasan 2016). Apart from its growth-inhibitory activity, capsaicin mitigates prostate cancer invasion and migration, melanoma and cholangiocarcinoma as well as fibrosarcoma (Hwang et al. 2011; Lee et al. 2014; Venier et al. 2015). Shin et al. (2008) found that capsaicin has a positive effect on B16-F10 mouse melanoma cells, which are highly metastasic, as well as inhibits the migration of melanoma cells.

The combination of capsaicin with conventional chemotherapy drugs or radiotherapy can improve the sensitivity, reduce the side effects and enhance the tolerance of patients to cancer treatment. The development of capsaicin-loaded nanoparticles may provide a very promising approach to chemotherapy for malignant tumors (Zhang et al. 2020).

Al-Samydai et al. (2021) improved the pharmacokinetic effects of capsaicin by its loading via nanoliposomes. Capsaicin-loaded nanoliposomes demonstrated important *in vitro* anticancer capacity against different cancers cell lines, with higher affinity against cancer cells than capsaicin itself. At the same time, Bhagwat et al. (2021) observed that capsaicin loaded self-nano emulsifying drug delivery system (SNEDDS) improved the wettability, solubility and bioavailability of capsaicin but also reduced gastric irritation to a larger extent. Further, the potential of capsaicin against colorectal cancer cells was found to be improved. Conclusively, SNEDDS remarkably improved both *in vitro* and *in vivo* effectiveness of capsaicin.

Capsaicin triggers apoptosis in small cell lung cancer by increasing the intracellular calcium level and elevating calcium-activated protease (calpain) activity (Friedman et al. 2017). It is also claimed that capsaicin has an antiproliferative effect on prostate and colorectal cancer cells by suppressing the cell growth and inhibiting Wnt/β-catenin signaling pathways in prostate cancer stem cells and down-regulating the expression levels of cyclin D1 and the caspase-like activity of 20S proteasomes in colorectal cancer cells (Zhu et al. 2020).

8.5.4 Urinary System Disorders

Topical application of capsaicin induces early neurogenically mediated cellular microcirculatory inflammatory reactions via the activation of the TRPV1 receptor and the release of CGRP and SP from sensory nerves in the bladder. Co-administration of SP and CGRP receptor antagonists may ameliorate microcirculatory inflammatory changes elicited by capsaicin in the urinary bladder (Járomi et al. 2018). The vesical instillation of 1 mM capsaicin in people with hypertensive bladder results in healthier urinary frequency and reduced incontinence (Patowary et al. 2017).

Prostaglandin E_2 (PGE_2) instilled into the bladder generates symptoms of urinary urgency in healthy women and reduces bladder capacity and urethral pressure in humans and female rats. Systemic capsaicin desensitization, which causes degeneration of C-fibers, prevented PGE_2-mediated reductions in bladder capacity (Maggi et al. 1988). Hokanson et al. (2021) also instilled PGE_2 in female rats after capsaicin desensitization. After capsaicin injection (125 mg/kg sc), rats underwent cystometric and urethral perfusion testing. Capsaicin-desensitized rats exhibited a reduction in bladder capacity, urethral perfusion pressure and bladder compliance. Urethral relax-ation/weakness and/or increased detrusor pressure as a result of decreased compliance may con-tribute to urinary urgency and highlight potential targets for new therapies for overactive bladder. Additionally, the diuretic effect of capsaicin in healthy and diabetic rats was evaluated by Ríos-Silva et al. (2019). They observed an increase in the urinary epidermal growth factor levels and a decrease of the urinary neutrophil gelatinase-associated lipocalin concentration. Thus, capsaicin my provide a larger capacity of the bladder, lower voiding frequency and greater intravesical pressure threshold for micturition.

Nerve growth factor (NGF) has been implicated as an important mediator to induce C-fiber bladder afferent hyperexcitability. NGF neutralization by anti-NGF antibody treatment reversed the spinal cord injury-induced increase in the number of action potentials and the reduction in spike thresholds and A-type K+ current density in mouse capsaicin-sensitive bladder afferent neurons. Thus, NGF plays an important and direct role in hyperexcitability of capsaicin-sensitive C-fiber bladder afferent neurons due to the reduction in A-type K+ channel activity in spinal cord injury (Shimizu et al. 2018).

Zheng et al. (2016) determined that TRPV1 behaves as a tumor suppressor by inducing apoptosis in bladder cancer cells. When TRPV1 was activated by capsaicin, it induced growth inhibition of 5637 cells. Thus, activation of TRPV1 may be applied as a novel strategy to treat bladder cancer or enhance the therapeutic efficacy of traditional chemotherapeutic drugs.

8.6 CONCLUSIONS AND FUTURE PERSPECTIVES

Most genetic materials of *C. annuum*, widely known as hot pepper and chili, originated in Latin America, from where this crop was distributed to the world after the arrival of the Spanish con-querors at the beginning of the 16th century. History shows that India, China and in general South East Asia also played an important role in its domestication. In the initial days, its fruits were used as vegetables and for condiments; later on, it was also utilized as a colorant and food ingredient and for medical uses.

Capsaicin is the most important active ingredient; and capsaicinoids in general, with their spicy flavor, play important roles in nutritional, nutraceutical, medical and industrial uses. Therapeutical applications and medical properties of capsaicin and capsaicinoid metabolites may be mainly con-centrated in various fields such as antiobesity effects; antioxidant activity and cardiovascular pro-tection; antimicrobial, antifungal and antiviral activities; urinary system disorders and renal failure; and anticancer and analgesic activities. In addition to scientific data demonstrating the outstanding healthy effects of capsaicin and capsaicinoids on different maladies, future and deeper clinical investigations, *in vitro* and *in vivo* performance, are needed to determine with doubt-free assess-ments the safety and efficacy of these compounds as therapeutic agents beneficial to human health worldwide.

In the last two decades, molecules without the pungent effect have been produced and tested. They may have a future for societies in which the hot flavor is not acceptable, and sensory and other chemical and physicochemical properties may be provided. At the same time, genetic and molecular agriculture is receiving strong attention in several countries, and different types of metabolites with various levels of pungency and with nutraceutical, pharma and industrial uses acceptable to the consumer and to society in general are expected to be produced. This strategy will increase all nutraceutical, medical and industrial potentialities of this crop useful to the societies of the 21st century.

8.7 ACKNOWLEDGEMENTS

The authors would like to acknowledge financial support from Consejo Nacional de Humanidades, Ciencia y Tecnología (CONAHCyT-México).

REFERENCES

Al-Samydai A, Alshaer W, Al-Dujaili EAS, Azzam H, Aburjai T (2021) Preparation, characterization, and anticancer effects of capsaicin-loaded nanoliposomes. *Nutrients* 13:3995. https://doi.org/10.3390/nu13113995

Baenas N, Belović M, Ilic N, Moreno DA, Garcia-Viguera C (2019) Industrial use of pepper (*Capsicum annum* L.) derived products: Technological benefits and biological advantages. *Food Chem* 274:872–885. https://doi.org/10.1016/j.foodchem.2018.09.047

Barbero GF, Liazid AG, Palma M, Vera JC, Barroso CG (2014) Evolution of total and individual capsaicinoids in peppers during ripening of the Cayenne pepper plant (*Capsicum annuum* L.). *Food Chem* 153:200–206. https://doi.org/10.1016/j.foodchem.2013.12.068

Basith S, Cui M, Hong S, Choi S (2016) Harnessing the therapeutic potential of capsaicin and its analogues in pain and other diseases. *Molecules* 21:966. https://doi.org/10.3390/molecules21080966

Baskaran P, Covington K, Bennis J, Mohandass A, Lehmann T, Thyagarajan B (2018) Binding efficacy and thermogenic efficiency of pungent and nonpungent analogs of capsaicin. *Molecules* 23:3198. https://doi.org/10.3390/molecules23123198

Bhagwat DA, Swami PA, Nadaf SJ, Choudhari PB, Kumbar VM, More HN, Killedar SG, Kawtikwar PS (2021) Capsaicin loaded solid SNEDDS for enhanced bioavailability and anticancer activity: *In-vitro, in-silico,* and *in-vivo* characterization. *J Pharm Sci* 110:280–291. https://doi.org/10.1016/j.xphs.2020.10.020

Brederson J-D, Kym PR, Szallasi A (2013) Targeting TRP channels for pain relief. *Eur J Pharmacol* 716:61–76. https://doi.org/10.1016/j.ejphar.2013.03.003

Chapa-Oliver AM, Mejia-Teniente L (2016) Capsaicin: From plants to a cancer-suppressing agent. *Molecules* 21:931. https://doi.org/10.3390/molecules21080931

Chinn MS, Sharma-Shivappa RR, Cotter JL (2011) Solvent extraction and quantification of capsaicinoids from *Capsicum chinense. Food Bioprod Process* 89:340–345. https://doi.org/10.1016/j.fbp.2010.08.003

Cui M, Gosu V, Basith S, Hong S, Choi S (2016) Polymodal transient receptor potential vanilloid type 1 nocisensor: Structure, modulators, and therapeutic applications. *Adv Protein Chem Struct Biol* 104:81–125. https://doi.org/10.1016/bs.apcsb.2015.11.005

Dairam A, Fogel R, Daya S, Limson JL (2008) Antioxidant and iron-binding properties of curcumin, capsaicin, and S-allylcysteine reduce oxidative stress in rat brain homogenate. *J Agric Food Chem* 56:3350–3356. https://doi.org/10.1021/jf0734931

Delgado-Vargas F, Jiménez-Aparicio A, Paredes-López O (2000) Natural pigments: Carotenoids, anthocyanins and betalains—characteristics, biosynthesis, processing and stability. *Crit Rev Food Sci Nutr* 40:173–289. https://doi.org/10.1080/10408690091189257

De Petrocellis L, Guida F, Moriello AS, De Chiaro M, Piscitelli F, de Novellis V, Maione S, Di Marzo V (2011) N-palmitoyl-vanillamide (palvanil) is a non-pungent analogue of capsaicin with stronger desensitizing capability against the TRPV1 receptor and anti-hyperalgesic activity. *Pharmacol Res* 63:294–299. https://doi.org/10.1016/j.phrs.2010.12.019

Diaz-Garcia CM, Morales-Lazaro SL, Sanchez-Soto C, Velasco M, Rosenbaum T, Hiriart M (2014) Role for the TRPV1 channel in insulin secretion from pancreatic beta cells. *J Membr Biol* 247:479–491. https://doi.org/10.1007/s00232-014-9658-8

Du HY, Shen ZJ, Li Y (2013) Microwave-assisted extraction of capsaicin from chili pepper powder. *Adv Mater Res* 634–638:1591–1594. https://doi.org/10.4028/www.scientific.net/AMR.634-638.1591

Friedman JR, Perry HE, Brown KC, Gao Y, Lin J, Stevenson CD, Hurley JD, Nolan NA, Akers AT, Chen YC, Denning KL, Brown LG, Dasgupta P (2017) Capsaicin synergizes with camptothecin to induce increased apoptosis in human small cell lung cancers via the calpain pathway. *Biochem Pharmacol* 129:54–66. https://doi.org/10.1016/j.bcp.2017.01.004

Gašparini D, Ljubičić R, Mršić-Pelčić J (2020) Capsaicin—potential solution for chronic pain treatment. *Psychiatr Danub* 32:420–428

Georgescu SR, Sarbu MI, Matei C, Ilie MA, Caruntu C, Constantin C, Tampa M (2017) Capsaicin: Friend or foe in skin cancer and other related malignancies? *Nutrients* 9:1365. https://doi.org/10.3390/nu9121365

Giuffrida D, Dugo P, Torre G, Bignardi C, Cavazza A, Corradini C, Dugo G (2013) Characterization of 12 *Capsicum* varieties by evaluation of their carotenoid profile and pungency determination. *Food Chem* 140:794–802. https://doi.org/10.1016/j.foodchem.2012.09.060

Govindarajan VS (1985) Capsicum—production, technology, chemistry and quality. Part 1: History, botany, cultivation, and primary processing. *Crit Rev Food Sci Nutr* 22:109–176. https://doi.org/10.1080/10408398509527412

Han J, Zhang S, Liu X, Xiao C (2020) Fabrication of capsaicin emulsions: Improving the stability of the system and relieving the irritation to the gastrointestinal tract of rats. *J Sci Food Agric* 100:129–138. https://doi.org/10.1002/jsfa.10002

Haramizu S, Kawabata F, Ohnuki K, Inoue N, Watanabe T, Yazawa S, Fushiki T (2011) Capsiate, a non-pungent capsaicin analog, reduces body fat without weight rebound like swimming exercise in mice. *Biomed Res* 32:279–284. https://doi.org/10.2220/biomedres.32.279

Hernández-Pérez T, Gómez-García MR, Valverde ME, Paredes-López O (2020) *Capsicum annuum* (hot pepper): An ancient Latin-American crop with outstanding bioactive compounds and nutraceutical potential. A review. *Comp Rev Food Sci Food Saf* 19:2972–2993. https://doi.org/10.1111/1541-4337.12634

Hokanson JA, Langdale CL, Milliken PH, Sridhar A, Grill WM (2021) Effects of intravesical prostaglandin E2 on bladder function are preserved in capsaicin-desensitized rats. *Am J Physiol Renal Physiol* 320:F212–F223. https://doi.org/10.1152/ajprenal.00302.2020

Huang W, Cheang WS, Wang X, Lei L, Liu Y, Ma KY, Zheng F, Huang Y, Chen Z-Y (2014) Capsaicinoids but not their analogue capsinoids lower plasma cholesterol and possess beneficial vascular activity. *J Agric Food Chem* 62:8415–8420. https://doi.org/10.1021/jf502888h

Hurley JD, Akers AT, Friedman JR, Nolan NA, Brown KC, Dasgupta P (2017) Non-pungent long chain capsaicin-analogs arvanil and olvanil display better anti-invasive activity than capsaicin in human small cell lung cancers. *Cell Adhes Migr* 11:80–97. https://doi.org/10.1080/19336918.2016.1187368

Hwang YP, Yun HJ, Choi JH, Han EH, Kim HG, Song GY, Kwon K, Jeong TC, Jeong HG (2011) Suppression of EGF-induced tumor cell migration andmatrixmetalloproteinase-9 expression by capsaicin via the inhibition of EGFR-mediated FAK/Akt, PKC/Raf/ERK, p38 MAPK, and AP-1 signaling. *Mol Nutr Food Res* 55:594–605. https://doi.org/10.1002/mnfr.201000292

Issa AY, ALSalamat HA, Awad WB, Haddaden RM, Aleidi SM (2021) The impact of pharmaceutical care on the efficacy and safety of transdermal glucosamine sulfate and capsaicin for joint pain. *Int J Clin Pharm* 43:101–106. https://doi.org/10.1007/s11096-020-01113-1

Janssens PLHR, Hursel R, Martens EAP, Westerterp-Plantenga MS (2013) Acute effects of capsaicin on energy expenditure and fat oxidation in negative energy balance. *PLoS ONE* 8:e67786. https://doi.org/10.1371/journal.pone.0067786

Járomi P, Garab D, Hartmann P, Bodnár D, Nyíri S, Sántha P, Boros M, Jancsó G, Szabó A (2018) Capsaicin-induced rapid neutrophil leukocyte activation in the rat urinary bladder microcirculatory bed. *Neurourol Urodyn* 37:690–698. https://doi.org/10.1002/nau.23376

Jeon YH (2015) Herpes Zoster and postherpetic neuralgia: Practical consideration for prevention and treatment. *Korean J Pain* 28:177–184. https://doi.org/10.3344/kjp.2015.28.3.177

Kang JH, Kim CS, Han IS, Kawada T, Yu R (2007) Capsaicin, a spicy component of hot peppers, modulates adipokine gene expression and protein release from obese-mouse adipose tissues and isolated adipocytes, and suppresses the inflammatory responses of adipose tissue macrophages. *FEBS Lett* 581:4389–4396. https://doi.org/10.1016/j.febslet.2007.07.082

Kim C-S, Kawada T, Kim B-S, Han I-S, Choe S-Y, Kurata T, Yu R (2003) Capsaicin exhibits anti-inflammatory property by inhibiting IkB-a degradation in LPS-stimulated peritoneal macrophages. *Cell Signal* 15:299–306. https://doi.org/10.1016/s0898-6568(02)00086-4

Kim SH, Hwang JT, Park HS, Kwon DY, Kim MS (2013) Capsaicin stimulates glucose uptake in C2C12 muscle cells via the reactive oxygen species (ROS)/AMPK/p38 MAPK pathway. *Biochem Biophys Res Commun* 439:66–70. https://doi.org/10.1016/j.bbrc.2013.08.027

Kwon KT, Uddin MS, Jung GW, Sim JE, Lee SM, Woo HC, Chun BS (2011) Solubility of red pepper (*Capsicum annum*) oil in near- and supercritical carbon dioxide and quantification of capsaicin. *Korean J Chem Eng* 28:1433–1438. https://doi.org/10.1007/s11814-010-0515-x

Lee E, Jung DY, Kim JH, Patel PR, Hu X, Lee Y, Azuma Y, Wang HF, Tsitsilianos N, Shafiq U, Kwon JY, Lee YJ, Lee KW, Kim JK (2015) Transient receptor potential vanilloid type-1 channel regulates diet induced obesity, insulin resistance, and leptin resistance. *FASEB J* 29:3182–3192. https://doi.org/10.1096/fj.14-268300

Lee GR, Jang SH, Kim CJ, Kim AR, Yoon DJ, Park NH, Han IS (2014) Capsaicin suppresses the migration of cholangiocarcinoma cells by down-regulating matrix metalloproteinase-9 expression via the AMPK-NF-kappaB signaling pathway. *Clin Exp Metastasis* 31:897–907. http://doi.org/10.1007/s10585-014-9678-x

Lee GR, Shin MK, Yoon DJ, Kim AR, Yu R, Park NH, Han IS (2013) Topical application of capsaicin reduces visceral adipose fat by affecting adipokine levels in high-fat diet induced obese mice. *Obesity* 21:115–122. https://doi.org/10.1038/oby.2012.166

Lu M, Chen C, Lan Y, Xiao J, Li R, Huang J, Huang Q, Cao Y, Ho CT (2020) Capsaicin-the major bioactive ingredient of chili peppers: Bio-efficacy and delivery systems. *Food Funct* 11:2848–2860. https://doi.org/10.1039/d0fo00351d

Ludy MJ, Mattes RD (2011) The effects of hedonically acceptable red pepper doses on thermogenesis and appetite. *Physiol Behav* 102:251–258. https://doi.org/10.1016/j.physbeh.2010.11.018

Maggi CA, Giuliani S, Conte B, Furio M, Santicioli P, Meli P, Gragnani L, Meli A (1988) Prostanoids modulate reflex micturition by acting through capsaicin-sensitive afferents. *Eur J Pharmacol* 145:105–112. https://doi.org/10.1016/0014-2999(88)90221-x

Marrelli M, Menichini F, Conforti F (2016) Hypolipidemic and antioxidant properties of hot pepper flower (*Capsicum annuum* L.). *Plant Foods Hum Nutr* 71:301–306. https://doi.org/10.1007/s11130-016-0560-7

Matharu M (2010) Cluster headache. *BMJ Clin Evid* 2010:1212.

Merritt JC, Richbart SD, Moles EG, Cox AJ, Brown KC, Miles SL, Finch PT, Hess JA, Tirona MT, Valentovic MA, Dasgupta P (2022) Anti-cancer activity of sustained release capsaicin formulations. *Pharmacol Ther* 238:108177. https://doi.org/10.1016/j.pharmthera.2022.108177

Morales-Soriano E, Ugás R (2022) Nutritional attributes and effect of processing on Peruvian chili peppers. In: Repo-Carrasco-Valencia R, Tomás MB (eds) *Native Crops in Latin America. Biochemical, Processing and Nutraceutical Aspects.* CRC Press, Boca Raton, FL, p. 161.

Narang N, Jiraungkoorskul W, Jamrus P (2017) Current understanding of antiobesity property of capsaicin. *Pharmacogn Rev* 11:23–26. https://doi.org/10.4103/phrev.phrev_48_16

Papoiu ADP, Yosipovitch G (2010) Topical capsaicin. The fire of a 'hot' medicine is reignited. *Expert Opinion Pharmacother* 11:1359–1371. https://doi.org/10.1517/14656566.2010.481670

Pasierski M, Szulczyk B (2022) Beneficial effects of capsaicin in disorders of the central nervous system. *Molecules* 27:2484. https://doi.org/10.3390/molecules27082484

Patowary P, Pathak MP, Zaman K, Raju PS, Chattopadhyay P (2017) Research progress of capsaicin responses to various pharmacological challenges. *Biomed Pharmacother* 96:1501–1512. https://doi.org/10.1016/j.biopha.2017.11.124

Perry L, Dickau R, Zarrillo S, Holst I, Pearsall DM, Piperno DR, Zeidler JA (2007) Starch fossils and the domestication and dispersal of chili peppers (*Capsicum* spp. L.) in the Americas. *Science* 315:986–988. https://doi.org/10.1126/science.1136914

Perucka I, Oleszek W (2000) Extraction and determination of capsaicinoids in fruit of hot pepper *Capsicum annuum* L. by spectrophotometry and high-performance liquid chromatography. *Food Chem* 71:287–291. https://doi.org/10.1016/S0308- 8146(00)00153-9

Privitera R, Anand P (2021) Capsaicin 8% patch Qutenza and other current treatments for neuropathic pain in chemotherapy-induced peripheral neuropathy (CIPN). *Curr Opin Support Palliat Care* 15:125–131. https://doi.org/10.1097/SPC.0000000000000545

Qin A, Kalemkerian GP (2018) Treatment options for relapsed small-cell lung cancer: What progress have we made? *J Oncol Pract* 14:369–370. https://doi.org/10.1200/JOP.18.00278

Richards BL, Whittle SL, Buchbinder R (2012) Neuromodulators for pain management in rheumatoid arthritis. *Cochrane Database Syst Rev* 1:CD008921. https://doi.org/10.1002/14651858.CD008921.pub2

Richbart SD, Friedman JR, Brown KC, Gadepalli RS, Miles SL, Rimoldi JM, Rankin GO, Valentovic MA, Tirona MT, Finch PT, Hess JA, Dasgupta P (2021) Nonpungent N-AVAM capsaicin analogues and cancer therapy. *J Med Chem* 64:1346–1361. https://doi.org/10.1021/acs.jmedchem.0c01679

Ríos-Silva M, Santos-Álvarez R, Trujillo X, Cárdenas-María RY, López-Zamudio M, Bricio-Barrios JA, Caridad Leal C, Saavedra-Molina A, Huerta-Trujillo M, Espinoza-Mejía K, Miguel Huerta M (2019) Effects of chronic administration of capsaicin on biomarkers of kidney injury in male Wistar rats with experimental diabetes. *Molecules* 24:36. https://doi.org/10.3390/molecules24010036

Rosca AE, Iesanu MI, Zahiu CDM, Voiculescu SE, Paslaru AC, Zagrean AM (2020) Capsaicin and gut microbiota in health and disease. *Molecules* 25:5681. https://doi.org/10.3390/molecules25235681

Saito M (2015) Capsaicin and related food ingredients reducing body fat through the activation of TRP and brown fat thermogenesis. *Adv Food Nutr Res* 76:1–28. https://doi.org/10.1016/bs.afnr.2015.07.002

Salgado-Roman M, Botello-Álvarez E, Rico-Martínez R, Jiménez-Islas H, Cárdenas-Manríquez M, Navarrete-Bolaños JL (2008) Enzymatic treatment to improve extraction of capsaicinoids and carotenoids from chili (*Capsicum annuum*) fruits. *J Agric Food Chem* 56:10012–10018. https://doi.org/10.1021/jf801823m

Scoville WL (1912) Note on capsicum. *J Am Pharm Assoc* 1:453–454. https://doi.org/10.1002/jps.3080010520

Shanmugham V, Subban R (2022) Comparison of the anti-obesity effect of enriched capsanthin and capsaicin from *Capsicum annuum* L. fruit in obesity-induced C57BL/6J mouse model. *Food Technol Biotechnol* 60:202–212. https://doi.org/10.17113/ftb.60.02.22.7376

Sharma SK, Vij AS, Sharma M (2013) Mechanisms and clinical uses of capsaicin. *Eur J Pharmacol* 720:55–62. https://doi.org/10.1016/j.ejphar.2013.10.053

Shimizu T, Majima T, Suzuki T, Shimizu N, Wada N, Kadekawa K, Takai S, Takaoka E, Kwon J, Kanai AJ, de Groat WC, Tyagi P, Saito M, Yoshimura N (2018) Nerve growth factor-dependent hyperexcitability of capsaicin-sensitive bladder afferent neurones in mice with spinal cord injury. *Exp Physiol* 103:896–904. https://doi.org/10.1113/EP086951

Shin DH, Kim OH, Jun HS, Kang MK (2008) Inhibitory effect of capsaicin on B16-F10 melanoma cell migration via the phosphatidylinositol 3-kinase/Akt/Rac1 signal pathway. *Experimental and Molecular Medicine*, 40(5), 486–94.

Shook RP, Hand GA, Paluch AE, Wang X, Moran R, Hébert JR, Jakicic JM, Blair SN (2015) High respiratory quotient is associated with increases in body weight and fat mass in young adults. *Eur J Clin Nutr* 70:1197–1202. https://doi.org/10.1038/ejcn.2015.198

Smeets AJ, Janssens PL, Westerterp-Plantenga MS (2013) Addition of capsaicin and exchange of carbohydrate with protein counteract energy intake restriction effects on fullness and energy expenditure. *J Nutr* 143:442–447. https://doi.org/10.3945/jn.112.170613

Srinivasan K (2016) Biological activities of red pepper (*Capsicum annuum*) and its pungent principle capsaicin: A review. *Crit Rev Food Sci Nutr* 56:1488–1500. https://doi.org/10.1080/10408398.2013.772090

Stellari GM, Mazourek M, Jahn MM (2010) Contrasting modes for loss of pungency between cultivated and wild species of Capsicum. *Heredity* 104:460–471. https://doi.org/10.1038/hdy.2009.131

Tremblay A, Arguin H, Panahi S (2016) Capsaicinoids: A spicy solution to the management of obesity? *Int J Obesity* 40:1198–1204. https://doi.org/10.1038/ijo.2015.253

Uarrota VG, Maraschin M, de Bairros Ade F, Pedreschi R (2021) Factors affecting the capsaicinoid profile of hot peppers and biological activity of their non-pungent analogs (capsinoids) present in sweet peppers. *Crit Rev Food Sci Nutr* 61:649–665. https://doi.org/10.1080/10408398.2020.1743642

Varghese S, Kubatka P, Rodrigo L, Gazdikova K, Caprnda M, Fedotova J, Zulli A, Kruzliak P, Büsselberg D (2017) Chili pepper as a body weight-loss food. *Int J Food Sci Nutr* 68:392–401. https://doi.org/10.108 0/09637486.2016.1258044

Venier NA, Yamamoto T, Sugar LM, Adomat H, Fleshner NE, Klotz LH, Venkateswaran V (2015) Capsaicin reduces the metastatic burden in the transgenic adenocarcinoma of the mouse prostate model. *Prostate* 75:1300–1311. https://doi.org/10.1002/pros.23013

Wang Y, Tang C, Tang Y, Yin H, Liu X (2020) Capsaicin has an anti-obesity effect through alterations in gut microbiota populations and short-chain fatty acid concentrations. *Food Nutr Res* 64. https://doi. org/10.29219/fnr.v64.3525

Xiang Y, Xu X, Zhang T, Wu X, Fan D, Hu Y, Ding J, Yang X, Lou J, Du Q, Xu J, Xie R (2022) Beneficial effects of dietary capsaicin in gastrointestinal health and disease. *Exp Cell Res* 417:113227. https://doi. org/10.1016/j.yexcr.2022.113227

Yoneshiro T, Aita S, Kawai Y, Iwanaga T, Saito M (2012) Nonpungent capsaicin analogs (capsinoids) increase energy expenditure through the activation of brown adipose tissue in humans. *Am J Clin Nutr* 95:845–850. https://doi.org/10. 3945/ajcn.111.01860

Yuan LJ, Qin Y, Wang L, Zeng Y, Chang H, Wang J, Wang B, Wan J, Chen SH, Zhang QY, et al. (2015) Capsaicin-containing chili improved postprandial hyperglycemia, hyperinsulinemia, and fasting lipid disorders in women with gestational diabetes mellitus and lowered the incidence of large-for-gestational-age newborns. *Clin Nutr* 35:388–393. https://doi.org/10.1016/j.clnu.2015.02.011

Yue L, Zhang F, Wang ZX (2012) Ultrasound-assisted extraction of capsaicin from red peppers and mathematical modeling. *Sep Sci Technol* 47:124–130. https://doi.org/10.1080/01496395.2011.604064

Zhang S, Wang D, Huang J, Hu Y, Xu Y (2020) Application of capsaicin as a potential new therapeutic drug in human cancers. *J Clin Pharm Ther* 45:16–28. https://doi.org/10.1111/jcpt.13039

Zheng L, Chen J, Ma Z, Liu W, Yang F, Yang Z, Wang K, Wang X, He D, Li L, Zeng J (2016) Capsaicin enhances anti-proliferation efficacy of pirarubicin via activating TRPV1 and inhibiting PCNA nuclear translocation in 5637 cells. *Mol Med Rep* 13:881–887. https://doi.org/10.3892/mmr.2015.4623

Zhu M, Yu X, Zheng Z, Huang J, Yang X, Shi H (2020) Capsaicin suppressed activity of prostate cancer stem cells by inhibition of Wnt/β-catenin pathway. *Phytother Res* 34:817–824. https://doi.org/10.1002/ptr.6563

9 Medicinal Bioactive Secondary Metabolites of Indian Chili *(Capsicum sp.)*

Debosree Ghosh, Partha Sarathi Singha and Ramkrishna Ghosh

CONTENTS

9.1 INTRODUCTION

Chilies are essential spices of Indian cuisine. They add hotness to dishes. They grow easily and extensively all over India. The fruit is consumed as a spice vegetable. Various metabolites of chili have been reported to have potential bioactive properties. They have been found to be working against various disease conditions. These secondary metabolites are by-products of primary metabolic processes. Studies report that these secondary metabolites in chili plants do not have any direct influence on the growth and development of the plant but indirectly contribute to promote good health and proper growth of chili plant. The secondary metabolites (SMs) play important and significant roles in protecting the plant. In simple words, they constitute the defense component of the plant. Many phytohormones have been recognized to be part of the secondary metabolites of the plants, and they have roles in protecting the plant against stress (Reddy et al., 2022). Studies report antioxidant properties of these SMs, by virtue of which they protect the plant against oxidative stress (Reddy et al., 2022). Chili has been reported to have immense medicinal properties (Ghosh et al., 2016). Studies show that the content of SMs varies in different stages of maturation of the chili fruit. Studies conducted at various stages of ripening reveals that the content of one of the major SM of high medicinal value in chili namely capsaicin was present in highest concentration in the highly ripened fruit (Nugroho, 2016). The time taken for ripening of chili fruit varies depending on the variety, the soil they grow in, the climatic condition and the season (http://www.celkau.in/crops/vegetables/Chilli/chilli.aspx). Content of SMs like capsaicin is also reported to be affected by the type of fertilizers used and by the growing medium. The distribution of the SMs is reported to be different in different parts of the chili fruit.

Besides capsaicin, other well-known medicinal SMs reported to be present in chili fruit are various other capsaicinoids, flavonoids and various vitamins like carotenoids and tocopherols etc. Capsaicin belongs to the category of capsaicinoids and makes up the highest percentage of total

 DOI: 10.1201/9781003378259-9

capsaicinoids, i.e. 69%. Other natural capsaicinoids recognized are dihydrocapsaicin (22%), nordi-hydrocapsaicin (7%), homocapsaicin (1%) and homodihydrocapsaicin (1%) (Reyes-Escogido Mde et al., 2011). Each part of the chili fruit is known to have a different concentration of capsaicin. The highest concentration of capsaicin has been reported to be present in the joint between the whitish pith where the seeds are attached and the membrane lining of the chili wall. Capsaicinoids have been reported to have potent medicinal properties like anticancer, antioxidant, anti-inflammatory, antiobesity, analgesic etc. (Luo et al., 2011). Based on the property of capsaicin-triggered TRPV1 signaling and desensitization processes, capsaicin is recommended as a very potent therapeutic agent against pain (Basith et al., 2016). Chili fruit has also been in use in traditional medicine for ages for mitigating certain pathological conditions like migraine, altered muscle tone, dyspepsia, anorexia etc. (Bal et al., 2022). High flavonoid content (> 100 mg/100 gm) is reported in Indian red chili powder (Nair et al., 1998). Certain spice herbs with pronounced antioxidant potential and other medicinal properties have also been found to be rich in flavonoid contents, which is known to tremendously contribute to the medicinal potential and therapeutic ability of the plants (Ghosh et al., 2017). Flavonoids have been reported to have a wide range of therapeutic potential and medicinal properties. Flavonoid-containing plant extracts, by virtue of their antioxidant potential, have been found to be cardioprotective, neuroprotective, hepatoprotective and nephroprotective (Ghosh et al., 2012a, b). Flavonoids are reported to be anticancer, antioxidant, antiviral and anti-inflammatory (Ullah et al., 2020). Thus, Indian chilies are sources of several such potent medicinal SMs, which, if recognized and isolated and further evaluated, may serve as potential core molecules for development of various therapeutic potentials against certain abnormal and serious health conditions like cancer, oxidative damage–induced ailments, inflammation, arthritis, cardiovascular disorders etc.

9.2   SECONDARY METABOLITES AS BIOACTIVE PHYTOCOMPOUNDS

In response to environmental biotic stress, the plants synthesize secondary metabolites as a second step of metabolism from primary metabolites (Guerriero et al., 2018). These secondary metabolites have several types of medicinal properties (Guerriero et al., 2018). Plants also produce secondary metabolites for various other purposes like as components of lignified cell walls of vascular tissues of plants attracting pollinators, for establishing symbiosis, etc. (Guerriero et al., 2018; Ncube and Van Staden, 2015). Alkaloids, terpenoids, phenolic compounds and sulphur-containing compounds are the four basic classes of plant secondary metabolites (Guerriero et al., 2018). These secondary metabolites are not essential for growth of plants but play important roles in the adaptation and survival of the plants. They are reported to act as signaling molecules during environmental interaction and stress (Pratiwi et al., 2018). These plant metabolites are known to have significant medicinal effects at the right doses and toxic effects in heavy or excess doses. They are reported to have extensive use in the pharmaceutical and cosmetics industries. Though many of these natural medicinal products have been replaced by synthetic drugs, the natural products with medicinal properties still remain irreplaceable because they come with minimum side effects. On the other hand, side effects of synthetic drugs are prominent and have adverse effects on patients' health. The secondary metabolites of plants have antioxidant, anticancer, anti-inflammatory, antipyretic, antimicrobial, antidiabetic and many other important medicinal uses and health benefits (Teoh, 2015). The plant secondary metabolites are also used in regenerative medicines (Umashanka, 2020). Various secondary metabolites from plants are used for treating conditions like bone disorders, arthritis, cancer, atherosclerosis etc. (Teoh, 2015; Umashanka, 2020). Uses of various secondary metabolites in various health conditions are illustrated in Figure 9.1.

The secondary metabolites of plants are reported to be the prime ones for the pharmacological actions of the medicinal plants. These secondary compounds have been in use to treat various health conditions for ages in Ayurveda and Unani. Secondary metabolites of plants are classified into different categories depending on their chemical structure and properties. These different types of secondary metabolites exert their pharmacological activities through various mechanisms. Studies

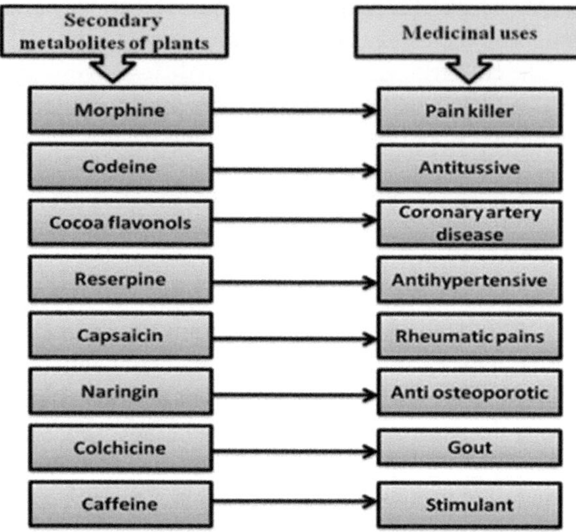

FIGURE 9.1   Various secondary metabolites of plants and their medicinal uses.

report that the target molecules on which the secondary metabolites act are either transcription factors, enzymes, hormones, receptors, ion channels or nucleic acids in some cases (Wink, 2015; Hussein & El-Anssary, 2019). Some secondary metabolites are also known to interact directly with the protein or lipid component of the cell membrane (Wink, 2015). The molecular targets for specific secondary metabolites are specific. For instance, the molecular targets for alkaloids are the receptors of neurotransmitters. On the other hand, SMs like phenolics and terpenoids act less specifically and form hydrogen bonds or hydrophobic bonds or ionic bonds with proteins, and thus, their primary molecular targets are the protein molecules. The brief mechanism of action of these phenolics and terpenoid class of plant secondary metabolites is by altering the three-dimensional structure of the proteins and thereby impacting and changing their activities. The best part of the medicinal potentials of the secondary metabolites of plants is that some of them have multi-target mechanisms of action, as a result of which they have the potential of acting against several types of disorders and health ailments (Wink, 2015). Structurally diverse plant secondary metabolites have also been evaluated for their medicinal potential against SARS CoV 2. The studies reveal that certain secondary metabolites of the plant have the ability to act against the deadly SARs CoV 2 virus (Puttaswamy et al., 2020). Searches for new secondary metabolites from plant sources and exploration of their medicinal properties are ongoing research projects around the world with the hope of finding new drug molecules that may be utilized for developing some potentially life-saving drugs.

9.3   SECONDARY METABOLITES OF INDIAN CHILIES AND THEIR MEDICINAL PROPERTIES

Different varieties of chilies are grown indigenously in different parts of India (Table 9.1). They differ in shape, size and intensity of hotness. Also, the phytoconstituents of each are different in quality and quantity. The secondary metabolites of chilies are reported to vary in different types of chilies. The one common and significant medicinal component of chili is capsaicin. This compound is also a secondary metabolite of chili and has pronounced pharmacological activities. Thus, the key bioactive ingredient of chili is capsaicin (Lu et al., 2020). Capsaicin has various medicinal uses. It is primarily known for its anti-inflammatory activities and is used in treating rheumatic arthritis. Studies reveal that capsaicin and its analogues have been extensively explored by researchers and are covered by more than a thousand patents. Capsaicin is reported to have analgesic, antioxidant, anti-inflammatory, antiobesity, antipruritic, anticancer, antiapoptotic, and neuro-protective functions (Basith et al., 2016). Studies also show that capsaicin is potent in treating conditions of metabolic syndrome and vascular diseases and also has gastro-protective potential (Basith et al., 2016). Capsaicin is also beneficial in pain management (Anand & Bley, 2011). Capsaicin is an alkaloid of the capsinoid family. Other members of this family are capsaicin, homodihydrocapsaicin, homocapsaicin, dihydrocapsaicin and nordihydrocapsaicin. All of these have potent medicinal properties (Singh et al., 2022).

9.3.1 ALKALOIDS FROM CHILI

Capsaicin is the main alkaloid found in almost all types of chili and is responsible for the spicy hot taste of chili (Voahanginirina, 2018; Tang et al., 2021). Chemically, it is *trans*-8-methyl-*N*-vanillyl-6-nonenamide. Other alkaloids reported from chili are homocapsaicin, nordihydrocapsaicin, dihydrocapsaicin and homodihydro-capsaicin. It is a unique alkaloid found in all genuses of *Capsicum* (Reyes-Escogido Mde et al., 2011). It is synthesized naturally in the placenta of the chili fruit. Capsaicin belongs to the class of capsinoids. These capsinoids are synthesized by the enzymatic condensation of a compound called vanillylamine and fatty acid chains of varying chain length, which are elongated by an enzyme called fatty acid synthase (Reyes-Escogido Mde et al., 2011). Capsaicin has been reported to have an affinity for the transient receptor potential vanilloid 1 (TRPV1). This receptor is primarily expressed in the sensory neurons (Cortright & Szallasi, 2004; Tominaga & Tominaga, 2005). The receptor is expressed in the small fibers of nociceptive neurons, in the brain, bladder, mast cells, keratinocytes of epidermis, kidneys, intestines, glial cells, liver, polymorphonuclear granulocytes and macrophages (Devesa et al., 2011). The TRPV1 receptor has a heat-sensitive subunit that is actually responsible for the burning sensation caused by capsaicin of chili. As capsaicin binds to TRPV1, there occurs an increase in the calcium level inside the cell, which, in turn, triggers release of various neuropeptides like substance P and the calcium gene-related peptide (CGRP). Capsaicin, when it comes in contact with the sensory neurons, produces burning sensation, inflammation and pain (Bevan & Szolcsanyi, 1990). Capsaicin is reported to impart an analgesic effect when applied locally on skin due to depletion of substance P (Fattori et al., 2016). Hence, capsaicin is extensively used as an anti-inflammatory agent and as an analgesic in various pathological conditions Other capsinoids from Indian chilies include dihydrocapsaicin nordihydrocapsaicin, homodihydrocapsaicin and homocapsaicin etc. All of these have pronounced medicinal properties (Antonious, 2018). Capsinoids have been in use in traditional medicine for ages. They were used to treat cough, toothache, sore throat etc. and for immunomodulation and also as antioxidants (Sanati et al., 2018). Capsaicin is known to have antiviral activities, and by virtue of this, the compound is expected to be effective against the SARS CoV 2 virus (Bal et al., 2022; Bousquet et al., 2021). Dihydrocapsaicin is known to have antibacterial activities against bacteria like *Helicobacter pylori* (Park et al., 2014).

9.3.2 TERPENOIDS

Indian chilies are known to produce terpenoids as one of their secondary metabolites. Terpenoids from plant sources are known to have pharmacological activities and clinical uses (Singh & Sharma, 2015; Masyita et al., 2022). Terpenes are basically a class of hydrocarbons. Different derivatives of terpenes, mainly oxygenated derivatives of terpenes, are widely distributed in nature in plants, microorganisms, fungi, insects and marine organisms. The main classes of terpenes are monoterpenes (e.g. limonene), sesquiterpenes (e.g. elemene), diterpenes (e.g. camphene) and polyterpenes (Fan et al., 2023). Terpenoid metabolism has been investigated in the chromoplast of *Capsicum annum* (Camara et al., 1983). Studies show that alpha-tocopherol synthesis occurs from 2,3-dimethylphytylquinol and S-adenosyl-l-methionine and was achieved using chromoplasts from *Capsicum annuum* fruit.

The enzyme involved in the cyclization process is 2,3-dimethyl-phytylquinol cyclase, and the one involved in the process of methylation is S-adenosyl methionine:gamma-tocopherol methyltransferase. Both the enzymes are reported to be localized in the chromoplast membrane fraction i.e. the envelopes and/or a-chlorophyll lamellae (Camara et al., 1982). Terpenoids are reported to have antitumor, antibacterial, anti-inflammatory, antiviral, antimalarial effects (Yang et al., 2020). They are also known to have cardio-protective and hypoglycemic activities. Indian chili is known to be rich in carotenoids. Carotenoids are a group of tetraterpenoids, and they form a large group. These carotenoids are widely distributed in nature, and they impart red, orange or yellow colours to various plant components (Nisar et al., 2015). Carotenoids are not produced by animals. In living organisms, these carotenoids play very significant roles. They are reported to have anticarcinogenic activities and are also potent antioxidants. The carotenoids also serve as the precursors for the synthesis of vitamin A and retinoic acid, which plays significant physiological roles in humans and animals (Giuffrida et al., 2013). Colour of ripened chili is mainly imparted by the varieties of carotenoids in it. Studies show that the content of carotenoids depends on certain factors like the cultivar, the soil it is grown in and the storage of the products (Okunlola et al., 2017). Ripened red chilies primarily are rich in carotenoids besides other bioactive secondary metabolites. Carotenoids contribute effectively to the medicinal properties of the red chili. These carotenoids are secondary metabolites of the *Capsicum* genus. The profile of carotenoids produced in the plant depends and varies depending on the environmental stress factors the plant gets exposed to. The carotenoids in the plant body help them to combat and survive in the stressful environment (Lee et al., 2005). These pigments, i.e. carotenoids, are produced by all photosynthetic organisms (Amengual, 2019). The initial precursor of carotenoids like phytoene and phytofluene are colourless, and they are abundantly distributed in our diet (Melendez-Martinez et al., 2019). Being colourless, these carotenes were not considered for studies for a long time and remained ignored in the carotenoid group. With time, studies reveal that these carotenes are potent bioactive substances. They have been reported to accumulate in the skin and have protective effects on skin against UV radiation and ageing and thus help to maintain good health of the integumentary system (Takemura et al., 2015; Antonioa et al., 2018). Carotenoids and retinoids are considered essential nutrients for humans. Carotenoids can be broadly classified into two classes: xanthophylls and carotenes. Carotenes are composed of compounds with linear or cyclicized hydrocarbons. Examples of carotenes are α-carotene and β-carotene. Examples of xanthophylls include carotenoids with oxygenated functions such as hydroxy-, keto-, methoxy-, epoxy and carbonyl, as in lutein, violaxanthin, neoxanthin and zeaxanthin etc. (Pereira et al., 2021).

9.3.3 TERPENOID-ALKALOIDS

Terpenoid and alkaloids constitute two different categories of secondary metabolite of plants including that of the chilies grown on the Indian subcontinent. Terpenoid-alkaloids are a different group which are formed from the same prenyl units which forms the terpene frameworks. During biosynthesis of terpenoid-alkaloids, a nitrogen atom or atoms are introduced as β-aminoethanol, ethylamine or methylamine. The result of this kind of biosynthesis is the formation of unique compounds with unpredictable structures (Cherney & Baran, 2001). These compounds are reported to be biologically active.

9.3.4 POLYPHENOLS

Indian chilies are rich in various polyphenolic compounds. These have antioxidant and anti-inflammatory activities (Perucka & Materska, 2007; Mateos et al., 2013; Abbas et al., 2017). Polyphenols are the most abundantly distributed bioactive molecule in nature. These compounds are widely distributed in plants and have immense medicinal properties. Polyphenols are broadly categorized into two general classes: flavonoids and phenolic acids. Flavonoids are further divided into flavones, flavonones, flavonols, flavanols and isoflavones. Phenolic acids are generally classified into

TABLE 9.1

Secondary Metabolites of Indian Chilies

Sl. No.	Common name of the Chili	Other local names	Scientific name of the Chili	Indigenous to the Geographical Region	Chemical composition (including Secondary Metabolites with medicinal potentials)	References
1.	Ghost Pepper	Bhut Jolokia, Naga chili or Raja mircha	*Capsicum chinense*	North-East Region of India	Capsinoids like capsaicin, dihydrocapsaicin, nordihydrocapsaicin etc., carbohydrate, soluble and insoluble dietary fibre, minerals like iron, zinc, copper, phosphorus, magnesium, Calcium etc., Polyphenols, vitamin A, C etc.	Voahanginirina, 2018; Tang et al., 2021; Oney-Montalvo et al., 2020; Malakar et al., 2018; Aqsa et al., 2021; https://spiceitupp.com/kashmiri-chillies/; https://www.onmanorama.com/food/features/2021/04/07/indian-chillies-you-must-know-about.html;Ayob et al., 2021.
2.	Kashmiri Chilies	Kashmiri lal mirch, Kashmiri mirch	*Capsicum annuum*	Kashmir, Jammu	Capsinoids like capsaicin, dihydrocapsaicin, nordihydrocapsaicin etc., carbohydrate, soluble and insoluble dietary fibre, minerals like iron, zinc, copper, phosphorus, magnesium, calcium etc. Saturated and unsaturated fatty acids, vitamin B complex, vitamin C, vitamin A, phenolic acids like gallic acid etc.	Ghosh et al., 2016; Tang et al., 2021; Oney-Montalvo et al., 2020; Malakar et al., 2018; Aqsa et al., 2021; https://spiceitupp.com/kashmiri-chillies/; https://www.onmanorama.com/food/features/2021/04/07/indian-chillies-you-must-know-about.html;Ayob et al., 2021.
3.	Guntur Chili	Sannam chili, Guntur sannam	*Capsicum annuum*	Andhra Pradesh	Capsaicinoids and capsinoids like capsaicin, vitamins like vitamin C etc., phenolics and flavonoids	Ghosh et al., 2016; https://www.gitagged.com/gunturchilli/blog/#:~:text=Guntur%20chilli%20is%20hot%20and,(11.98g% 2F100 g); Phanindra & Mounika, 2020; Reyes-Escogido Mde et al., 2011;
4.	Dalle Khursani	Cherry pepper	*Capsicum annum*	Sikkim	Capsaicinoids and capsinoids like capsaicin and dihydrocapsaicin etc., carotenoids and phenolic compounds. High in vitamin A, E and potassium	Ghosh et al., 2016; https://pragyanxetu.com/2022/08/18/dalle-khursani/
5.	Kanthari Chili	Bird's eye chili of Kerala, Kanthari Mulaku	*Capsicum frutescens*	Kerala	Capsaicinoids and capsinoids like capsaicin, vitamins like vitamin A, E, C etc., niacin etc., sodium, potassium etc. minerals	https://www.timesfoodie.com/nutritional-facts/birds-eye-chilli-kanthari-mulaku-health-benefits/88002587.cms; Dubey et al., 2015

(Continued)

TABLE 9.1 (*Continued*)
Secondary Metabolites of Indian Chilies

Sl. No.	Common name of the Chili	Other local names	Scientific name of the Chili	Indigenous to the Geographical Region	Chemical composition (including Secondary Metabolites with medicinal potentials)	References
6.	Byadagi Chili	Tabasco	*Capsicum annuum*	Karnataka	Capsaicinoids and capsinoids like capsaicin etc., vitamins like vitamin A, E, C, folate etc., minerals like manganese, copper, molybdenum etc., phenolics and flavonoids	Ghosh et al., 2016; https://www.potsandpans.in/blogs/articles/byadagi-chilli-health-benefits-uses-and-important-facts
7.	Ramnad Mundu	Gundu, Mundu chili, Fat chili	*Warangal chappatta*	Tamil Nadu	Capsaicinoids and capsinoids like capsaicin etc., vitamins like vitamin A, E, C, etc., minerals, phenolics and flavonoids	Ghosh et al., 2016; https://issuu.com/infogitagged/docs/7_most_famous_chilli_pepper_in_india.docx/s/10788789
8.	Jwala Chili	Finger hot Indian chili, Indian Jwala	*Capsicum annuum*	Gujarat	Capsaicin and other capsaicinoids, phenolics and flavonoids, vitamins like C, E, K etc., and minerals; high protein	Ghosh et al., 2016; https://mygardenlife.com/plant-library/chili-pepper-indian-jwala-*Capsicum-annuum*
9.	Dhani	Bird's eye chili of northeast	*Capsicum frutescens*	Manipur	Capsaicin and other capsaicinoids, phenolics and flavonoids, vitamins like C, E, K etc., and minerals	Ghosh et al., 2016; https://www.onmanorama.com/food/features/2021/04/07/indian-chillies-you-must-know-about.html
10.	Madras Puri	Gundu Milagai	*Capsicum annuum*	Andhra Pradesh	Capsaicin and other types of capsinoids, flavonoids, phenolics, vitamins and minerals	Ghosh et al., 2016; https://pramoda.co.in/types-of-indian-chillies/;
11.	Khola Chili	Canacona chili	*Capsicum annuum*	Goa	Capsaicinoids and capsinoids like capsaicin etc., vitamins like vitamin A, E, C, etc., minerals like potassium, manganese, iron, magnesium; phenolics and flavonoids	Ghosh et al., 2016; https://sahasa.in/2021/06/29/khola-chilli-canacona-chilli/
12.	Bhiwapur Chili	Doda chill	*Capsicum annuum*	Maharashtra	Capsaicin, chilies are excellent source of vitamin A, B, B6, C, molybdenum, very high potassium, carbohydrates, dietary fibers, fats, proteins, rich in phosphorus, calcium, rich in iron, magnesium, potassium	https://ipindia.gov.in/writereaddata/Portal/IPOJournal/1_322_1/Journal_88.pdf; Michael et al., 2020; https://niftem.ac.in/newsite/pmfme/wp-content/uploads/2022/08/mizochilliwriteup.pdf
13.	Mizo chili	Mizoram's Bird Eye Chili (MZBEC)/ Hmarchate/ Vaihmarchate	*Capsicum frutescence*	Mizoram	Capsaicin, ascorbic acid, vitamin A, B1 and B2, iron, calcium, magnesium, phosphorus, protein and beta carotene	https://ipindia.gov.in/writereaddata/Portal/IPOJournal/1_4739_1/Journal_121.pdf; https://ipindia.gov.in/writereaddata/Portal/Images/pdf/Journal_140.pdf

hydroxybenzoic and hydroxycinnamic acids (Waksmundzka-Hajnos et al., 2008). Chilies are reported to contain flavonoids, which effectively contributes to their medicinal properties. Flavonoids from plants are a broad class containing more than 7,000 secondary metabolites (Lee et al., 1995). Higher organisms like humans cannot synthesize these flavonoids in their bodies. On the other hand, these flavonoids have potent antioxidant and other medicinal activities. Hence, flavonoids from plant sources serve as essential medicinal agents and aids in healing various health ailments (Lee et al., 1995). Flavonoids are known to have potential medicinal properties like anti-inflammatory, antihypertensive and antiviral activities (Rodriguez-Mateos et al., 2014). Flavonoids are low-molecular-weight compounds that share a common skeletal structure containing diphenylpropanes (C6-C3-C6), along with two phenyl rings (A and B), and those are bridged by a C ring of heterocyclic pyran. Studies report quercetin and luteolin as the two primary flavonoids from pepper (Lee et al., 1995; Fujiwake et al., 1982). Other flavonoids reported from chili are catechin, luteolin, epicatechins, rutin and kaempferol (Bal et al., 2022). In chili, flavonoids remain mostly accumulated in the peel (Lee et al., 2005). Studies show that flavonoids play important roles in modulating signaling of lipid kinases and proteins in cells (Mehmet et al., 2015). Studies also establish the fact that certain flavonoids have anticancer activities and are helpful in treating breast cancer, with the flavonoid apigenin also effective in inhibiting protein kinase (Mosmann, 1983).

Phenolic compounds in chili contribute to the antioxidant properties of chili (Ruanma et al., 2010). The phenolic compounds are reported to contribute a hydrogen atom and accept the electron and thus act as an antioxidant by neutralizing the free radicals. Studies show that the antioxidant potency of chili is dependent on the content of the phenolic compounds in it (Medina-Juárez et al., 2012). The concentration and content of flavonoids and phenolic acids in chili is reported to increase with maturity of the chili. These flavonoids and phenolic acids are mostly concentrated in the pericarp of the chili, and they are known to be potent in rendering protection against pathological conditions like cancer, oxidative stress, diabetes, neurological disorders etc. (Salehi et al., 2018).

9.4 CONCLUSIONS AND FUTURE PERSPECTIVES

Plant secondary metabolites serve as excellent medicinal agents against various health ailments. Most of the secondary metabolites are produced in plants in response to environmental stress, and they help the plant to survive in adverse conditions. They are also essential to maintain the immunity and health of the plant. Most of these secondary metabolites from plants have been found to be effective in combating several disease conditions in humans. Chilies grown around the world are used extensively as a popular spice in cooking. These chilies have been recognized for ages as rich source of powerful medicinal compounds. These medicinal compounds are mostly produced as secondary metabolites in the chili plant and accumulate in different parts of the plant. The edible part of the plant, the fruit, the chili, is also rich in various medicinal secondary metabolites. The different varieties of chilies grown in different parts of India vary in their composition qualitatively and quantitatively and have varied taste, hotness and medicinal potency. Knowledge of these secondary metabolites in Indian chilies will help to identify them as potential lead molecules for developing drugs against various diseases. The literature is limited, and there are few studies regarding the phytochemical analysis of the different species of chilies grown in different parts of India. The field needs further extensive research work to know and identify the phytochemical composition of the Indian chilies. Detailed research and detailed knowledge of the phytochemicals present in the Indian chilies will be helpful to utilize them for developing pharmaceutical formulations against various diseases. Capsaicin, being the prime component of almost all *Capsaicin* genuses, is extensively studied and is reported to be highly potent against various health ailments like diabetes, cancer, inflammation, skin problems, arthritis and many more. There are several other potent medicinal secondary metabolites in chili, which need much research work before they can be recognized for their potency against various diseases.

9.5   ACKNOWLEDGEMENT

Dr. DG acknowledges the Department of Physiology, Govt. General Degree College, Kharagpur II West Bengal, India. Dr. PSS acknowledges the Department of Chemistry, Govt. General Degree College, Kharagpur II West Bengal, India. Mr. RG acknowledges the Department of Botany, Govt. General Degree College, Kharagpur II West Bengal, India.

REFERENCES

Abbas, M., Saeed, F., Anjum, F.M., Muhammad, A., Tabussam, T., Muhammad, S., Bashir, A.I., Shahzad, H., Hafiz, A.R.S. 2017. Natural polyphenols: An overview. *International Journal of Food Properties*. 20(8), 1689–1699. http://doi.org/10.1080/10942912.2016.1220393.

Amengual, J. 2019. Bioactive properties of carotenoids in human health. *Nutrients*. 11(10), 2388. http://doi.org/10.3390/nu11102388

Anand, P., Bley, K. 2011. Topical capsaicin for pain management: Therapeutic potential and mechanisms of action of the new high-concentration capsaicin 8% patch. *British Journal of Anaesthesia*. 107(4), 490–502. http://doi.org/10.1093/bja/aer260

Ananthan, R., Subhash, K., Longvah, T. 2018. Assessment of nutrient composition and capsaicinoid content of some red chilies. *International Conference on Food and Nutrition Technology IPCBEE*. 72. http://doi.org/10.7763/IPCBEE.2014.V72.1

Antonioa, A.S., Wiedemanna, L.S.M., Veiga Junior, V.F. 2018. The genus *Capsicum*: A phytochemical review of bioactive secondary metabolites. *RSC Advances*. 8, 25767–25784.

Antonious, G.F. 2018. Capsaicinoids and vitamins in hot pepper and their role in disease therapy [Internet]. In *Capsaicin and Its Human Therapeutic Development*. InTech. http://doi.org/10.5772/intechopen.78243

Aqsa, A., Waqas, A., Nauman, K. 2021. Phytochemical constituents and biological properties of domesticated capsicum species: A review. *Bioactive Compounds in Health and Disease*. 4, 201–225. http://doi.org/10.31989/bchd.v4i9.837

Ayob, O., Hussain, P.R., Suradkar, P., Naqash, F., Rather, S.A., Joshi, Z.R.S., Azad, A.A. 2021. Evaluation of chemical composition and antioxidant activity of Himalayan Red chilli varieties. *LWT*. 146, 111413.

Bal, S., Sharangi, A.B., Upadhyay, T.K., Khan, F., Pandey, P., Siddiqui, S., Saeed, M., Lee, H.J., Yadav, D.K. 2022. Biomedical and antioxidant potentialities in chilli: Perspectives and way forward. *Molecules*. 27(19), 6380. http://doi.org/10.3390/molecules27196380

Basith, S., Cui, M., Hong, S., Choi, S. 2016. Harnessing the therapeutic potential of capsaicin and its analogues in pain and other diseases. *Molecules*. 21(8), 966. http://doi.org/10.3390/molecules21080966

Bevan, S., Szolcsanyi, J. 1990. Sensory neuron-specific actions of capsaicin: Mechanisms and applications. *Trends in Pharmacological Sciences*. 11, 330–333.

Bousquet, J., Czarlewski, W., Zuberbier, T. 2021. Spices to control COVID-19 symptoms: Yes, but not only . . . . *International Archives of Allergy and Immunology*. 182(6), 489–495. http://doi.org/10.1159/000513538.

Camara, B., Bardat, F., Dogbo, O., Brangeon, J., Monéger, R. 1983. Terpenoid metabolism in plastids: Isolation and biochemical characteristics of capsicum annuum chromoplasts. *Plant Physiology*. 73(1), 94–99. http://doi.org/10.1104/pp.73.1.94.

Camara, B., Bardat, F., Seye, A., D'Harlingue, A., Monéger, R. 1982. Terpenoid metabolism in plastids: Localization of alpha-tocopherol synthesis in capsicum chromoplasts. *Plant Physiology*. 70(5), 1562–1563. http://doi.org/10.1104/pp.70.5.1562

Capsaicin and Capsaicinoids. https://birdhousechillies.com/2019/03/12/capsaicin-and-capsaicinoids/

Cherney, E.C., Baran, P.S. 2001. Terpenoid-alkaloids: Their biosynthetic twist of fate and total synthesis. *Israel Journal of Chemistry*. 51(3–4), 391–405. http://doi.org/10.1002/ijch.201100005

Chilli (*Capsicum annum* L.). Available from: http://www.celkau.in/crops/vegetables/Chilli/chilli.aspx [Accessed on 02.06.2023].

Cortright, D.N., Szallasi, A. 2004. Biochemical pharmacology of the vanilloid receptor TRPV1. *European Journal of Biochemistry*. 271, 1814–1819. http://doi.org/10.1111/j.1432-1033.2004.04082.x

Devesa, I., Planells-Cases, R., Fernández-Ballester, G., González-Ros, J.M., Ferrer-Montiel, A., Fernández-Carvajal, A. 2011. Role of the transient receptor potential vanilloid 1 in inflammation and sepsis. *Journal of Inflammation Research*. 4, 67–81. https://doi.org/10.2147/JIR.S12978

Dubey, R.K., Singh, V., Upadhyay, G., Pandey, A.K., Prakash, D. 2015. Assessment of phytochemical composition and antioxidant potential in some indigenous chilli genotypes from North East India. *Food Chemistry*. 188, 119–125. http://doi.org/10.1016/j.foodchem.2015.04.088

Fan, M., Yuan, S., Li, L., Zheng, J., Zhao, D., Wang, C., Wang, H., Liu, X., Liu, J. 2023. Application of terpenoid compounds in food and pharmaceutical products. *Fermentation*. 9, 119. https://doi.org/10.3390/fermentation9020119

Fattori, V., Hohmann, M.S., Rossaneis, A.C., Felipe A Pinho-Ribeiro, F. A., Verri, W. A. 2016. Capsaicin: Current understanding of its mechanisms and therapy of pain and other pre-clinical and clinical uses. *Molecules*. 21(7), 844. https://doi.org/10.3390/molecules21070844

Fujiwake, H., Suzuki, T., Iwai, K. 1982. Intracellular distribution of enzymes and intermediates involved in biosynthesis of capsaicin and its analogues in Capsicum fruits. *Agricultural and Biological Chemistry*. 46, 2685–2689. http://doi.org/10.1271/bbb1961.46.2685

Ghosh, D., Firdaus, S.B., Mitra, E., Dey, M., Bandyopadhyay, D. 2012a. Protective effect of aqueous leaf extract of *Murraya koenigi* against lead induced oxidative stress in rat liver, heart and kidney: A dose response study. *Asian Journal of Pharmaceutical and Clinical Research*. 5(4), 54–58.

Ghosh, D., Mitra, E., Firdaus, S.B., Dey, M., Ghosh, A.K., Chattopadhyay, A., Bandyopadhya, D. 2012b. In vitro studies on the antioxidant potential of the aqueous extract of Curry leaves (*Murraya koenigii L.*) collected from different parts of the state of West Bengal. *Indian Journal of Physiology and Allied Sciences*. 66(3), 77–95.

Ghosh, S., Singha, P.S., Ghosh, D. 2017. Leaves of coriandrum sativum as an indigenous medicinal spice herb of India: A mini review. *International Journal of Pharmaceutical Sciences Review and Research*. 45(2), Article No. 20, 110–114.

Ghosh, S., Syamal, A.K., Ghosh, D. 2016. Medicines from chillies. *Asian Journal of Pharmaceutical and Clinical Research*. 9(5), 1–3.

Giuffrida, D., Dugo, P., Torre, G., Bignardi, C., Cavazza, A., Corradini, C., Dugo, G. 2013. Characterization of 12 Capsicum varieties by evaluation of their carotenoid profile and pungency determination. *FoodChemistry*. 140, 794–802.

Government of India. Geographical Indications. Journal No. 88. July 28, 2016. Available from: https://ipindia.gov.in/writereaddata/Portal/IPOJournal/1_322_1/Journal_88.pdf [Accessed on 11.06.2023].

Government of India. Geographical Indications. Journal No. 121. April 26, 2019. Available from: https://ipindia.gov.in/writereaddata/Portal/IPOJournal/1_4739_1/Journal_121.pdf [Accessed on 11.06.2023].

Government of India. Geographical Indications. Journal No. 141. August 31, 2020. Available from: https://ipindia.gov.in/writereaddata/Portal/Images/pdf/Journal_140.pdf.

Guerriero, G., Berni, R., Muñoz-Sanchez, J.A., Apone, F., Abdel-Salam, E.M., Qahtan, A.A., Alatar, A.A., Cantini, C., Cai, G., Hausman, J.F., Siddiqui, K.S., Hernández-Sotomayor, S.M.T., Faisal, M. 2018. Production of plant secondary metabolites: Examples, tips and suggestions for biotechnologists. *Genes (Basel)*. 9(6), 309. http://doi.org/10.3390/genes9060309

Handbook of Processing of Mizo Chilli. *National Institute of Food Technology Entrepreneurship and Management*. Ministry of Food Processing Industries. Government of India. Available from: https://niftem.ac.in/newsite/pmfme/wp-content/uploads/2022/08/mizochilliwriteup.pdf [Accessed on 11.06.2023].

https://issuu.com/infogitagged/docs/7_most_famous_chilli_pepper_in_india.docx/s/10788789 [Accessed on 11.06.2023].

https://mygardenlife.com/plant-library/chili-pepper-indian-jwala-capsicum-annuum [Accessed on 11.06.2023].

https://pragyanxetu.com/2022/08/18/dalle-khursani/ [Accessed on 11.06.2023].

https://pramoda.co.in/types-of-indian-chillies/ [Accessed on 11.06.2023].

https://sahasa.in/2021/06/29/khola-chilli-canacona-chilli/ [Accessed on 11.06.2023].

https://spiceitupp.com/kashmiri-chillies/ [Accessed on 02.06.2023].

https://www.gitagged.com/gunturchilli/blog/#:~:text=Guntur%20chilli%20is%20hot%20and,(11.98g%2F100g) [Accessed on 11.06.2023].

https://www.onmanorama.com/food/features/2021/04/07/indian-chillies-you-must-know-about.html [Accessed on 02.06.2023].

https://www.potsandpans.in/blogs/articles/byadagi-chilli-health-benefits-uses-and-important-facts [Accessed on 11.06.2023].

https://www.timesfoodie.com/nutritional-facts/birds-eye-chilli-kanthari-mulaku-health-benefits/88002587.cms [Accessed on 11.06.2023].

Hussein, R.A., El-Anssary, A.A. 2019. Plants secondary metabolites: The key drivers of the pharmacological actions of medicinal plants [Internet]. *Herbal Medicine*. Intech Open. http://doi.org/10.5772/intechopen.76139

Lee, J.J., Crosby, K.M., Pike, L.M., Yoo, K.S., Leskovar, D.I. 2005. Impact of genetic and environmental variation on development of flavonoids and carotenoids in pepper (*Capsicum* spp.). *Scientia Horticulturae.* 106, 341–352. http://doi.org/10.1016/j.scienta.2005.04.008

Lee, Y., Howard, L.R., Villalón, B. 1995. Flavonoids and antioxidant activity of fresh pepper (Capsicum annuum) cultivars. *Journal of Food Science.* 60(3), 473–476. https://doi.org/10.1111/j.1365-2621.1995.tb09806.x

Lu, M., Chen, C., Lan, Y., Xiao, J., Li, R., Huang, J., Huang, Q., Cao, Y., Ho, C.T. 2020. Capsaicin-the major bioactive ingredient of chili peppers: Bio-efficacy and delivery systems. *Food Function.* 11(4), 2848–2860. http://doi.org/10.1039/d0fo00351d

Luo, X.J., Peng, J., Li, Y.J. 2011. Recent advances in the study on capsaicinoids and capsinoids. *European Journal of Pharmacology.* 650, 1–7. http://doi.org/10.1016/j.ejphar.2010.09.074

Malakar, S., Sarkar, S., Kumar, N., Jaganmohan, R. 2018. Studies of biochemical characteristics and identification of active phyto-compounds of king chili (*Capsicum chinense* Jacq.) using GC-MS. *Journal of Pharmacognosy and Phytochemistry.* 7(3), 3100–3104.

Masyita, A., Mustika Sari, R., Dwi Astuti, A., Yasir, B., Rumata, N. R., Emran, T. B., Nainu, F., Simal-Gandra, J. 2022. Terpenes and terpenoids as main bioactive compounds of essential oils, their roles in human health and potential application as natural food preservatives. *Food Chemistry.* X(13), 100217. https://doi.org/10.1016/j.fochx.2022.100217

Mateos, R.M., Jiménez, A., Román, P., Romojaro, F., Bacarizo, S., Leterrier, M., Gómez, M., Sevilla, F., Del Río, L.A., Corpas, F.J., Palma, J.M. 2013. Antioxidant systems from pepper (Capsicum annuum L.): Involvement in the response to temperature changes in ripe fruits. *International Journal of Molecular Sciences.* 14(5), 9556–9580. http://doi.org/10.3390/ijms14059556

Medina-Juárez, L.Á., Molina-Quijada, D.M., Del-Toro-Sánchez, C.L., González-Aguilar, G.A., Gámez-Meza, N. 2012. Antioxidant activity of peppers (*Capsicum annuum* L.) extracts and characterization of their phenolic constituents. *Interciencia.* 37, 588–593.

Mehmet, B., Metin, Y., Gulhan, A., Omer, T., Oruc, A. 2015. Effect of capsaicin on transcription factor in 3T3-L1 cell line. *Eastern Journal of Medicine.* 20, 34–45.

Melendez-Martinez, A.J., Stinco, C.M., Mapelli-Brahm, P. 2019. Skin carotenoids in public health and nutri-cosmetics: The emerging roles and applications of the UV radiation-absorbing colourless carotenoids phytoene and phytofluene. *Nutrients.* 11, 1093. http://doi.org/10.3390/nu11051093

Michael, P.M., Lalbiaknunga, J., Lalnunmawia, F. 2020. Mizo chilli (*Capsicum frutescens*): A potential source of capsaicin with broad-spectrum ethno pharmacological applications. *JPP.* 9(5), 670–672. https://doi.org/10.22271/phyto.2020.v9.i5j.12306

Mosmann, T. 1983. Rapid colorimetric assay for cellular growth and survival: Application to proliferation and cytotoxicity assays. *Journal of Immunological Methods.* 65, 55–63. http://doi.org/10.1016/0022-1759(83)90303-4

Nair, S., Nagar, R., Gupta, R. 1998. Antioxidant phenolics and flavonoids in common Indian foods. *Journal of the Association of Physicians of India.* 46(8), 708–710.

Ncube, B., Van Staden, J. 2015. Tilting plant metabolism for improved metabolite biosynthesis and enhanced human benefit. *Molecules.* 20, 12698–12731. http://doi.org/10.3390/molecules200712698

Nisar, N.L., Li, S. Lu, S., Khin, N.C., Pogson, B.J. 2015. Carotenoid metabolism in plants. *Molecular Plant Pathology.* 8, 68–82.

Nugroho, L.H. 2016. Red pepper (*Capsicum spp.*) fruit: A model for the study of secondary metabolite product distribution and its management. *AIP Conference Proceedings.* 1744, 020034.

Okunlola, G.O., Olatunji, O.A., Akinwale, R.O., Tariq, A., Adelusi, A.A. 2017. Physiological response of the three most cultivated pepper species (*Capsicum* spp.) in Africa to drought stress imposed at three stages of growth and development. *Scientia Horticulturae.* 224, 198–205.

Oney-Montalvo, J.E., Avilés-Betanzos, K.A., Ramírez-Rivera, E.D.J., Ramírez-Sucre, M.O., Rodríguez-Buenfil, I.M. 2020. Polyphenols content in Capsicum chinense Fruits at different harvest times and their correlation with the antioxidant activity. *Plants.* 9, 1394. https://doi.org/10.3390/plants9101394.

Park, S.Y., Kim, J.Y., Lee, S.M., Jun, C.H., Cho, S.B., Park, C.H., Joo, Y.E., Kim, H.S., Choi, S.K., Rew, J.S. 2014. Capsaicin induces apoptosis and modulates MAPK signaling in human gastric cancer cells. *Molecular Medicine Reports.* 9, 499–502. http://doi.org/10.3892/mmr.2013.1849

Pereira, A.G., Otero, P., Echave, J., Carreira-Casais, A., Chamorro, F., Collazo, N., Jaboui, A., Lourenço-Lopes, C., Simal-Gandara, J., Prieto, M.A. 2021. Xanthophylls from the sea: Algae as source of bioactive carotenoids. *Marine Drugs.* 19(4), 188. https://doi.org/10.3390/md19040188

Perucka, I., Materska, M. 2007. Antioxidant vitamin contents of Capsicum annum fruit extract as affected by processing and varietal effects. *Acta Scientiarum Polonorum, Technologia Alimentaria.* 6(4), 67–74.

Phanindra, N.M., Mounika, R.N. 2020. Quality assessment of selected commercial brand of chilli powder in Andhrapradesh region. *IOSR Journal of Pharmacy.* 10(1), 47–63.

Pratiwi, R., Hanafi, M., Artanti, N., Pratiwi, R. 2018. Bioactivity of antibacterial compounds produced by endophytic actinomycetes from neesia altissima. *Journal of Tropical Life Science.* 8, 37–42. http://doi.org/10.11594/jtls.08.01.07

Puttaswamy, H., Gowtham, H.G., Ojha, M.D., Yadav, A., Choudhir, G., Raguraman, V., Kongkham, B., Selvaraju, K., Shareef, S., Gehlot, P., Ahamed, F., Chauhan, L. 2020. *In silico* studies evidenced the role of structurally diverse plant secondary metabolites in reducing SARS-CoV-2 pathogenesis. *Scientific Reports.* 10(1), 20584. http://doi.org/10.1038/s41598-020-77602-0

Reddy, G.D.R., Prakash, A.B., Vardan, N.H., Bhattacharyya, S., Bhati, K. 2022. Influence of secondary metabolites against biotic stress in Chilli. *Pharma Innovation.* 11(5), 2419–2422.

Reyes-Escogido Mde, L., Gonzalez-Mondragon, E.G., Vazquez-Tzompantzi, E. 2011. Chemical and pharmacological aspects of capsaicin. *Molecules.* 16(2), 1253–1270. http://doi.org/10.3390/molecules16021253

Rodriguez-Mateos, A., Vauzour, D., Krueger, C.G., Shanmuganayagam, D., Reed, J., Calani, L., Mena, P., Del Rio D., Crozier, A. 2014. Bioavailability, bioactivity and impact on health of dietary flavonoids and related compounds: An update. *Archives of Toxicology.* 88, 1803–1853.

Ruanma, K., Shank, L., Chairote, G. 2010. Phenolic content and antioxidant properties of green chilli paste and its ingredients. *Maejo International Journal of Science and Technology.* 4, 193–200.

Salehi, B., Mishra, A.P., Shukla, I., Sharifi-Rad, M., Contreras, M., Segura-Carretero, A., Fathi, H., Nasrabadi, N.N., Kobarfard, F., Sharifi-Rad, J. 2018. Thymol, thyme, and other plant sources: Health and potential uses. *Phytotherapy Research.* 32, 1688–1706. http://doi.org/10.1002/ptr.6109

Sanati, S., Razavi, B.M., Hosseinzadeh, H. 2018. A review of the effects of Capsicum annuum L. and its constituent, capsaicin, in metabolic syndrome. *Iranian Journal of Basic Medical Sciences.* 21, 439–448. http://doi.org/10.22038/IJBMS.2018.25200.6238

Singh, B., Sharma, R.A. 2015. Plant terpenes: Defense responses, phylogenetic analysis, regulation and clinical applications. *3 Biotech.* 5(2), 129–151. http://doi.org/10.1007/s13205-014-0220-2

Singh, S., Uddin, M., Khan, M.M.A., Singh, S., Chishti, A.S., Bhat, U.H. 2022. Therapeutic properties of Capsaicin: A medicinally important bio-active constituent of chilli pepper. *Asian Journal of Pharmaceutical and Clinical Research.* 15(7), 47–58. http://doi.org/10.22159/Ajpcr.2022.V15i7.44405

Takemura, M., Maoka, T., Misawa, N. 2015. Biosynthetic routes of hydroxylated carotenoids (xanthophylls) in Marchantia polymorpha, and production of novel and rare xanthophylls through pathway engineering in Escherichia coli. *Planta.* 241(3), 699–710. http://doi.org/10.1007/s00425-014-2213-0

Tang, Y., Zhang, G., Yang, T., Yang, S., Aisimutuola, P., Wang, B., Li, N., Wang, J., Yu, Q. 2021. Biochemical variances through metabolomic profile analysis of Capsicum chinense Jacq. during fruit development. *Folia Horticulturae.* 33(1), 17–26. https://doi.org/10.2478/fhort-2021-0001

Teoh, E.S. 2015. Secondary metabolites of plants. *Medicinal Orchids of Asia.* 5, 59–73. http://doi.org/10.1007/978-3-319-24274-3_5

Tominaga, M., Tominaga, T. 2005. Structure and function of TRPV1. *Pflugers Archiv.* 451, 143–150. http://doi.org/10.1007/s00424-005-1457-8

Ullah, A., Munir, S., Badshah, S.L., Khan, N., Ghani, L., Poulson, B.G., Emwas, A.H., Jaremko, M. 2020. Important flavonoids and their role as a therapeutic agent. *Molecules.* 25(22), 5243. http://doi.org/10.3390/molecules25225243

Umashanka, D.D. 2020. Plant secondary metabolites as potential usage in regenerative medicine. *JPHYTO.* 9(4), 270–273.

Voahanginirina, R. 2018. Influence of the use of Capsaicin on the storage of Rice Grain. *Open Access Library Journal.* 5, 1–13. http://doi.org/10.4236/oalib.1104833

Waksmundzka-Hajnos, M., Sherma, J., Kowalska, T. 2008. *Thin layer chromatography in phytochemistry.* CRC Press, Boca Raton, p. 896. https://doi.org/10.1201/9781420046786

Wink, M. 2015. Modes of action of herbal medicines and plant secondary metabolites. *Medicines (Basel).* 2(3), 251–286. http://doi.org/10.3390/medicines2030251

Yang, W., Chen, X., Li, Y., Guo, S., Wang, Z., Yu, X. 2020. Advances in pharmacological activities of terpenoids. *Natural Product Communications.* 15(3). http://doi.org/10.1177/1934578X20903555

10 Recent Trends in Pepper-Derived Products for Food Matrices

Somnath Basak and Rekha S. Singhal

CONTENTS

10.1 INTRODUCTION

Peppers are an important fruit (i.e. berries) of the *Capsicum* plant which belong to the *Solanaceae* family. Peppers consist of a variety of species, with the most widely cultivated being *Capsicum annuum, Capsicum frutescens*, and *Capsicum chinense. Capsicum annum* L. is one of the most investigated species of the peppers (Parvez, 2017). The word *pepper* derives its name from pungency. However, there are sweeter varieties of pepper also known as bell peppers or sweet peppers, also properly known as *Capsicum* in Indian English. The word chilli is usually employed for hotter and more pungent varieties. While strong flavor and intense color are the prominent reasons for use of peppers in food matrices, the strong pungency often limits its applications. Therefore, newer and novel products are being developed for the same. Peppers are known for several health benefits due to the presence of several bioactive compounds such as carotenoids, capsaicin, ascorbic acid, and several other antioxidant compounds (Saleh et al., 2018). Apart from nutraceutical compounds, peppers are also storehouses of several vitamins and minerals. The presence of capsaicinoids is responsible for therapeutic properties of peppers. Capsaicin has been found to alleviate arthritis and diabetic neuropathy. Peppers have chemopreventive potential along with antioxidant, anti-inflammatory, cardioprotective, and hypocholesterolemic effects (Srinivasan, 2016). Therefore, pepper has found its place in traditional medicine.

Peppers are highly perishable and require adequate processing to achieve shelf stability. Some of the conventional processing treatments involve blanching, osmotic dehydration, steaming, canning, roasting, and stir-frying, which have a profound effect on the physical and nutritional profile of peppers. Capsanthin, capsorubin, and cryptocapsin are the pigment compounds responsible for the red color during maturation of peppers (Gomes et al., 2014). The yellow orange colors arise from the zeaxanthin, β-carotene and cryptoxanthin (Melgar-Lalanne et al., 2017). These natural pigments can be an excellent alternative to the synthetic food colorants and additives which are added to foods to achieve color and flavor. The pepper-based natural pigments are lipophilic and cannot

DOI: 10.1201/9781003378259-10"

be employed directly in aqueous foods. They are also highly labile to oxidation and light, which propels academia and industry to devise solutions to overcome these limitations.

Newer strategies of microencapsulation of *Capsicum* oleoresin have also been investigated in recent times. Newer and novel wall materials are being explored for their ability to encapsulate pepper oleoresins and essential oils. da Silva Anthero et al. (2022) incorporated malt along with modified corn starch and gum arabic as encapsulating agents for the spray-drying of *Capsicum* oleoresin. The developed oleoresin microcapsules reduced weight gain in mice when they were fed a high-fat diet. Interestingly, the addition of red pepper extract powder into the yogurt maintained the initial counts of lactic acid bacteria better than the unfortified yogurt (Šeregelj et al., 2019. Pepper-based pigments and oleoresins can be used as colorants and active ingredients in cosmetic formulations. A novel Carbopol-based gel with 50% capsaicin released capsaicin within 52 hours with improved antibacterial and antifungal activities (Goci et al., 2021).

Green peppers are usually consumed fresh, while red pepper is usually consumed fresh or added as a spice powder in several cuisines. However, oleoresins and essential oils have gained significant importance in recent times due to the higher quality and ease of incorporation in several food formulations. However, the pungency of capsaicin is still a limitation. Several newer fractionation technologies are being researched times for selective removal of capsaicinoids while maintaining a high yield. This chapter discusses the recent trends and diverse products of pepper and their preparation strategies. The products can help in the formulation of nutraceutical products and can also be used as food ingredients for achieving flavor and color in a food system.

10.2 OLEORESINS AND ESSENTIAL OILS

Pepper oleoresins and essential oils are an important product category segment in peppers, which are incorporated in food formulations as functional ingredients. The oleoresins and essential oils are also added in edible films and coatings to improve the shelf life of fresh produce. The reduction of salts in foods has been one of the major ongoing efforts of academia and industry. The salt levels can be decreased in foods if the perception of saltiness is increased. The pungent Sichuan pepper oleoresin (0.41 g/L) enhanced the perception of saltiness at low to strong NaCl solutions, which helped in the reduction of added NaCl by 38.61 and 39.06% for the hypersensitive and semisensitive groups, respectively (Zhang et al., 2020).

Supercritical carbon dioxide ($scCO_2$) is one of the most used methods for the extraction of oleoresins from peppers (Figure 10.1). The conventional solvent extraction is being intensified by microwave, ultrasound, and ohmic heating, leading to lower solvent requirements. Pretreatments of the samples by a pulsed electric field can result in membrane permeabilization, which can affect the heat and mass transfer processes, thereby positively affecting extractability of oleoresins and essential oils (Melgar-Lalanne et al., 2017). The $scCO_2$ extraction has been reported to be one of the most economical methods for the extraction of oleoresins from malagueta peppers at al large scale of production (*Capsicum frutescens*) (Aguiar et al., 2018). The traditional water extraction is performed by maceration or hydrodistillation. The aqueous extraction can be intensified by cavitation, pressurized hot water extraction, or pulsed electric field (Melgar-Lalanne et al., 2017). Cold pressing of the *Capsicum* spp. and maceration of the chilli powder in vegetable oils are other traditional methods of achieving functional extracts (Melgar-Lalanne et al., 2017).

Uquiche et al. (2022) developed a densification treatment using extrusion of red pepper to improve the extractability and yield of red pepper oleoresin. On the contrary, the pretreatment of extrusion of red pepper had a significant negative effect on the yield, carotenoid content, and antioxidant capacity of the oleoresin. The thermal and shear degradation induced by extrusion decreased the extractable color and antioxidant capacity of peppers. The increase in bulk density of red pepper due to extrusion reduced the porosity and increased the resistance of mass transfer from the solid raw material to the $scCO_2$ phase. Capsanthin-rich avocado oil was obtained by simultaneous $scCO_2$ extraction (400 bar|50°C) of avocado and red bell pepper (Barros et al., 2016). The oil had the potential to be

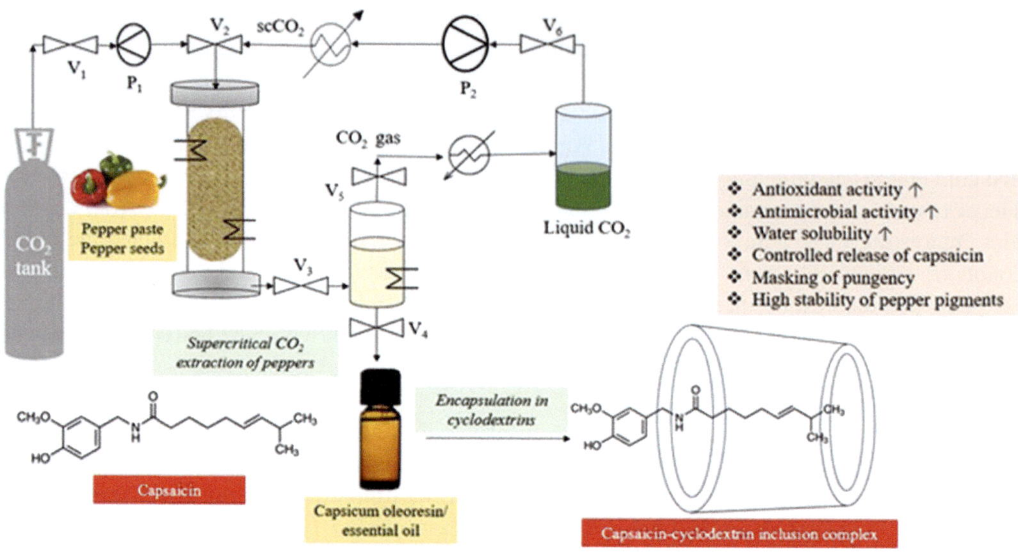

FIGURE 10.1   The process of scCO₂ extraction of pepper essential oil and its incorporation in cyclodextrin and its potential positive implications on the functional, physical, and sensory properties of the extract.

used as a food colorant with biological activity. Due to the use of avocado oil as a co-solvent, a high recovery (50%) was observed. The freeze-dried avocado pulp containing 68% lipids and the red bell pepper pulp containing 625 µg/g capsanthin were packed separately in a single packed bed reactor. The scCO₂ first passed through the avocado pulp, followed by the flow of the lipid-rich extract into the pepper pulp. The red bell pepper pulp extracted solely through scCO₂ exhibited only 15% recovery of capsanthin. Araus et al. (2012) used triolein as a co-solvent during scCO₂ extraction of capsanthin. The solubility of triolein is nearly 100 times the solubility of capsanthin in scCO₂. This helped in bridging the solubility gap, and hence, the dissolution of capsanthin was increased three-fold irrespective of the working conditions of temperature and pressure. The authors attributed this trend to the nonpolar interactions between the 18-C fatty acid chains in triolein and the 22-C nonpolar core of the carotenoids.

The addition of ethanol as a co-modifier and tapping solvent facilitated the selective extraction of pigments from hot varieties of peppers, with minimal extraction of chlorophyll. The additional precipitation step after the scCO₂ extraction differentially isolated the pigments from the capsaicinoids. Therefore, irrespective of the hotness or pungency level of peppers, the pigments can be selectively fractionated from the peppers with minimal levels of chlorophyll and capsaicinoids using ethanol as a modifier in scCO₂ extraction followed by a precipitation step. Apart from scCO₂, subcritical dimethyl ether was found to give comparable yields to scCO₂ (Catchpole et al., 2003).

Other extraction techniques such as microwave-assisted extraction have been used to intensify the solvent extraction of bioactive compounds from different matrices. Although microwave extraction results in a shorter time of extraction with lower amounts of the solvent, the dielectric properties of the solvent and the pepper significantly affect the extractability of capsaicinoids. A polar solvent shows better extractability in microwave-assisted extraction. Gogus et al. (2015) extracted the volatiles of fresh and red commercial sweet red pepper pastes using microwave-assisted extraction in water and n-hexane. Despite n-hexane being a nonpolar solvent, the highest number of volatile compounds was obtained. Browning was, however, observed in both the pastes. Interestingly, the yield using n-hexane was twofold higher than that extracted in water. Ultrasound has also been coupled with scCO₂ extraction in several studies (Dias et al., 2016; Santos et al., 2015). The coupling of the two techniques increased the yield and rate of extraction of capsaicin and phenolics from dedo de moça pepper (*Capsicum baccatum* L. var. *pendulum*) (Dias et al., 2016). Ultrasound-induced disturbances in the pepper matrix facilitate the release of extractable matter from pepper and improve the yield of capsaicinoids (Santos et al., 2015).

Due to the antimicrobial activity of essential oils, the edible films and coatings are loaded with essential oils to inhibit the growth of spoilage bacteria and pathogens, which helps in improving the shelf life of fresh produce. Wang et al. (2022a) incorporated Sichuan pepper (*Zanthoxylum armatum* DC.) essential oil in starch films. The addition of essential oil at 2.5% positively increased the barrier to water vapor permeability and inhibited *S. aureus* and *E. coli*. Such films with a high light transmittance and antimicrobial activity can be used as an edible packaging in high-fat products. These essential oil–incorporated films can be a great alternative to plastic-based food packaging.

The addition of essential oils into cooking oils can have a dual positive impact, i.e. the quality improvement of frying oil and the food product which is being processed or fried. Salazar et al. (2012) established that the addition of piquin pepper (*Capsicum annuum* L. var. *Aviculare*) essential oil mitigated the formation of acrylamide in potato and tortilla chips during frying in soybean oil. This also had a positive effect on the sensory aspects of the chips. Yang et al. (2010) reported that the addition of red pepper (*Capsicum annuum*) seed oil enhanced the thermal oxidative stability of vegetable oil during frying. To summarize, pepper oleoresins and essential oils have a wide range of functional properties, which can be incorporated in edible films, frying oils, and food matrices.

10.3 INCLUSION COMPLEXES

As already discussed, pepper oleoresins and essential oils can be encapsulated as guest molecules in a cyclodextrin inclusion complex. This has a positive influence on the physicochemical properties (Figure 10.2). Teixeira et al. (2013) synthesized an inclusion complex using hydroxypropyl β-cyclodextrin and pepper oleoresin, achieving an entrapment efficiency of 78.3% at a 1:1 stoichiometry. The pepper oleoresin- cyclodextrin nanocapsule exhibited greater solubility, antioxidant capacity, and antimicrobial activity against *Escherichia coli* K12 and *Salmonella enterica* serovar Typhimurium LT2. The nanoencapsulated pepper oleoresin could inhibit *Salmonella* at a lower concentration than the free extract. The kneading method is the most popular method for producing nano-sized particles. The displacement of water molecules from the cyclodextrin cavity due to the hydrophobic interactions and the increase in van der Waals forces and hydrogen bonds due to interactions between the guest molecule and the host cavity drive the formation of the inclusion complex (Teixeira et al., 2013). Pestovsky and Martínez-Antonio (2018) immobilized inclusion complexes of capsaicin with β-cyclodextrin on the surface of gold nanoparticles. It was interesting to note that the immobilization of capsaicin on the Au nanoparticles without the presence of β-cyclodextrin did not result in a prolonged release, indicating that the encapsulation of capsaicin in the β-cyclodextrin cavity played an important role in the immobilization process. When the Au nanoparticles were replaced with chitosan nanoparticles in the presence of β-cyclodextrin, the release kinetics were less pronounced in comparison to the gold nanoparticles.

Several pigments are usually hydrophobic, and the fabrication of the inclusion complexes can be an efficient method for improving the solubility in water. Petito et al. (2016) complexed red bell pepper carotenoids with 2-hydroxypropyl-β-cyclodextrin using ultrasonic homogenization and observed the enhancement of solubility by up to 660 times. This can help in the incorporation of fat-soluble pigments in water-based food matrices. Domínguez-Cañedo and Guevera (2011) encapsulated habanero chilli (*Capsicum chinense*) oleoresin in β-cyclodextrin using the precipitation method, and observed improved stability of the carotenoid content and color of the encapsulated powder.

Venkataswamy et al. (2019) developed a metal organic framework using γ-cyclodextrin and entrapped capsaicin in it. This helped in overcoming the pungency and solubility issues of the capsaicinoids. The vapor diffusion of ethanol into the synthesis solution containing γ-cyclodextrin,

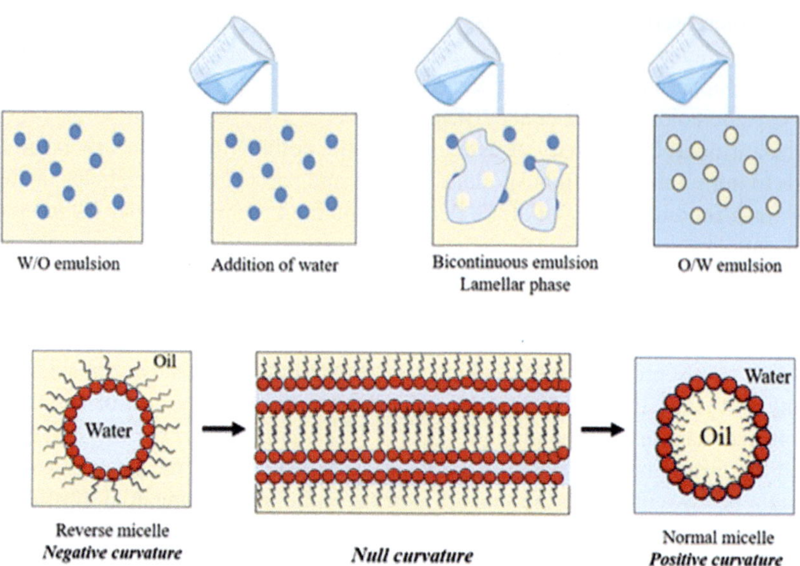

FIGURE 10.2 The emulsion phase inversion method for the fabrication of O/W emulsions incorporating pepper essential oils or oleoresins.

KOH, and capsaicin helped in the synthesis of capsaicin-encapsulated metallic organic frameworks. Instead of the conventional kneading method, Li et al. (2022) developed a process in which *Zanthoxylum bungeanum* resin microcapsules were encapsulated in β-cyclodextrin using the ultrasonic-assisted molecular embedding technique. The core:wall ratio of 1:3.47 (w/w) was optimized, and the ultrasonic treatment of 240 W for 11.6 min was used for the fabrication. This process of encapsulation could mask the bitterness and astringency of the resin and also improved the thermal stability. An encapsulation efficiency of 82% was achieved. The type of cyclodextrin can also affect the dissolution and release behavior of the bioactive compounds. The α-cyclodextrin-based inclusion complex exhibited greater solubility than the β- and γ-cyclodextrin-based complex. These pepper-based inclusion complexes can be incorporated in food matrices such as salad dressings and mayonnaise to achieve a desirable flavor profile. For instance, Gomes et al. (2014) prepared inclusion complexes of red bell pepper pigments with β-cyclodextrin for the potential application as a natural colorant in yogurt. Two different approaches, namely magnetic stirring and ultrasonic homogenization, were used. Ultrasonic homogenization was found to be a superior method for better encapsulation of the bell pepper pigment. When incorporated in yogurt, the color indices were much more stable than the crude red bell pepper extract. This helped in simulating the papaya-like flavor in yogurt. The stability of the colors induced by the inclusion complexes was also reported by Lobo et al. (2018). The authors prepared inclusion complexes using β-cyclodextrin and yellow pepper pigment (2:1). The complexes were incorporated in isotonic drinks, which were subjected to illumination (1400 lx) at ambient conditions for 21 days. The inclusion of the yellow bell pepper pigment can be a great substitute for synthetic food colors. Protection against photodegradation and aerobic oxidation of the pigments can be easily facilitated by the encapsulation of these natural pepper-based pigments in cyclodextrins.

Inclusion of oleoresins into cyclodextrins also stabilizes their functional properties (Table 10.1). Two different approaches of freeze-drying and kneading were explored. Encapsulation efficiencies of 79.3% and 90.2% were observed for freeze-dried and kneaded inclusion complexes, respectively (Ozdemir et al., 2018). The inclusion complexes fabricated by kneading exhibited greater inhibitory activity

TABLE 10.1 Recent Studies Undertaken on the Formation of Inclusion Complexes of Capsaicin for Improved Delivery of Capsaicin

Pepper variety	Conditions of the formation of inclusion complex	Key conclusions	Reference
Sichuan pepper (*Zanthoxylum bungeanum* Maxim)	Ultrasonic-assisted molecular embedding (240 W|11.6 min) β-cyclodextrin:pepper resin = 3.47:1 (w/w)	• Encapsulation efficiency of 81.94% • Enhancement of heat stability of Sichuan pepper resin • Masking of the bitterness and pungency achieved by the formation of inclusion complex • Smooth and surface spherical particles obtained	Li et al. (2022)
—	Capsaicin loading onto high-amylose corn starch (HACS) Capsaicin content: 44% Capsaicin:HACS = 1:10 (w/w)	• Nano-sized delivery of capsaicin in helical V-amylose architectures • High level of crystallinity observed • Swelling of the starch molecules and the leaching of the capsaicin from the host HACS molecule • A higher particle surface diameter acts as a driving force for sustained release. • A gradual and sustained release of capsaicin in the stomach, up to 28% • A sustained release of ~45% in the intestine • A higher capsaicin bioaccessibility	Isaschar-Ovdat et al. (2021)
Yellow bell pepper (*Capsicum annuum* L.)	Ultrasonic homogenization (50 W|15 min) and kneading (20 min) Extract: β-cyclodextrin = 1:2	• Lutein, zeaxanthin, α-cryptoxanthin, α-carotene and β-carotene are the major pigments. • Inclusion efficiency of 57% • High stability of the yellow color in the isotonic beverage • A great substitute to the synthetic colors used in food matrices	Lobo et al. (2018)
Red bell pepper (*Capsicum annuum* L.)	Ultrasonic homogenization 100 W|5 min) Extract: 2-hydroxypropyl β-cyclodextrin = 1:4, 1:6,1:8, 1:10	• Increase in solubility of the carotenoids in water by more than 660 times • Modified cyclodextrins have less susceptibility to aggregation • Higher mean particle diameter compared to the control	Petito et al. (2016)

against *Listeria monocytogenes* than the inclusion complex produced by freeze-drying. Hence, these inclusion complexes can be a functional additive that can help in inducing antioxidant activity and inhibition of foodborne pathogens. Rakmai et al. (2017) observed that the inclusion complex of the encapsulated pepper essential oil in hydroxypropyl-β-cyclodextrin had fourfold higher antimicrobial activity against *Staphylococcus aureus* and *Escherichia coli* in comparison to the free pepper essential oil. The active components of the oil were also protected against the effect of light. The increased water solubility enhances the accessibility of the essential oils to the primitive sites of the cell, thereby improving the antimicrobial activity. To summarize, the inclusion complexes are a great tool for improving the stability and solubility of bioactive compounds and pigments in pepper essential oils and oleoresins.

10.4 ESSENTIAL OIL–BASED EMULSIONS

Another approach for improving the bioavailability of fat-soluble bioactives is the development of emulsions to encapsulate the bioactive of interest. These emulsion-based systems are important for the formulation of pain balms, analgesic creams, chest rubs, and face serums. The thermodynamically stable microemulsions can be a great vehicle for the delivery of capsaicin and other phenolic compounds, which protect against UV-induced damage and counteract free radicals. This can be a potential strategy in the formulation of cosmetic formulations targeted for antiaging and skin care (Baenas et al., 2019). Emulsion-based gels or emulgels are being developed for sustained delivery of capsaicin. Phytantriol- and glycerol monooleate–based cubosomes were successful in transdermal and targeted delivery of capsaicin, which was stable to high temperature and strong light (Peng et al., 2015). An oil-in-water (O/W) nanoemulsion of olive oil containing capsaicin was formulated into a topical cream and a gel with enhanced analgesic and anti-inflammatory effects (Ghiasi et al., 2019). Interestingly, the emulgel outperformed the emulsion-based cream in terms of transdermal delivery of capsaicin.

The two major components for encapsulation of fat-soluble bioactive into the micro- or nanodroplets are the surfactant and oil. Apart from food, these emulsion-based systems also have several cosmetic applications. Hien et al. (2021) loaded pepper essential oil into a nanoemulsion by an emulsion phase inversion (EPI) method. A 10% (v/v) of essential oil and a surfactant (Tween 80) oil ratio of 2 were found to be the most suitable for the development of the nanoemulsion. The EPI method for the fabrication of oil-in-water nanoemulsion involves the conversion of the water-in-oil emulsion to the oil-in-water phase. The technique majorly involves the addition of a nonionic surfactant to the essential oil phase followed by addition of water (the continuous phase) in small amounts. Since water is in lower amounts, the emulsion is a water-in-oil (W/O) system with an inverted micelle system. An increase in the addition of water converts the W/O emulsion to O/W emulsion and decreases the surfactant concentration. This converts the geometric packing to a bilayer gel system (Figure 10.2). The O/W emulsion prepared by the EPI method had a homogeneous and low droplet size (<100 nm) (Hien et al., 2021). The nanoemulsions could be stabilized only under frozen storage (-15°C for 24 h followed by thawing at room temperature for a period of 5 weeks. The volatile compounds were significantly retained. The stability of the phytochemicals in the nanoemulsion manifested the capacity of the formulation to confer antimicrobial properties and inhibit oxidative rancidity in fats. The bioaccessibility of carotenoids from yellow peppers was explored by incorporating them in nanoemulsions. The type of oil also affected the lipid digestibility, microstructural properties, and carotenoid bioaccessibility of the excipient emulsions (Liu et al., 2015). The bioaccessibility increased in the order: indigestible orange oil < medium-chain triglycerides < long-chain triglycerides. The microfluidization technique was the most efficient method for the preparation of the emulsions. The emulsions prepared in orange juice had the highest mean particle diameters compared to the triglycerides. Orange oil is a relatively polar lipid, and diffusion of substantial amounts of oil can occur from small droplets to large droplets. This diffusion increases the size of the droplets and enhances the coalescence of the oil droplets, which was evidenced from visual appearance of the emulsions. Jimenez et al. (2018) established that nanoemulsions of pepper essential oil prepared by ultrasound and high-pressure homogenization had antimicrobial properties against *Listeria monocytogenes* and *Escherichia coli*. Interestingly, the emulsions exhibited higher physicochemical stability and antimicrobial activity than essential oils. These emulsion-based systems have been reported to have positive health benefits. Red bell pepper extract was solubilized in soybean oil and water using soy lecithin as an emulsifier and a range of wall materials. The maximum encapsulation was observed for bovine gelatin (64%), whey protein isolate (57.6%), whey protein concentrate (56.6%), and calcium caseinate (54%). The emulsions were freeze-dried to yield a powder which had intense color and were stable to sedimentation for 48 hours (Petito et al., 2022). This improved the dispersibility of carotenoid in water. Incorporating essential oils into nano-emulsions can improve the applicability of hydrophobic and oil-soluble pigments in water-based food matrices. The high pH stability of the emulsions was beneficial for their incorporation in soy sauce and cooking wine (Shi et al., 2022). The subsequent section discusses the essential oil powders, which are mainly produced by the process of spray-drying or freeze-drying.

10.5 ESSENTIAL OIL–BASED POWDER

These essential oil powders have the potential to act as natural preservatives in food matrices. The pepper essential oil powders have been reported to exhibit antibacterial activity against *Staphylococcus aureus*, *Bacillus subtilis*, *Listeria monocytogenes*, and *Listeria innocua* (Pereira et al., 2020). The O/W emulsions of essential oils are usually dried to convert them to a free-flowing powder by the process of spray-drying. Pereira et al. (2019) optimized the process of the encapsulation of pink pepper essential oil. The O/W emulsions were stabilized using two different approaches, one using soy protein isolate as the single layer and the other one using a protein/pectin double layer. These emulsions were spray-dried at an inlet temperature of 140°C, outlet temperature of 98°C, and emulsion feed rate of 2 mL/min. The double-layer emulsions were much less cohesive, retained greater oil, adsorbed less water, and exhibited higher wetting time in water due to less hygroscopicity. The addition of pectin in the protective layer prevented agglomeration of the resultant microcapsule powders and facilitated higher flowability. Interestingly, the powders were spherical, with no presence of pores. The absence of pores preserved the volatile compounds which were encapsulated in the microcapsules. No cracks were observed, which led to a lower permeability to gases, thereby preserving the quality of the essential oil. Sichuan pepper essential oil was encapsulated in quinoa protein isolate-gum arabic coacervates, which imparted great thermal and ionic stability. This is also beneficial for a controlled and slower release of essential oil, which helps in greater absorption during oral and gastric digestion. Therefore, it can be summarized that encapsulation of pepper essential oils in powders and microcapsules can help in imparting improved functional properties and a controlled release of the essential oil, apart from protection against harsh conditions of gastric digestion and food processing (Chen et al., 2023).

10.6 PEPPER-BASED BEVERAGE

With the increase in the demand of functional beverages, pepper-based beverages can be a good functional plant-based beverage which can be used to tackle metabolic diseases due to the presence of polyphenols and other functional compounds. An interesting clinical study was undertaken by Zanzer et al. (2018) in which the effect of a pepper-based beverage on the postprandial glycemia and overall gastrointestinal well-being of the individuals was investigated. A randomized crossover intervention study was performed on 16 healthy subjects (10 men and 6 women). While the pepper-based beverage did not have any effect on insulin sensitivity and β-cell function, nor any changes in the glycemic response and gut and thyroid hormones, it lowered hunger and the desire to eat. No well-defined mechanism could be identified for the appetite-reducing ability of the pepper-based beverage. The oral administration of sweet pepper juice also had a significant protective effect on

UV-B induced skin and DNA damage (Truong et al., 2022). The downregulation of the expression of pro-inflammatory cytokines such as interleukin (IL)-17, IL-23, IL-1β, and cyclooxygenase-2 was facilitated by the inhibition of the NF-κB pathway. Oral supplementation of the juices enhanced the expression of the primary antioxidant enzymes such as catalase, superoxide dismutase-2, and glutathione peroxidase. The UV-B induced photoaging of the skin was reduced due to regulation of the genes involved in melloproteinase-2, 3, and 9 and collagen type 1 α 1.

A randomized control study performed by Nagasukeerthi et al. (2017) established the role of bell pepper juice in the management of type 2 diabetes mellitus, along with integrated yoga therapy. Pepper juice has also been reported to be helpful in reducing weight gain of mice when they were placed on a high-fat diet (Kim et al., 2015). Newer and novel nonthermal treatments such as pulsed electric field (PEF) can help in improving the yield and properties of the pepper juice. A higher turbidity and viscosity along with a lower pH and electrical conductivity of the juice was observed in the juice expelled from the PEF-treated bell pepper (Rybak et al., 2020). A higher extraction of ascorbic acid was also observed. The pretreatment of 1 kJ/kg resulted in a lower hygroscopicity of the spray-dried powder in comparison to the untreated control.

Pepper juice has been incorporated in several food matrices such as soybean curd for tofu (Hwang et al., 2011a) and noodles (Hwang et al., 2011b), where it exerted a positive influence on the physical and sensory properties of the food. Although pepper juice is well established as a functional food, the high pungency and bitterness of the juice is often rejected by consumers, leading to underutilization. Fukao et al. (2019) could successfully reduce bitterness by more than 83% in ripe red bell pepper juice using a synthetic styrene-divinylbenzene adsorbent resin (pore radius ≥ 250 Å). This was attributed to the removal of quercetin-3-O-rhamnoside. Simultaneously, a high retention of vitamin B6 was also observed.

Fermentation of the juice is the best approach to reduce pungency in red and green pepper juices. *Bacillus licheniformis* SK1230 is the most preferred microorganism which is responsible for the degradation of capsaicinoids. Cho et al. (2015) suggested a fermentation period of 5 days for green and red pepper juices with minimal loss of antioxidants and total phenols. The process of fermentation using lactic acid bacteria can be used to develop probiotic juices, which are gaining huge attention. *Levilactobacillus brevis* and *Limosilactobacillus fermentum* were used for yielding pickled juices with high pigment and lactic acid bacteria content (Janiszewska-Turak et al., 2022). Incorporation of this fermented pepper juice in yogurt improved the antimicrobial, antioxidant, and antiobesity properties of the product. A high sensory acceptability with reduced pungency and enhanced acidity was observed due to the addition of hot pepper juice to the yogurt (Kang et al., 2018). The yogurt inhibited hypertrophy and fat accumulation in organs. This prevented the accumulation of hepatic lipid and lowered cholesterol (Yeon et al., 2015).

Apart from the fruit, a pepper leaf–based beverage was developed using two lactic acid bacteria, namely *Lactobacillus homohiochii* JBCC 25 and *L. homohiochii* JBCC 46, which were isolated from naturally fermented vinegar (Song et al., 2014). A high content of γ-aminobutyric acid and inhibitory activity against α-glucosidase were observed in the fermented leaf juice. Fermentation resulted in reduced bitterness, increased hot taste, and refreshment value. The subsequent section discusses fermentation as a processing method to create a diverse range of products with improved sensory and functional properties (Table 10.2).

10.7 FERMENTED WHOLE PEPPERS

Fermentation is one of the most important methods which has been found to positively affect the quality of peppers by increasing the polyphenol content and reducing the pungency. Yu et al. (2014) established that the addition of the fermented red pepper to stirred yogurt at a level of 0.05% benefitted the sensory scores of yogurt as compared to the nonfermented pepper.

Li et al. (2023) observed that the addition of *P. ethanolidurans* M1117 to peppers decreased the salinity, pH, capsaicinoids, and nitrite content of the chilli pepper compared to the spontaneously fermented pepper. Sichuan paocai is a fermented chilli pepper used as a side dish or an appetizer. The addition of this culture helped in enhancing the taste of the peppers due to the higher amounts of terpenes and other metabolites. This increase can be correlated to the metabolites of gluconeogenesis, citrate cycle, and biosynthesis of terpenoids by *P. ethanolidurans*, thereby establishing

---

TABLE 10.2

A Summary of the Recent Trends in the Fermentation of Pepper to Improve Its Functional and Sensory Properties

Pepper variety	Conditions of fermentation	Key conclusions	Reference
Fresh chili pepper (*Capsicum annuum* L.)	Fermentation broth of ginger and garlic with salt (6% w/v) and sugar (1% w/v) [A] Inoculation with 1% (v/v) *Pediococcus ethanolidurans* M1117 (108 log cfu mL-1) [B] Spontaneously fermented pepper	• Salinity reached the lowest level on 6th day. • Overall salinity [A]<[B] • Faster maturation after inoculation with starter culture • Nitrite peak observed on the 3rd day ([A]<[B]) • Hardness and fracturability ↓ ([A]>[B]) • [A] peppers turned yellow. • L* ↓ and a* ↑ • Better brittleness and mouthfeel of [A] compared to [B] • Capsaicin and dihydrocapsaicin ↓ • Linalyl acetate, linalyl formate, ethyl hexanoate, terpinen-4-ol, γ-terpinene, β-myrcene, eucalyptol, D-limonene, linalool, and diallyl disulphide ↑ ([A]>[B]) • Ethyl hexanoate, β-sesquiphellandrene, and zingiberene ↓ • Sweet and tasteless free amino acids ↑ ([A]>[B]). • *Pediococcus* most dominant in [A] • *Erwinia* most dominant in [B] • Pseudomonas present in both [A] and [B] • Inoculation results in better food safety of fermented pepper.	Li et al. (2023)
Habanero pepper (*Capsicum chinense* L.)	5% to 25% salt And 1.5 L of water to 400 g chopped pepper portions followed by natural fermentation	• Average pH during the fermentation period: 4.62–4.92 • pH of fermented peppers with ≤10 % salt lower than ≥15% salt • Higher LAB counts in ≤15% salt after 50 d of fermentation. • Majority of strains belong to *Klebsiella oxytoca* • Reduction in pH via the formation of organic acids • Enterobacteriaceae is spontaneously fermenting peppers irrespective of salt concentration. • No significant difference in color, aroma, texture, and overall acceptability at all salt concentrations • Pepper fermented with no salt addition showed higher aroma, color, overall acceptability, and texture scores than the other treatments. • Tris (tert-butyldimethylsilyloxy) arsine and dimethyl sulfide ↓ with increasing concentration of salt • Development of adequate starter cultures important	Aryee et al. (2022)
Chinese chilli pepper (*Capsicum frutescens* L.)	Pickling in brine (6% salt, 0.84% acetic acid, 0.42% citric acid, 0.6% calcium chloride, and 0.17% sodium erythorbate) for 3 months	• Hardness and chewiness ↓ • L* ↓ and a*, b* ↑ • Bright green to olive-brown color observed • More severe change observed for green pepper than red pepper • Conversion of chlorophyll to pheophytin • Capsaicinoid content and pungency ↑ during ripening and ↓ during fermentation • Ascorbic acid, dehydroascorbic acid, and total vitamin C ↓ • Free amino acids ↑	Ye et al. (2022)

(Continued)

TABLE 10.2 *(Continued)*

A Summary of the Recent Trends in the Fermentation of Pepper to Improve Its Functional and Sensory Properties

Pepper variety	Conditions of fermentation	Key conclusions	Reference
Fresh chilli pepper (*Capsicum frutescens* L.)	Crushing of peppers followed by addition of 10% (w/w) salt in pickle jars for 32 days at 30°C in absence of light (natural spontaneous fermentation)	• Fungal diversity ↑ followed by a ↓ • *Hyphopichia* (46.94%) and *Kodamaea* (17.09%) dominated during the fermentation. • *Meyerozyma* undetected prior to fermentation period but abundance ↑ during the fermentation period • *Rosenbergiella* and *Staphylococcus* were the most dominant. • Acetic and tartaric acid ↑ • Malic and succinic acid ↓ • Pepper-like, fruity, sour plum-like, pickled, and spicy flavors detected after fermentation • *Hyphopichia* positively correlated with flavor formation	Xu et al. (2021)
Fresh Yanhong pepper (*Capsicum annum* L.)	Addition of 10% NaCl (w/v) and 0.5% CaCl2 (w/v) to the minced peppers followed by the inoculation of 5% (w/w) LAB (107 CFU mL−1)	• Hardness, cell wall material, sodium carbonate-soluble, and chelate-soluble pectin content ↓ • Degree of esterification ↓ • Water-soluble pectin ↑ • Rhamnose (Rha) ↑; Arabinose (Ara) and galactose (Gal) molar ratios↓ • Extensive deopolymerization into low M_w • Degradation of pectin RG-I backbone and branched chains	Xu et al. (2021)
Chilli pepper (*Capsicum annum* L.)	Chopping of pepper followed by spontaneous fermentation	• 42 hydrocarbons, 34 esters, 16 alcohols, 21 aldehydes, 8 acids, 4 ketones, 4 ethers • Contents of esters • Ethyl palmitate and ethyl linoleate are the main esters. • Pepper seeds the major source of palmitic acid, oleic acid, and linoleic acid • β-ionone ↑ • High level of bacterial diversity at the beginning of fermentation, with Proteobacteria being the most prominent • *Weissella* played an important role in the initial stage. • Proteobacteria ↓ and Firmicutes ↑ • *Lactobacillus* was the most dominant during fermentation.	Wang et al. (2019)
Guajillo Pepper (*Capsicum annum* L.)	Addition of brine (1.5% sodium chloride and 8% glucose) followed by spontaneous fermentation	• *Hanseniaspora opuntiae, Pichia kudriavzevii,* and *Wickerhamomyces anomalus* were the most prominent yeasts. • Adhesion capacity of these yeasts similar to that of the probiotic yeast (*Saccharomyces boulardii*) • Co-aggregation with pathogenic microorganisms observed • All the yeast strains exhibited antioxidant, hypocholestrolemic, and antibacterial activity against *Salmonella enterica* ser. Typhimurium ATCC 14028 and *Candida albicans* ENCBDM2. • Protective effect on the viability of *Lactobacillus casei* Shirota in gastric and intestinal conditions	Lara-Hidalgo et al. (2018)
Red pepper (*Capsicum annum* L.)	Chopping and grinding, followed by the addition of *L. parabuchneri* (2% w/w)	• Benzene, lactones, pyrazines (2-methoxy-3-(2-methylpropyl) pyrazine), and terpenes ↑ • Aldehydes, sulfur-containing compounds, and ketones ↓ • Degradation compounds of carotenoids (β-ionone, β-cyclocitral, α-ionone, and β-damascenone) ↑ • Significant degradation of carotenoids observed during fermentation • Alcohols were produced by the lactic acid bacteria. • Hydrolysis of glycosidic forms resulted in the formation of terpenes.	Lee et al. (2018)

its potential as a starter culture for Sichuan pickles. While *Pediococcus* was the most abundant in *P. ethanolidurans*-fermented chilli pepper, *Erwinia* was the most dominant in spontaneously fermented chilli pepper. Interestingly, the sulphide content of spontaneously-fermented chilli pepper was higher than the chilli fermented by starter culture. This can be attributed to the inhibition of the release of the garlic flavor by the decomposition of the sulfide-containing amino acids. The release of sweet amino acids such as methionine, serine, and proline was facilitated by the growth of *P. ethanolidurans*. The importance of the bacterial and fungal communities for the fermentation of chili pepper was discussed by Xu et al. (2021). The two dominant bacterial genera were *Rosenbergiella* and *Staphylococcus*, while *Kodamaea* and *Hyphopichia* were the most abundant fungi during fermentation. The *Hyphopichia* cluster positively correlated with flavor formation of peppers. The microbiota of spontaneous fermentation of jalapeno pepper (*Capsicum annum* L.) was majorly comprised of lactic acid bacteria (*Weissella cibaria, Lactobacillus plantarum, Lactobacillus paraplantarum*, and *Leuconostoc citreum*) and yeasts (*Hanseniaspora pseudoguilliermondii* and *Kodamaea ohmeri*). The lactobacilli exhibited high halotolerance in the order of: *L. citreum* > *L. plantarum* > *W. cibaria*. The lactobacilli flourished well in the presence of phenylpropanoids such as ferulic acid and *p*-coumaric acid (González-Quijano et al., 2018). The yeasts isolated from fermented guajillo pepper exhibited great probiotic potential (Lara-Hidalgo et al., 2018). The three yeasts isolated from the spontaneous fermentation had mucin adhesion similar to the commercial probiotic yeast *Saccharomyces boulardii*. Furthermore, the yeast *Pichia kudriavzevii* protected *Lactobacillus casei* Shirota in gastric and intestinal conditions.

Ye et al. (2022) reported that terpenes, esters, and aldehydes were the discriminating volatile compounds in *paojiao* (pickled *Capsicum frutescens* L.) during ripening. The *Lactobacillus* in fermented pepper increased the content of carotenoids in the fermented pickle. While ripening increased the ascorbic acid content in pepper, the process of fermentation decreased the ascorbic acid content. Due to the low stability of ascorbic acid, it can be easily degraded to dehydroascorbic acid by the deteriorative enzymes produced during fermentation. The highly acidic environment generated during the lactic acid fermentation can degrade capsaicin and dihydrocapsaicin. An increase in umami amino acids (Asp and Glu) was also observed due to the secretion of peptidases by the lactic acid bacteria. The odour compounds which had a significant impact on the olfactory profile were isopentyl 8-methylnon-6-enoate, (*E,E*)-2,4-heptadienal, (*E*)-2-heptenal, β-myrcene, D-limonene, (*Z*)-β-ocimene, (E)-β-ocimene, linalool, neryl oxide, and α-terpineol. The inoculation of red pepper with *Lactobacillus parabuchneri* increased the content of pyrazines, lactones, and terpenes. The aldehydes, ketones, and sulphur-containing amino acids decreased during fermentation. The degradation products of several carotenoids (β-ionone, β-cyclocitral, α-ionone, and β-damascenone) significantly increased (Lee et al., 2018). Di Cagno et al. (2009) developed a process of fermented red and yellow peppers. A mixed starter culture (*Lactobacillus curvatus, Lactobacillus plantarum*, and *Weisella confusa*) was used as the autochthonous starter for the purpose. The process involved blanching (85°C│2 min) followed by fermentation in 1% w/v brine at 35°C for 15 hours and heat treatment (85°C│15 min). As the fermentation progressed, the counts of the starter cultures increased, while that of enterobacteria and yeasts were inhibited. The spontaneously fermented peppers had a lower acidity and were contaminated with enterobacteria and yeasts during storage. The peppers suspended in sunflower oil and fermented using starter cultures had higher values of firmness and other sensory attributes. The peppers could be stored at 25°C for 30 days.

Apart from fermented whole peppers, pepper paste or minced pepper-based fermented products are traditional pepper-based products that have a characteristic aroma which is developed during fermentation. *Cladosporium* and *Hansenpora* were found to significantly impact the flavour profile of the pepper paste. While alkane volatiles were generated in the early stages of fermentation (3–5 days), alcohols were dominant in the middle stage (7–17 days). Ester volatiles were produced in the last stage (17–20 days) of fermentation (Ma et al., 2022). Often, minced peppers suffer from quality deterioration during fermentation, and hence, they are subjected to pasteurization. Cho et al. (2016) developed an ohmic heating process as an alternative to the thermal pasteurization of the fermented pepper paste. The ohmic heating process (5 kHz|60 V) resulted in a 99.7% reduction in vegetative *Bacillus* strains compared to only 81.9% reduction achieved by conduction heating (100°C|8 min). The physicochemical properties of the paste produced by ohmic heating was comparable to those of the fresh samples and higher than that of the conventionally heated samples. High-pressure processing was also found to be a good alternative to thermal pasteurization of fermented minced pepper. A high degree of capsaicin and firmness retention, with low levels of biogenic amines, was associated with the HPP-treated fermented minced pepper (Li et al., 2016). Vacuum impregnation of fermented minced pepper with $CaCl_2$ and pectin methylesterase can help in maintaining the firmness even after fermentation, thereby positively affecting sensory and consumer acceptance (Wang et al., 2022).

10.8 PEPPER-BASED PROTEIN

The by-product of the pepper oil extraction, i.e. the pepper seed meal, is rich in protein. Paprika seeds contain approximately 24% protein and higher levels of lysine, threonine, and total aromatic amino acids as per the reference of FAO/WHO guidelines. Firatligil-Durmus and Evranuz (2010) extracted proteins with water at pH 8.8 at 31°C for 20 minutes and observed an yield of 12.24 g protein/100 g of red pepper seed flour. The red pepper seed meal protein, extracted from the by-product of the red pepper seed oil, was used for the fabrication of a composite film using gelatin by Lee et al. (2016). The film prepared from a combination of 4% pepper seed protein, 1% gelatin, and 0.5% oregano oil was used to wrap fatty tuna meat. The microbial populations of *Listeria monocytogenes* and *Salmonella typhimurium* were reduced by 0.73 and 0.89 log CFU/mL after 12 days of refrigerated storage. Interestingly, the film also decreased the degree of lipid oxidation in fatty tuna, thus leading to an improved quality of tuna fish at the end of the storage. The waste *Capsicum* (*Capsicum frutescens*) powder generated during capsanthin extraction was used as a substrate to produce single-cell protein using *Candida utilis* 1769 (Zhao et al., 2010). Therefore, the production of single-cell protein can be an efficient strategy for the valorization of red pepper seed meal and produce high-quality, protein-rich animal feed. The peptides isolated from chilli pepper seeds have reported biological activities. The lipid transfer proteins are a type of antimicrobial peptide synthesized by plants. Chilli pepper seeds are known to have antifungal peptides. The peptide isolated by Cruz et al. (2010) was closer to 9 kDa and showed high sequence homology to the lipid transfer protein. The peptide inhibited the growth of *Fusarium oxysporum*, *Colletotrium lindemunthianum*, *Saccharomyces cerevisiae*, *Pichia membranifaciens*, *Candida tropicalis*, and *Candida albicans*. To summarize, pepper proteins hold immense potential to be incorporated in food systems, and more research needs to be undertaken to help in the valorization of pepper seed meal to comply with the principles of circular economy.

10.9 SUMMARY

Peppers are commercially important fruits which are exploited in food matrices for their characteristic pungency and heat, which is attributed to the presence of capsaicin. The development of new and novel products from pepper has gained significant attention in recent times. Pepper oleoresins and essential oils can be extracted from pepper pastes and seeds, which can be encapsulated as inclusion complexes to improve their stability, reduce their pungency, and facilitate a controlled release in the gastrointestinal tract. The development of pepper oil–based emulsions and powders helps in the incorporation of the lipophilic bioactive compounds and lipids into aqueous food systems. Fermentation is an effective method for the reduction of pungency of peppers and also has the potential to impart probiotic properties to the fermented juice and whole peppers. The enzymes released during the process can improve the sensory profile of peppers by influencing the texture and sensory profile of peppers. Pepper seed meal is a proteinaceous by-product obtained after the extraction of essential oil from pepper seeds. The valorization of this pepper seed meal can open new avenues for the development of novel products by keeping in mind the principles of circular economy.

REFERENCES

Araus, K. A., del Valle, J. M., Robert, P. S., & de la Fuente, J. C. (2012). Effect of triolein addition on the solubility of capsanthin in supercritical carbon dioxide. *The Journal of Chemical Thermodynamics, 51*, 190–194.

Aryee, A. N., Owusu-Kwarteng, J., Senwo, Z., & Alvarez, M. N. (2022). Characterizing fermented habanero pepper (*Capsicum chinense* L). *Food Chemistry Advances, 1*, 100137.

Baenas, N., Belović, M., Ilic, N., Moreno, D. A., & García-Viguera, C. (2019). Industrial use of pepper (Capsicum annum L.) derived products: Technological benefits and biological advantages. *Food Chemistry, 274*, 872–885.

Barros, H. D., Coutinho, J. P., Grimaldi, R., Godoy, H. T., & Cabral, F. A. (2016). Simultaneous extraction of edible oil from avocado and capsanthin from red bell pepper using supercritical carbon dioxide as solvent. *The Journal of Supercritical Fluids, 107*, 315–320.

Catchpole, O. J., Grey, J. B., Perry, N. B., Burgess, E. J., Redmond, W. A., & Porter, N. G. (2003). Extraction of chili, black pepper, and ginger with near-critical CO2, propane, and dimethyl ether: Analysis of the extracts by quantitative nuclear magnetic resonance. *Journal of Agricultural and Food Chemistry, 51*(17), 4853–4860.

Chen, K., Zhang, M., Mujumdar, A. S., & Wang, M. (2023). Quinoa protein isolate-gum Arabic coacervates cross-linked with sodium tripolyphosphate: Characterization, environmental stability, and Sichuan pepper essential oil microencapsulation. *Food Chemistry, 404*, 134536.

Cho, S., Kim, J. M., Yu, M. S., Yeon, S. J., Lee, C. H., & Kim, S. K. (2015). Fermentation of hot pepper juice by Bacillus licheniformis to reduce pungency. *Journal of the Korean Society for Applied Biological Chemistry, 58*(4), 611–616.

Cho, W. I., Yi, J. Y., & Chung, M. S. (2016). Pasteurization of fermented red pepper paste by ohmic heating. *Innovative Food Science & Emerging Technologies, 34*, 180–186.

Cruz, L. P., Ribeiro, S. F., Carvalho, A. O., Vasconcelos, I. M., Rodrigues, R., Cunha, M. D., & Gomes, V. M. (2010). Isolation and partial characterization of a novel lipid transfer protein (LTP) and antifungal activity of peptides from chilli pepper seeds. *Protein and Peptide Letters, 17*(3), 311–318.

da Silva Anthero, A. G., Moya, A. M. T. M., Torsoni, A. S., Cazarin, C. B. B., & Hubinger, M. D. (2022). Characterization of Capsicum oleoresin microparticles and in vivo evaluation of short-term capsaicin intake. *Food Chemistry: X, 13*, 100179. https://doi.org/10.1016/j.fochx.2021.100179

de Aguiar, A. C., Osorio-Tobón, J. F., Silva, L. P. S., Barbero, G. F., & Martínez, J. (2018). Economic analysis of oleoresin production from malagueta peppers (*Capsicum frutescens*) by supercritical fluid extraction. *The Journal of Supercritical Fluids, 133*, 86–93. https://doi.org/10.1016/j.supflu.2017.09.031

Dias, A. L., Arroio Sergio, C. S., Santos, P., Barbero, G. F., Rezende, C. A., & Martínez, J. (2016). Effect of ultrasound on the supercritical CO2 extraction of bioactive compounds from dedo de moça pepper (*Capsicum baccatum* L. var. pendulum). *Ultrasonics Sonochemistry, 31*, 284–294. https://doi.org/10.1016/j.ultsonch.2016.01.013

Di Cagno, R., Surico, R. F., Minervini, G., De Angelis, M., Rizzello, C. G., & Gobbetti, M. (2009). Use of autochthonous starters to ferment red and yellow peppers (*Capsicum annum* L.) to be stored at room temperature. *International Journal of Food Microbiology, 130*(2), 108–116. https://doi.org/10.1016/j.ijfoodmicro.2009.01.019

Domínguez-Cañedo, I. L., & Beristain-Guevara, C. I. (2011). Microencapsulation of Habanero chilli (*Capsicum chinense*) oleoresin in β-cyclodextrin and antioxidant activity during storage. In *European Drying Conference-EuroDrying 2011. Palma. Balearic Island, Spain, 26–28 October 2011.* https://www.academia.edu/6846684/MICROENCAPSULATION_OF_HABANERO_CHILLI

Firatligil-Durmus, E., & Evranuz, O. (2010). Response surface methodology for protein extraction optimization of red pepper seed (*Capsicum frutescens*). *LWT-Food Science and Technology, 43*(2), 226–231.

Fukao, M., Shirono, H., Takada, W., Moriuchi, T., & Fukaya, T. (2019). Debittering of red bell pepper (*Capsicum annuum*) juice retaining a high vitamin B6 content, using a styrene-divinylbenzene adsorbent resin. *Food Science and Technology Research, 25*(1), 57–63.

Ghiasi, Z., Esmaeli, F., Aghajani, M., Ghazi-Khansari, M., Faramarzi, M. A., & Amani, A. (2019). Enhancing analgesic and anti-inflammatory effects of capsaicin when loaded into olive oil nanoemulsion: An in vivo study. *International Journal of Pharmaceutics, 559*, 341–347.

Goci, E., Haloci, E., Di Stefano, A., Chiavaroli, A., Angelini, P., Miha, A., . . . & Marinelli, L. (2021). Evaluation of in vitro capsaicin release and antimicrobial properties of topical pharmaceutical formulation. *Biomolecules, 11*(3), 432.

Gogus, F., Ozel, M. Z., Keskin, H., Yanık, D. K., & Lewis, A. C. (2015). Volatiles of fresh and commercial sweet red pepper pastes: Processing methods and microwave assisted extraction. *International Journal of Food Properties, 18*(8), 1625–1634.

Gomes, L. M. M., Petito, N., Costa, V. G., Falcão, D. Q., & de Lima Araújo, K. G. (2014). Inclusion complexes of red bell pepper pigments with β-cyclodextrin: Preparation, characterisation and application as natural colorant in yogurt. *Food Chemistry, 148*, 428–436. https://doi.org/10.1016/j.foodchem.2012.09.065

González-Quijano, G. K., Dorantes-Alvarez, L., Hernández-Sánchez, H., Jaramillo-Flores, M. E., de Jesús Perea-Flores, M., Vera-Ponce de León, A., & Hernández-Rodríguez, C. (2018). Halotolerance and survival kinetics of lactic acid bacteria isolated from jalapeño pepper (*Capsicum annuum* L.) fermentation. *Journal of Food Science, 79*(8), M1545–M1553. https://doi.org/10.1111/1750-3841.12498

Hein, L. T. M., Dao, D. T. A. (2021). Antibacterial activity of black pepper essential oil nanoemulsion formulated by emulsion phase inversion method. *Current Research in Nutrition and Food Science, 10*(1), 311–320.

Hwang, I. G., Hwang, Y., Kim, H. Y., Lee, J., Jeong, H. S., & Yoo, S. M. (2011a). Quality characteristics of tofu (soybean curd) added with Cheongyang hot pepper (*Capsicum annuum* L.) juice. *Journal of the Korean Society of Food Science and Nutrition, 40*(7), 999–1005.

Hwang, I. G., Kim, H. Y., Hwang, Y., Jeong, H. S., & Yoo, S. M. (2011b). Quality characteristics of wet noodles combined with Cheongyang hot pepper (*Capsicum annuum* L.) juice. *Journal of the Korean Society of Food Science and Nutrition, 40*(6), 860–866.

Isaschar-Ovdat, S., Shani-Levi, C., & Lesmes, U. (2021). Capsaicin stability and bio-accessibility affected by complexation with high-amylose corn starch (HACS). *Food & Function, 12*(15), 6992–7000.

Janiszewska-Turak, E., Tracz, K., Bielińska, P., Rybak, K., Pobiega, K., Gniewosz, M., . . . & Gramza-Michałowska, A. (2022). The impact of the fermentation method on the pigment content in pickled beetroot and red bell pepper juices and freeze-dried powders. *Applied Sciences, 12*(12), 5766.

Jiménez, M., Domínguez, J. A., Pascual-Pineda, L. A., Azuara, E., & Beristain, C. I. (2018). Elaboration and characterization of O/W cinnamon (*Cinnamomum zeylanicum*) and black pepper (*Piper nigrum*) emulsions. *Food Hydrocolloids, 77*, 902–910.

Kim, N. H., & Park, S. H. (2015). Evaluation of green pepper (*Capsicum annuum* L.) juice on the weight gain and changes in lipid profile in C57BL/6 mice fed a high-fat diet. *Journal of the Science of Food and Agriculture, 95*(1), 79–87.

Lara-Hidalgo, C. E., Dorantes-Álvarez, L., Hernández-Sánchez, H., Santoyo-Tepole, F., Martínez-Torres, A., Villa-Tanaca, L., & Hernández-Rodríguez, C. (2018). Isolation of yeasts from guajillo pepper (*Capsicum annuum* L.) fermentation and study of some probiotic characteristics. *Probiotics and Antimicrobial Proteins, 11*(3), 748–764. https://doi.org/10.1007/s12602-018-9415-x

Lee, J. H., Yang, H. J., Lee, K. Y., & Song, K. B. (2016). Physical properties and application of a red pepper seed meal protein composite film containing oregano oil. *Food Hydrocolloids, 55*, 136–143.

Lee, S. M., Lee, J. Y., Cho, Y. J., Kim, M. S., & Kim, Y. S. (2018). Determination of volatiles and carotenoid degradation compounds in red pepper fermented by *Lactobacillus parabuchneri*. *Journal of Food Science, 83*(8), 2083–2091. https://doi.org/10.1111/1750-3841.14221

Li, J., Hou, X., Jiang, L., Xia, D., Chen, A., Li, S., . . . & Zhang, Z. (2022). Optimization and characterization of Sichuan pepper (Zanthoxylum bungeanum Maxim) resin microcapsule encapsulated with β-cyclodextrin. *LWT- Food Science and Technology, 171*, 114120.

Li, J., Zhao, F., Liu, H., Li, R., Wang, Y., & Liao, X. (2016). Fermented minced pepper by high pressure processing, high pressure processing with mild temperature and thermal pasteurization. *Innovative Food Science & Emerging Technologies, 36*, 34–41.

Li, Y., Luo, X., Long, F., Wu, Y., Zhong, K., Bu, Q., . . . & Gao, H. (2023). Quality improvement of fermented chili pepper by inoculation of *Pediococcus ethanolidurans* M1117: Insight into relevance of bacterial community succession and metabolic profile. *LWT-Food Science and Technology, 179*, 114655. https://doi.org/10.1016/j.lwt.2023.114655

Liu, X., Bi, J., Xiao, H., & McClements, D. J. (2015). Increasing carotenoid bioaccessibility from yellow peppers using excipient emulsions: Impact of lipid type and thermal processing. *Journal of Agricultural and Food Chemistry, 63*(38), 8534–8543.

Lobo, F. A. T., Silva, V., Domingues, J., Rodrigues, S., Costa, V., Falcão, D., & de Lima Araújo, K. G. (2018). Inclusion complexes of yellow bell pepper pigments with β-cyclodextrin: Preparation, characterisation and application as food natural colorant. *Journal of the Science of Food and Agriculture, 98*(7), 2665–2671. https://doi.org/10.1002/jsfa.8760

Locali-Pereira, A. R., Lopes, N. A., Menis-Henrique, M. E. C., Janzantti, N. S., & Nicoletti, V. R. (2020). Modulation of volatile release and antimicrobial properties of pink pepper essential oil by microencapsulation in single-and double-layer structured matrices. *International Journal of Food Microbiology, 335*, 108890.

Ma, D., Li, Y., Chen, C., Fan, S., Zhou, Y., Deng, F., & Zhao, L. (2022). Microbial succession and its correlation with the dynamics of volatile compounds involved in fermented minced peppers. *Frontiers in Nutrition, 9*, 1041608.

Melgar-Lalanne, G., Hernández-Álvarez, A. J., Jiménez-Fernández, M., & Azuara, E. (2017). Oleoresins from Capsicum spp.: Extraction methods and bioactivity. *Food and Bioprocess Technology, 10*, 51–76. https://doi.org/10.1007/s11947-016-1793-z

Nagasukeerthi, P., Mooventhan, A., & Manjunath, N. K. (2017). Short-term effect of add on bell pepper (*Capsicum annuum* var. grossum) juice with integrated approach of yoga therapy on blood glucose levels and cardiovascular functions in patients with type 2 diabetes mellitus: A randomized controlled study. *Complementary Therapies in Medicine, 34*, 42–45. https://doi.org/10.1016/j.ctim.2017.07.011

Ozdemir, N., Pola, C. C., Teixeira, B. N., Hill, L. E., Bayrak, A., & Gomes, C. L. (2018). Preparation of black pepper oleoresin inclusion complexes based on beta-cyclodextrin for antioxidant and antimicrobial delivery applications using kneading and freeze drying methods: A comparative study. *LWT, 91*, 439–445.

Parvez, G. M. (2017). Current advances in pharmacological activity and toxic effects of various Capsicum species. *International Journal of Pharmaceutical Sciences and Research, 8*(5), 1900–1912. http://doi.org/10.13040/IJPSR.0975-8232

Peng, X., Zhou, Y., Han, K., Qin, L., Dian, L., Li, G., Pan, X. & Wu, C. (2015). Characterization of cubosomes as a targeted and sustained transdermal delivery system for capsaicin. *Drug Design, Development and Therapy*, 4209–4218.

Pereira, A. R. L., Cattelan, M. G., & Nicoletti, V. R. (2019). Microencapsulation of pink pepper essential oil: Properties of spray-dried pectin/SPI double-layer versus SPI single-layer stabilized emulsions. *Colloids and Surfaces A: Physicochemical and Engineering Aspects, 581*, 123806.

Pestovsky, Y. S., & Martínez-Antonio, A. (2018). Gold nanoparticles with immobilized β-cyclodextrin-capsaicin inclusion complex for prolonged capsaicin release. In *IOP Conference Series: Materials Science and Engineering* (Vol. 389, No. 1, p. 012030). IOP Publishing.

Petito, N. D. L., da Silva Dias, D., Costa, V. G., Falcão, D. Q., & de Lima Araujo, K. G. (2016). Increasing solubility of red bell pepper carotenoids by complexation with 2-hydroxypropyl-β-cyclodextrin. *Food Chemistry, 208*, 124–131. https://doi.org/10.1016/j.foodchem.2016.03.122

Petito, N. D. L., Devens, J. M., Falcão, D. Q., Dantas, F. M. L., Passos, T. S., & Araujo, K. G. D. L. (2022). Nanoencapsulation of red bell pepper carotenoids: Comparison of encapsulating agents in an emulsion based system. *Colorants, 1*(2), 132–148.

Rakmai, J., Cheirsilp, B., Mejuto, J. C., Torrado-Agrasar, A., & Simal-Gándara, J. (2017). Physico-chemical characterization and evaluation of bio-efficacies of black pepper essential oil encapsulated in hydroxypropyl-beta-cyclodextrin. *Food Hydrocolloids, 65*, 157–164.

Rybak, K., Samborska, K., Jedlinska, A., Parniakov, O., Nowacka, M., Witrowa-Rajchert, D., & Wiktor, A. (2020). The impact of pulsed electric field pretreatment of bell pepper on the selected properties of spray dried juice. *Innovative Food Science & Emerging Technologies, 65*, 102446.

Salazar, R., Arámbula-Villa, G., Hidalgo, F. J., & Zamora, R. (2012). Mitigating effect of piquin pepper (*Capsicum annuum* L. var. Aviculare) oleoresin on acrylamide formation in potato and tortilla chips. *LWT-Food Science and Technology, 48*(2), 261–267.

Saleh, B. K., Omer, A., & Teweldemedhin, B. J. M. F. P. T. (2018). Medicinal uses and health benefits of chili pepper (Capsicum spp.): A review. *MOJ Food Process Technology, 6*(4), 325–328. https://doi.org/10.15406/mojfpt.2018.06.00183

Santos, P., Aguiar, A. C., Barbero, G. F., Rezende, C. A., & Martínez, J. (2015). Supercritical carbon dioxide extraction of capsaicinoids from malagueta pepper (*Capsicum frutescens* L.) assisted by ultrasound. *Ultrasonics Sonochemistry, 22*, 78–88.

Šeregelj, V., Tumbas Šaponjac, V., Lević, S., Kalušević, A., Ćetković, G., Čanadanović-Brunet, J., Nedović, V., Stajčić, S., Vulić, J., & Vidaković, A. (2019). Application of encapsulated natural bioactive compounds from red pepper waste in yogurt. *Journal of Microencapsulation, 36*(8), 704–714. https://doi.org/10.1080/02652048.2019.1668488

Shi, Y., Zhang, M., Chen, K., & Wang, M. (2022). Nano-emulsion prepared by high pressure homogenization method as a good carrier for Sichuan pepper essential oil: Preparation, stability, and bioactivity. *LWT-Food Science and Technology, 154*, 112779.

Song, Y. R., Shin, N. S., & Baik, S. H. (2014). Physicochemical and functional characteristics of a novel fermented pepper (Capsicum annuum L.) leaves-based beverage using lactic acid bacteria. *Food Science and Biotechnology, 23*, 187–194. https://doi.org/10.1007/s10068-014-0025-4

Srinivasan, K. (2016). Biological activities of red pepper (*Capsicum annuum*) and its pungent principle capsaicin: A review. *Critical Reviews in Food Science and Nutrition, 56*(9), 1488–1500. https://doi.org/10.1080/10408398.2013.772090

Teixeira, B. N., Ozdemir, N., Hill, L. E., & Gomes, C. L. (2013). Synthesis and characterization of nanoencapsulated black pepper oleoresin using hydroxypropyl beta-cyclodextrin for antioxidant and antimicrobial applications. *Journal of Food Science, 78*(12), N1913–N1920.

Truong, V. L., Rarison, R. H. G., & Jeong, W. S. (2022). Protective effects of orange sweet pepper juices prepared by high-speed blender and low-speed masticating juicer against UVB-induced skin damage in SKH-1 hairless mice. *Molecules (Basel, Switzerland), 27*(19), 6394. https://doi.org/10.3390/molecules27196394

Uquiche, E., Millao, S., & del Valle, J. M. (2022). Extrusion affects supercritical CO_2 extraction of red pepper (*Capsicum annuum* L.) oleoresin. *Journal of Food Engineering, 316*, 110829.

Venkataswamy, G., Meena, N. kiran, Naik, K. K., & Harohally, N. (2019). Capsaicin Entrapment in Metal-Organic Framework Material Derived from γ-Cyclodextrin. *ChemRxiv.* doi:10.26434/chemrxiv.10484192.v1 [preprint and has not been peer-reviewed].

Wang, J., Wang, R., Xiao, Q., Liu, C., Deng, F., & Zhou, H. (2019). SPME/GC-MS characterization of volatile compounds of Chinese traditional-chopped pepper during fermentation. *International Journal of Food Properties, 22*(1), 1863–1872.

Wang, Y., Luo, J., Hou, X., Wu, H., Li, Q., Li, S., . . . & Zhang, Z. (2022a). Physicochemical, antibacterial, and biodegradability properties of green Sichuan pepper (*Zanthoxylum armatum* DC.) essential oil incorporated starch films. *LWT-Food Science and Technology, 161*, 113392.

Wang, Y., Qin, K., Chen, F., Jiang, L., Zhou, H., Ding, S., & Wang, R. (2022b). Texture improvement of fermented minced pepper under vacuum impregnation with pectin methylesterase and $CaCl_2$ during fermentation. *International Journal of Food Science & Technology, 57*(6), 3477–3489.

Xu, H., Chen, Y., Ding, S., Qin, Y., Jiang, L., Zhou, H., . . . & Wang, R. (2021). Changes in texture qualities and pectin characteristics of fermented minced pepper during natural and inoculated fermentation process. *International Journal of Food Science & Technology, 56*(11), 6073–6085.

Yang, C. Y., Mandal, P. K., Han, K. H., Fukushima, M., Choi, K., Kim, C. J., & Lee, C. H. (2010). Capsaicin and tocopherol in red pepper seed oil enhances the thermal oxidative stability during frying. *Journal of Food Science and Technology, 47*, 162–165.

Ye, Z., Shang, Z., Li, M., Zhang, X., Ren, H., Hu, X., & Yi, J. (2022). Effect of ripening and variety on the physiochemical quality and flavor of fermented Chinese chili pepper (Paojiao). *Food Chemistry, 368*, 130797. https://doi.org/10.1016/j.foodchem.2021.130797

Yeon, S. J., Hong, G. E., Kim, C. K., Park, W. J., Kim, S. K., & Lee, C. H. (2015). Effects of yogurt containing fermented pepper juice on the body fat and cholesterol level in high fat and high cholesterol diet fed rat. *Korean Journal for Food Science of Animal Resources, 35*(4), 479–485. https://doi.org/10.5851/kosfa.2015.35.4.479

Yu, M. S., Kim, J. M., Lee, C. H., Son, Y. J., & Kim, S. K. (2014). Quality characteristics of stirred yoghurt added with fermented red pepper. *Korean Journal for Food Science of Animal Resources, 34*(4), 408–414. https://doi.org/10.5851/kosfa.2014.34.4.408

Zanzer, Y. C., Plaza, M., Dougkas, A., Turner, C., & Östman, E. (2018). Black pepper-based beverage induced appetite-suppressing effects without altering postprandial glycaemia, gut and thyroid hormones or gastrointestinal well-being: a randomized crossover study in healthy subjects. *Food & Function, 9*(5), 2774–2786.

Zhang, Q. B., Zhao, L., Gao, H. Y., Zhang, L. L., Wang, H. Y., Zhong, K., . . . & Xie, R. (2020). The enhancement of the perception of saltiness by Sichuan pepper oleoresin in a NaCl model solution. *Food Research International, 136*, 109581.

Zhao, G., Zhang, W., & Zhang, G. (2010). Production of single cell protein using waste capsicum powder produced during capsanthin extraction. *Letters in Applied Microbiology, 50*(2), 187–191.

11.1 INTRODUCTION

The genus *Capsicum* originated in humid areas of tropical and sub-tropical America and Central Asia and includes capsicums and chillies. Out of 200 species, the five most domesticated species are *C. chinense*, *C. annuum*, *C. pubescens*, *C. baccatum* and *C. frutescens* (Budzianowski, 2017). With the increase in food demand, the production of capsicum has also increased by 35% in the last few decades. China is the largest producer of capsicums in the world, followed by Turkey and Indonesia. The *Capsicum* genus is known for its pungent flavour, which is the most desirable trait of this fruit

DOI: 10.1201/9781003378259-11

by consumers (Mariano et al., 2022). The Scoville heat unit (SHU) is used to express pungency levels. The Carolina Reaper is on the top of the Scoville scale with more than 1,500,000 SHU. The pungency of *Capsicum* species is due to a group of non-volatile alkaloid compounds called capsaicinoids (Barbero et al., 2008). They are produced in the placenta of the fruit as the plant's natural defence mechanism against pests (Srinivasan, 2016). Capsaicin and dihydrocapsaicin are the two most abundant naturally occurring capsaicinoids and account for almost 90% of *Capsicum's* pungency (Sganzerla et al., 2014). Apart from the placenta, capsaicinoids can be found in different parts of capsicum plants, including the pericarp, seeds, leaves and stems (Estrada et al., 2002). The synthesis of these secondary metabolites can be influenced by interaction of developmental stages and environmental conditions. The capsaicinoid content in plants is regulated by factors like light, temperature, water, stress, nutrient supply, genetics and growth stage of plant (Lozada et al., 2021).

Capsaicinoids have gained attention because of their pharmacological properties with anti-cancer, anti-diabetic, anti-obesity, analgesic and anti-inflammatory activities (Luo et al., 2011). Other nutritional compounds present in *Capsicum* species are carbohydrates, proteins, calcium, magnesium, potassium and vitamins A, C and E, carotenoids, flavonoids and phenols. Capsaicinoids are extracted through several different techniques such as solvent extraction, Soxhlet extraction, ultrasound-assisted extraction, supercritical fluid extraction and microwave-assisted extraction. The chapter's end also discusses the importance of reducing food waste and improving sustainable production of capsicums and chillies. Bioactive compounds can be extracted from capsicum fruit and plant waste to reduce its environmental impact and work more strategically towards achieving sustainable developmental goals.

11.2   THE GENUS *CAPSICUM*

The *Capsicum* is a genus of *Solanaceae*—an important plant family of Eudicots. Other prominent members of the *Solanaceae* family are important horticultural and industrial crops such as potato (*Solanum tuberosum*), tomato (*Solanum lycopersicum*), eggplant (*S. melongena*) and tobacco (*Nicotiana tabacum*). The *Capsicum* genus originated in humid, tropical and sub-tropical regions of America and Central Asia and most likely from Bolivia, as the country has 27 known species (Walsh and Hoot, 2001). Archaeological evidence has proved their use in Mexico, northern Central America, the Caribbean, lowland Bolivia, the northern lowland Amazonia and the mid-elevation southern Andes since 6000 BC (Basu and De, 2003; Perry et al., 2007). The *Capsicum* genus has over 200 species, out of which there are 5 commercially cultivated and 25 wild and semi-cultivated species (Budzianowski, 2017). The domesticated species are *C. chinense*, *C. annuum*, *C. pubescens*, *C. baccatum* and *C. frutescens*. *Capsicum annuum* is considered to be one of the oldest cultivated plants in America (Sawatdeenarunat et al., 2016) and among the most-used condiments across the globe (Perantoni et al., 2018).

Capsicum species have different names depending upon the region and culture (Basu and De, 2003). Previously, capsicum and chilli fruits were confused with black pepper (*Piper nigrum*) because of similar taste and spice. Therefore, the *Capsicum* was incorrectly given the name "pepper" (Walsh and Hoot, 2001). However, black pepper (*Piperaceae* family) and capsicums are not related by either appearance or phylogeny. In 1543, the botanical term *Capsicum* was suggested for the first time by Fuchs, and it was then adopted by Linneo in 1753 (Tripodi and Kumar, 2019). Now, the fruits are commonly known as chilli, red chilli, hot pepper, chilli pepper, bell pepper, paprika, sweet pepper and cayenne (Basu and De, 2003). Generally, the term "chilli pepper" is used for species with small and spicy fruits, e.g. *C. chinense* and *C. frutescens*, whereas the term "sweet pepper" is used for species with larger fruits and less or no spice, e.g. bell varieties of *Capsicum annuum* (Tripodi and Kumar, 2019).

Capsicum annuum is the most widely cultivated *Capsicum* species in the world. They are usually grown as annual vegetables despite their natural perennial character (Tripodi and Kumar, 2019). It consists of both pungent and non-pungent species. The plants have herb or sub-shrub growth patterns with different sizes, shapes and colours of fruits (Tripodi and Kumar, 2019). *C. annuum* includes numex, jalapeño, sweet pepper or bell varieties, cayenne pepper, etc. The fruit colour of different varieties varies from red to green, yellow, purple and orange. The jalapeño variety is the consumer favourite in Mexico and contributes 31% to the total pepper production of the country (Sánchez-Toledano et al., 2021) (Figure 11.1).

Capsicum frutescence is commonly known as red pepper, goat pepper and chillies. It is a hardy perennial shrub that grows up to 2 m in height with a wild or semi-wild nature (De, 2003). *C. frutescence* originated from South America and is now cultivated across American, Asian and African countries. The fruits are usually smaller in size (less than 2 cm) (Arancon et al., 2013). Chilli fruits are usually sun dried and then marketed as dry chilli fruit and powder. Another common use is to make flavouring pickles (De, 2003) (Figure 11.1).

Known as habanero-type pepper, *Capsicum chinense* is another commonly cultivated species in America, Africa and Asia. The fruits of this species have high pungency (exceptional heat) levels (Tripodi and Kumar, 2019). It is reported as the most common species in Brazil because there is high genetic diversity in the Amazon basin. Moreover, it is very popular among consumers across the tropical region (Heinrich et al., 2015). The common cultivars of *Capsicum chinense* include Carolina reaper, habanero, Scotch bonnet and bhut jolokia.

Capsicum baccatum and *Capsicum pubescens* species are cultivated in Central and South America, mainly in Bolivia and Argentina. These two species show phenotypically different characters compared to other cultivated *Capsicum* species. The former has coloured spots on corolla (yellow or green) (Tripodi and Kumar, 2019), while the latter has purplish flowers and dark-coloured seeds (De, 2003). The most popular cultivars of *C. baccatum* in South America are Bishop's crown and ají amarillo chilli (DeWitt and Bosland, 2009).

11.3 GLOBAL PRODUCTION OF CAPSICUMS AND CHILLIES

Capsicum spp. are semi-perennial bushy plants and can grow up to 0.75 to 1.8 m height in cultivated locations (Batiha et al., 2020), but they are mostly commercially cultivated as annual plants. It is a warm-season crop with a typical growing period of over 5 months (Pinto et al., 2016) in the field and over 9 months in greenhouses (Zhao et al., 2021). The germination temperature required by seeds of *Capsicum* is 25°C to 30°C. Capsicums and chillies require optimal temperatures ranging between 16°C and 21°C for fruit setting. For better quality fruit development, the ideal temperature ranges are 15°C to 17°C at night and 24°C to 30°C during the day (Pinto et al., 2016). In general, chilli plants are more heat tolerant than capsicum plants (Burt, 2005). If planted very early in the season or at lower temperature, seedling emergence can be delayed, resulting in poor plant growth

FIGURE 11.1   (a) *Capsicum annuum* var. annuum, (b) *Capsicum frutescens* (Source: Australian Plant Image Index).

(Bosland and Votava, 2012). Capsicums are highly susceptible to frost, resulting in retarded and stunted plant growth. When the temperature falls below 15°C or rises above 32°C, capsicum plant growth can be retarded, which may lead to conditions like blossom end rot, termination of fruit-set, malformed (cracked, small or hardened) fruits and lower yields (Thuy and Kenji, 2015). For commercial quality production, a well-drained, soft sandy loam soil with pH range of 6.5 to 7.5 is considered suitable in the field (Wallender and Tanji, 2011), and a hydroponic solution with electric conductivity (EC) of ~2.3 and pH ~6.1 is best for greenhouse production (Zhao et al., 2021). Fruits can be harvested at several stages depending upon the variety and desired stage of maturity. Considering season and maturity, green fruits can be harvested almost 75 days after transplanting, while to harvest fruits at fully ripen stage (red or other colours) requires another 15 to 20 days (Govindarajan and Salzer, 1985).

The production of capsicums and chillies has been increased worldwide. According to FAO data, there has been a 35% increase in world capsicum production. In 2021, the global production of fresh capsicums and chillies was 36,286,643 tonnes (FAO, 2022), and its yield increased from 15.5 tonnes per hectare in 2008 to 18.5 tonnes per hectare in 2018 (Sánchez-Toledano et al., 2021). The increase in yield is much needed, as it is believed that there will be a 4% increase in food demand per capita in the next decade. The production of dry chillies and capsicums was 4,839,309 tonnes globally (FAO, 2022). Asia is the number one producer of capsicums and chillies, followed by America, Europe and Africa with 65%, 13.3%, 11.9% and 10.1% contributions, respectively. The top 10 countries producing green *Capsicum* and *Pimenta* species along with their area of harvest are listed in Table 11.1. Figure 11.2 shows the increase in world production of chillies and capsicums from 2001 to 2021 (FAO, 2022).

11.4   THE SCIENCE OF PUNGENCY IN *CAPSICUM*

The most popular trait of *Capsicum* for culinary purposes is the pungency or spiciness of the fruits. The spicy flavour in food is admired globally, and it becomes one of the fundamentals in food flavouring (Mariano et al., 2022). The pungency is expressed in Scoville heat units (SHU) and

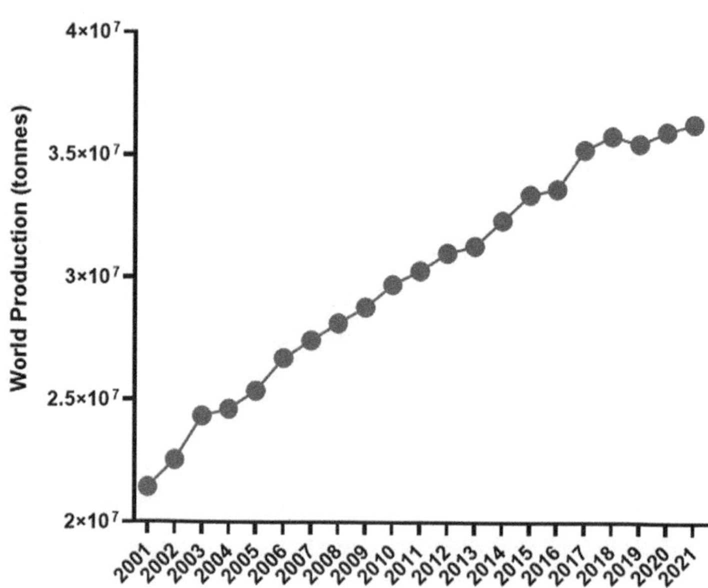

FIGURE 11.2   Global production of capsicums and chillies from 2001 to 2021.

TABLE 11.1

Top 10 *Capsicum*-Producing Countries + Australia for Year 2020–2021 (FAO, 2022)

Top 10 countries along with Australia	Average production of green chillies and peppers (*Capsicum* spp. and *Pimenta* spp.) in tonnes	Area harvested in hectares (ha)
China, mainland	16,725,716	752,121
Türkiye	2,864,100	80,239
Indonesia	2,759,806	321,923
Mexico	2,701,293	147,808
Spain	1,492,205	22,240
Egypt	864,563.6	47,892
Nigeria	758,359.3	101,335
United States of America	548,472.5	16,916
Algeria	534,745.5	11,190
Netherlands	435,000	1,630
Australia	55,150	2,396

traditionally calculated through the Scoville organoleptic test (Scoville, 1912). This test is conducted by diluting the solution of pepper extract with sugar water. Scoville heat units refers to the dilution factor required for making pepper extract and sugar solution (Lau et al., 2015). The chillies with SHU higher than 1 million are referred to as "superhot" (Bosland et al., 2015). The hottest chillies in the world belong to varieties of the *C. chinense* species. The Carolina reaper is on top of the Scoville scale with > 1,500,000 SHU, followed by ghost pepper and Infinity chilli with SHUs of 750,000 to 1,500,000 (Bosland and Baral, 2007). Carolina reaper is a crossbreed between two very hot chillies, la soufriere pepper from Saint Vincent and a naga pepper from Pakistan. A newly developed hotter cultivar named Pepper X was reported to have ~3 million SHU (Buchanan, 2020). It is important to quantify the pungent compounds of *Capsicum* fruits, capsaicinoids, for marketing and consumer purposes, as it is necessary to identify safety of chillies and their products as food ingredients. Ultra-high-performance liquid chromatography (UHPLC), high-performance liquid chromatography (HPLC), gas chromatography and gas liquid chromatography are being used to identify capsaicinoids in capsicum tissues (Lau et al., 2015). The diluted solution is then tested by trained volunteers, hence making the results arguable because different people have different tolerance to hotness of chillies (Kachoosangi et al., 2008).

11.5 CAPSAICINOIDS

The unique pungent character of the *Capsicum* genus is mainly due to a group of secondary metabolites called capsaicinoids (Barbero et al., 2008). They are primarily produced in plants as a defence mechanism and to repel pests, herbivores and fungal attack (Srinivasan, 2016). These compounds are synthesised through the phenylalanine ammonia-lyase (PAL) pathway, and phenylalanine is the precursor. Different enzymes are involved in the process of capsaicinoid synthesis, and capsaicin synthase plays the key role (Sung et al., 2005) by acting on the fatty acid chain with the help of Mg^{2+}, ATP, coenzyme A and vanillylamine. Vanillylamine, derived from phenylalanine, produces vinylamine, which is then condensed to a branched-chain fatty acid from 9 to 11 carbon atoms to create a capsaicinoid molecule (Chapa-Oliver and Mejía-Teniente, 2016). The hotness of the fruit is related to the bond between the vanillyl ring and the acyl chain (Reyes-Escogido et al., 2011).

There are more than 20 natural compounds in the group of capsaicinoids. The difference in structure is due to the presence or absence of unsaturation (double bonds) and the number of carbons on the lateral chain (Reyes-Escogido et al., 2011). Capsaicin and dihydrocapsaicin are the most abundantly found members, contributing to ~90% of chillies' pungency (Sganzerla et al., 2014). Scoville units for capsaicin and dihydrocapsaicin are 16×10^6 and 15×10^6, respectively (Srinivasan, 2016). Other less abundant compounds are nordihydrocapsaicin, norcapsaicin, homocapsaicin, homodihydrocapsaicin and nonivamide (Barbero et al., 2014). The structures of different capsaicinoids are shown in Figure 11.3 (Yasin et al., 2023).

Capsaicin is the most important and hottest compound of this pungent group. It is a colourless, odorless, lipophilic alkaloid with molecular formula $C_{18}H_{27}NO_3$. Its chemical name is *trans*-8-methyl-*N*-vanillyl-6-nonenamide. It is not soluble in water but is soluble in oil, fat and alcohol. It is a derivative of homovanillic acid, and by changing the acid portion, analogues of capsaicin are developed with different pungency levels (Al-Snafi, 2015). For example, in a study, two pungent and one low-pungent compounds were obtained by changing acyl chain length and chemical substitutes in the aromatic ring (Castillo et al., 2007). Dihydrocapsaicin is chemically known as 8-methyl-*N*-vanillylnonanamide, and it is the second-most-abundant member of the capsaicinoids.

11.6 PRODUCTION OF CAPSAICINOIDS

The primary production site for capsaicinoids is the placenta of the fruit. Mostly, it is concentrated on the epidermis of the fruit ribs where seeds are attached. The epidermal cells are of the interlocular septum that undergo morphological changes during fruit development, and the presence of osmiophilic granules in these cells makes them associated with capsaicinoid biosynthesis (Suzuki and Iwai, 1984). These blisters are formed when the cuticle layer of the cell wall is lifted to fill subcuticular cavities with capsaicinoids (Rowland et al., 1983). The biosynthetic pathway of capsaicinoids was identified such that these compounds are synthesised from the two arms: the aromatic component from L-phenylalanine to vanillylamine and C9-C11 branched-chain fatty acid moieties from L-valine or L-leucine (Ishikawa, 2003). The epidermal layer of the placenta has been found to be the accumulation site of capsaicinoids during all stages of fruit maturation, and concentration is higher in the placenta than in the pericarp of the fruit (example in Table 11.2). Apart from the placenta, the pericarp and seeds also have capsaicinoids in them. Smaller amounts are also found

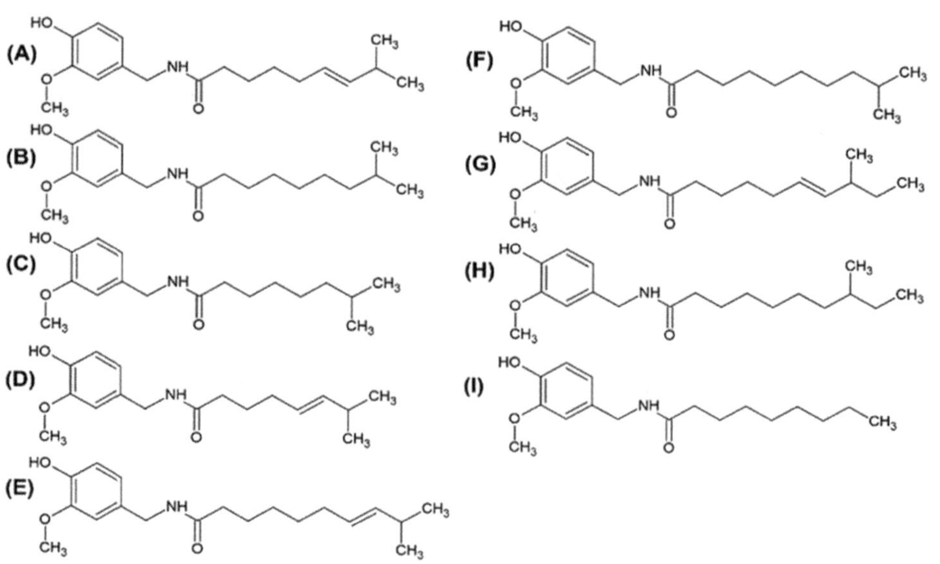

FIGURE 11.3 Structures of capsaicinoids: (A) capsaicin, (B) dihydrocapsaicin, (C) nordihydrocapsaicin, (D) norcapsaicin, (E) homocapsaicin I, (F) homodihydrocapsaicin I, (G) homocapsaicin II, (H) homodihydrocapsaicin II and (I) nonivamide.

TABLE 11.2

Capsaicinoid Concentration in Different Parts of Capsicum Plant

Specie or cultivar	Part of plant/fruit	Capsaicin µg/g dry weight	Dihydrocapsaicin µg/g dry weight	References
Capsicum annuum	Bottom leaves	1.04	16.34	(Estrada et al., 2002)
Padrón Pepper	Top leaves	19.99	20.00	
	Bottom stem	Not detected	5.50	
	Top stem	7.00	7.25	
Capsicum chinense	Pericarp	610.9	167.6	(Gavilán Guillen et al., 2018)
	Seed	336.6	95.1	
	Placenta	3,451.6	1,008.7	
Capsicum baccatum	Pericarp	355.7	78.6	(Gavilán Guillen et al., 2018)
	Seed	423.0	103.8	
	Placenta	802.8	198.8	

in vegetative parts of the plant like stem and leaves, as mentioned in Table 11.2 (Estrada et al., 2002). The fruit starts to accumulate capsaicinoids in early stages of development, usually 10 days after pollination (Fayos et al., 2017). Capsaicinoids reach maximum levels when fruit maturity is at maximum (40 to 50 days after pollination) and then decreases, and almost 60% capsaicinoids are degraded in later stages (Estrada et al., 2002).

Researchers are trying to produce natural and synthetic, less pungent analogues of capsaicinoids to increase the clinical application of these unique compounds (Molinillo et al., 2010). Enzymatic and chemical synthesis are being conducted for this purpose, with the former being preferred because of its toxic reagent–free process. Capsaicin synthesis was performed using amidation reactions catalysed with different lipases. The study used vanillylamine condensed with fatty acid derivatives and obtained 40% to 59% of capsaicin yield and 2% to 44% yield of capsaicin analogues (Kobata et al., 1999). Capsaicinoids are produced at a set moderate pressure and temperature (140°C and 170°C) by using chlorinated fatty acids and amines chlorinated in the industry (Kaga et al., 1996). *In vitro* synthesis has also been introduced for capsaicinoid production as an alternative to chemical and enzymatic synthesis. Tissue culture or cell suspension cultures are used for these purposes. Phenylalanine, ferulic acid and vanillylamine can be used as intermediates to improve the efficiency of the biosynthetic pathway (Sudhakar Johnson et al., 1996). For example, a 45% increase of capsaicin yield was observed when produced through callus culture. The application of 100 µM of vanillin, vanillylamine, *p*-coumaric acid and ferulic acid in the cell culture also enhanced the capsaicin content by 7.5, 5.2 and 2.5 times, respectively (Pandhair and Gosal, 2009). Hence, artificial and *in vitro* synthesis of capsaicinoids can be investigated to improve the production quantity and to obtain the desirable traits in these pungent compounds.

## 11.7	FACTORS AFFECTING CAPSAICINOID PRODUCTION

Capsaicinoid production is influenced by various environmental conditions of the regions where plants are cultivated (Lozada et al., 2021). It is an interaction between plant growth and development and environmental factors (Vázquez-Espinosa et al., 2020, Tito et al., 2018). Biotic stresses such as pests and diseases and abiotic factors such as temperature, water availability, nutrients, light (solar radiation), moisture, salinity and carbon dioxide, as well as growth stage and genetics, are the factors that can cause variation in capsaicinoid concentrations (Uarrota et al., 2021, Luna-Ruiz et al., 2018). The crops grown in open-field conditions are affected by the environmental conditions more than plants grown in controlled systems like greenhouses (Uarrota et al., 2021). Previous studies have also reported that pungency of *Capsicum* fruits is affected by the growing conditions of the plant (Harvell and Bosland, 1997). The environmental conditions can affect the biosynthesis, metabolism and accumulation of metabolites in capsicum plants. It is reported that capsaicinoid concentration was decreased under high temperature and light stress (Gurung et al., 2011). Therefore, environmental parameters play crucial roles in determining the plant growth and fruit yield and the production of secondary metabolites like capsaicinoids in *Capsicum* species.

11.7.1 GENETICS

The pungency trait of capsicums can vary with the genetic makeup of the plants. Significant differences were observed between genotypes, within the genotypes and with genotype–environment interaction (Zewdie and Bosland, 2000). Genotype and ripening relation studies concluded that concentrations of all the three capsaicinoids (capsaicin, dihydrocapsaicin and nordihydrocapsaicin) were low at the mature green stage in all the observed cultivars (Gnayfeed et al., 2001).

The *Capsicum* genome is approximately 3.5 Gbp in size. Most *Capsicum* species are diploid ($n = 12$) in nature, with 24 chromosomes in each nucleus, e.g. the domesticated species *C. annuum, C. frutescens, C. chinense* and *C. baccatum*. However, some wild species like *C. buforum, C. campylopodium* and *C. cornutum* have 26 chromosomes ($n = 13$), which is supposed to be the result of long history of evolution (Moscone et al., 2007). There is one tetraploid species ($n = 2\times = 24$), which is *C. annuum* var. *glabriusculum*, with 48 chromosomes (Tripodi and Kumar, 2019). The studies show that there is a difference of two chromosomal interchanges between *C. annuum* and *C. chinense* (Bosland and Votava, 2012). The genus is divided in 11 clades based on morphological characters, provenance and phylogenetic relationships (Carrizo García et al., 2016). The species with cultivation and breeding importance are from three main groups (Carrizo García et al., 2016). The first group is *Annuum*, with three domesticated (*C. annuum, C. frutescens* and *C. chinense*) and two wild species (*C. annuum* var. *glabriusculum* and *C. galapagoense*). The second one is *Baccatum*, with three forms of varieties (var. *baccatum*, var. *pendulum* and var. *umbilicatum*) and two wild species (*C. chacoense* and *C. praetermisssum*). The third one is *Pubescens*, which only has homonymous domesticated species (Carrizo García et al., 2016). Some of the differences shown by three main *Capsicum* species are presented in Table 11.3 (Eshbaugh, 1976; Moscone et al., 2007). For

TABLE 11.3

Taxonomic Traits of Some *Capsicum* Species (Eshbaugh, 1976; Moscone et al., 2007)

Species	Growth form	Calyx	Flower colour	No. of flowers per node	Fruit shape & colour	Seeds	Chromosome no (2n)	Pollen fertility	Capsaicinoids concentration
C. annuum (Sweet capsicum)	Herb or subshrub (1–2m)	Present	Large white flowers	1	Shape varies; green, red, violet, yellow or orange	Yellow smooth seeds	24	94.6%	0.17–15.36 mg/g of dried fruit (Othman et al., 2011)
C. chinense (hot chillies)	Herb or shrub (0.5–2m)	Absent	Dull white	2–3	Spherical or conical; red, orange, yellow	Yellow smooth seeds	24	97.0%	1.464–2.01 mg/g of fresh fruit (de Aguiar et al., 2016)
C. frutescens (piri piri)	Herb or shrub (1–2m)	Absent	Greenish white	2–3	Elongate; red	Yellow smooth seeds	24	19.8%	1.28–1.59 mg/g of fresh fruit (de Aguiar et al., 2016)

breeders to effectively target traits such as improved yield, pungency, fruit colour, size, shape, water use efficiency and disease resistance, it is very important to elucidate the genome of all *Capsicum* species and the pangenome of each of the commercially important *Capsicum* species.

11.7.2 LIGHT

Light quality and intensity are two of the most important environmental factors that can alter the metabolic pathways in plants. Apart from being the energy source for photosynthesis and carbon assimilation, light also has a key role in growth and development of plants (Uarrota et al., 2021; He et al., 2023; Chavan et al., 2020). For example, eggplants were grown under smart glass film with 85% of UV (221–279 nm), 26% of red (600–699 nm) and 58% far-red (710–850 nm) blocked. The smart glass film significantly reduced the mean season PAR by 34%, thus leading to the reduction of fruit yield by 32%. However, it didn't alter the fruit quality and increased the eggplant sweetness (Chavan et al., 2020). A study on transcriptome analysis of *Capsicum annuum* fruits grown under light-blocking films (LBF) showed that LBF altered the metabolic pathways like photosynthetic activity, carotenoid composition and differentially expressed genes (He et al., 2023). Therefore, it is an important factor in production and accumulation of capsaicinoids (Iwai et al., 1979). By altering the light intensity, capsaicinoid concentration can either increase or decrease, which varies between species, degree of shading and other parameters such as temperature and relative humidity (Jeeatid et al., 2017). The interaction of temperature, light intensity and relative humidity was also examined in the study. In *C. chinense* Jacq., the capsaicinoids production was improved (4,820 and 349 mg/plant) when grown at low temperature and 50% and 70% shading, respectively. But it was decreased under higher radiation and temperature. However, capsaicinoid production was elevated with reduced light intensity (713–783 μmol/m^2/s) and high relative humidity (~80%) (Jeeatid et al., 2018a). It has been proposed that concentration of capsaicinoids decreased with the increase of light intensity, as *Capsicum sp.* belongs to the nightshade family (prefer to grow in shady areas and flower at night). Thus, it is suggested that shading can enhance the accumulation of these secondary compounds (Gurung et al., 2011; Jeeatid et al., 2017). Thus, the study identified that light intensity between 700 and 950 μmol/m^2/s is best for higher production of capsaicinoids (Jeeatid et al., 2017). However, some researchers reported that shading has no effect on the accumulation of capsaicinoids (on pungency of *Capsicum* fruits) (Valiente-Banuet and Gutiérrez-Ochoa, 2016); rather light has a positive effect on it (Arce-Rodríguez and Ochoa-Alejo, 2019). Therefore, it is concluded that light conditions and intensity can affect the synthesis and accumulation of capsaicinoids in *Capsicum* species.

11.7.3 WATER AVAILABILITY

Several studies reported that water stress is directly proportional to the pungency of *Capsicum* fruits, but it is a variety-dependent trait (Jeeatid et al., 2018b; Sung et al., 2005; Pimchan et al., 2012). Water-deficient plants had smaller fruits, while the placental proportion was not disturbed. Therefore, increased placental proportion resulted in a higher amount of capsaicinoids per fruit, as placenta is the production site for these compounds (Sung et al., 2005). The concentration of capsaicinoids increased in all the nine cultivars when subjected to water deficiency (Pimchan et al., 2012). In addition, the contents of two most abundant capsaicinoids, capsaicin and dihydrocapsaicin, were increased in habanero chilli under water stress conditions (Ruiz-Lau et al., 2011).

The enhancement of capsaicinoid concentrations due to water shortage is likely to be due to the stress increasing the activity of enzymes phenylalanine ammonia-lyase and cinnamate 4-hydroxylase for biosynthesis of capsaicinoids and stress reduced their degradation by peroxidases (Contreras-Padilla and Yahia, 1998). However, in a wild native *Capsicum* species, *Capsicum chacoense* of Bolivia, capsaicinoids increased even under high frequency and quantity of irrigation (Haak et al., 2012). In a study in Bhutan and Thailand, 14 chilli cultivars were evaluated under high rainfall and elevation. The results showed increased concentrations of capsaicinoids in all the 14 varieties

(Gurung et al., 2011). Another study on high relative humidity caused 66.7% variation in total capsaicinoid content with an increase in their concentration (Jeeatid et al., 2018a). It can be inferred that pungency may be enhanced by water stress or waterlogging, depending upon the ecological and physiological factors.

11.7.4 MINERAL NUTRIENTS

Capsaicinoid concentrations in capsicum and chilli plants may be affected by altering the concentrations of available minerals to the plants. Nitrogen and potassium are the two main components studied mostly in this regard. Nitrogen availability is the key factor because the synthesis of capsaicin requires three amino acids (phenylalanine, valine and leucine). Higher nitrogen concentration in the growing medium is directly linked to increased capsaicin content in capsicums cultivated in soil (Johnson and Decoteau, 1996), hydroponics (Aldana-Iuit et al., 2015) or *in vitro* (Monforte-González et al., 2010). However, higher potassium applications were observed to have a negative impact on capsaicin concentration in *C. chinense*. It not only reduced capsaicin concentrations but also lowered the nitrogen content in leaves (Medina-Lara et al., 2008). During hydroponic cultivation of capsicums and chillies, smaller capsaicin concentration in the placenta was found by lower concentrations of nitrogen in the media. However, the change in potassium did not cause any changes in capsaicinoid accumulation (Monforte-González et al., 2010). Hence, it can be concluded from the available data that pungency of capsicums and chillies can be manipulated with agronomical management of nutrients applied to the crops.

11.7.5 TEMPERATURE

The biosynthetic mechanisms of most of the tropical and sub-tropical fruits undergo variations in response to temperature alterations. These changes may include a higher respiration rate, ethylene formation, increased electrolyte leakage, toxin accumulations and, finally, cell collapse (Finger and Pereira, 2016). The activity of enzymes responsible for the synthesis of capsaicinoids in capsicums may be affected by introducing plants to short-term temperature stress (high or low temperatures). The capsaicin content of six cultivars of green and ripened capsicums were observed under both high (29.9°C) and control temperatures (24.1°C) (Rahman et al., 2012). The high-temperature treatment was found to be more effective on capsaicinoid content enhancement than the control. Moreover, ripe fruits under high temperatures were found to be richer in secondary metabolites than in the green fruits (Arce-Rodríguez and Ochoa-Alejo, 2019). It is suggested that pungency will be different for the same varieties grown in different temperature conditions such as hot and semi-dry versus cool and wet. The capsaicinoids will be higher in fruits grown in a dry, semi-arid region as compared to the cooler environment (Guzmán and Bosland, 2017). A research study was performed in the Yucatan region in Mexico on three "superhot" chilli varieties, namely bhut jolokia, Trinidad moruga scorpion and Carolina reaper. The spiciness of these fruits in SHU was elevated by 48%, 15% and 27%, respectively, when grown in the Yucatan region under high temperature (maximum 37°C), low rainfall, high relative humidity (86%) and alkaline soil (Munoz-Ramírez et al., 2018). Moreover, the Scoville unit of the Carolina reaper increased from 2.2 million to 3 million, which means that edaphoclimatic conditions of Yucatan are more suitable for chilli cultivation specifically for heat purposes (Lozada et al., 2021). In that study, the Carolina reaper also came second to the Trinidad moruga scorpion variety in SHU terms, though it holds the world record for being the hottest chilli in the world grown under optimal climate conditions. Hence, temperature is an important factor to control the pungency of *Capsicum* fruits.

11.8 GROWTH AND DEVELOPMENTAL STAGES

The *Capsicum* plant starts to produce capsaicinoids in the early stages of fruit development, mostly 10 days after pollination (Fayos et al., 2017). The concentration tends to increase with the fruit

TABLE 11.4

Taxonomic Traits of Some *Capsicum* Species (Eshbaugh, 1976; Moscone et al., 2007)

Species	Growth form	Calyx	Flower colour	No. of flowers per node	Fruit shape & colour	Seeds	Chromosome no (2n)	Pollen fertility	Capsaicinoids concentration
C. annuum (Sweet capsicum)	Herb or subshrub (1–2m)	Present	Large white flowers	1	Shape varies; green, red, violet, yellow or orange	Yellow smooth seeds	24	94.6%	0.17–15.36 mg/g of dried fruit (Othman et al., 2011)
C. chinense (hot chillies)	Herb or shrub (0.5–2m)	Absent	Dull white	2–3	Spherical or conical; red, orange, yellow	Yellow smooth seeds	24	97.0%	1.464- 2.01 mg/g of fresh fruit (de Aguiar et al., 2016)
C. frutescens (piri piri)	Herb or shrub (1–2m)	Absent	Greenish white	2–3	Elongate; red	Yellow smooth seeds	24	19.8%	1.28 -1.59 mg/g of fresh fruit (de Aguiar et al., 2016)

development till 40 to 50 days after pollination and then starts to decrease (Barbero et al., 2014). *Capsicum* ripened fruits are comprised of more capsaicinoids than immature ones (Bae et al., 2014). Six chilli varieties from Malaysia were compared for their content of capsaicinoids and antioxidant properties (Alam et al., 2018). The results showed that Cili Padi Putih has the highest concentration of capsaicinoids (3.41 ± 0.03 mg/g) among all the six varieties, and the highest antioxidant concentrations are observed in the Cili Padi Ragnip variety. Despite the fact that fruits of the latter variety were harvested at the fully ripened stage, it had lower capsaicinoid concentration (3.35 ± 0.01 mg/g). A study analysed capsaicinoids at different plant stages, i.e. 16 to 48 days after anthesis. Capsaicin and dihydrocapsaicin both were identified in higher concentrations during later developmental stages (Jang et al., 2015). In another study, different capsaicinoid concentration was calculated for green and red fruit maturity stages of different *Capsicum annuum* cultivars (Bronowicka Ostra, Cyklon, Tornado and Tajfun). All the cultivars except for Tornado had higher capsaicin concentration at the fully ripened stage (0.53, 0.34, 0.035 and 0.047 mg/g dry weight, respectively) as compared to ripening stage (0.44, 0.29, 0.043 and 0.029 mg/g dry weight, respectively) (Materska and Perucka, 2005).

11.9 PROPERTIES OF CAPSAICINOIDS AND CAPSINOIDS

In plants, the primary role of capsaicinoids is to provide defense against insects and rodents. These compounds also facilitate seed dispersal of wild pepper plants by mediating interactions with birds, which are immune to the burn of hot capsicums (Mares-Quiñones and Valiente-Banuet, 2019). In addition to the pungent flavor, *Capsicum* genuses have many biochemical and pharmacological properties. They have anti-microbial, antioxidant, anti-inflammatory, anti-allergic and anti-carcinogenic abilities, which make these fruits unique to study (Lee et al., 2005). For instance, ripened chilli fruits are rich with provitamin A that provides immunity against arthritis, cardiovascular diseases, cancer and the aging process (Nishino et al., 2009).

The character of *Capsicum* species of producing pungent compounds makes them a unique member of the *Solanaceae* family. Along with producing a heat sensation in food, the capsaicinoids also have many medicinal, chemical, agricultural and cosmetic uses (Chapa-Oliver and Mejía-Teniente, 2016). They are being studied as antioxidants and radioprotectors in physiological systems (Gangabhagirathi and Joshi, 2015). Capsaicinoids are also identified with anti-mutagenic and anti-tumoural properties (Surh and Lee, 1995), and they are of great interest to researchers recently (Clark and Lee, 2016). They are used as topical pain relief for diseases like toothache, arthritis, osteoarthritis, post-herpetic neuralgia and other neurological conditions (Batiha et al., 2020). Capsaicinoids also aid in weight control, diabetes, gastrointestinal and gastrovascular problems, cough and sore throat (Luo et al., 2011; Batiha et al., 2020). Because of these pharmaceutical properties of capsaicinoids, research is being done to improve their production in *Capsicum* by manipulation of cultivation technologies, chemical and enzymatic synthesis, or *in-vitro* through cell or tissue culture (Reyes-Escogido et al., 2011).

Capsinoids are non-pungent analogues of capsaicinoids. Capsinoids have an ester bond instead of an amide bond in their structure linking vanillyl and acyl chains (Snitker et al., 2009). Capsinoids lack pungency due to their hydrolysis while crossing the oral mucosa, but their activity on transient receptor potential vanilloid 1 (TRPV1) receptors is retained (Snitker et al., 2009). Capsinoids also have many biological, anti-tumour, antioxidant and anti-obesity properties. Their clinical application in cancer prevention and weight loss is considered to be more beneficial than capsaicinoids because they are less harmful than their pungent analogues at higher doses (Antonio et al., 2018). Capsinoids enhance lipid metabolism, thus decreasing fat accumulation. Capsinoids are also found effective in prevention of cardiovascular and gastrointestinal problems (Antonio et al., 2018). Capsinoids decreased insulin levels, body weight, visceral fat accumulation and glucose tolerance in diabetic rats, hence improving their glucose metabolism (Kwon et al., 2013). A study in Std ddY mice with capsiate (a member of the capsinoid group) resulted in increased metabolic and fat oxidation rate due to regulation of mRNA level of uncoupling proteins and thyroid hormones, which have a major role in energy expenditure and thermoregulation (Masuda et al., 2003).

11.10 EXTRACTION OF CAPSAICINOIDS

11.10.1 Methods of Extraction

Capsaicinoids are extracted from capsicum and chilli fruits by various techniques. The conventional methods are maceration, solvent extraction, Soxhlet extraction and enzymatic extraction. The old methods require large amount of solvent, higher temperature and long time of extraction. Therefore, modern and advanced extraction techniques are now being investigated which are more time, energy and solvent efficient. These are ultrasound-assisted extraction, solid-phase microextraction, supercritical fluid extraction, pressurised liquid and microwave-assisted extraction (Sganzerla et al., 2014).

The ultrasound-assisted extraction technique is suggested to be simple and low cost among other modern techniques (Deng et al., 2012). Microwave- and ultrasound-assisted extractions are modern methods which combine the old solvent extraction techniques with the new microwave and ultrasound treatments. Ultrasound-assisted extraction was compared with microwave-assisted extraction and the Soxhlet procedure for capsaicinoid extraction from dry chillies (Chuichulcherm et al., 2013). The results concluded that ultrasound-assisted extraction is the most efficient extraction technique. However, the capsaicin yield from microwave-assisted extraction was higher than ultrasound-assisted extraction (5.28 and 4.01 mg/g dried chili), but the latter required less energy consumption and time. Supercritical fluid extraction has also been investigated by researchers as a modern technique for capsaicinoid extraction. When coupled with carbon dioxide, it is more efficient and reduces the degradation of secondary compounds due to light or high temperature during the extraction process (Wu et al., 2016). The final product has fewer solvent residues and provides higher yield by consuming less energy, time and organic solvent (Santos et al., 2015). The capsaicinoids yield of *Capsicum* species obtained by different extraction techniques is shown in Table 11.5.

TABLE 11.5

Capsaicinoid Yield through Different Extraction Techniques

Name of Technique	Extraction factors	Source	Capsaicinoids yield	References
Soxhlet extraction	Methanol, 2 hours, 1 atm	Jamaican hot red fruit	50 g/kg of dry ground fruit	(Bajer et al., 2015)
		Yellow Habanero fruit	0.95 g/kg of dry grounded fruit	
	Ethyl acetate, 6 hours, 25°C	Malagueta pepper	3.36 g/kg of dried sample	(Santos et al., 2015)
Ultrasound-assisted extraction	Hydroethanolic solution, 1 hour, 37°C, 40KHz	Habanero chilli	19 mg/g extract	(Martins et al., 2017)
	Methanol, 10 minutes, 50°C, 1atm	Cayenne	713 µmol/kg of fresh pepper	(Barbero et al., 2008)
	Olive oil, 20 minutes, 28.5/31.5 kHz	Red hot chilli pepper powder	169.9 mg/kg pepper	(Paduano et al., 2014)
	Ethanol, 5–30 minutes, 40kHz, 600W	*Capsicum frutescens* L.	4 mg/g dried chilli	(Chuichulcherm et al., 2013)
Microwave-assisted extraction	Ethanol, 5 minutes, 125°C, 1 atm	Cayenne	717 µmol/kg of fresh pepper	(Barbero et al., 2006)
		Long marble pepper	564 µmol/kg fresh pepper	
	Olive oil, 60s, 500W	Red hot chilli powder	164.7 mg/kg pepper	(Paduano et al., 2014)
	90, 320, 360, 600W	*Capsicum frutescens*	5.2 mg/dried chilli	(Chuichulcherm et al., 2013)
Microwave + Aqueous two-phase extraction	20% of ethanolic extract, 25% of NaH$_2$PO$_4$	Cumari-do-Para (*Capsicum chinense*)	Extraction efficiency: 85.6%	(Cienfuegos et al., 2017)
Pressurised liquid extraction	200°C, 20MPa, 10 + 20 minutes of static extraction	*Capsicum chinense; Capsicum annuum*	Capsaicin 20µg/g; 264µg/g	(Bajer et al., 2015)
Superfluid extraction	Acetonitrile, acetic acid, water, 40°C, 15MPa	Malagueta pepper (*Capsicum frutescens* L.)	115 mg/g extract	(de Aguiar et al., 2018)
	40°C, 15MPa	Malagueta pepper	Yield 76.1%	(Santos et al., 2015)
	90.2 kg/h CO$_2$ flow, 35°C, 10MPa	Habanero chili	Yield 92%	(Rocha-Uribe et al., 2014)
	CO$_2$, 40°C, 25MPa	Dedo de moça pepper	1.25 g/kg of raw material	(Dias et al., 2016)
	5% methanol	Green	60.33 ng/g	(Hamada et al., 2019)
		Yellow	31.79ng/g	
		Red bell pepper	35.38 ng/g	
Supercritical fluid extraction + ultrasound-assisted extraction	40°C, 15MPa, 150 mins 360W, 60 mins	Malagueta pepper (*Capsicum frutescens* L.)	Yield 9%	(Santos et al., 2015)

The efficiency of an extraction method depends on factors like temperature, time, solvent type, solvent percentage and sample-to-solvent ratio. A variety of different solvents has been used for capsaicinoid extraction. Some commonly used solvents are methanol, ethanol, acetonitrile, ethyl acetate, isopropanol, acetone, chloroform and hexane etc. A previous study suggested acetone to be an efficient solvent for capsaicinoids extraction from dehydrated capsicum fruits (Attuquayefio and Buckle, 1987). Later on, methanol, ethanol and acetonitrile were proposed to be more efficacious solvents as compared to acetone (Barbero et al., 2006). Another study observed ethanol and acetonitrile as better solvents for extraction of capsaicin from fresh chilli samples, whereas acetone was better for dried samples (Chen and Kang, 2013). Ethanol is the only organic solvent which is found to be non-toxic for capsaicinoid extraction (Stoica et al., 2016). A study investigated six solvents for optimisation of extraction method for capsaicin and reported methanol with the highest extraction yield of 123.47 µg/g. Extraction yield with ethanol was second highest, followed by isopropanol, acetonitrile, ethyl acetate and acetone, with yields as 110.84 µg/g, 99.10 µg/g, 81.22 µg/g, 77.59 µg/g and 71.62 µg/g, respectively (Mokhtar et al., 2016). Methanol and ethanol were found to be the best solvents for dihydrocapsaicin extraction as well. Water is not considered a good solvent for these pungent metabolites (Barbero et al., 2008). The addition of water to methanol or other solvents can improve their extraction yield (Rostagno et al., 2003). For example, 5%, 10%, 25% and 50% of water was added to methanol for capsaicinoids extraction. The results reported a higher yield with 25% water as compared to the pure solvent, and the yield was improved by 54% and 62% for capsaicin and dihydrocapsaicin, respectively (Mokhtar et al., 2016).

The temperature of the extraction procedure also participates in determining the capsaicinoid yield. It is believed that high temperatures will improve the efficacy of an extraction method. Extraction time is also crucial, as the optimum time is required to obtain highest yield. A study tested different times and temperatures for extraction of capsaicinoids. The result reported 100°C for 25 minutes provided the highest capsaicinoid amount, with 75% methanol (Mokhtar et al., 2016). The capsaicinoids started degrading when the time and temperature increased from 20 minutes and 100°C, respectively (Barbero et al., 2008; Supalkova et al., 2007).

11.11 PROPERTIES OF CAPSAICIN

Capsicums are being used for their medicinal benefits in different countries. They were used for treating asthma, coughs and sore throat by the Mayas and for toothaches by the Aztecs in old civilisations (Saleh et al., 2018).

Capsaicin is the principal component responsible for pungency and irritation caused by consuming the *Capsicum* fruits. Capsaicin is reported to control the appetite and increase the body temperature. This temperature regulation can help in weight loss (Backonja et al., 2010). Transient receptor potential vanilloid subtype 1 (TRPV1) is the receptor channel for capsaicin and can be controlled through various mechanisms of low pH, temperature, osmotic sensing, pressure and molecules (Bae et al., 2016). These channels are widely distributed in the brain, sensory nerves, dorsal root ganglia, bladder, gut and blood vessels (Geppetti and Trevisani, 2004). TRPV1 is mainly responsible for controlling sensory neurons. When capsaicin makes a bond with these sensory neurons, a sense of pain, heat and inflammation is produced. Topical creams and heat patches of capsaicin are used for treatment of neuropathic pain. Capsaicin provides cardiometabolic protection through activation of TRPV1 channels in different organs and tissues. Thus, helps in treatment of cardiometabolic diseases like hypertension, obesity, dyslipidaemia, diabetes and atherosclerosis (Geppetti and Trevisani, 2004; Sun et al., 2016). However, the daily dosage of capsaicin required for cardiometabolic protection is still under investigation (Sun et al., 2016).

Capsaicin is also investigated as an antimicrobial agent for many foodborne pathogens such as *Listeria monocytogenes, Helicobacter pylori, Pseudomonas aeruginosa, Botrytis cinerea, Aspergillus niger* and *Penicillium expansum* (Omolo et al., 2014; Tayseer et al., 2020). It is also observed as an antivirulence agent against *Vibrio cholerae, Staphylococcus aureus* and

Porphyromonas gingivalis (Zhou et al., 2014). Capsaicin is also known for its antioxidative, hypotensive and anticancer effects. A study performed in rats showed that dietary capsaicin can be used for prevention of Alzheimer's disease (Xu et al., 2017). Capsaicin extract from habanero pepper, administered as a diet supplement for 6 months, was tested on 60 dogs for anticancer evaluation. The tumour size decreased by 5% to 50% in 15 dogs and was unaffected in 29 dogs, whereas it increased by 10% to 30% in 5 dogs. However, the preliminary observations suggested that dogs were tolerant to capsaicin. This can be a basis for future capsaicin studies in dog oncology (Adaszek et al., 2017). In a study, the effect of capsaicin and dihydrocapsaicin on number of white blood cells, lymphocytes and NK cells in rats was studied. The effect of dihydrocapsaicin was relatively higher on these cells as compared to capsaicin.

Dihydrocapsaicin has been reported to provide functional recovery from neurological problems after cerebral ischemia and reperfusion (Janyou et al., 2020). Dihydrocapsaicin has a similar role as capsaicin in performing bioactive actions, but it is considered to be stronger for producing hypothermia than capsaicin (Akimoto et al., 2009). In a study, the effect of capsaicin and dihydrocapsaicin on number of white blood cells, lymphocytes and NK cells in rats was studied. The effect of dihydrocapsaicin was relatively higher on these cells as compared to capsaicin (Akimoto et al., 2009).

11.12 POSTHARVEST TECHNOLOGY TO INCREASE PRODUCTION OF CAPSAICINOIDS

Capsicums and chillies are perishable fruits which require careful postharvest handling. Proper postharvest processes are essential to maintain the quality of the product despite its consumption form (fresh or processed). All types of *Capsicum* spp. fruits are suitable for fresh use, but some are widely famous for their fresh consumption, for example, jalapeno, New Mexican green chilli and pimiento. The grade 1 quality fresh *Capsicum* spp. is described as one with firm fruit, bright colour, thick flesh and firm green calyx (Lownds et al., 1993). Moreover, it should be free from disease, bruises and abrasions. Immature fruits usually are soft and thin walled, pliable and pale-green coloured (Bosland and Votava, 2012). In summer, harvest of capsicums and chillies should be done in the early morning, and fruits should be removed from direct sunlight as soon as possible (Boyette et al., 1990), as they are very likely to dehydrate and get sunscald and heat damage (Lownds et al., 1993). Quality losses can occur if *Capsicum* spp. fruits are not cooled within 1 to 2 hours of harvesting or not kept refrigerated for a few days. This is due to temperatures above 21°C enhancing respiration and ethylene production of the fruit. Therefore, refrigeration helps to prolong the shelf life and quality of fruits by reduced water loss, ethylene production, colour change and disease attack (Bosland and Votava, 2012). The optimum temperature for storage is 7°C to 10°C with 85% to 90% relative humidity for up to 3 weeks (Kader, 2002). Prior to storage, precooling of fruits is achieved through room cooling or the more efficient technique of forced-air cooling.

Capsicum spp. fruits are highly sensitive to freeze injury and chilling injury, which occur at 0°C and 4°C, respectively (Bosland and Votava, 2012). The first injury symptom appears as dot pitting, which changes into sheet pitting, sunken lesions or even tissue collapse if the disorder continues for longer periods (Lim et al., 2007), and chilling injury is both cultivar and maturity dependent (Lim et al., 2007). It has been observed that ripened capsicums are less susceptible to chilling stress than the mature green fruit (Lim et al., 2007). Moreover, prevention of water loss during storage is another crucial factor of capsicum postharvest because water loss causes shrivelling and softening of fruit and initiates fruit senescence. Both chilling injury and water loss are considered to be directly related to each other (Smith et al., 2006). Hence, storage of *Capsicum* spp. fruits at a moderately low temperature to prevent chilling injury is an important postharvest process. Controlled atmosphere storage and moderate atmosphere storage are two techniques for *Capsicum* spp. fruits postharvest storage, with the latter proven to be more beneficial in improving the shelf life of the

TABLE 11.6

Postharvest Treatments for Capsicums and Chillies

	Treatments	Purpose
Chemical	Gaseous ozone	To decrease bacterial and fungal contamination
	Methyl jasmonate and methyl salicylate	To reduce pathogen infection and chilling injury, to prolong shelf life
	Brassinolide	To prevent chilling injury and improve quality (minimising decline in chlorophyll and vitamin C)
	Glycine betaine	To reduce cellular leakage and membrane damage
	Calcium chloride	To maintain firmness and quality
Physical	Heat shock	To prevent microbial rot and chilling injury, to maintain quality
	Intermittent warming	To reduce chilling injury and improve storage life
	Low-temperature conditioning	To withstand cold temperatures
	Irradiation with ultraviolet light	To improve natural pathogen resistance, reduce fungal infestation, storage quality
	Gamma irradiation	To improve hygienic quality and shelf life

fruit (O'Donoghue et al., 2018). Postharvest processing of *Capsicum* spp. fruits is done through various techniques such as canning, brining and pickling, freezing, fermentation, dehydration and powdering, preparation of oleoresin extracts and capsaicinoid extraction (Bosland and Votava, 2012). Some postharvest treatments are listed in Table 11.6.

11.13 METABOLOMICS IN CAPSICUMS

The term "metabolomics" was first introduced in 1998, defining it as the approach to analyse the change of relative metabolite concentrations due to the alteration of gene expression (Oliver et al., 1998). Metabolomics is the metabolic profiling of cells, tissues or organisms in relation to genetic variation or external stimuli (Idle and Gonzalez, 2007). Four approaches are being used in metabolomics: target analysis, metabolite profiling, metabolomics and metabolic fingerprinting (Fiehn, 2002). It is considered to be a combination of analytical technologies to identify and quantify the cellular metabolites with application of statistical and multivariant methods for data extraction, mining and interpretation (Roessner and Bowne, 2009).

The biosynthesis of capsaicinoids in plants is divided into two pathways, phenylpropanoid and fatty acid metabolism. The former is responsible for phenolic structure and the latter for the fatty acids of the capsaicinoid molecules (Ochoa-Alejo and Gómez-Peralta, 1993). Previous research work suggested that capsaicinoid levels increase by applying hydric stress to these plants. It alters the phenylpropanoid pathway and increases the activity of several enzymes like phenylalanine ammonia-lyase, cinnamic acid-4-hydroxylase and capsaicin synthase, which are all responsible for capsaicinoids synthesis (Estrada et al., 1999).

Capsicums are classified as non-climacteric fruits because they do not show a sudden rise in ethylene production at the onset of ripening (Pretel et al., 1995). Some hot chillies are respiratory climacteric, while Mexican varieties are non-climacteric (Gross et al., 1986). There are different metabolites responsible for colour, flavour, aroma and nutrition of capsicums and chillies. *Capsicum* sp. are enriched with antioxidants like flavonoids, carotenoids, vitamin C and capsaicinoids, and these generally increase with the maturity of fruit. Metabolomic study by metabolite profiling (GC-MS) of capsicum tissue was initially done to identify expression of metabolites during ripening (Aizat et al., 2014). The enzymatic and LC-MS targets were then used to further quantify more classes of metabolites in additional ripening stages. Pathways of different capsicum metabolites like monosaccharides (glucose, fructose), organic acids (malate, citrate), an amine (putrescine) and amino acids (glutamine, isoleucine, serine, valine) have been observed by metabolomics. The study identified 81 unique known compounds in capsicum (Aizat et al., 2014). The two main observations described from the study were the modification of sugars and downstream compounds and the changes of metabolites related to the ethylene pathway (Aizat et al., 2014).

For the biochemical analysis of *Capsicum annuum* fruit, non-targeted metabolomic analysis was done on six different developmental stages which are 16, 25, 36, 38, 43 and 48 days postanthesis (Jang et al., 2015). It was observed that distinct changes occurred in the metabolites, gene expressions and antioxidant activities during early (16–25 days postanthesis), breaker (36–38 days postanthesis) and later (43–48 days postanthesis) stages. Capsaicin and dihydrocapsaicin were found in higher concentrations during later stages. Pathway analysis showed metabolite–gene interactions during biosynthesis of compounds like amino acids, capsaicinoids, fatty acids and flavonoids. The main compounds found in *Capsicum* fruits are capsaicinoids, carotenoids, phenolics, ascorbic acid and some minor compounds like alcohols, terpenes, aldehydes, ketones, fatty acids and fatty acid esters (Saha et al., 2015).

11.13.1 Carotenoids

Capsicum varieties can be differentiated visually based on their colour and shape Mostly, the unripened fruits are green, yellow or white and later turn to red, dark red, brown or, in some cases, almost black upon maturation (Ha et al., 2007). Fruit maturation and ripening is a complex process regulated by many genes and proteins (Klie et al., 2014; He et al., 2023). In *Capsicum* species, the process involves certain modifications in metabolism, respiration, volatile compounds, chlorophyll content (into carotenoids), protein alterations, flavour and changes in total soluble contents (Mateos et al., 2013). These species are carotenogenic, which means they produce carotenoids upon ripening (Figure 11.4). In most species, the green chlorophyll disappears and is converted into two different red carotenoids by capsanthin–capsorubin synthase. The two pigments are keto-carotenoids capsanthin and capsorubin, which are formed from the appropriate 5,6-epoxycarotenoids (antheraxanthin and violaxanthin) by a pinacol rearrangement (Deli and Molnar, 2002). The yellow and orange pigments are mostly obtained from β-carotene, zeaxanthin, lutein, β-cryptoxanthin, violaxanthin and/or capsolutein (Topuz et al., 2009). Carotenoids are fat-soluble antioxidants, and they are important for health of the epithelial cell layer in humans (Byers and Perry, 1992). The bright colours of the fruits are used by the plants to attract birds in order to spread seeds for reproductive success.

Carotenoids are pigments responsible for fruit colour and are sometimes an indication of nutrition and fruit quality in *Capsicum* fruits. Total carotenoids can vary among fruits of different cultivars and can be affected by plant stage and environmental conditions (Russo, 2012). The biosynthesis of carotenoids is regulated during fruit ripening, and it is observed to increase up to 90% in red capsicum genotypes. The ripening phase begins when green chloroplasts of immature fruit start to convert into red chromoplasts. The carotenoid change during several fruit ripening stages is shown in Figure 11.4 (Russo, 2012).

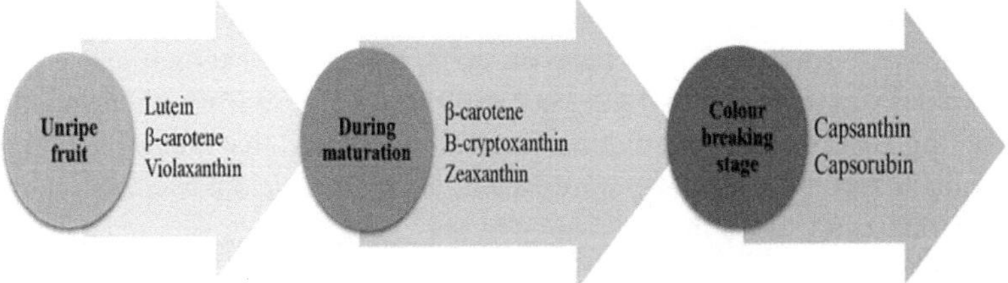

FIGURE 11.4 Carotenoid changes during fruit ripening.

The synthesis of carotenoids in *Capsicum* fruits can be affected by the light environment. Carotenoids in red and orange *Capsicum* varieties are subjected to be changed by red, blue and UV lights. A study observed the effect of light-blocking films on orange capsicum cultivars (O06614, *Capsicum annuum* L.) at the mature green and mature ripe stages (35 and 65 days after pollination, respectively) (He et al., 2022). The targeted metabolite analysis showed that these films increased lutein but suppressed the ratio of zeaxanthin and neoxanthin at 35 days after pollination. The study identified differentially expressed genes DEGs that are responsible for the plant adaptation to the altered light conditions. Moreover, it was observed that light-blocking films downregulated the UV light receptor UVR8, light-signalling DEGs and transcription factors phytochrome-interacting factor 4 (PIF4) at 65 days after pollination (He et al., 2022). Total carotenoids in red (Gina) and orange (O06614) varieties grown under smart glass film were not affected by the reduced red (-26%), far-red (-58%) and overall (-19%) of photosynthetically active radiation (PAR, 380–699 nm) (He et al., 2022; Chavan et al., 2020).

11.13.2 PHENOLICS

Phenolic compounds are important because of their antioxidant, antiradical and signaling properties in plant and human cells (Oboh and Rocha, 2007). They also help in prevention of diseases like cancer and cardiovascular and neurodegenerative problems (Ghasemnezhad et al., 2011). Different *Capsicum* species have varying amounts of phenolics depending upon plant maturity and colour. A study recorded the highest phenolic compounds in red ripe capsicum cultivars of Malagueta and Cambuci as compared to their green unripe fruits (Ghasemnezhad et al., 2011). This shows that phenolic concentration can vary with developmental stages of capsicum fruits. In another study, 20 *Capsicum* varieties were compared, and their phenolic contents ranged from 0.35 (±0.03) mg GAE/g for Cambuci Verde to 3.06 (±0.12) mg GAE/g for Naga Jolokia (highest among all the varieties). Moreover, it was also observed, based on the correlation between capsaicinoids and phenolics, that the two groups of metabolites (capsaicinoids vs. phenolic compounds) are independent of each other. For example, Biquinho pepper did not have capsaicinoids, but it had the third-highest phenolic compounds in the study (de Aguiar et al., 2016). The most abundantly found phenolics in green capsicum are myricetin (658 µg/g) and pyrogallol (572 µg/g), while pyrogallol (757 µg/g) was highest in red capsicum. The orange capsicums were found to have gallic acid (900 µg/g) and chlorogenic acid (117 µg/g) in abundance (Anaya-Esparza et al., 2021).

11.13.3 ASCORBIC ACID (VITAMIN C)

Ascorbic acid, known as vitamin C, is essential for the human body to maintain immunity and normal well-being. It is abundantly found in fruits and vegetables. *Capsicum* fruits are rich sources of ascorbic acid and concentration varies due to genome, age and maturity of fruit (Lau et al., 2015). Fresh capsicum fruits and vegetables (mainly fresh green chillies) are reported to have higher Vitamin C content as compared to highly processed and aged chillies (Zeng et al., 2005; Paul and Ghosh, 2012). The ascorbic acid oxidises to dehydroascorbic acid and irreversible 2,3-diketo-L-gulonic acid with time and thermal processing of food. It has been quantified in capsicum and chillis in several studies using electrochemical techniques, especially HPLC and reagent-assisted UV-vis spectroscopy (Zeng et al., 2005). In a recent study, two capsicum varieties, namely Red (Gina) and Orange (O06614), were cultivated under smart glass film (SGF) in a glasshouse. These films alter the light quality and quantity by blocking heat-generating radiation. The results showed that altered light had no effect on fruit development and ripening. However, it reduced the ascorbic acid concentration in fruits within the acceptable market rang, i.e. 40 mg to 45 mg (He et al., 2022).

The fruits of capsicum and chillies have nutritional and medicinal importance because of their rich metabolic profile. Phytochemicals such as carotenoids, ascorbic acid, phenolics and capsaicinoids are the primary compounds in this regard (Cisternas-Jamet et al., 2020). The metabolites also play role as plants' natural defence against insects, pathogens and microorganisms, and they help the plant to respond under stress conditions. Therefore, the metabolic pathways and their responses to different environmental conditions are of keen interest to researchers to improve the capsicums yield and economic value.

11.14 SOLUTIONS TOWARD REDUCING CAPSICUM AND CHILLI WASTE

The food waste issue is increasing worldwide, while a large population is still experiencing hunger in some developing countries. Even after fulfilling the nutritional requirement, most of the food is not consumed in time, and it is treated as waste (Viaggi, 2018). Food loss is identified as the commodities lost in the supply chain up to retail. Food waste refers to the food and inedible parts which are wasted by retail and consumers, including households and food service (Forbes et al., 2021). Food production is one of the largest users of freshwater, land and energy resources. It is responsible for 70% of global water withdrawal (Guo et al., 2023). Almost one-third of all the food produced is lost or wasted, which accounts for 1.3 billion tonnes of food and $940 billion annually (Gustafsson et al., 2013). This accounts for almost 65 kg of food waste by a person every year. The waste then increases the demand for more food production and hence the environmental impact related to its production. The destination of the wasted food is landfills, sewer, litter, compost, aerobic digestion or co-anerobic digestion etc. According to a 2019 report, the total estimated global food waste was 931 million tonnes from all three sectors, which are household, food services and retail (responsible for 61%, 26% and 13%, respectively) (Forbes et al., 2021). Globally, food waste is the third-largest source of greenhouse gas emissions after China and the United States. It also contributes towards 10% of the total emission (UNEP, 2021). This greenhouse gas emission is worse than emission from airplanes, plastic production and even oil extraction, which is 1.9%, 3.8% and 3.8%, respectively. Moreover, food dumped in landfills produces methane gas 28 times stronger than carbon dioxide (IPCC, 2019). Therefore, it is important to prevent the food waste and losses to improve food security, the environment and economic development (Malorgio and Marangon, 2021). The food waste data index includes both edible and non-edible food items. Mostly in less developed (low- or middle-income) countries like Sub-Saharan Africa, the household waste usually comprises of leftovers and inedible parts of food, e.g. fruit peels, vegetable seeds, potato peels etc. (Mucyo, 2013). Sustainable Development Goal (SDG) 12.3 is focused on food waste and its inedible portion that exits the supply chain and is thus lost or wasted. Another better solution for food waste is valorisation or recycling of the waste. Waste valorisation means to reuse, recycle or compost the waste material to produce a more valuable product. The valorised products from the wastes can include nutraceuticals, useful materials for a variety of applications and biofuels (Arancon et al., 2013).

More than 40% of food waste occurs due to mishandling of fresh fruits and vegetables during storage and postharvest processing (Sawatdeenarunat et al., 2016). Fruit and vegetable waste accounts for 12% and 25% of the total food wasted by a single person every year in the world (Chen et al., 2020). Despite the increase in *Capsicum* production worldwide, the annual losses are estimated to be 40% (Anaya-Esparza et al., 2021). Almost 30% of the fresh capsicum fruits are discarded as non-marketable products or by-products. The waste produced after industrial processing includes stalk, peels, seeds and unused flesh (Sandoval-Castro et al., 2017). This crop waste is a raw source of bioactive compounds like capsaicinoids, phenols, flavonoids and carotenoids in the case of *Capsicum* species.

11.15 SUSTAINABLE USE OF WASTE OF CAPSICUM CROPS

Seeds from capsicums and chillies are now being investigated to reduce and reuse waste because they are enriched with bioactive compounds. One of the main reasons is to reduce waste material produced as a result of processing of these fruits by industries (Silva et al., 2013). For instance, non-fermented pickle slices are the main industrialised product of jalapeño pepper, and almost 500 tonnes of by-products are generated in Mexico (Sandoval-Castro et al., 2017). These by-products include seeds and placenta from fruits, which are discarded as waste after scalding. It is a source of overall economic losses due to vegetable waste (23% of total cost). A study analysed these by-products and found them to be effective source of phenolics, capsaicinoids (40% less than fresh fruit) and other antioxidants (Sandoval-Castro et al., 2017).

The seeds of *C. annuum* (bell pepper) were reported to be enriched with sterols and triterpenes like botulin (161 mg/kg) and campesterol (54.1 mg/kg). Myristic, pentadecylic and palmitoleic acids were also identified in seeds. Moreover, these compounds were observed to have antioxidant properties and acetylcholinesterase-inhibitory effects (Silva et al., 2013). Capsicum seeds were reported to have higher antioxidant activity as compared to the fruits. The study used DPPH, ABTS and reducing power methods to analyse the antioxidant potential of seeds and fruit pulp. The results showed that seed extract had the highest antioxidant capacity (11.32, 89.25 and 9.94 µmol TE/g, respectively) while pulp extract had 2.28, 17.17 and 3.99 µmol TE/g, respectively. The total phenolic content of seeds was also higher (409 mg gallic acid equivalent/g) than fruit pulp (119 mg gallic acid equivalent/g) (Sora et al., 2015). A study obtained Hevein-like antimicrobial peptides from leaves of capsicum plants that have antifungal and antibacterial traits (Games et al., 2016). The extracts of *Capsicum annuum* inhibited the growth of fungi *Penicillium expansum* and *Debaryomyces hansenii*, which are potential food pathogens and toxins (Nazzaro et al., 2009). Cayenne pepper extract had antimicrobial activity against natural microflora, coliforms, moulds and *Staphylococcus aureus* in the Kareish cheese (Wahba et al., 2010). The treatment of minced beef with *C. annuum* extract inhibited the growth of *S. typhimurium* (Careaga et al., 2003). Peppers are also used as a natural colour enhancer in the food industry. Therefore, products from *Capsicum* fruits and waste can be used in the pharmaceutical, cosmetic and food industries.

11.16 CONCLUSIONS AND FUTURE PERSPECTIVES

The *Capsicum* genus (capsicums and chillies) is enriched with a group of compounds called capsaicinoids, which are responsible for spiciness of these fruits. Capsaicin and dihydrocapsaicin are the major members of the capsaicinoid group, contributing ~90% of the spiciness to this genus. These compounds being produced and stored in the placenta and other parts of the plant such as the stem, leaves, pericarp and seeds, making these parts valuable in obtaining these health-beneficial compounds. Capsaicinoids can be extracted through several different techniques, making them viable options for nutraceutical, pharmacological, cosmetic and food industry applications. It is essential to investigate different *Capsicum* species and varieties to obtain higher capsaicinoid content under various environmental factors.

Food waste has become a major global challenge, thus making it important to identify approaches to valorise large volumes of food waste. Fruits and vegetables are an important part of food that is wasted; thus, any measures, such as isolation of nutrients and non-nutritive components, will help in reducing the volume of food waste along with reducing environmental impacts. This recycling of components from fruits and vegetables will also help in reducing the loss of energy and resources used to produce food. On farms, production losses and plant part waste can also be recycled for isolation of health-beneficial components. *Capsicum* spp. can be major crops for such recycling, because these are short-shelf-life crops, and any measures such as extraction of capsaicinoids and developing nutraceuticals can help farmers and relevant industries with new sources of income.

REFERENCES

ADASZEK, Ł., GADOMSKA, D., STANIEC, M., GOŁYŃSKI, M., ŁYP, P., ZIĘTEK, J., RÓŻAŃSKA, D., ORZELSKI, M., ŚMIECH, M. & WINIARCZYK, S. 2017. Clinical assessment of the anti-cancer activity of the capsaicin-containing habanero pepper extract in dogs–preliminary study. *Med Weter*, 73, 404–411.

AIZAT, W. M., DIAS, D. A., STANGOULIS, J. C. R., ABLE, J. A., ROESSNER, U. & ABLE, A. J. 2014. Metabolomics of capsicum ripening reveals modification of the ethylene related-pathway and carbon metabolism. *Postharvest Biol Technol*, 89, 19–31. http://doi.org/10.1016/j.postharvbio.2013. 11.004.

AKIMOTO, S., TANIHATA, J., KAWANO, F., SATO, S., TAKEI, Y., SHIRATO, K., SOMEYA, Y., NOMURA, S., TACHIYASHIKI, K. & IMAIZUMI, K. 2009. Acute effects of dihydrocapsaicin and capsaicin on the distribution of white blood cells in rats. *J Nutr Sci Vitaminol*, 55, 282–287. https://doi. org/10.3177/jnsv.55.282.

ALAM, M. A., SYAZWANIE, N. F., MAHMOD, N. H., BADALUDDIN, N. A., MUSTAFA, K. A., ALIAS, N., ASLANI, F. & PRODHAN, M. A. 2018. Evaluation of antioxidant compounds, antioxidant activities and capsaicinoid compounds of Chili (*Capsicum* sp.) germplasms available in Malaysia. *J Appl Res Med Aromatic Plants*, 9, 46–54. http://doi.org/10.1016/j.jarmap.2018.02.001.

ALDANA-IUIT, J. G., SAURI-DUCH, E., MIRANDA-HAM, M. D. L., CASTRO-CONCHA, L. A., CUEVAS-GLORY, L. F. & VÁZQUEZ-FLOTA, F. A. 2015. Nitrate promotes capsaicin accumulation in *Capsicum chinense* immobilized placentas. *BioMed Res Int*, 2015, 794084. http://doi.org/ 10.1155/2015/794084.

AL-SNAFI, A. E. 2015. The pharmacological importance of *Capsicum* species (*Capsicum annuum* and *Capsicum frutescens*) grown in Iraq. *J Pharmaceut Biol*, 5, 124–142.

ANAYA-ESPARZA, L. M., MORA, Z. V.-D. L., VÁZQUEZ-PAULINO, O., ASCENCIO, F. & VILLARRUEL-LÓPEZ, A. 2021. Bell peppers (*Capsicum annuum* L.) losses and wastes: Source for food and pharmaceutical applications. *Molecules*, 26, 5341. http://doi.org/10.3390/molecules26175341.

ANTONIO, A. S., WIEDEMANN, L. S. M. & VEIGA JUNIOR, V. F. 2018. The genus *Capsicum*: A phytochemical review of bioactive secondary metabolites. *RSC Adv*, 8, 25767–25784. http://doi.org/10.1039/ C8RA02067A.

ARANCON, R. A. D., LIN, C. S. K., CHAN, K. M., KWAN, T. H. & LUQUE, R. 2013. Advances on waste valorization: New horizons for a more sustainable society. *Energy Sci Eng*, 1, 53–71. http://doi. org/10.1002/ese3.9.

ARCE-RODRÍGUEZ, M. L. & OCHOA-ALEJO, N. 2019. Biochemistry and molecular biology of capsaicinoid biosynthesis: Recent advances and perspectives. *Plant Cell Rep*, 38, 1017–1030. http://doi. org/10.1007/s00299-019-02406-0.

ATTUQUAYEFIO, V. K. & BUCKLE, K. A. 1987. Rapid sample preparation method for HPLC analysis of capsaicinoids in capsicum fruits and oleoresins. *J Agric Food Chem*, 35, 777–779. http://doi.org/10.1021/ jf00077a032.

BACKONJA, M. M., MALAN, T. P., VANHOVE, G. F. & TOBIAS, J. K. 2010. NGX-4010, a high-concentration capsaicin patch, for the treatment of postherpetic neuralgia: A randomized, double-blind, controlled study with an open-label extension. *Pain Med*, 11, 600–608. http://doi.org/10.1111/j.1526-4637.2009.00793.x.

BAE, C., ANSELMI, C., KALIA, J., JARA-OSEGUERA, A., SCHWIETERS, C. D., KREPKIY, D., WON LEE, C., KIM, E.-H., KIM, J. I. & FARALDO-GOMEZ, J. D. 2016. Structural insights into the mechanism of activation of the TRPV1 channel by a membrane-bound tarantula toxin. *Elife*, 5, e11273. http:// doi.org/10.7554/eLife.11273.

BAE, H., JAYAPRAKASHA, G. K., CROSBY, K., YOO, K. S., LESKOVAR, D. I., JIFON, J. & PATIL, B. S. 2014. Ascorbic acid, capsaicinoid, and flavonoid aglycone concentrations as a function of fruit maturity stage in greenhouse-grown peppers. *J Food Comp Anal*, 33, 195–202. http://doi.org/10.1016/j. jfca.2013.11.009.

BAJER, T., BAJEROVÁ, P., KREMR, D., EISNER, A. & VENTURA, K. 2015. Central composite design of pressurised hot water extraction process for extracting capsaicinoids from chili peppers. *J Food Comp Anal*, 40, 32–38. http://doi.org/10.1016/j.jfca.2014.12.008.

BARBERO, G. F., LIAZID, A., PALMA, M. & BARROSO, C. G. 2008. Fast determination of capsaicinoids from peppers by high-performance liquid chromatography using a reversed phase monolithic column. *Food Chem*, 107, 1276–1282. http://doi.org/10.1016/j.foodchem.2007.06.065.

BARBERO, G. F., PALMA, M. & BARROSO, C. G. 2006. Determination of capsaicinoids in peppers by microwave-assisted extraction–high-performance liquid chromatography with fluorescence detection. *Anal Chim Acta*, 578, 227–233. http://doi.org/10.1016/j.aca.2006.06.074.

BARBERO, G. F., RUIZ, A. G., LIAZID, A., PALMA, M., VERA, J. C. & BARROSO, C. G. 2014. Evolution of total and individual capsaicinoids in peppers during ripening of the Cayenne pepper plant (*Capsicum annuum* L.). *Food Chem*, 153, 200–206. http://doi.org/10.1016/j.foodchem.2013.12.068.

BASU, S. K. & DE, A. K. 2003. Capsicum: Historical and botanical perspectives. In: DE, A. K. (ed.) *Capsicum*. London, UK: CRC Press.

BATIHA, G. E.-S., ALQAHTANI, A., OJO, O. A., SHAHEEN, H. M., WASEF, L., ELZEINY, M., ISMAIL, M., SHALABY, M., MURATA, T., ZARAGOZA-BASTIDA, A., RIVERO-PEREZ, N., MAGDY BESHBISHY, A., KASOZI, K. I., JEANDET, P. & HETTA, H. F. 2020. Biological properties, bioactive constituents, and pharmacokinetics of some *Capsicum* spp. and capsaicinoids. *Int J Mol Sci*, 21, 5179. http://doi.org/10.3390/ijms21155179.

BOSLAND, P. W. & BARAL, J. B. 2007. 'Bhut Jolokia'—The world's hottest known chile pepper is a putative naturally occurring interspecific hybrid. *HortScience*, 42, 222–224. http://doi.org/10.21273/HORTSCI.42.2.222.

BOSLAND, P. W., COON, D. & COOKE, P. H. 2015. Novel formation of ectopic (nonplacental) capsaicinoid secreting vesicles on fruit walls explains the morphological mechanism for super-hot chile peppers. *J Am Soc Hort Sci*, 140, 253–256. http://doi.org/10.21273/JASHS.140.3.253.

BOSLAND, P. W. & VOTAVA, E. J. 2012. *Peppers: Vegetable and Spice Capsicums.* Wallingford, UK: CABI.

BOYETTE, M., WILSON, L. G. & ESTES, E. 1990. *Postharvest Cooling and Handling of Peppers: Postharvest Cooling and Handling of North Carolina Fresh Produce.* Raleigh, NC: The North Carolina State Extension Publications.

BUCHANAN, M. 2020. Some like it hot. *Nat Phys*, 16, 112. http://doi.org/10.1038/s41567-019-0766-3.

BUDZIANOWSKI, W. M. 2017. High-value low-volume bioproducts coupled to bioenergies with potential to enhance business development of sustainable biorefineries. *Renew Sustain Energy Rev*, 70, 793–804. http://doi.org/10.1016/j.rser.2016.11.260.

BURT, J. 2005. Growing capsicums and chillies. Department of Primary Industries–Vegetable Resource Database. Pub. DAWA. *Farmnote*, 64, 99.

BYERS, T. & PERRY, G. 1992. Dietary carotenes, vitamin C, and vitamin E as protective antioxidants in human cancers. *Annu Rev Nutr*, 12, 139–159. http://doi.org/10.1146/annurev.nu.12.070192.001035.

CAREAGA, M., FERNÁNDEZ, E., DORANTES, L., MOTA, L., JARAMILLO, M. E. & HERNANDEZ-SANCHEZ, H. 2003. Antibacterial activity of capsicum extract against *Salmonella typhimurium* and *Pseudomonas aeruginosa* inoculated in raw beef meat. *Int J Food Microbiol*, 83, 331–335. http://doi.org/10.1016/s0168-1605(02)00382-3.

CARRIZO GARCÍA, C., BARFUSS, M. H. J., SEHR, E. M., BARBOZA, G. E., SAMUEL, R., MOSCONE, E. A. & EHRENDORFER, F. 2016. Phylogenetic relationships, diversification and expansion of chili peppers (*Capsicum*, Solanaceae). *Ann Bot*, 118, 35–51. http://doi.org/10.1093/aob/mcw079.

CASTILLO, E., LÓPEZ-GONZÁLEZ, I., DE REGIL-HERNÁNDEZ, R., REYES-DUARTE, D., SÁNCHEZ-HERRERA, D., LÓPEZ-MUNGUÍA, A. & DARSZON, A. 2007. Enzymatic synthesis of capsaicin analogs and their effect on the T-type Ca^{2+} channels. *Biochem Biophys Res Commun*, 356, 424–430. http://doi.org/10.1016/j.bbrc.2007.02.144.

CHAPA-OLIVER, A. M. & MEJÍA-TENIENTE, L. 2016. Capsaicin: From plants to a cancer-suppressing agent. *Molecules*, 21, 931. https://doi.org/10.3390/molecules21080931.

CHAVAN, S. G., MAIER, C., ALAGOZ, Y., FILIPE, J. C., WARREN, C. R., LIN, H., JIA, B., LOIK, M. E., CAZZONELLI, C. I., CHEN, Z. H., GHANNOUM, O. & TISSUE, D. T. 2020. Light-limited photosynthesis under energy-saving film decreases eggplant yield. *Food Energy Secur*, 9, e245. http://doi.org/10.1002/fes3.245.

CHEN, C., CHAUDHARY, A. & MATHYS, A. 2020. Nutritional and environmental losses embedded in global food waste. *Resour Conserv Recycl*, 160, 104912. http://doi.org/10.1016/j.resconrec.2020.104912.

CHEN, L. & KANG, Y.-H. 2013. Anti-inflammatory and antioxidant activities of red pepper (*Capsicum annuum* L.) stalk extracts: Comparison of pericarp and placenta extracts. *J Funct Foods*, 5, 1724–1731. http://doi.org/10.1016/j.jff.2013.07.018.

CHUICHULCHERM, S., PROMMAKORT, S., SRINOPHAKUN, P. & THANAPIMMETHA, A. 2013. Optimization of capsaicin purification from *Capsicum frutescens* Linn. with column chromatography using Taguchi design. *Ind Crops Prod*, 44, 473–479. http://doi.org/10.1016/j.indcrop.2012.10.007.

CIENFUEGOS, N. E. C., SANTOS, P. L., GARCÍA, A. R., SOARES, C. M. F., LIMA, A. S. & SOUZA, R. L. 2017. Integrated process for purification of capsaicin using aqueous two-phase systems based on ethanol. *Food Bioprod Process*, 106, 1–10. http://doi.org/10.1016/j.fbp.2017.08.005.

CISTERNAS-JAMET, J., SALVATIERRA-MARTÍNEZ, R., VEGA-GÁLVEZ, A., STOLL, A., URIBE, E. & GOÑI, M. G. 2020. Biochemical composition as a function of fruit maturity stage of bell pepper (*Capsicum annuum*) inoculated with *Bacillus amyloliquefaciens*. *Sci Hortic*, 263, 109107. http://doi.org/10.1016/j.scienta.2019.109107.

CLARK, R. & LEE, S.-H. 2016. Anticancer properties of capsaicin against human cancer. *Anticancer Res*, 36, 837–843.

CONTRERAS-PADILLA, M. & YAHIA, E. M. 1998. Changes in capsaicinoids during development, maturation, and senescence of chile peppers and relation with peroxidase activity. *J Agric Food Chem*, 46, 2075–2079. http://doi.org/10.1021/jf970972z.

DE, A. K. 2003. *Capsicum: The Genus Capsicum.*London, UK: CRC Press. https://doi.org/10.1201/9780 203381151.

DE AGUIAR, A. C., COUTINHO, J. P., BARBERO, G. F., GODOY, H. T. & MARTÍNEZ, J. 2016. Comparative study of capsaicinoid composition in capsicum peppers grown in Brazil. *Int J Food Prop*, 19, 1292–1302. http://doi.org/10.1080/10942912.2015.1072210.

DE AGUIAR, A. C., OSORIO-TOBÓN, J. F., SILVA, L. P. S., BARBERO, G. F. & MARTÍNEZ, J. 2018. Economic analysis of oleoresin production from malagueta peppers (*Capsicum frutescens*) by supercritical fluid extraction. *J Supercrit Fluids*, 133, 86–93. http://doi.org/10.1016/j.supflu.2017.09.031.

DELI, J. & MOLNAR, P. 2002. Paprika carotenoids: Analysis, isolation, structure elucidation. *Curr Org Chem*, 6, 1197–1219. http://doi.org/10.2174/1385272023373608.

DENG, X.-Y., GAO, K., HUANG, X. & LIU, J. 2012. Optimization of ultrasonic-assisted extraction procedure of capsaicinoids from Chili peppers using orthogonal array experimental design. *Afr J Biotech*, 11, 13153–13161. http://doi.org/10.5897/AJB12.752.

DEWITT, D. & BOSLAND, P. W. 2009. *The Complete Chile Pepper Book: A Gardener's Guide to Choosing, Growing, Preserving, and Cooking.* Portland, OR: Timber Press.

DIAS, A. L. B., ARROIO SERGIO, C. S., SANTOS, P., BARBERO, G. F., REZENDE, C. A. & MARTÍNEZ, J. 2016. Effect of ultrasound on the supercritical CO2 extraction of bioactive compounds from dedo de moça pepper (*Capsicum baccatum* L. var. pendulum). *Ultrasonics Sonochemistry*, 31, 284–294. http://doi.org/10.1016/j.ultsonch.2016.01.013.

ESHBAUGH, W. H. 1976. Genetic and biochemical systematic studies of chili peppers (*Capsicum*-Solanaceae). *Bull Torrey Bot Club*, 102, 396–403. http://doi.org/10.2307/2484766.

ESTRADA, B., BERNAL, M. A., DÍAZ, J., POMAR, F. & MERINO, F. 2002. Capsaicinoids in vegetative organs of *Capsicum annuum* L. in relation to fruiting. *J Agric Food Chem*, 50, 1188–1191. http://doi.org/10.1021/jf011270j.

ESTRADA, B., POMAR, F., DıAZ, J., MERINO, F. & BERNAL, M. A. 1999. Pungency level in fruits of the Padrón pepper with different water supply. *Sci Hortic*, 81, 385–396. http://doi.org/10.1016/S0304-4238(99)00029-1.

FAO. 2022. Crops and livestock products. *Statistics Division.* Food Agriculture Organization of the United Nations, Rome, Italy.

FAYOS, O., DE AGUIAR, A. C., JIMÉNEZ-CANTIZANO, A., FERREIRO-GONZÁLEZ, M., GARCÉS-CLAVER, A., MARTÍNEZ, J., MALLOR, C., RUIZ-RODRÍGUEZ, A., PALMA, M., BARROSO, C. G. & BARBERO, G. F. 2017. Ontogenetic variation of individual and total capsaicinoids in Malagueta peppers (*Capsicum frutescens*) during fruit maturation. *Molecules*, 22, 736. http://doi.org/10.3390/molecules22050736.

FIEHN, O. 2002. Metabolomics—the link between genotypes and phenotypes. *Plant Mol Biol*, 48, 155–171. https://doi.org/10.1023/A:1013713905833.

FINGER, F. L. & PEREIRA, G. M. 2016. Physiology and postharvest of pepper fruits. In: DO RÊGO, E. R., DO RÊGO, M. M. & FINGER, F. L. (eds.) *Production and Breeding of Chilli Peppers (Capsicum spp.).* Cham: Springer. https://doi.org/10.1007/978-3-319-06532-8_2.

FORBES, H., QUESTED, T. & CLEMENTINE, O. C. 2021. *UNEP Food Waste Index Report 2021.* Nairobi, Kenya: United Nations Environment Programme.

GAMES, P. D., DASILVA, E. Q. G., BARBOSA, M. D. O., ALMEIDA-SOUZA, H. O., FONTES, P. P., DEMAGALHÃES-JR, M. J., PEREIRA, P. R. G., PRATES, M. V., FRANCO, G. R., FARIA-CAMPOS, A., CAMPOS, S. V. A. & BARACAT-PEREIRA, M. C. 2016. Computer aided identification of a Hevein-like antimicrobial peptide of bell pepper leaves for biotechnological use. *BMC Genom*, 17, 999. http://doi.org/10.1186/s12864-016-3332-8.

GANGABHAGIRATHI, R. & JOSHI, R. 2015. Antioxidant activity of capsaicin on radiation-induced oxidation of murine hepatic mitochondrial membrane preparation. *Res Rep Biochem*, 5, 163–171. http://doi.org/10.2147/RRBC.S84270.

GAVILÁN GUILLEN, N., TITO, R. & GAMARRA MENDOZA, N. 2018. Capsaicinoids and pungency in *Capsicum chinense* and *Capsicum baccatum* fruits. *Pesqui Agropecu Trop*, 48, 237–244. http://doi.org/10.1590/1983-40632018v4852334.

GEPPETTI, P. & TREVISANI, M. 2004. Activation and sensitisation of the vanilloid receptor: Role in gastrointestinal inflammation and function. *Br J Pharmacol*, 141, 1313–1320. http://doi.org/10.1038/sj.bjp.0705768.

GHASEMNEZHAD, M., SHERAFATI, M. & PAYVAST, G. A. 2011. Variation in phenolic compounds, ascorbic acid and antioxidant activity of five coloured bell pepper (*Capsicum annuum*) fruits at two different harvest times. *J Funct Foods*, 3, 44–49. http://doi.org/10.1016/j.jff.2011.02.002.

GNAYFEED, M. H., DAOOD, H. G., BIACS, P. A. & ALCARAZ, C. F. 2001. Content of bioactive compounds in pungent spice red pepper (paprika) as affected by ripening and genotype. *J Sci Food Agric*, 81, 1580–1585. http://doi.org/10.1002/jsfa.982.

GOVINDARAJAN, V. S. & SALZER, U. J. 1985. Capsicum-production, technology, chemistry, and quality part 1: History, botany, cultivation, and primary processing. *Crit Rev Food Sci Nutr*, 22, 109–176. https://doi.org/10.1080/10408398509527412.

GROSS, K. C., WATADA, A. E., KANG, M. S., KIM, S. D., KIM, K. S. & LEE, S. W. 1986. Biochemical changes associated with the ripening of hot pepper fruit. *Physiol Plant*, 66, 31–36. http://doi.org/10.1111/j.1399-3054.1986.tb01227.x.

GUO, Y., TAN, H., ZHANG, L., LIU, G., ZHOU, M., VIRA, J., HESS, P. G., LIU, X., PAULOT, F. & LIU, X. 2023. Global food loss and waste embodies unrecognized harms to air quality and biodiversity hotspots. *Nat Food*, 4, 686–698. http://doi.org/10.1038/s43016-023-00810-0.

GURUNG, T., TECHAWONGSTIEN, S., SURIHARN, B. & TECHAWONGSTIEN, S. 2011. Impact of environments on the accumulation of capsaicinoids in *Capsicum* spp. *HortScience*, 46, 1576–1581. http://doi.org/10.21273/HORTSCI.46.12.1576.

GUSTAFSSON, J., CEDERBERG, C. & SONESSON, U. 2013. *The Methodology of the FAO Study: Global Food Losses and Food Waste-Extent, Causes and Prevention—FAO, 2011*. Göteborg, Sweden: The Swedish Institute for Food and Biotechnology.

GUZMÁN, I. & BOSLAND, P. W. 2017. Sensory properties of chile pepper heat–and its importance to food quality and cultural preference. *Appetite*, 117, 186–190. http://doi.org/10.1016/j.appet.2017.06.026.

HA, S.-H., KIM, J.-B., PARK, J.-S., LEE, S.-W. & CHO, K.-J. 2007. A comparison of the carotenoid accumulation in *Capsicum* varieties that show different ripening colours: Deletion of the capsanthin-capsorubin synthase gene is not a prerequisite for the formation of a yellow pepper. *J Exp Bot*, 58, 3135–3144. http://doi.org/10.1093/jxb/erm132.

HAAK, D. C., MCGINNIS, L. A., LEVEY, D. J. & TEWKSBURY, J. J. 2012. Why are not all chilies hot? A trade-off limits pungency. *Proc Biol Sci*, 279, 2012–2017. http://doi.org/10.1098/rspb.2011.2091.

HAMADA, N., HASHI, Y., YAMAKI, S., GUO, Y., ZHANG, L., LI, H. & LIN, J.-M. 2019. Construction of online supercritical fluid extraction with reverse phase liquid chromatography–tandem mass spectrometry for the determination of capsaicin. *Chinese Chem Lett*, 30, 99–102. http://doi.org/10.1016/j.cclet.2018.10.029.

HARVELL, K. P. & BOSLAND, P. W. 1997. The environment produces a significant effect on pungency of chillies. *HortScience*, 32, 1292. http://doi.org/10.21273/HORTSCI.32.7.1292.

HE, X., CHAVAN, S. G., HAMOUI, Z., MAIER, C., GHANNOUM, O., CHEN, Z. H., TISSUE, D. T. & CAZZONELLI, C. I. 2022. Smart glass film reduced ascorbic acid in red and orange capsicum fruit cultivars without impacting shelf life. *Plants (Basel)*, 11, 985. http://doi.org/10.3390/plants11070985.

HE, X., SOLIS, C. A., CHAVAN, S. G., MAIER, C., WANG, Y., LIANG, W., KLAUSE, N., GHANNOUM, O., CAZZONELLI, C. I., TISSUE, D. T. & CHEN, Z.-H. 2023. Novel transcriptome networks are associated with adaptation of capsicum fruit development to a light-blocking glasshouse film. *Front Plant Sci*, 14, 1280314. http://doi.org/10.3389/fpls.2023.1280314.

HEINRICH, A. G., FERRAZ, R. M., RAGASSI, C. F. & REIFSCHNEIDER, F. J. 2015. Characterization and evaluation of self-fertilized progenies of biquinho salmon pepper. *Hortic Brazil*, 33, 465–470. http://doi.org/10.1590/S0102-053620150000400010.

IDLE, J. R. & GONZALEZ, F. J. 2007. Metabolomics. *Cell Metab*, 6, 348–351. http://doi.org/10.1016/j.cmet.2007.10.005.

IPCC. 2019. Summary for policymakers.In: SHUKLA, P.R., SKEA, J., CALVO BUENDIA, E., MASSON-DELMOTTE, V., PÖRTNER, H.-O., ROBERTS, D. C., ZHAI, P., SLADE, R., CONNORS, S., VAN DIEMEN, R., FERRAT, M., HAUGHEY, E., LUZ, S., NEOGI, S., PATHAK, M., PETZOLD, J., PORTUGAL PEREIRA, J., VYAS, P., HUNTLEY, E., KISSICK, K., BELKACEMI, M., MALLEY, J. (eds.) *Climate Change and Land: An IPCC Special Report on Climate Change, Desertification, Land Degradation, Sustainable Land Management, Food Security, and Greenhouse Gas Fluxes in Terrestrial Ecosystems*. London, UK: Intergovernmental Panel on Climate Change. https://doi.org/10.1017/9781009157988.001.

ISHIKAWA, K. 2003. Chapter 5. Biosynthesis of capsaicinoids in *Capsicum*. In: DE, A. K. (ed.) *Capsicum*. London, UK: CRC Press.

IWAI, K., SUZUKI, T. & FUJIWAKE, H. 1979. Formation and accumulation of pungent principle of hot pepper fruits, capsaicin and its analogues, in *Capsicum annuum* var. *annuum* cv. Karayatsubusa at different growth stages after flowering. *Agric Biol Chem*, 43, 2493–2498. http://doi.org/10.1080/00021369.1979.10863843.

JANG, Y. K., JUNG, E. S., LEE, H.-A., CHOI, D. & LEE, C. H. 2015. Metabolomic characterization of hot pepper (*Capsicum annuum* "CM334") during fruit development. *J Agric Food Chem*, 63, 9452–9460. http://doi.org/10.1021/acs.jafc.5b03873.

JANYOU, A., WICHA, P., SEECHAMNANTURAKIT, V., BUMROONGKIT, K., TOCHARUS, C., SUKSAMRARN, A. & TOCHARUS, J. 2020. Dihydrocapsaicin-induced angiogenesis and improved functional recovery after cerebral ischemia and reperfusion in a rat model. *J Pharm Sci*, 143, 9–16. http://doi.org/10.1016/j.jphs.2020.02.001.

JEEATID, N., SURIHARN, B., TECHAWONGSTIEN, S., CHANTHAI, S., BOSLAND, P. W. & TECHAWONGSTIEN, S. 2018a. Evaluation of the effect of genotype-by-environment interaction on capsaicinoid production in hot pepper hybrids (*Capsicum chinense* Jacq.) under controlled environment. *Sci Hortic*, 235, 334–339. http://doi.org/10.1016/j.scienta.2018.03.022.

JEEATID, N., TECHAWONGSTIEN, S., SURIHARN, B., BOSLAND, P. W. & TECHAWONGSTIEN, S. 2017. Light intensity affects capsaicinoid accumulation in hot pepper (*Capsicum chinense* Jacq.) cultivars. *Hortic Environ Biote*, 58, 103–110. http://doi.org/10.1007/s13580-017-0165-6.

JEEATID, N., TECHAWONGSTIEN, S., SURIHARN, B., CHANTHAI, S. & BOSLAND, P. W. 2018b. Influence of water stresses on capsaicinoid production in hot pepper (*Capsicum chinense* Jacq.) cultivars with different pungency levels. *Food Chem*, 245, 792–797. http://doi.org/10.1016/j.foodchem.2017.11.110.

JOHNSON, C. D. & DECOTEAU, D. R. 1996. Nitrogen and potassium fertility affects jalapeño pepper plant growth, pod yield, and pungency. *HortScience*, 31, 1119–1123. http://doi.org/10.21273/HORTSCI.31.7.1119.

KACHOOSANGI, R. T., WILDGOOSE, G. G. & COMPTON, R. G. 2008. Carbon nanotube-based electrochemical sensors for quantifying the 'heat' of chilli peppers: The adsorptive stripping voltammetric determination of capsaicin. *Analyst*, 133, 888–895. http://doi.org/10.1039/B803588A.

KADER, A. A. 2002. *Postharvest Technology of Horticultural Crops*. Davis, CA: University of California Agriculture and Natural Resources.

KAGA, H., GOTO, K., TAKAHASHI, T., HINO, M., TOKUHASHI, T. & ORITO, K. 1996. A general and stereoselective synthesis of the capsaicinoids via the orthoester Claisen rearrangement. *Tetrahedron*, 52, 8451–8470. http://doi.org/10.1016/0040-4020(96)00414-0.

KLIE, S., OSORIO, S., TOHGE, T., DRINCOVICH, M. F., FAIT, A., GIOVANNONI, J. J., FERNIE, A. R. & NIKOLOSKI, Z. 2014. Conserved changes in the dynamics of metabolic processes during fruit development and ripening across species. *Plant Physiol*, 164, 55–68. http://doi.org/10.1104/pp.113.226142.

KOBATA, K., KOBAYASHI, M., TAMURA, Y., MIYOSHI, S., OGAWA, S. & WATANABE, T. 1999. Lipase-catalyzed synthesis of capsaicin analogs by transacylation of capsaicin with natural oils or fatty acid derivatives in n-hexane. *Biotechnol Lett*, 21, 547–550. http://doi.org/10.1023/A:1005567923159.

KWON, D. Y., KIM, Y. S., RYU, S. Y., CHA, M.-R., YON, G. H., YANG, H. J., KIM, M. J., KANG, S. & PARK, S. 2013. Capsiate improves glucose metabolism by improving insulin sensitivity better than capsaicin in diabetic rats. *J Nutr Biochem*, 24, 1078–1085. http://doi.org/10.1016/j.jnutbio.2012.08.006.

LAU, B. B. Y., PANCHOMPOO, J. & ALDOUS, L. 2015. Extraction and electrochemical detection of capsaicin and ascorbic acid from fresh chilli using ionic liquids. *N J Chem*, 39, 860–867. http://doi.org/10.1039/C4NJ01416B.

LEE, J. J., CROSBY, K. M., PIKE, L. M., YOO, K. S. & LESKOVAR, D. I. 2005. Impact of genetic and environmental variation on development of flavonoids and carotenoids in pepper (*Capsicum* spp.). *Sci Hortic*, 106, 341–352. http://doi.org/10.1016/j.scienta.2005.04.008.

LIM, C. S., KANG, S. M., CHO, J. L., GROSS, K. C. & WOOLF, A. B. 2007. Bell pepper (*Capsicum annuum* L.) fruits are susceptible to chilling injury at the breaker stage of ripeness. *HortScience*, 42, 1659–1664. http://doi.org/10.21273/HORTSCI.42.7.1659.

LOWNDS, N. K., BANARAS, M. & BOSLAND, P. W. 1993. Relationships between postharvest water loss and physical properties of pepper fruit (*Capsicum annuum* L.). *HortScience*, 28, 1182–1184. http://doi.org/10.21273/HORTSCI.28.12.1182.

LOZADA, D. N., COON, D. L., GUZMÁN, I. & BOSLAND, P. W. 2021. Heat profiles of 'superhot' and New Mexican type chile peppers (*Capsicum* spp.). *Sci Hortic*, 283, 110088. http://doi.org/10.1016/j.scienta.2021.110088.

LUNA-RUIZ, J. D. J., NABHAN, G. P. & AGUILAR-MELÉNDEZ, A. 2018. Shifts in plant chemical defenses of chile pepper (*Capsicum annuum* L.) due to domestication in Mesoamerica. *Front Ecol Evol*, 6, 48. http://doi.org/10.3389/fevo.2018.00048.

LUO, X.-J., PENG, J. & LI, Y.-J. 2011. Recent advances in the study on capsaicinoids and capsinoids. *Eur J Pharmacol*, 650, 1–7. http://doi.org/10.1016/j.ejphar.2010.09.074.

MALORGIO, G. & MARANGON, F. 2021. Agricultural business economics: The challenge of sustainability. *Agric Food Econ*, 9, 6. http://doi.org/10.1186/s40100-021-00179-3.

MARES-QUIÑONES, M. D. & VALIENTE-BANUET, J. I. 2019. Horticultural aspects for the cultivated production of Piquin peppers (*Capsicum annuum* L. var. *glabriusculum*)—A review. *HortScience*, 54, 70–75. http://doi.org/10.21273/HORTSCI13451-18.

MARIANO, B. J., DE OLIVEIRA, V. S., CHÁVEZ, D. W. H., CASTRO, R. N., RIGER, C. J., MENDES, J. S., DA COSTA SOUZA, M., FRANKLAND SAWAYA, A. C. H., SAMPAIO, G. R., DA SILVA TORRES, E. A. F. & TATIANA, S. 2022. Biquinho pepper (*Capsicum chinense*): Bioactive compounds, *in vivo* and *in vitro* antioxidant capacities and anti-cholesterol oxidation kinetics in fish balls during frozen storage. *Food Biosci*, 47, 101647. http://doi.org/10.1016/j.fbio.2022.101647.

MARTINS, F. S., BORGES, L. L., RIBEIRO, C. S. C., REIFSCHNEIDER, F. J. B. & CONCEIÇÃO, E. C. 2017. Novel approaches to extraction methods in recovery of capsaicin from habanero pepper (CNPH 15.192). *Pharmacogn Mag*, 13, S375–S379. https://doi.org/10.4103/0973-1296.210127.

MASUDA, Y., HARAMIZU, S., OKI, K., OHNUKI, K., WATANABE, T., YAZAWA, S., KAWADA, T., HASHIZUME, S.-I. & FUSHIKI, T. 2003. Upregulation of uncoupling proteins by oral administration of capsiate, a nonpungent capsaicin analog. *J Appl Physiol (1985)*, 95, 2408–2415. http://doi.org/10.1152/japplphysiol.00828.2002.

MATEOS, R. M., JIMÉNEZ, A., ROMÁN, P., ROMOJARO, F., BACARIZO, S., LETERRIER, M., GÓMEZ, M., SEVILLA, F., DEL RÍO, L. A., CORPAS, F. J. & PALMA, J. M. 2013. Antioxidant systems from pepper (*Capsicum annuum* L.): Involvement in the response to temperature changes in ripe fruits. *Int J Mol Sci*, 14, 9556–9580. http://doi.org/10.3390/ijms14059556.

MATERSKA, M. & PERUCKA, I. 2005. Antioxidant activity of the main phenolic compounds isolated from hot pepper fruit (*Capsicum annuum* L.). *J Agric Food Chem*, 53, 1750–1756. http://doi.org/10.1021/jf035331k.

MEDINA-LARA, F., ECHEVARRÍA-MACHADO, I., PACHECO-ARJONA, R., RUIZ-LAU, N., GUZMÁN-ANTONIO, A. & MARTINEZ-ESTEVEZ, M. 2008. Influence of nitrogen and potassium fertilization on fruiting and capsaicin content in habanero pepper (*Capsicum chinense* Jacq.). *HortScience*, 43, 1549–1554. http://doi.org/10.21273/HORTSCI.43.5.1549.

MOKHTAR, M., RUSSO, M., CACCIOLA, F., DONATO, P., GIUFFRIDA, D., RIAZI, A., FARNETTI, S., DUGO, P. & MONDELLO, L. 2016. Capsaicinoids and carotenoids in *Capsicum annuum* L.: Optimization of the extraction method, analytical characterization, and evaluation of its biological properties. *Food Anal Methods*, 9, 1381–1390. http://doi.org/10.1007/s12161-015-0311-7.

MOLINILLO, J. M. G., DOMINGUEZ, F. A. M., MONTOYA, R. M. V., LOVILLO, M. P., BARROSO, C. G. & BARBERO, G. F. 2010. *Method for the chemical synthesis of capsinoids*. US20100256413A1.

MONFORTE-GONZÁLEZ, M., GUZMÁN-ANTONIO, A., UUH-CHIM, F. & VÁZQUEZ-FLOTA, F. 2010. Capsaicin accumulation is related to nitrate content in placentas of habanero peppers (*Capsicum chinense* Jacq.). *J Sci Food Agric*, 90, 764–768. http://doi.org/10.1002/jsfa.3880.

MOSCONE, E. A., SCALDAFERRO, M. A., GRABIELE, M., CECCHINI, N. M., SÁNCHEZ GARCÍA, Y., JARRET, R., DAVIÑA, J. R., DUCASSE, D. A., BARBOZA, G. E. & EHRENDORFER, F. 2007. The evolution of chili peppers (*Capsicum* - Solanaceae): A cytogenetic perspective. *Acta Hortic*, 745, 137–170. http://doi.org/10.17660/ActaHortic.2007.745.5.

MUCYO, S. 2013. *Analysis of Key Requirements for Effective Implementation of Biogas Technology for Municipal Solid Waste Management in Sub-Saharan Africa: A Case Study of Kigali City, Rwanda.* Dundee, Scotland: University of Abertay Dundee.

MUNOZ-RAMÍREZ, L. S., PENA-YAM, L. P., AVILÉS-VINAS, S. A., CANTO-FLICK, A., GUZMÁN-ANTONIO, A. A. & SANTANA-BUZZY, N. 2018. Behavior of the hottest chili peppers in the world cultivated in Yucatan, Mexico. *HortScience*, 53, 1772–1775. http://doi.org/10.21273/HORTSCI13574-18.

NAZZARO, F., CALIENDO, G., ARNESI, G., VERONESI, A., SARZI, P. & FRATIANNI, F. 2009. Comparative content of some bioactive compounds in two varieties of *Capsicum annuum* L. Sweet pepper and evaluation of their antimicrobial and mutagenic activities. *J Food Biochem*, 33, 852–868. http://doi.org/10.1111/j.1745-4514.2009.00259.x.

NISHINO, H., MURAKOSHI, M., TOKUDA, H. & SATOMI, Y. 2009. Cancer prevention by carotenoids. *Arch Biochem Biophys*, 483, 165–168. http://doi.org/10.1016/j.abb.2008.09.011.

OBOH, G. & ROCHA, J. B. T. 2007. Distribution and antioxidant activity of polyphenols in ripe and unripe tree pepper (*Capsicum pubescens*). *J Food Biochem*, 31, 456–473. http://doi.org/10.1111/j.1745-4514.2007.00123.x.

OCHOA-ALEJO, N. & GÓMEZ-PERALTA, J. E. 1993. Activity of enzymes involved in capsaicin biosynthesis in callus tissue and fruits of chili pepper (*Capsicum annuum* L.). *J Plant Physiol*, 141, 147–152. http://doi.org/10.1016/S0176-1617(11)80751-0.

O'DONOGHUE, E. M., BRUMMELL, D. A., MCKENZIE, M. J., HUNTER, D. A. & LILL, R. E. 2018. Sweet capsicum: Postharvest physiology and technologies. *New Zealand J Crop Hortic Sci*, 46, 269–297. http://doi.org/10.1080/01140671.2017.1395349.

OLIVER, S. G., WINSON, M. K., KELL, D. B. & BAGANZ, F. 1998. Systematic functional analysis of the yeast genome. *Trends Biotechnol*, 16, 373–378. http://doi.org/10.1016/s0167-7799(98)01214-1.

OMOLO, M. A., WONG, Z.-Z., MERGEN, A. K., HASTINGS, J. C., LE, N. C., REILAND, H. A., CASE, K. A. & BAUMLER, D. J. 2014. Antimicrobial properties of chili peppers. *J Infect Dis Ther*, 2, ISSN: 2332–0877.

OTHMAN, Z. A. A., AHMED, Y. B. H., HABILA, M. A. & GHAFAR, A. A. 2011. Determination of capsaicin and dihydrocapsaicin in *Capsicum* fruit samples using high performance liquid chromatography. *Molecules*, 16, 8919–8929. http://doi.org/10.3390/molecules16108919.

PADUANO, A., CAPORASO, N., SANTINI, A. & SACCHI, R. 2014. Microwave and ultrasound-assisted extraction of capsaicinoids from chili peppers (*Capsicum annuum* L.) in flavored olive oil. *J Food Res*, 3, 51–59. http://doi.org/10.5539/jfr.v3n4p51.

PANDHAIR, V. & GOSAL, S. S. 2009. Capsaicin production in cell suspension cultures derived from placenta of *Capsicum annuum* L. fruit. *Indian J Agric Biochem*, 22, 78–82.

PAUL, R. & GHOSH, U. 2012. Effect of thermal treatment on ascorbic acid content of pomegranate juice. *Indian J Biotechnol*, 11, 309–313.

PERANTONI, I. C. R., DOS SANTOS SOUSA, R. & SILVA, N. B. 2018. Gênero Capsicum no mercado do porto, em Cuiabá-MT. *Biodiversidade*, 17, 71–78.

PERRY, L., DICKAU, R., ZARRILLO, S., HOLST, I., PEARSALL, D. M., PIPERNO, D. R., BERMAN, M. J., COOKE, R. G., RADEMAKER, K., RANERE, A., RAYMOND, J. S., SANDWEISS, D. H., SCARAMELLI, F., TARBLE, K. & ZEIDLER, J. A. 2007. Starch fossils and the domestication and dispersal of chili peppers (*Capsicum* spp. L.) in the Americas. *Science*, 315, 986–988. http://doi.org/10.1126/science.1136914.

PIMCHAN, P., TECHAWONGSTIEN, S., CHANTHAI, S. & BOSLAND, P. W. 2012. Impact of drought stress on the accumulation of capsaicinoids in *Capsicum* cultivars with different initial capsaicinoid levels. *HortScience*, 47, 1204–1209. http://doi.org/10.21273/HORTSCI.47.9.1204.

PINTO, C. M. F., DOS SANTOS, I. C., DE ARAUJO, F. F. & DA SILVA, T. P. 2016. Pepper importance and growth (*Capsicum* spp.). In: RAMALHO DO RÊGO, E., MONTEIRO DO RÊGO, M. & FINGER, F. L. (eds.) *Production and Breeding of Chilli Peppers (Capsicum spp.).* Cham: Springer International Publishing. https://doi.org/10.1007/978-3-319-06532-8_1.

PRETEL, M. T., SERRANO, M., AMOROS, A., RIQUELME, F. & ROMOJARO, F. 1995. Non-involvement of ACC and ACC oxidase activity in pepper fruit ripening. *Postharvest Biol Technol*, 5, 295–302. http://doi.org/10.1016/0925-5214(94)00025-N.

RAHMAN, M. J., INDEN, H. & HOSSAIN, M. M. 2012. Capsaicin content in sweet pepper (*Capsicum annuum* L.) under temperature stress. *Acta Hortic*, 936, 195–201. http://doi.org/10.17660/ActaHortic.2012.936.23.

REYES-ESCOGIDO, M. D. L., GONZALEZ-MONDRAGON, E. G. & VAZQUEZ-TZOMPANTZI, E. 2011. Chemical and pharmacological aspects of capsaicin. *Molecules*, 16, 1253–1270. http://doi.org/10.3390/molecules16021253.

ROCHA-URIBE, J. A., NOVELO-PÉREZ, J. I. & ARACELI RUIZ-MERCADO, C. 2014. Cost estimation for CO_2 supercritical extraction systems and manufacturing cost for habanero chili. *J Supercrit Fluids*, 93, 38–41. http://doi.org/10.1016/j.supflu.2014.03.014.

ROESSNER, U. & BOWNE, J. 2009. What is metabolomics all about? *Biotechniques*, 46, 363–365. http://doi.org/10.2144/000113133.

ROSTAGNO, M. A., PALMA, M. & BARROSO, C. G. 2003. Ultrasound-assisted extraction of soy isoflavones. *J Chromatogr A*, 1012, 119–128. http://doi.org/10.1016/S0021-9673(03)01184-1.

ROWLAND, B. J., VILLALON, B. & BURNS, E. E. 1983. Capsaicin production in sweet bell and pungent jalapeno peppers. *J Agric Food Chem*, 31, 484–487. http://doi.org/10.1021/jf00117a005.

RUIZ-LAU, N., MEDINA-LARA, F., MINERO-GARCÍA, Y., ZAMUDIO-MORENO, E., GUZMÁN-ANTONIO, A., ECHEVARRÍA-MACHADO, I. & MARTÍNEZ-ESTÉVEZ, M. 2011. Water deficit affects the accumulation of capsaicinoids in fruits of *Capsicum chinense* Jacq. *HortScience*, 46, 487–492. http://doi.org/10.21273/HORTSCI.46.3.487.

RUSSO, V. M. 2012. *Peppers: Botany, Production and Uses.* Wallingford, UK: CABI.

SAHA, S., WALIA, S., KUNDU, A., KAUR, C., SINGH, J. & SISODIA, R. 2015. Capsaicinoids, tocopherol, and sterols content in chili (*Capsicum* sp.) by gas chromatographic-mass spectrometric determination. *Int J Food Prop*, 18, 1535–1545. http://doi.org/10.1080/10942912.2013.833222.

SALEH, B. K., OMER, A. & TEWELDEMEDHIN, B. 2018. Medicinal uses and health benefits of chili pepper (*Capsicum* spp.): A review. *MOJ Food Process Technol*, 6, 325–328.

SÁNCHEZ-TOLEDANO, B. I., CUEVAS-REYES, V., KALLAS, Z. & ZEGBE, J. A. 2021. Preferences in 'Jalapeño' pepper attributes: A choice study in Mexico. *Foods*, 10, 3111. http://doi.org/10.3390/foods10123111.

SANDOVAL-CASTRO, C. J., VALDEZ-MORALES, M., OOMAH, B. D., GUTIÉRREZ-DORADO, R., MEDINA-GODOY, S. & ESPINOSA-ALONSO, L. G. 2017. Bioactive compounds and antioxidant activity in scalded Jalapeño pepper industrial byproduct (*Capsicum annuum*). *J Food Sci Technol*, 54, 1999–2010. http://doi.org/10.1007/s13197-017-2636-2.

SANTOS, P., AGUIAR, A. C., BARBERO, G. F., REZENDE, C. A. & MARTÍNEZ, J. 2015. Supercritical carbon dioxide extraction of capsaicinoids from malagueta pepper (*Capsicum frutescens* L.) assisted by ultrasound. *Ultrason Sonochemistry*, 22, 78–88. http://doi.org/10.1016/j.ultsonch.2014.05.001.

SAWATDEENARUNAT, C., NGUYEN, D., SURENDRA, K., SHRESTHA, S., RAJENDRAN, K., OECHSNER, H., XIE, L. & KHANAL, S. K. 2016. Anaerobic biorefinery: Current status, challenges, and opportunities. *Bioresourc Technol*, 215, 304–313. http://doi.org/10.1016/j.biortech.2016.03.074.

SCOVILLE, W. L. 1912. Note on capsicums. *J Am Pharmaceut Assoc*, 1, 453–454. http://doi.org/10.1002/jps.3080010520.

SGANZERLA, M., COUTINHO, J. P., DE MELO, A. M. T. & GODOY, H. T. 2014. Fast method for capsaicinoids analysis from *Capsicum chinense* fruits. *Food Res Int*, 64, 718–725. http://doi.org/10.1016/j.foodres.2014.08.003.

SILVA, L. R., AZEVEDO, J., PEREIRA, M. J., VALENTÃO, P. & ANDRADE, P. B. 2013. Chemical assessment and antioxidant capacity of pepper (*Capsicum annuum* L.) seeds. *Food Chem Toxicol*, 53, 240–248. http://doi.org/10.1016/j.fct.2012.11.036.

SMITH, D. L., STOMMEL, J. R., FUNG, R. W. M., WANG, C. Y. & WHITAKER, B. D. 2006. Influence of cultivar and harvest method on postharvest storage quality of pepper (*Capsicum annuum* L.) fruit. *Postharvest Biol Technol*, 42, 243–247. http://doi.org/10.1016/j.postharvbio.2006.06.013.

SNITKER, S., FUJISHIMA, Y., SHEN, H., OTT, S., PI-SUNYER, X., FURUHATA, Y., SATO, H. & TAKAHASHI, M. 2009. Effects of novel capsinoid treatment on fatness and energy metabolism in humans: Possible pharmacogenetic implications. *Am J Clin Nutr*, 89, 45–50. http://doi.org/10.3945/ajcn.2008.26561.

SORA, G. T., HAMINIUK, C. W., DA SILVA, M. V., ZIELINSKI, A. A., GONÇALVES, G. A., BRACHT, A. & PERALTA, R. M. 2015. A comparative study of the capsaicinoid and phenolic contents and *in vitro* antioxidant activities of the peppers of the genus *Capsicum*: An application of chemometrics. *J Food Sci Technol*, 52, 8086–8094. http://doi.org/10.1007/s13197-015-1935-8.

SRINIVASAN, K. 2016. Biological activities of red pepper (*Capsicum annuum*) and its pungent principle capsaicin: A review. *Crit Rev Food Sci Nutr*, 56, 1488–1500. http://doi.org/10.1080/10408398.2013.772090.

STOICA, R.-M., MOSCOVICI, M., TOMULESCU, C. & BABEANU, N. 2016. Extraction and analytical methods of capsaicinoids—A review. *Sci Bull Series F. Biotechnol*, 20, 93–98.

SUDHAKAR JOHNSON, T., RAVISHANKAR, G. A. & VENKATARAMAN, L. V. 1996. Biotransformation of ferulic acid and vanillylamine to capsaicin and vanillin in immobilized cell cultures of *Capsicum frutescens*. *Plant Cell Tissue Organ Cult*, 44, 117–121. http://doi.org/10.1007/BF00048188.

SUN, F., XIONG, S. & ZHU, Z. 2016. Dietary capsaicin protects cardiometabolic organs from dysfunction. *Nutrients*, 8, 174. https://doi.org/10.3390/nu8050174

SUNG, Y., CHANG, Y.-Y. & TING, N.-L. 2005. Capsaicin biosynthesis in water-stressed hot pepper fruits. *Bot Bull Acad Sin*, 46, 35–42. http://doi.org/10.7016/BBAS.200501.0035.

SUPALKOVA, V., STAVELIKOVA, H., KRIZKOVA, S., ADAM, V., HORN, A., HAVEL, L., RYANT, P., BABULA, P. & KIZEK, R. 2007. Study of capsaicin content in various parts of pepper fruit by liquid chromatography with electrochemical detection. *Acta Chim Slov*, 54, 55–59.

SURH, Y.-J. & LEE, S. S. 1995. Capsaicin, a double-edged sword: Toxicity, metabolism, and chemopreventive potential. *Life Sci*, 56, 1845–1855. http://doi.org/10.1016/0024-3205(95)00159-4.

SUZUKI, T. & IWAI, K. 1984. Chapter 4. Constituents of red pepper species: Chemistry, biochemistry, pharmacology, and food science of the pungent principle of *Capsicum* species. In: BROSSI, A. (ed.) *The Alkaloids: Chemistry and Pharmacology*. Cambridge, MA: Academic Press. https://doi.org/10.1016/S0099-9598(08)60072-3.

TAYSEER, I., ABURJAI, T., ABU-QATOUSEH, L., AL-KARABIEH, N., AHMED, W. & AL-SAMYDAI, A. 2020. *In vitro* anti-*Helicobacter pylori* activity of capsaicin. *J Pure Appl Microbiol*, 14, 279–286. https://doi.org/10.22207/JPAM.14.1.29.

THUY, T. L. & KENJI, M. 2015. Effect of high temperature on fruit productivity and seed-set of sweet pepper (*Capsicum annuum* L.) in the field condition. *J Agric Sci Technol Hue Uni J Sci*, 5, 515–520. http://doi.org/10.17265/2161-6256/2015.12.010.

TITO, R., VASCONCELOS, H. L. & FEELEY, K. J. 2018. Global climate change increases risk of crop yield losses and food insecurity in the tropical Andes. *Glob Chang Biol*, 24, e592–e602. http://doi.org/10.1111/gcb.13959.

TOPUZ, A., FENG, H. & KUSHAD, M. 2009. The effect of drying method and storage on color characteristics of paprika. *LWT Food Sci Technol*, 42, 1667–1673. http://doi.org/10.1016/j.lwt.2009.05.014.

TRIPODI, P. & KUMAR, S. 2019. The capsicum crop: An introduction. In: RAMCHIARY, N. & KOLE, C. (eds.) *The Capsicum Genome. Compendium of Plant Genomes.* Cham: Springer. https://doi.org/10.1007/978-3-319-97217-6_1.

UARROTA, V. G., MARASCHIN, M., DE BAIRROS, A. D. F. M. & PEDRESCHI, R. 2021. Factors affecting the capsaicinoid profile of hot peppers and biological activity of their non-pungent analogs (capsinoids) present in sweet peppers. *Crit Rev Food Sci Nutr*, 61, 649–665. http://doi.org/10.1080/1040839 8.2020.1743642.

UNEP DTU PARTNERSHIP. 2021. Reducing consumer food waste using green and digital technologies. UNEP DTU Partnership, Copenhagen and Nairobi. https://www.unep.org/resources/publication/reducing-consumer-food-waste-using-green-and-digital-technologies

VALIENTE-BANUET, J. I. & GUTIÉRREZ-OCHOA, A. 2016. Effect of irrigation frequency and shade levels on vegetative growth, yield, and fruit quality of piquin pepper (*Capsicum annuum* L. var. *glabriusculum*). *HortScience*, 51, 573–579. http://doi.org/10.21273/HORTSCI.51.5.573.

VÁZQUEZ-ESPINOSA, M., OLGUÍN-ROJAS, J. A., FAYOS, O., GONZÁLEZ-DE-PEREDO, A. V., ESPADA-BELLIDO, E., FERREIRO-GONZÁLEZ, M., BARROSO, G. C., BARBERO, G. F., GARCÉS-CLAVER, A. & PALMA, M. 2020. Influence of fruit ripening on the total and individual capsaicinoids and capsiate content in Naga Jolokia peppers (*Capsicum chinense* Jacq.). *Agronomy*, 10, 252. http://doi.org/10.3390/agronomy10020252.

VIAGGI, D. 2018. *The Bioeconomy: Delivering Sustainable Green Growth.* Wallingford, UK: CABI.

WAHBA, N. M., AHMED, A. S. & EBRAHEIM, Z. Z. 2010. Antimicrobial effects of pepper, parsley, and dill and their roles in the microbiological quality enhancement of traditional Egyptian Kareish cheese. *Foodborne Pathog Dis*, 7, 411–418. http://doi.org/10.1089/fpd.2009.0412.

WALLENDER, W. W. & TANJI, K. K. 2011. *Agricultural Salinity Assessment and Management.* Reston, VA: American Society of Civil Engineers (ASCE).

WALSH, B. M. & HOOT, S. B. 2001. Phylogenetic relationships of *Capsicum* (Solanaceae) using DNA sequences from two noncoding regions: The chloroplast *atpB-rbcL* spacer region and nuclear *waxy* introns. *Int J Plant Sci*, 162, 1409–1418. http://doi.org/10.1086/323273.

WU, J., GE, F., WANG, D. & XU, X. 2016. Combination of supercritical fluid extraction with high-speed countercurrent chromatography for extraction and isolation of ethyl *p*-methoxycinnamate and ethyl cinnamate from *Kaempferia galanga* L. *Separation Sci Technol*, 51, 1757–1764. http://doi.org/10.1080/01 496395.2016.1176046.

XU, W., LIU, J., MA, D., YUAN, G., LU, Y. & YANG, Y. 2017. Capsaicin reduces Alzheimer-associated tau changes in the hippocampus of type 2 diabetes rats. *PLoS ONE*, 12, e0172477. https://doi.org/10.1371/journal.pone.0172477.

YASIN, M., LI, L., DONOVAN-MAK, M., CHEN, Z.-H. & PANCHAL, S. K. 2023. *Capsicum* waste as a sustainable source of capsaicinoids for metabolic diseases. *Foods*, 12, 907. http://doi.org/10.3390/foods12040907.

ZENG, W., MARTINUZZI, F. & MACGREGOR, A. 2005. Development and application of a novel UV method for the analysis of ascorbic acid. *J Pharm Biomed Anal*, 36, 1107–1111. http://doi.org/10.1016/j.jpba.2004.09.002.

ZEWDIE, Y. & BOSLAND, P. W. 2000. Evaluation of genotype, environment, and genotype-by-environment interaction for capsaicinoids in *Capsicum annuum* L. *Euphytica*, 111, 185–190. http://doi.org/10.1023/A:1003837314929.

ZHAO, C., CHAVAN, S., HE, X., ZHOU, M., CAZZONELLI, C. I., CHEN, Z.-H., TISSUE, D. T. & GHANNOUM, O. 2021. Smart glass impacts stomatal sensitivity of greenhouse *Capsicum* through altered light. *J Exp Bot*, 72, 3235–3248. http://doi.org/10.1093/jxb/erab028.

ZHOU, Y., GUAN, X., ZHU, W., LIU, Z., WANG, X., YU, H. & WANG, H. 2014. Capsaicin inhibits *Porphyromonas gingivalis* growth, biofilm formation, gingivomucosal inflammatory cytokine secretion, and *in vitro* osteoclastogenesis. *Eur J Clin Microbiol Infect Dis*, 33, 211–219. http://doi.org/10.1007/s10096-013-1947-0.

12 Post-Harvest Handling and Processing of Green and Red Indian Chillies

A.J. Sachin, P. Preethi, S.V.R. Reddy and P. Naresh

CONTENTS

ABBREVIATIONS

Symbols	Explanations
%	percentage or per cent
/	per
>	more than
±	plus or minus
≤	less than or equal to
$^{\circ}$C	degrees Celsius
μm	micrometer
cm	centimeter
CO_2	carbon dioxide
cv.	cultivar
d.b.	dry basis
et al.	coworkers
g	gram
GA_3	gibberlic acid
h	hours
kg	kilogram
kW	kilowatt
L	litre
m2	square meter
m3	cubic meter
mg	milligram
O_2	oxygen
ppm	parts per million

DOI: 10.1201/9781003378259-12

RH   relative humidity
s   seconds
w.b.   wet basis

12.1 INTRODUCTION

The chillies, or *Capsicum annuum* L., are frequently used as a savoury food additive and ingredient to give dishes and products a fiery flavour and colour. Green chillies are part of Indian cuisine. Though the quantity may vary, none of the recipes are completed without green chillies. The green chillies are preferred to be used fresh and hence are refrigerated or wrapped with damp muslin cloths, or the green chilli paste is preserved with salt and sesame oil. One of the most important aspects of red chillies is colour, which is caused by carotenoids such capsanthin and capsorubin. The components of violaxanthin that give them their yellow-orange hue include capsanthin-5,6 epoxide, zeaxanthin, lutein, cryptoxanthin, and β-carotene. Drying red pepper is one of the most often-used and reasonably priced methods to extend its shelf life (Arslan and Ozcan, 2011). This is accomplished by reducing the moisture content to a low activity level, which limits the growth of microorganisms and several moisture-mediated degradation events (Anoraga et al., 2018). When it comes to rural and other locations with sporadic or limited energy supply, it is more practical than sun drying. For this reason, solar dryer is an excellent substitute for sun drying when producing dried goods of superior quality. At present, convective hot air drying is the most popular method of drying chillies. However, long-term exposure to high air temperatures can cause significant deterioration of product quality, including colour loss, heat-sensitive nutrient reduction, and decreased antioxidant and rehydration capacity (Gupta et al., 2002).

Packaging and storage play an important role in post-harvest handling of dried chillies. The elements that cause or contribute to food deterioration during storage are influenced by packaging, which is a significant component in determining the stability of foods. The air within a package is determined by its type, and this has been shown to have an impact on a variety of factors, including microbial activity and the rate and degree of nutrient loss. Hence, the goal of this review study is to offer a thorough understanding of the developments in various green and dried red chillies' post-harvest unit operations (Alsebaeai et al., 2017; Jalgaonkar et al., 2022). This chapter discusses the post-harvest procedures (harvesting, drying, packing, and storing) of both green and dried chillies.

12.2 GREEN CHILLIES

12.2.1 HARVESTING, PACKAGING, AND TRANSPORTATION

Green chillies attain harvestable maturity after 75 to 85 days of transplantation depending on variety. Hand picking followed by packaging in wet gunny bags and transportation in closed or open wagons are the current practices followed in many parts of India. Chilli fruits contain high moisture content at maturity, and significant loss of moisture, mechanical injuries, and microbial spoilages occur at every point after harvest. At farm level, the freshly harvested green chillies can be stored up to 6 days in a zero-energy cool chamber against 2 days in farm conditions (Shil et al., 2018). The storage life of existing practice is 8 days, which can be extended up to 12 days when the chillies are coated with 5% shellac and packed in 18 μm antifog film, under ambient conditions with maximum temperature of 26°C. This combination minimizes the weight loss and retains freshness and bio-compound activity (Chitravathi et al., 2014). This treatment, combined with modified atmospheric packaging (MAP), gives 48 days of storage at 8°C (Chitravathi et al., 2015; Chitravathi et al., 2016). According to Ranjeet et al. (2014), active MAP at 4.5% O_2, 7.8% CO_2 level, and polypropylene film (38 μm) packages extended the storage life up to 7 weeks at 8°C.

Anthracnose is a major after-harvest disease in chillies and drastically affects the quality during storage. Washing green chillies with ozonated water (30 mg/L) before packing profoundly reduces the microbial population on the fruit surface and extends the storage life up to 36 days (Chitravathi et al., 2015). Plant growth regulators and polyamines have moderately improved the post-harvest character of chillies and other horticultural crops. Chilli fruits pretreated with 2 ppm GA_3 stored at 4°C retain maximum colour and regulated senescence-related enzyme activity (Panigrahi et al., 2017). Green bell pepper is highly susceptible to high temperature storage, which results in shrinking, rotting, and loss of colour. External application of spermidine and putrescine at the rate of 20 μm each followed by low-temperature storage (4°C) extends the storage life up to 40 days (Patel et al., 2019).

12.2.2 Processing and Value Addition

12.2.2.1 Dehydrated Green Chillies: Whole, Flakes, and Powder

Dehydration is one of the oldest and simplest method of processing in fruits and vegetables. The de-stemmed and top cut green chillies is retained as whole or made into thin slices and dried under open sun. After removal of moisture (88–90%), this can be stored in air tight container either as whole or fine powder or flakes form. Capsaicin content in chilli is an important parameter to determine the quality of the produce. In that way, blanching green chillies with 85–90°C for 2 minutes, treating with 0.01% potassium metabi sulfite followed by tray drying at 50°C retain maximum capsaicin, color, and quality. Meanwhile, this treatment maintains the quality of green chillies powder up to 60 days under ambient conditions (Pavani et al., 2018). Beside its cost in effectiveness, the finest particles size of green chillies powder was obtained from lyophilize drying method. Moreover, the Scanning Electron Microscope and X-ray diffraction images represent inequality among different methods of drying in the backbone structure of green chillies powder particles (Muneer et al., 2018). Alsebaeai et al. (2017) found that the flexible packaging foil is the best suitable packing material for green chillies powder where the maximum retention of physicochemical properties was recorded under 5°C. The varieties with more dry matter are preferred for dry processing to obtain more recovery, e.g., CA-960 & LVA-655.

12.2.2.2 Brining and Pickling

Green chillies are washed in water, air dried, and destemmed, and the top and bottom portion are cut. They are then cured in 10% salt solution for 2½ to 3 days. After that, oil, turmeric powder, and other suitable ingredients (depending upon local palate) are added and mixed thoroughly to get the end product. However, green chillies brined at 10% salt solution and in 8% salt + 1% sugar are highly suitable for pickling. Pickles made up of vinegar-cured, brined green chillies account for taste, flavor, texture and color, and those parameters were not changed significantly at room temperature (22°C–28°C) up to 12 months of storage (Devi, 2018).

12.2.2.3 Green Chilli Paste

Chopping and pulping method: This method involves chopping of fresh fruits followed by boiling. The fresh fruits are de-stemmed and the top and bottom are cut after harvest. They are chopped into small pieces, followed by boiling and filtration using a muslin cloth or filtration unit.

Grinding and filtration method: The de-stemmed, top-and-bottom-cut chillies are ground before boiling. The ground chilli paste is filtered through a filtration machine or muslin cloth, taken to the boiling. After proper boiling, the fruit pulp is stored in an airtight container for further use.

12.2.2.4 Sauce and Ketchup

Sauce and ketchup are semisolid products with a flowable consistency. These both can differentiate by total soluble solids (TSS) content, in which the former and later consist of 30 and 28°Brix, respectively. Green chilli sauce can be prepared either homogenous or mixed with other vegetables like tomato. This can be added with vinegar and salt for preservation, which is the common household preparation. This base can be used for preparation of several dishes or side dish for diverse cuisines. De-stemmed and top-cut green chillies are boiled till soft, cooled, pulped, and mixed with macerated spices and condiments (onion, ginger, and garlic powder) and salt for taste. Sodium benzoate is added as a preservative, and the whole is homogenized for even consistency. The suitable packing material for longer storage life (6 months) of green chilli sauce is glass containers, which can be even extended (8 months) under refrigerated conditions. Green chilli sauce mixed with honey can be stored in glass containers for about 8 months with least changes in capsaicin, and other bioactive compounds under room temperature (Kaur et al., 2021).

12.2.2.5   Other Industrial Uses

12.2.2.2.5.1   Green Chilli Oleoresins

Oils and oleoresins from chillies are used to spice snack items. The active ingredient in chillies, capsaicin, is employed in pharmaceutical and cosmetic formulations because it works well as a counter-irritant. Capsaicin overcomes the storage of chillies, and it can be stored for a prolonged period. It can be made available in low-growing chilli areas. Oleoresin can be extracted using organic solvents at small scale industrial level. For large scale, enzymatic extraction is a recommended method. In that case, β-Glucanase (endo-1, 3(4)- based enzyme for pretreatment yields higher-quality oleoresin with better recovery percentage (Baby and Ranganathan, 2016).

12.3   DRY CHILLIES

12.3.1   DRYING

Chilli fruits contain high moisture content at maturity and are perishable. Thus, appropriate processing and storage technologies are important for farmers, processors, and consumers. Since the initial moisture content of the freshly harvested red chillies fruits is very high (300%–400% d.b.), it is difficult to process them or store under such high-moisture conditions. Hence, their storage life was limited to 2 to 3 days with about 12% to 15% cumulative losses. Thus, it is important to reduce their moisture content and improve aeration in storage after harvest to reduce the growth of microbial flora, resulting in loss of quality and spoilage (Singh and Alam, 1982). Chilli fruit dehydration is influenced by a number of factors, such as (i) raw material quality and nature; (ii) preparation method (whole, sliced, or dried); (iii) pre-treatments (pricking, alkali/antioxidant treatment, etc.); (iv) lading density; (v) dehydration time, temperature, and method; (vi) critical temperature of dehydration, etc. (Pruthi, 2003). There are various methods available for reducing the moisture content in the chillies, some of which are discussed in what follows.

12.3.1.1   Traditional Sun Drying

Traditional method of drying in which the red ripe chillies fruits are directly dried in the sun without any pre-treatments (Kyi et al., 2020). This is the most widely practiced method not only in India but also throughout Asia, Africa, and Central and South America. Most of the time, the harvested fruits are heaped under indoor conditions for allowing complete color development of partially ripe fruits if there are any in the lot. A partially ripe fruit develops white patches on the pericarp and reduces the market value of the whole lot. During the traditional drying process, the fruits are spread thinly over the concrete floors/terraces/flat ground and turned frequently to attain uniform drying. This process takes about 2 to 3 weeks depending on the prevailing weather conditions. Generally, the fruits are heaped and covered with tarpaulin sheets/gunny bags during nights for achieving moisture equilibrium. Since there are no pucca structures/mats used for this drying, there are greater chances of contamination by dust, dirt, and other foreign materials (Gupta et al., 2018; Kyi et al., 2020).

12.3.1.2   Improved Sun Drying Methods

Numerous attempts have been made by various scientists to improve the traditional and widely adopted sun drying process during the post-independence era. Laul et al. (1970) tried various ways for improving the traditional sun drying by imposing various pre-treatments *viz.* pricking (making longitudinal slit); blanching (98.5°C for 3 min); and chemical treatment with a solution containing potassium carbonate (2.5 %) and olive oil (1.5 %) for 5 min. Among all these pre-treatments, pricked samples dried in 12 days, blanched ones in 7 days, and chemical dip samples in 6 days, while the control samples took 15 days for complete drying. The color was not much affected with pricking, but it was intensified with blanching. The chemical dip treatment samples exhibited sharp color with glossy appearance, which could be due to the use of olive oil.

CSIR-Central Food Technology Research Institute (CFTRI), Mysore, has developed a new, improved technique which exhibited faster rate of drying and thus reducing the drying period to 1 week compared to 2 to 3 weeks by traditional methods. This method requires relatively less space since it uses multi-tiered wire nets with fruit spread at 5 to 10 kg/m^2. Fresh chillies are dipped in a water-based emulsion (Dipsol) that contains gum acacia (0.1%), refined groundnut oil (1.0%), potassium carbonate (2.5%), and butylated hydroxyl anisole (0.001%) as part of the procedure. One quintal of fresh chillies requires about 15 liters of emulsion for dipping treatment. The retention of color and pungency was better with this treatment. The whole dry chilli recovery was 2% higher than with traditional methods due to reduced breakage and loss of seeds.

Singh and Alam (1982) experimented with drying of chillies on different surfaces *viz.* mud floor and tarpaulin with a fruit spread of 3–6 kg/m^2. During March, the mud floor and tarpaulin surfaces took 15.8 and 10.5 days, respectively, for complete drying process. They suggested tarpaulin surfaces for drying since they are more profitable and practical than dirt floor drying. Similarly, Pruthi (2003) experimented sun drying over concrete floors with black tar coating and wire mesh elevations (0.5–1.0 m). The drying rates were significantly high when dried directly over the black tar-coated concrete floors, while the fruits spread over wire mesh took 16 to 23 hours for reducing the moisture content to desired level of 5%.

The effect of few physical treatments *viz.* pricking, destalking, and chopping on the drying characteristics of red chillies was characterized by Chandy et al. (1992). The chopping treatment was found to be best among all, as it took only 18 hours for complete drying, while pricking and de-stalking took 20 and 50 hours, respectively, with 54 hours' drying time for control fruits. Mangaraj et al. (2001) dried chillies cv. Jwala over cemented floor under open sun with and without punching. The punched fruits took 102 hours, while unpunched fruits took 150 hours for reduction of moisture from 300% to 8% to 9% (d.b.). Gupta et al. (2017) studied the drying behavior of red chillies cv. Punjab Sindhuri using different physical treatments and drying techniques. Mechanical drying at 50°C gave the best results, followed by drying at 60°C in terms of drying behavior, while the product quality was compromised at 70°C and under sun drying. However, improved sun drying was better than sun drying. In general, slit chillies dried faster than whole chillies (Gupta et al., 2018).

12.3.1.3   Solar Drying

Solar drying is an improvement over the sun drying techniques with various advantages on final product quality as well as sustainability. The chillies were dried by Garg and Krishnan (1974) in a solar cabinet dryer with an average air temperature of 22.8°C. They found that the solar drier reduced the drying time and produced superior-quality produce. Solar radiation is a renewable energy resource available for about 2600 to 3200 h/year throughout the country (Palaniappan and Subramanian, 1997). Other advantages include reduced drying time and a better-quality product free from dirt, dust, and insect infestation. Though the initial investments are little high, the operating and maintenance costs are relatively low.

Red chillies' moisture content was reduced from 14% to 18% (w.b.) to 8% (w.b.) by using a roof-integrated solar air collector constructed on a corrugated aluminium roof. This design was helpful in reducing the fuel consumption to 10%, and the product quality was substantially improved to facilitate better powdering. In this solar air collector, the temperature was raised by 15°C to 46°C above the ambient temperature, and the economic breakeven point was observed to be 2 years (Palaniappan and Subramanian, 1997).

ICAR-Central Arid Zone Research Institute, Jodhpur, has designed a commercial solar drier with inclined box-type driers with area of 10 m^2. It has the capacity to dry about 80 to 100 kg of green/red chillies and reduce moisture content from 85% to 6% (w.b.) in 3.5 to 4 days. The color of dried chillies remained consistent after drying (Thanvi and Pande, 1987). ICAR-Central Institute of Agricultural Engineering, Bhopal, developed a solar cabinet drier 1980s, and its efficacy on chilli drying was studied by Singh and Alam (1982). They spread the red chilli fruits at 3, 9, and 12 kg/m^2 during the months of November, January, and March, which took 4.2, 5.5, and 3 days, respectively. This is found to be highly economical for large quantities of materials spread at 6 kg/m^2.

For the purpose of commercially drying red chillies, the Sarder Patel Renewable Energy Research Institute in Gujrat, India, constructed and installed a forced-circulation indirect solar drier. It had a drying chamber that could hold about 65 kg to 70 kg of chillies per batch and could reduce the moisture content from 12 % to 4 % (d.b.) in 2 to 5 hours of operation at 60°C. This had the capacity to dry about 200 kg chillies per day in three batches and 5000 kg per month. It could be effectively operated for 8 to 9 months in a year, and the end product quality was very good.

Joy et al. (2001) used a German-made solar drier for drying of red chillies spread in a single layer over a clean perforated mat placed inside the drying tunnel of the solar tunnel dryer. The end product contained higher capsaicin content and color values compared to those dried using conventional methods. Dhanore and Jibhakate (2014) developed a solar tunnel drier for drying red chilli fruits. It was prepared using a Galvanized Iron sheet and has a wooden frame covered with polythene and facilitated with a chimney. The red chillies' moisture content reduced from 75% to 5% and was found to save time compared to conventional sun drying.

In another experiment, Kumar et al. (2021) compared drying in solar dryer with open sun drying of red chillies and found that the former method reduced the moisture content from 79% (w.b.) to10% in 55 hours, while the later method did so in 124 hours, indicating a 55% difference in drying time. The thermal efficiency of the dryer was 16.25%, and specific energy consumption was 6.06 kW/kg. Drying in a solar dryer also produced better-quality produce. Therefore, solar tunnel drying is a better alternative to sun drying.

12.3.1.4 Mechanical Drying

This method of drying chillies has the special advantage of producing better and higher-quality produce. Compared to traditional sun drying, it takes less time and minimizes crop losses. In certain countries, the chilli fruits are dried in tobacco barns, heated buildings, or more usually in tunnel dryers/stainless steel continuous belt/belt-trough driers by exposing the fruits to forced hot air at temperatures of 50°C to 60°C and reducing the moisture content to 7% to 8% (Pruthi, 2003). Some processors adopt a two-stage method for chilli processing in which the fruits are initially dried to a moisture content of 12% to 20%, followed by storage at 0°C, and in the second step the fruits are dried to a final moisture content ≤ 7% to 8% during grinding.

Dhanegopal et al. (1988) conducted research on the drying properties of chillies. The investigation, which was carried out at an airflow rate of 0.064 m^3/s, discovered that when temperature increased from 40°C to 60°C, the total amount of time needed to reach equilibrium moisture content (8.9% d.b.) dropped from 40 hours to 18 hours. After the seed moisture content reached 25%, the drying process was subject to a decreasing rate period (d.b.). This may be the result of less liquid diffusing through the seed than evaporating off the surface below a moisture content of 25% (d.b.). As drying air temperature increased, the heat utilization factor dropped.

A waste-fired drier was used for drying red chillies of cv. 'G$_3$'at Punjabrao Deshmukh Krishi Vidyapeeth, Akola. It could accommodate 200 kg of fresh chillies, and the average air velocity maintained was 1.22 m/s. The heat energy was obtained by burning agricultural waste. The hot air of 53°C ± 2°C circulated over 10-cm-bed-thick fruits gave better drying, avoiding the loss of capsaicin (Phirke et al., 1992). Comparative studies of sun drying, improved sun drying, and mechanical drying at 50°C, 60°C, and 70°C indicated mechanical drying at 50°C was better than the other methods at retaining quality attributes such as ascorbic acid, capsaicin content, oleoresin content, and coloring matter (Gupta et al., 2017, 2018). When the product was mechanically and sun dried at 70°C, its quality decreased. But when it came to maintaining product quality, enhanced sun drying outperformed sun drying.

12.3.2 PACKAGING AND STORAGE

After drying or curing, fruits are cleaned of extraneous matter, and the damaged or discolored pods are discarded prior to packaging and storage. Generally, the dried chillies are packed in gunny bags each containing 25 kg of dried chillies for domestic marketing. However, the quality deterioration is more with this practice owing to improper packaging and storage. Fruits are generally stored from 1 to 3 months at growers' level, but they are stored longer in storage houses at traders' level, where fumigation is needed to control insect and pest infestation. A series of studies on packaging and storage of different varieties of Guntur chillies in flexible consumer packages was done by Mahadevaiah et al. (1976). They found that the moisture content (> 15%) is critical for mould growth in stored dried chillies. The discoloration of stored red chillies was greatly influenced by their moisture content and storage temperature. Packaging studies revealed that 200-gauge low- and high-density polyethylene (HDPE) films are suitable for packing of whole chillies in the units of 250 g under tropical conditions. Under high temperature (38°C) and RH (92%), the fruits turned intensely dark due to non-enzymatic browning reactions, while with lower storage temperature (27°C) and RH (65%), the color value decreases slightly, and the pods retained relatively brighter colors.

Efforts were made by Naik et al. (2001) for development of suitable consumer packaging for Guntur and Byadagi chilli varieties. They found that the critical moisture content for facilitation of mould growth in them is 14% and 11.2%, respectively. However, for storage of these varieties, an initial moisture content of 10.45% at 64% RH and 9.6% at 57 % RH was found to be critical. Under tropical conditions, 300-gauge high-density polyethylene was found to be highly suitable for packaging of Guntur and Byadagi chillies in unit packs of 250 g. The metallized polyester polythene (MPP) packaging retained more color than HDPE in both conditions.

It has been reported that chillies stored for household consumption and as a source of seed for the next crop are unable to grow on due to low moisture content and highly pungent odour. Generally, capsaicin content decreases by 60% to 70% over the course of a year when stored in non-airtight containers, regardless of the storage conditions and temperatures whereas, airtight containers significantly preserve capsaicin levels. Samples collected from local markets contained 50% less capsaicin than samples stored in airtight packages for a year. These reports indicate that storage conditions and packaging more strongly affect quality than the drying temperature. For long-term and better-quality storage, chillies should be stored in airtight packages (Yogeesha and Gowda, 2003).

When sun-dried chillies with 11% moisture content packed in gunny bags were stored under ambient and cold storage conditions, all varieties showed a decline in biochemical properties irrespective of the storage conditions (Jyothi et al., 2008). However, cold storage was better at retaining all biochemical properties. There are also differences between varieties in retention of quality attributes in storage. The influence of vacuum packaging and long-term storage on quality in red chillies fruits was studied by Chetti et al. (2014). Dried chillies were packed in vacuum-packed and jute bags and stored at two moisture levels (10% and 12%) under ambient (25 ± 2°C) and cold environments (4 ± 1°C) with and without light for a period of 2 years. Among all the treatment combinations, red chillies with 12% moisture stored in vacuum packages exhibited better quality parameters compared to those with 10% moisture. Also, the vacuum-packed fruits stored under cold temperature were found to have the lowest percentage decline in various quality parameters.

12.3.3 SORTING AND GRADING

Chilli fruits are sorted and graded according to colour and size. Fruits are sorted or graded by the help of hand either on a moving conveyer or on a grading table. Chillies are graded with the following specifications.

Criteria	Grade 1	Grade 2
Length	Uniform (> 10 cm)	
Uniformity of variety	5% maximum	5% maximum
Freshness	5% maximum	5% maximum
Dirt, foreign particles	10% maximum	15% maximum
Damage, injury, misshaped	5% maximum	5% maximum

12.4 CONCLUSION

Proper handling, packing, shipping, and storage minimize chilli post-harvest losses. Recent developments in drying processes and dehydration techniques have made it possible to prepare a variety of dried and dehydrated chilies. This chapter covered a range of historical and contemporary drying techniques. Technology for processing and preservation aids in saving extra fruit during glut years. Technology contributes to an increase in chilli exports in the form of preserved and with added-value products. The technical tasks that must be included in the value chain of chillies are harvesting at the right maturity level, clearing the field heat, suitable packaging, sound-absorbing and ventilated trucks, field drying, storage, appropriate distribution, marketing, and processing.

REFERENCES

Alsebaeai, M., Chauhan, A. and Arvind, H.S. 2017. Effect of storagibility on the shelf life of green chillies powder using different packaging materials. *International Journal of Innovative Science Engineering and Technology* 6(9): 18595–18602.

Anoraga, S.B., Sabarisman, I. and Ainuri, M. 2018, March. Effect of different pretreatments on dried chillies (*Capsicum annum* L.) quality. *IOP Conference Series: Earth and Environmental Science* 131(1): 012014.

Arslan, D. and Özcan, M.M. 2011. Dehydration of red bell-pepper (*Capsicum annuum* L.): Change in drying behavior, colour and antioxidant content. *Food and Bioproducts Processing* 89(4): 504–513.

Baby, K.C. and Ranganathan, T.V. 2016. Effect of enzyme pretreatment on yield and quality of fresh green chillies (*Capsicum annuum* L) oleoresin and its major capsaicinoids. *Biocatalysis and Agricultural Biotechnology* 7: 95–101.

Chandy, E., Ilyas, S.M., Samuel, D.V.K. and Singh, A. 1992, December. Effect of some physical treatments on drying characteristics of red chillies. In: *Proceedings of the International Agricultural Engineering Conference* (pp. 489–498). Asian Institute of Technology, Bangkok, Thailand.

Chetti, M.B., Deepa, G.T., Antony, R.T., Khetagoudar, M.C., Uppar, D.S. and Navalgatti, C.M. 2014. Influence of vacuum packaging and long term storage on quality of whole chillies (*Capsicum annuum* L.). *Journal of Food Science and Technology* 51: 2827–2832.

Chitravathi, K., Chauhan, O.P. and Raju, P.S. 2014. Postharvest shelf-life extension of green chillies (*Capsicum annuum* L.) using shellac-based edible surface coatings. *Postharvest Biology and Technology* 92: 146–148.

Chitravathi, K., Chauhan, O.P. and Raju, P.S. 2016. Shelf life extension of green chillies (*Capsicum annuum* L.) using shellac-based surface coating in combination with modified atmosphere packaging. *Journal of Food Science and Technology* 53: 3320–3328.

Chitravathi, K., Chauhan, O.P., Raju, P.S. and Madhukar, N. 2015. Efficacy of aqueous ozone and chlorine in combination with passive modified atmosphere packaging on the postharvest shelf-life extension of green chillies (*Capsicum annuum* L.). *Food and Bioprocess Technology* 8: 1386–1392.

Devi, Y.P. 2018. Efficacy of preservatives on the shelf life and quality of green chillies pickle. *Journal of Krishi Vigyan* 7(Special): 48–52.

Dhanegopal, M.S., Shashidhara, S.D. and Kulakarni, G.N. 1988. Studies on drying characteristics of chillies. *Journal of Agricultural Engineering* 25(3): 72–75.

Dhanore, R.T. and Jibhakate, Y.M. 2014. A solar tunnel dryer for drying red chilly as an agricultural product. *International Journal of Engineering Research & Technology* 3(7): 310–314.

Garg, H.P., Krishnan, A. 1974. Solar Drying of Agricultural Products I. Drying of Chillies in a Solar Cabinet Dryer. *Annals of Arid Zone*, 13(4): 285–292.

Gupta, P., Ahmed, J., Shivhare, U.S. and Raghavan, G.S.V., 2002. Drying characteristics of red chillies. *Drying Technology* 20(10): 1975–1987.

Gupta, S., Sharma, S. R., Mittal, T. C., Jindal, S. K. and Gupta, S. K. 2017. Study of drying behaviour in red chillies. *Green Farming* 8(6): 1364–1369.

Gupta, S., Sharma, S. R., Mittal, T. C., Jindal, S. K. and Gupta, S. K. 2018. Effect of different drying techniques on the quality of red chilly powder. *Indian Journal of Ecology* 45(2): 402–405.

Jalgaonkar, K., Mahawar, M.K., Girijal, S. and Hp, G., 2022. Post-harvest profile, processing and value addition of dried red chillies (*Capsicum annum* L.). *Journal of Food Science and Technology* 61: 1–19.

Joy, C.M., George, P.P. and Jose, K.P. 2001. Solar tunnel drying of red chilies (Capsicum annum L.). *Journal of Food Science and Technology* 38(3): 213–216.

Jyothi, P., Jagdish, S., Ajay, V., Singh, A.K., Mathura, R. and Sanjeet, K. 2008. Evaluation of chillies (*Capsicum annum* L) genotypes for some quality traits. *Journal of Food Science and Technology* 45(5): 463.

Kaur, M., Aggarwal, P. and Kumar, V. 2021. Effect of chillies varieties and storage conditions on quality attributes of honey chillies sauce: A preservation study. *Journal of Food Processing and Preservation* 45(9): e15734.

Kumar, J.P., Paramaguru, P., Arumugam, T., Boopathi, N.M. and Venkatesan, K. 2021. Genetic divergence among Ramnad mundu chillies (*Capsicum annuum* L.) genotypes for yield and quality. *Electronic Journal of Plant Breeding* 12(1): 228–234.

Kyi, T.H., Soe, T.T., Kar, A and Lin, M., 2020. Study on different drying methods of red chillies (*Capsicum annuum* L.). *Acta Scientific Nutritional Health* 4(4): 62–66.

Laul, M.S., Bhale Rao, S.D., Rane, V.R. and Amla, B.L. 1970. Studies on the sun drying of chillies (Capsicum annum Linn.). *Indian Food Packer* 24(2): 22–28.

Mahadevaiah, B., Chang, K.S. and Balasubrahmanyam, N. 1976. Packaging and storage studies on dried ground and whole chillies (*Capsicum annuum*) in flexible consumer packages. *Indian Food Packer* 30(6): 33–40.

Mangaraj, S., A. Singh, D.V.R. Samual, et al. 2001. Comparative performance evaluation of different drying methods for chillies. *Journal of Food Science and Technology*, 38: 296–299.

Muneer, A.S., Al-Dalali, S. and Al Eryani, H. 2018. Microscopic characterizations of green chilli powder. *International Journal of Current Microbiology and Applied Sciences* 7(8): 2488–2497.

Naik, P.J., Nagalakshmi, S., Balasubramanyam, N., Dhanaraj, S. and Shankaracharya, N.B. 2001. Packaging and storage studies on commercial varieties of Indian chillies (*Capsicum annuum* L.). *Journal of Food Science and Technology* 38(3): 227–230.

Palaniappan, C. and Subramanian, S.V. 1997. Performance of 500 m^2 area solar air heater for spices drying. In *Renewable Energy Application to Industries*. Narosa Publishing House, New Delhi.

Panigrahi, J., Gheewala, B., Patel, M., Patel, N. and Gantait, S. 2017. Gibberellic acid coating: A novel approach to expand the shelf-life in green chillies (*Capsicum annuum* L.). *Scientia Horticulturae* 225: 581–588.

Patel, N., Gantait, S. and Panigrahi, J. 2019. Extension of postharvest shelf-life in green bell pepper (*Capsicum annuum* L.) using exogenous application of polyamines (spermidine and putrescine). *Food Chemistry* 275: 681–687.

Pavani, S., SudhaVani, V., DorajeeRao, A.V.D., Viji, C.P., Suneetha, D.S., Venkata, K., Subbaiah, G.N. and Sarada, P. 2018. Effect of pre-treatments and drying methods on quality of green chillies powder. *International Journal of Pure and Applied Bioscience* 6(2): 1153–1157.

Phirke, P.S., Umbarkar, S.P. and Tapre, A.B. 1992. Development and evaluation of a waste fired dryer for red chillies. *ASAE Paper 016002*. ASAE.

Pruthi, J.S., 2003. Advances in post-harvest processing technologies of Capsicum. In A. K. De (ed.), *Capsicum: The Genus Capsicum* (pp. 175–218). CRC Press, London.

Ranjeet, S., Giri, S. K. and Kotawaliwale, N. 2014. Shelf-life enhancement of green bell pepper (*Capsicum annuum* L.) under modified atmosphere storage. *Food Packaging and Shelf-Life*: 101–112.

Shil, S., Mandal, J. and Das, S.P. 2018. Evaluation of postharvest quality of four local chillies (*Capsicum frutescens*) genotypes of Tripura under zero energy cool chamber. *Journal of Pharmacognosy and Phytochemistry* 7(3): 3698–3702.

Singh, H.P. and Alam, A. 1982. Techno-economic study on chillies drying. *Journal of Agricultural Engineering* 19(1): 27–32.

Thanvi, K.P. and Pande, P.C. 1987. Development of a low-cost solar agricultural dryer for arid regions of India. *Energy in Agriculture* 6(1): 35–40.

Yogeesha, H.S. and Gowda, R. 2003. The storage of Capsicum. In A. K. De (ed.), *Capsicum: The Genus Capsicum* (pp. 234–242). CRC Press, London.

13 Influence of Packaging on Postharvest Quality of Peppers

Komal Sonawane and Ramanand Jagtap

CONTENTS

13.1 INTRODUCTION

Any dried part of a plant other than leaves that can be used for flavoring or seasoning but not as a main ingredient in foods are known as a spice. Green leaves with the same characteristics are known as herbs (Ahmad et al., 2021). The unique flavor and aroma of spices are because of their specific composition. However, these desirable attributes begin to diminish shortly after harvesting due to changes in metabolic activity and potential contamination. Contamination tends to increase as it progresses through various stages, including harvesting, processing, packaging, and storage (Subroto et al., 2023).

Spices are of two types: **dried** and **fresh**. Fresh spices are unprocessed and are obtained directly from the plant, whereas dried spices undergo drying—that is, dehydration—to remove moisture content (King, 2006).

Pepper comes under the category of dried spices. Moisture in spices promotes microbial growth, leading to contamination and spoilage. Drying preserves spices, inhibiting microbial growth and

DOI: 10.1201/9781003378259-13

improving handling, packaging, and transportation. Proper drying maintains spice quality and flavor for longer shelf life (Mandeel, 2005).

Packaging plays a crucial role in preserving the flavor of spices, extending their shelf life, and preventing contamination because of external factors. It safeguards spices from various reactions with the environment, ensuring their quality and freshness over an extended period. Some of the unfavorable reactions are listed as follows:

Oxidation: Pepper is rich in alkaloids, essential oils, pigments, ions, and sensory ingredients. Exposure to oxygen in the air can cause the breakdown of these compounds, leading to flavor and aroma loss as well as color changes. Additionally, air creates a favorable environment for microbial growth.

Photodegradation: The pigments present in pepper can undergo photodegradation in the presence of light, leading to color fading as well as flavor and aroma loss.

Heat generation: Spices, if exposed to sunlight for longer durations of time, can absorb heat, which can then accelerate chemical reactions.

Volatile components can evaporate, resulting in lower quality.

Therefore, all these factors, along with other basic and consumer requirements, must be considered while selecting the packaging material (Hammouti et al., 2019; King, 2006).

Pepper spices are available in various forms such as whole, ground, or powder. These are commonly packaged in airtight containers made of materials such as glass, plastic, or metal. Glass jars with screw-top lids are a popular choice because they provide good protection against moisture, light, and air. Plastic containers, such as high-density polyethylene (HDPE) or polypropylene (PP) bottles, are also commonly used due to their durability and lightweight nature. Metal tins or cans are preferred for commercial packaging, offering robust protection and long shelf life. The choice of packaging material depends on factors such as product quality, market demands, and sustainability considerations. Also, packaging material can also vary depending upon the size of pepper as change in size changes surface area exposed to atmosphere (Fávaro et al., 2007).

While these packaging materials provide effective protection, they do present certain challenges. Over time, moisture can penetrate packaging, while transparent materials like glass or certain polymers allow light to pass through. Despite proper sealing, some oxygen may remain inside, and the non-biodegradable nature of plastic packaging materials raises environmental concerns.

To address these challenges, researchers have explored advanced packaging solutions. These include the incorporation of oxygen absorbers or the use of vacuum sealing methods to minimize oxygen levels. The industry is also shifting towards more environmentally friendly packaging options. Moreover, researchers are now focusing on modified atmosphere packaging (MAP) techniques, which can alter internal atmosphere, which enhances the shelf life and preserves the quality of spices (Qu et al., 2022).

Thus, selection of appropriate packaging materials for spices requires a thorough understanding of their composition and potential reactions with external factors and packaging materials. This chapter aims to delve into the intricate world of pepper packaging, exploring the challenges, advancements, and innovative approaches to preserving the flavor, aroma, and quality of this beloved spice. By considering the composition of spices and their interactions with packaging materials, we can ensure the longevity and sensory experience of pepper and other spices in culinary applications.

13.2 PEPPER PACKAGING

13.2.1 Pepper Characteristics

Pepper, as a spice, contains various volatile compounds that contribute to its distinct flavor and aroma. These compounds are sensitive to external factors such as exposure to air, light, and moisture. Oxygen in the air can initiate the breakdown of these volatile compounds, leading to a loss of flavor and aroma. Light, especially ultraviolet (UV) light, can cause photodegradation of pigments present in pepper, resulting in color changes and potential flavor alterations. Moisture can accelerate the growth of microorganisms, leading to spoilage and further degradation of pepper quality (Ozturkoglu-Budak, 2017).

Additionally, pepper contains essential oils, alkaloids, and other sensory ingredients that contribute to its overall taste profile. These components can be susceptible to oxidation, evaporation, and degradation if not properly protected during packaging (Perumal et al., 2022).

Therefore, when designing packaging solutions for pepper, it is crucial to consider these characteristics and their vulnerabilities. The packaging should provide an effective barrier against oxygen, light, and moisture to minimize the impact on flavor, aroma, and color. Different packaging materials, such as opaque films, dark-colored bottles, or laminated structures with moisture-resistant properties, can be employed to provide the necessary protection.

Furthermore, the selection of packaging materials should also take into account the compatibility between the volatile compounds in pepper and the packaging materials. Some materials may interact with the compounds, leading to off flavors or other undesirable changes in the product.

By understanding the specific characteristics of pepper and their susceptibilities during packaging, suitable packaging solutions can be developed to preserve the products' quality, flavor, and appearance over an extended shelf life.

13.2.2 Contamination and Flavor Preservation in Pepper Packaging: From Harvesting to Processing

The journey of pepper from harvesting to processing involves several stages at which contamination can occur. During harvesting, pepper berries may come into contact with soil, dust, insects, and other environmental contaminants. Improper handling or unhygienic conditions during harvesting can introduce microbial contaminants, further compromising the quality of the pepper (Jayakumar et al., 2021).

After harvesting, the pepper berries undergo processing, which typically involves cleaning, drying, and sometimes grinding or crushing. During these processes, there is a potential for additional contamination to occur. Equipment, storage facilities, or processing surfaces that are not properly cleaned and sanitized can introduce microorganisms, molds, or other contaminants to the pepper. Contamination not only affects the safety of the product but also contributes to the loss of flavor. Microorganisms and enzymes present on the surface of the pepper berries can initiate chemical reactions that lead to the breakdown of flavor compounds. These reactions, known as enzymatic and microbial spoilage, can result in off flavors, loss of aroma, and overall deterioration in taste quality (Ozturkoglu-Budak, 2017).

To mitigate contamination and preserve the flavor of pepper, appropriate measures should be taken at each stage of the supply chain. This includes implementing good agricultural practices during harvesting, ensuring proper hygiene and sanitation during processing, and adopting suitable packaging techniques to prevent further contamination.

13.2.3 Oxygen Interaction with Packaging Material

An effective oxygen barrier is crucial in pepper packaging because it plays a vital role in preventing the breakdown of the spice's ingredients through oxidation, thereby preserving its flavor. Oxygen, present in the surrounding environment, can react with the components of pepper spices, leading to flavor deterioration and loss of aromatic compounds. Therefore, choice of packaging material should be made accordingly (Kim et al., 2023).

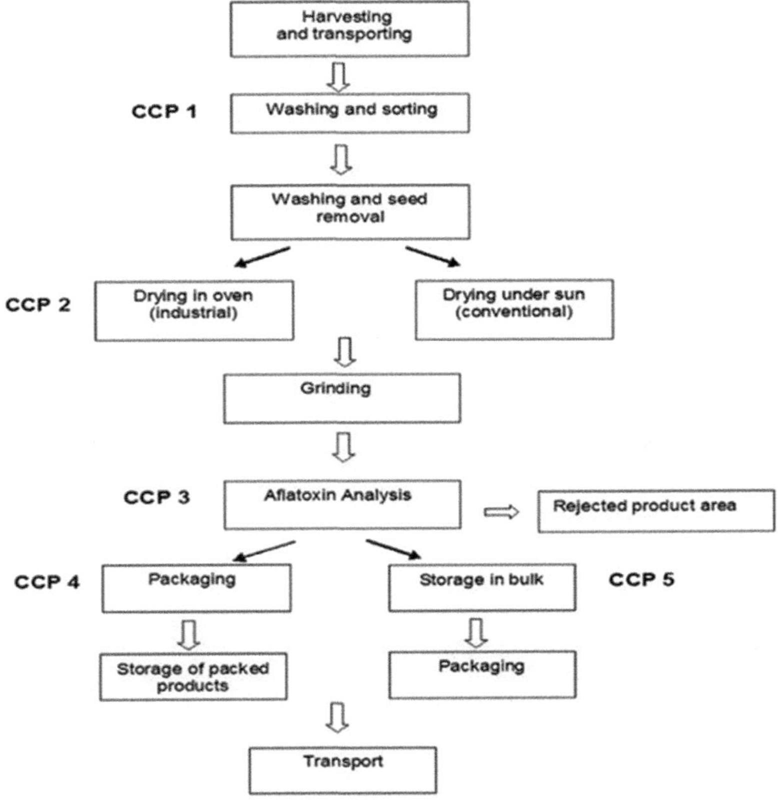

FIGURE 13.1 Different stages for pepper production (Ozturkoglu-Budak, 2017)

Studying the interaction of oxygen with polymeric packaging materials is of paramount importance when it comes to selecting the appropriate barrier properties (Zabihzadeh Khajavi et al., 2020a). This understanding plays a crucial role in guiding the development and decision-making process in the field of packaging.

By investigating the oxygen permeability of various polymeric materials, researchers and packaging engineers can assess their barrier performance. This involves measuring the rate at which oxygen molecules diffuse through the material and come into contact with the product. Understanding the permeation behavior helps in evaluating the effectiveness of different polymers in impeding oxygen ingress and preserving the desired properties of pepper spices.

13.2.4 Interaction of Pepper Ingredients with Packaging Material

Some packaging materials may react chemically with specific compounds present in pepper. For instance, certain metals, such as copper or aluminum, can react with phenolic compounds, causing color changes and off flavors. Similarly, interaction between pepper constituents, like ascorbic acid, and packaging materials can lead to oxidation reactions, impacting flavor and aroma (Singh et al., 2021; Wongsa et al., 2023).

Some of the ingredients in pepper can interact with the packaging material, which results in loss of flavor. Certain packaging materials, such as polymeric films, can adsorb volatile compounds from pepper, resulting in flavor and aroma changes. For instance, compounds like monoterpenes (e.g. limonene) and sesquiterpenes (e.g. beta-caryophyllene) present in pepper can be absorbed by plastics, potentially leading to off flavors or off odors (Arvanitoyannis & Kotsanopoulos, 2014).

It is important to note that the specific chemicals involved in these reactions can vary depending on the composition of the pepper and the packaging materials used. Different varieties of pepper may contain distinct volatile compounds, while packaging materials can have their own chemical components that interact with the volatile matter.

To mitigate these interactions, packaging material should be so chosen that it should have low absorption and permeability properties.

13.3 DEPENDABILITY OF OXYGEN PERMEATION ON STRUCTURE OF POLYMERIC PACKAGING MATERIAL

The oxygen permeability of a polymer packaging material is influenced by its structure and composition. The arrangement and characteristics of polymer chains play a significant role in determining the material's ability to impede the passage of oxygen molecules. The molecular structure of polymers can vary in terms of chain length, branching, crystallinity, and the presence of functional groups. These structural features directly impact the packing density and the formation of physical barriers within the material (Kausar, 2020).

In general, polymers with a highly organized and densely packed structure exhibit lower oxygen permeability (Ordoñez et al., 2022; Tang et al., 2023). This is because tightly packed polymer chains restrict the movement of oxygen molecules, reducing their ability to diffuse through the material. Crystalline regions within a polymer can act as additional barriers, hindering the permeation of gases.

On the other hand, polymers with a more amorphous structure tend to have higher oxygen permeability. The random arrangement of polymer chains in amorphous regions allows for greater free volume and facilitates the movement of oxygen molecules through the material.

Let's understand by the structures:

1 Polyethylene terephthalate (PET)

$$HOOC-C_6H_4-COOH + HOCH_2CH_2OH \longrightarrow [CO-C_6H_4-COOCH_2CH_2O] + 2H_2O$$

terephthalic acid   ethylene glycol   PET   water

2 Polyethylene (PE)

$$\left(\begin{array}{cc} H & H \\ | & | \\ -C & -C- \\ | & | \\ H & H \end{array} \right)_n$$

If we compare structure one and two, in the case of PET, ester bonds connect the terephthalic acid and ethylene glycol units, forming a linear chain of repeating units. This linear chain structure contributes to the crystalline nature of PET. The polymer chains can align and pack closely together in a regular pattern, creating regions of high order and density. PET has a small range of amorphous regions as well, which makes it flexible and transparent to be used as packaging material. Crystallinity of PET is also due to the presence of aromatic structures as they provide rigidity and planarity to the polymer chains, allowing them to stack more efficiently (Laria et al., 2023). On the other hand, PE is a polyolefin polymer composed of repeating units of ethylene. The molecular structure of PE is relatively simple, with only carbon and hydrogen atoms. It lacks aromatic rings or other rigid structures that promote strong intermolecular interactions. As a result, the polymer chains in PE can only form limited crystalline regions with less order and density compared to

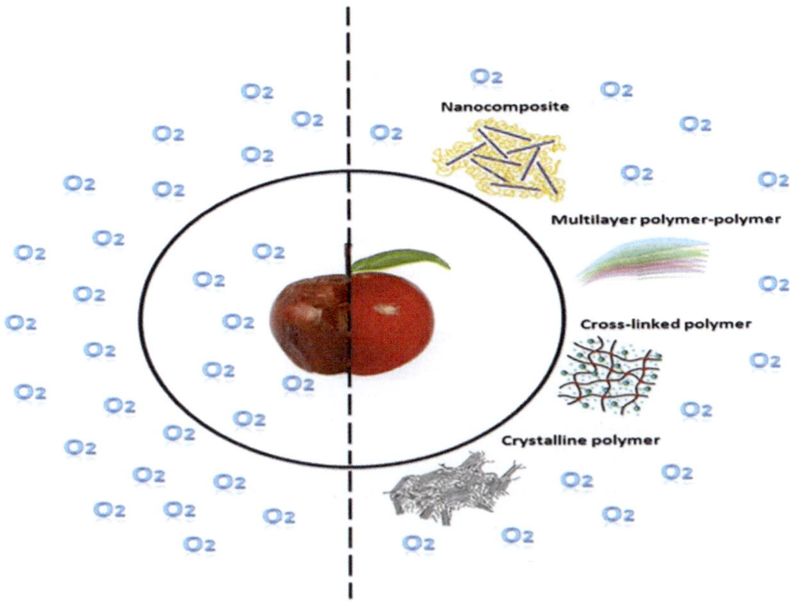

FIGURE 13.2   Effect of oxygen barrier on food (Zabihzadeh Khajavi et al., 2020b)

PET, and therefore, barrier properties provided by PET will be highly greater than polyethylene (Lamberti et al., 2020). Again, high-density polyethylene (HDPE) can provide better barrier properties than low-density polyethylene (LDPE) for the same reasons.

The presence of functional groups in the polymer structure can also influence oxygen permeability. Certain functional groups, such as polar groups or aromatic rings, can create intermolecular interactions that hinder the diffusion of oxygen.

In addition to the polymer structure, glass transition temperature (Tg),[1] humidity,[2] and thickness[3] of the packaging material can affect the oxygen permeability. Lower Tg, higher temperatures, and humidity levels can increase the mobility of polymer chains, leading to increased permeability. Thicker packaging materials generally offer higher oxygen barrier properties due to the longer path that oxygen molecules need to travel (Zabihzadeh Khajavi et al., 2020b).

13.4   TYPES OF PACKAGING MATERIAL

When it comes to packaging pepper, there is a wide range of packaging materials available, each with its own unique characteristics and suitability for preserving the quality and extending the shelf life of the spice. Choosing the right packaging material is crucial to ensure that the pepper remains fresh, flavorful, and protected from external factors. Let's delve into some commonly used packaging materials:

1 Glass Jars (Weber Macena et al., 2021): Glass jars offer excellent barrier properties, protecting pepper from oxygen, moisture, and light. They provide a transparent view of the product, allowing consumers to see the color and texture of the pepper. Glass is also inert and non-reactive, ensuring that the flavor and aroma of the spice remain unaffected. However, glass jars can be fragile and heavier compared to other packaging options
2 Plastic Pouches (Weber Macena et al., 2021): Plastic pouches, typically made from high-density polyethylene (HDPE) or polypropylene (PP), are widely used for packaging pepper due to their lightweight, cost-effectiveness, and flexibility. These pouches can be heat-sealed, ensuring good product integrity and preventing moisture and oxygen ingress.

However, plastic pouches may have lower barrier properties compared to glass or metal packaging, allowing a certain amount of oxygen and light transmission.

High-Density Polyethylene (HDPE) (Fávaro et al., 2007; Huang et al., 2019; Piringer, 1994):

- HDPE is a versatile plastic polymer known for its high strength-to-density ratio, making it sturdy and resistant to impact and tearing.
- It has excellent moisture barrier properties, protecting the pepper from humidity and moisture ingress that can lead to quality deterioration.
- HDPE is also resistant to chemicals, oils, and greases, which helps maintain the integrity of the packaging and prevents contamination.
- It is widely used for producing flexible pouches and bags, offering convenience, lightweight packaging, and ease of handling.
  However, HDPE has certain limitations:
- It has limited resistance to UV radiation, which means prolonged exposure to sunlight can degrade the material and potentially impact the quality of the packaged pepper.
- HDPE has relatively lower oxygen barrier properties compared to other packaging materials like glass or metal, which may allow a certain amount of oxygen permeation over time.
- The transparency of HDPE pouches can vary, with some options being more translucent than others, which may affect product visibility for consumers.

Polypropylene (PP) (Khalaj et al., 2016; Siracusa & Blanco, 2020):

- PP is a thermoplastic polymer known for its excellent chemical resistance and high-temperature tolerance.
- It offers good moisture barrier properties, helping to protect the pepper from moisture absorption that can lead to spoilage.
- PP has good impact strength and is resistant to cracking and stress, making it suitable for packaging that requires durability and protection during transportation and handling.
- It is commonly used for producing rigid containers, caps, and closures for pepper packaging.
  Similar to HDPE, PP also has limitations to consider:
- PP has lower oxygen barrier properties compared to materials like glass or metal, allowing some oxygen permeation over time.
- It may have lower resistance to UV radiation, which can cause degradation of the material and potentially impact the quality of the packaged pepper.
- PP can become brittle at low temperatures, so it may not be suitable for long-term storage in extremely cold environments.
- The transparency of PP containers may vary, with some options being more opaque, which may limit product visibility for consumers.

Both HDPE and PP offer advantages such as cost-effectiveness, flexibility in design, and ease of manufacturing. However, their limitations should be carefully considered when selecting the appropriate packaging material for pepper. Manufacturers may need to balance factors like barrier properties, durability, product visibility, and cost to ensure that the chosen material effectively preserves the quality and extends the shelf life of the packaged pepper (Fávaro et al., 2007).

3 Metal Tins (Deshwal & Panjagari, 2020): Metal tins, such as aluminum or tin-plated steel cans, offer excellent barrier properties against oxygen, light, and moisture. They provide sturdy protection, ensuring product integrity and preserving the flavor and aroma of the

pepper. Metal tins are also recyclable and have a longer shelf life compared to other packaging materials. However, metal packaging may be more expensive and can add weight to the overall product.

4 Laminated or Multilayered Films (Azevedo et al., 2022; Grzebieniarz et al., 2023): Laminated or multilayer films are flexible, impermeable and are commonly used in the form of pouches or sachets for packaging pepper. These films often consist of multiple layers, including a barrier layer, such as aluminum foil or metallized film, to provide high oxygen and moisture barrier properties. Flexible films offer convenience and easy handling and are lightweight, making them suitable for retail packaging. However, their transparency may vary, and they may not provide the same level of protection against light as opaque packaging materials.

5 Paper-Based Packaging (Oloyede & Lignou, 2021): Paper-based packaging, such as cartons or paper bags, can be used for pepper packaging, especially in bulk quantities. While paper may not offer the same level of barrier properties as glass, metal, or flexible films, it can provide a certain degree of protection against light and moisture. Paper-based packaging is often chosen for its eco-friendly and recyclable nature, as well as for its cost-effectiveness. However, it may not be suitable for long-term storage or extensive protection against oxygen and moisture.

FIGURE 13.3 Conventional packaging methods.

Choosing the appropriate packaging material depends on several factors, including the desired shelf life, product visibility, barrier properties required, and cost considerations. It is essential to evaluate the specific needs of the pepper product and select a packaging material that provides optimal protection, preserves the quality, and meets the preferences and expectations of consumers.

By carefully considering the advantages and limitations of each packaging material, pepper manufacturers can make informed decisions to ensure the longevity, safety, and sensory attributes of their products.

13.5 RECENT ADVANCES IN PEPPER PACKAGING

Conventional packaging methods for pepper had limitations, including limited shelf life, poor barrier properties, inadequate physical protection, lack of convenience, environmental concerns, and inadequate quality control. These challenges prompted the need for recent advances in pepper packaging. The advancements aim to address these issues by providing solutions such as improved barrier materials, innovative packaging designs for enhanced physical protection, convenience features for consumers, eco-friendly materials, and better-quality control and traceability mechanisms. By overcoming these limitations, the recent advances in pepper packaging strive to extend shelf life, maintain product quality, enhance consumer satisfaction, and promote sustainable packaging practices (Salama & Abdel Aziz, 2020).

13.5.1 SINGLE-LAYER AND DOUBLE-LAYER PACKAGING

Single-layer and double-layer packaging are common approaches used in the packaging of pepper to provide protection, maintain quality, and extend the shelf life of the spice.

Single-layer packaging, such as polyethylene (PE) bags or pouches, offers basic protection against moisture, light, and air (Tyagi et al., 2021a). However, these materials may have limited barrier properties against oxygen and light, which can lead to undesirable reactions in pepper. For example, exposure to oxygen can initiate the breakdown of certain compounds in pepper, such as essential oils and pigments, resulting in flavor and aroma loss. Light exposure can also cause the degradation of pigments, leading to color fading. Additionally, air can create a favorable environment for microbial growth, increasing the risk of contamination (Watada et al., 1987).

In contrast, double-layer packaging provides enhanced protection for pepper. One common configuration is a combination of polyethylene terephthalate (PET) film as the inner layer and a metalized or aluminum foil layer as the outer layer. This structure offers improved barrier properties against moisture, oxygen, and light. The impermeable nature of the foil layer prevents oxygen from entering the package, minimizing the degradation of flavor and aroma compounds. The metalized layer reflects light, reducing its impact on the pigments in pepper and helping to maintain color stability. This double-layer structure also provides a more effective barrier against microbial contamination (Anukiruthika et al., 2020a).

For example, the reaction between oxygen and the essential oils in pepper can lead to oxidation, causing a loss of flavor and aroma (Marques, 2010). The reaction between light and pigments in pepper, such as chlorophyll and carotenoids, can result in color changes. Double-layer packaging helps to minimize these reactions by creating a protective barrier, preserving the flavor, aroma, and color of the pepper.

The choice between single-layer and double-layer packaging depends on factors such as the desired shelf life, the level of protection required, and the specific characteristics of the pepper product. For longer shelf life and higher protection needs, double-layer packaging is often preferred. However, it is essential to consider the specific requirements of the pepper and select packaging materials and structures accordingly to ensure optimal preservation of its quality during storage and transportation.

By carefully selecting packaging materials and structures and understanding the reactions that can occur in pepper, producers can effectively maintain the flavor, aroma, and quality of the spice, providing consumers with a desirable and enjoyable culinary experience.

13.5.2 Multilayer Packaging

Multilayer packaging refers to the use of multiple layers of different materials in the construction of a packaging structure. This approach combines the desirable properties of each material to create a packaging solution that offers enhanced performance and functionality. In the context of pepper spice packaging, multilayer structures are commonly used to provide a combination of barrier properties, strength, and durability.

The selection of materials and the arrangement of layers in a multilayer packaging system depend on the specific requirements of pepper spice preservation, such as oxygen and moisture barrier, aroma retention, and protection against light and physical damage. Different materials, such as polymers, aluminum foil, paper, and adhesives, can be used to create an optimized multilayer structure.

For example, a typical multilayer packaging structure for pepper spices may consist of an outer layer of paper or polymer film for strength and printability, an inner layer of aluminum foil for excellent barrier properties against oxygen and light, and an innermost layer of polymer film for direct contact with the spices, providing additional barrier properties and maintaining the quality of the product (Anukiruthika et al., 2020b).

Different functions that each layer should provide depending on the specific requirement of spices packaging are as follows.

13.5.2.1 Outer Layer

Protection: The outermost layer acts as a protective barrier against external factors such as moisture, dust, and physical damage during handling, transportation, and storage.

Printing and Branding: It provides a surface for printing essential product information, branding, and visual appeal to attract consumers.

13.5.2.2 Oxygen Barrier Layer

Oxygen Protection: This layer acts as a barrier to prevent oxygen from entering the packaging and oxidizing the pepper spices. It helps maintain the freshness, flavor, and quality of the product.

Examples of materials: EVOH (ethylene vinyl alcohol), PVDC (polyvinylidene chloride), or nylon can be used as oxygen barrier layers (Tyuftin & Kerry, 2023).

13.5.2.3 Moisture Barrier Layer

Moisture Protection: The moisture barrier layer prevents moisture from permeating into the packaging, thereby reducing the risk of moisture absorption and potential degradation of the spices.

Examples of materials: Polyethylene terephthalate (PET), polypropylene (PP), or aluminum foil can be used as moisture barrier layers (Mujtaba et al., 2022).

13.5.2.4 Light Barrier Layer

Light Protection: This layer blocks or reduces the transmission of light to prevent photodegradation of the pepper spices, which can result in flavor loss and color changes (Kwon et al., 2018).

Examples of materials: Metallized films, opaque polymers (Paixão et al., 2019), or light-blocking additives can be used as light barrier layers (Kwon et al., 2018).

13.5.2.5 Heat Seal Layer

Sealing Function: The heat seal layer ensures proper sealing of the packaging, maintaining the integrity and freshness of the product.

Examples of materials: Polyethylene (PE) or polypropylene (PP) films are commonly used as heat seal layers (Bamps et al., 2023; Marangoni Júnior et al., 2020).

13.5.2.6   Inner Layer

Direct Contact: The inner layer comes in direct contact with the pepper spices, and therefore, it should be safe, food-grade, and compatible with the product.

Examples of materials: Food-grade polyethylene (PE), polypropylene (PP), or polyester (PET) films can be used as inner layers (Hu et al., 2017; Shin & Selke, 2014).

13.5.2.7   Adhesive Layer

Bonding: In multilayer structures, an adhesive layer is used to bond different layers together, ensuring the integrity and stability of the packaging.

Examples of materials: Adhesive resins or tie layers can be used for effective bonding (Bayer, 2021).

By combining these layers in a multilayer configuration, the packaging system effectively addresses multiple preservation factors, including oxygen permeation, moisture control, light exposure, and physical protection. This comprehensive approach helps to maintain the freshness, flavor, and overall quality of pepper spices over an extended period.

It's important to note that the specific choice and configuration of layers in a multilayer packaging system can vary based on the desired shelf life, product characteristics, and regulatory requirements. Manufacturers and packaging experts carefully evaluate and optimize the layer composition to provide the most effective protection while ensuring product safety and consumer satisfaction.

13.6   DISPOSAL CHALLENGES FOR MULTILAYER SYSTEM

While multilayer packaging systems offer excellent protection and preservation benefits for pepper spices, one of the challenges associated with these systems is the disposal of the packaging after use. The disposal problem arises due to the combination of different materials used in the layers, which

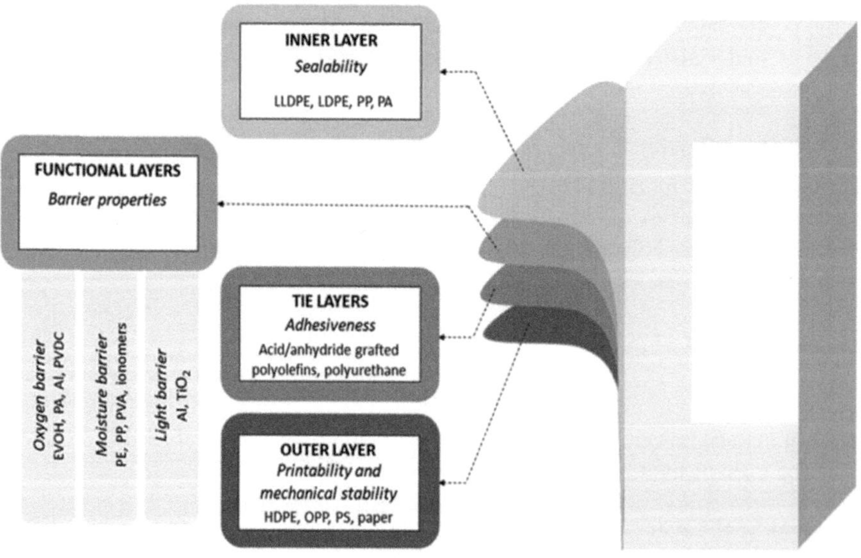

FIGURE 13.4   Various layers of multilayer packaging (Anukiruthika et al., 2020c)

TABLE 13.1

Few Examples of Multilayer Packaging Configurations

Multilayer system	Outer layer	Barrier layer	Barrier layer	Inner layer
PET/Aluminum Foil/PE (Domeño et al., 2017)	PET (polyethylene terephthalate) provides physical protection and printing surface.	Aluminum foil acts as an effective oxygen and light barrier.	PE (polyethylene) prevents moisture ingress.	PE ensures direct contact with the pepper spices.
PET/EVOH/PE (Viana Batista et al., 2023)	PET offers protection and printing surface.	EVOH (ethylene vinyl alcohol) provides excellent oxygen barrier properties.	PE prevents moisture permeation.	PE ensures safe contact with the pepper spices.
PP/Metallized Film/ PE (Baele et al., 2021)	PP (polypropylene) provides physical protection and printing surface.	Metallized film offers effective light-blocking properties.	PE acts as a moisture barrier.	PE ensures safe contact with the pepper spices.
Nylon/PVDC/PE (Leelaphiwat et al., 2018)	Nylon provides strength and protection.	PVDC (polyvinylidene chloride) acts as an excellent oxygen barrier.	PE prevents moisture ingress.	PE ensures safe contact with the pepper spices.
Paper/Aluminum Foil/PE (Patel et al., 2015)	Paper provides a natural and sustainable appearance	Aluminum foil acts as an effective oxygen and light barrier.	PE prevents moisture permeation.	PE ensures safe contact with the pepper spices.

can make recycling and waste management more complex (Soares et al., 2022). Here are some key points to consider regarding the disposal problem:

- Non-Biodegradable Materials: Multilayer packaging often incorporates non-biodegradable materials such as aluminum, metallized films, or certain plastics. These materials do not easily break down in the environment, leading to increased waste accumulation and potential environmental impacts (Maduwantha & Jayasinghe, 2023).
- Separation and Recycling: The presence of multiple layers, each made of different materials, makes it difficult to separate and recycle the packaging effectively. Recycling facilities may require specialized processes to separate the different layers and recycle them individually, which can be economically and technologically challenging (Anukiruthika et al., 2020d; Kaiser et al., 2017).
- Landfill and Incineration: In some cases, multilayer packaging may end up in landfills or incineration facilities, contributing to the overall waste generation and potential release of harmful pollutants into the environment.
- Sustainable Alternatives: To address the disposal problem, there is an increasing focus on the development and adoption of more sustainable packaging alternatives. These include mono-material packaging, compostable films, and bio-based polymers that offer improved recyclability and reduced environmental impact (Anukiruthika et al., 2020d).

It is crucial for packaging manufacturers, brand owners, and consumers to consider the environmental implications of packaging choices and work towards more sustainable solutions. This can involve exploring innovative materials, supporting recycling initiatives, and promoting responsible waste management practices. Information about sustainable solution is provided later in this chapter.

13.7 ACTIVE PACKAGING SYSTEMS

- Active packaging systems incorporate active components that interact with the environment to extend the shelf life of pepper (Dey & Neogi, 2019).
- Oxygen absorbers or scavengers are commonly used to reduce oxygen levels inside the packaging, minimizing oxidative reactions and maintaining pepper freshness (Dey & Neogi, 2019).
- Moisture absorbers or desiccants help control moisture content, preventing microbial growth and maintaining the quality of the pepper (Gaikwad et al., 2019).
- Ethylene absorbers can be employed to remove ethylene gas, which is produced by fruits and vegetables and can accelerate ripening and deterioration of pepper.
- Antimicrobial Agents: Active packaging systems may incorporate antimicrobial agents that release substances to inhibit the growth of microorganisms and prevent spoilage or foodborne illnesses. Examples include antimicrobial films or coatings containing substances like silver nanoparticles or essential oils (Vilas et al., 2020).
- Flavor and Aroma Enhancers: Some active packaging systems are designed to release specific flavors or aromas into the package, enhancing the sensory attributes of the product and providing a fresh experience for the consumer.

13.8 MODIFIED ATMOSPHERE PACKAGING (MAP)

Modified Atmosphere Packaging (MAP) is a packaging technique that involves altering the composition of the air surrounding the product to extend its shelf life (Church & Parsons, 1995). It is widely used in the food industry, including the packaging of pepper, to maintain quality and freshness (Smith et al., 2018). Here are some more technical details about MAP:

- Gas Composition: MAP typically involves the use of gases like nitrogen (N_2), carbon dioxide (CO_2), and oxygen (O_2) in varying proportions. The specific gas composition is determined based on the product characteristics and desired shelf life. Nitrogen is commonly used as an inert gas to displace oxygen, while carbon dioxide helps inhibit microbial growth and enzymatic activity (Gorris & Peppelenbos, 2020; Meredith et al., 2014).
- Gas Flush and Residual Oxygen: The packaging process involves flushing the package with the desired gas mixture to create the modified atmosphere. The goal is to minimize the residual oxygen content inside the package, as oxygen can lead to oxidation and spoilage. The residual oxygen level is typically measured in percentage (e.g. less than 1% O2) and tightly controlled to ensure product quality and safety (Khan & Mittal, 2020).
- Permeability and Barrier Properties: The packaging materials used for MAP must possess specific barrier properties to prevent gas permeation. Different materials have different permeability rates for gases, so the selection of packaging films or laminates is critical. High-barrier materials, such as certain plastics or aluminum foil, are commonly used to minimize gas exchange and maintain the desired gas composition inside the package (Syverud & Stenius, 2009).
- Gas Exchange Rate: While the goal is to create a modified atmosphere inside the package, it is important to consider the gas exchange rate. Some products, like fresh produce, may still require a certain level of oxygen for respiration (Deshwal et al., 2021). Therefore, the packaging system should be designed to allow controlled gas exchange, allowing the product to respire while maintaining the desired gas composition.
- Gas Sensors and Monitoring: To ensure the efficacy of MAP, gas sensors and monitoring systems are often employed. These sensors detect the gas composition inside the package and provide real-time data on gas levels (Puligundla et al., 2012). This allows for quality control and adjustments in the packaging process if needed (Shaalan et al., 2022).

- Temperature and Storage Conditions: Temperature control and proper storage conditions are essential in conjunction with MAP. Low temperatures help slow down microbial growth and enzymatic reactions, further extending the shelf life of the product (Illeperuma & Jayasuriya, 2002). Maintaining a suitable storage environment is crucial for maximizing the benefits of MAP.

Modified atmosphere packaging offers several advantages, including extended shelf life, reduced spoilage, and improved product quality. It allows for better preservation of flavors, colors, and nutritional properties of the packaged product. However, it requires careful design, gas selection, packaging materials, and monitoring to achieve optimal results for different products.

13.9 SUSTAINABLE AND ECO-FRIENDLY PACKAGING

Sustainable packaging is an important aspect of modern packaging practices, aimed at reducing environmental impact and promoting eco-friendly solutions. In addition to the active packaging systems mentioned earlier, several sustainable packaging options are gaining popularity:

- Biodegradable Packaging: Biodegradable materials, such as bio-based plastics or biopolymers derived from renewable resources like cornstarch or sugarcane, can be used as packaging materials (Ayu et al., 2020). These materials break down naturally over time, reducing waste accumulation (Usal et al., 2021).
- Compostable Packaging: Compostable packaging is designed to break down in industrial composting facilities, creating nutrient-rich soil. It offers an alternative to traditional plastics and reduces landfill waste. Compostable materials are often made from renewable resources like plant-based polymers (Ayu et al., 2020; Kostag & El Seoud, 2021).
- Recyclable Packaging: Packaging materials that can be easily recycled and converted into new products are considered sustainable. Materials like cardboard, paper, glass, and certain types of plastics, such as PET or HDPE, can be recycled (Popescu et al., 2020). Designing packaging with recyclability in mind, such as using mono-materials or minimizing the use of mixed materials, facilitates the recycling process (Arvanitoyannis & Bosnea, 2001).
- Reusable Packaging: Reusable packaging aims to reduce waste by providing a durable, long-lasting alternative to single-use packaging. Examples include glass jars, stainless steel containers, or fabric bags that can be refilled or returned for reuse (Camps-Posino et al., 2021; Coelho et al., 2020).
- Minimalist Packaging: Simplifying packaging design by reducing unnecessary layers, using less material, and optimizing the size and shape of packages can contribute to sustainability. This approach minimizes resource consumption and waste generation while still ensuring product protection and functionality (Ahvenainen, 2003).
- Renewable Energy Use: Sustainable packaging goes beyond material choices and can also involve using renewable energy sources during packaging production, such as solar or wind energy. Implementing energy-efficient manufacturing processes further reduces the environmental footprint (Aggarwal et al., 2018; Haghighi et al., 2020).
- Life Cycle Assessment: Conducting a life cycle assessment of packaging materials and processes helps evaluate their environmental impact throughout the entire life cycle, from sourcing and production to use and disposal (Pauer et al., 2019). This assessment enables informed decision-making and the identification of areas for improvement (Kakadellis & Harris, 2020).

It is crucial to consider a combination of these sustainable packaging practices to achieve optimal environmental outcomes. Each packaging solution should align with the specific needs and requirements of the product while striving for reduced waste generation, lower carbon footprint, and resource conservation (Oloyede & Lignou, 2021).

13.9.1 Biocompostable Packaging

Biocompostable packaging for pepper offers a sustainable alternative that combines effective flavor preservation with reduced environmental impact. These packaging materials are derived from renewable sources such as plant-based polymers, which can undergo decomposition and return to the environment through natural processes (Raźniewska, 2022). The technical properties of biocompostable packaging play a crucial role in ensuring the desired functionality and performance.

The composition of biocompostable packaging materials is carefully designed to provide barrier properties against moisture, oxygen, and light, similar to conventional packaging materials. Various biopolymers, such as polylactic acid (PLA), polyhydroxyalkanoates (PHA), and starch- based polymers, are commonly used in biocompostable packaging (Pal & Katiyar, 2017). These materials can be processed into films, coatings, or laminates to create a protective barrier around the pepper, preventing moisture loss, flavor degradation, and exposure to external contaminants.

Biocompostable packaging offers several technical advantages. It provides excellent moisture resistance, preventing the loss of essential oils and flavor compounds from pepper. The oxygen barrier properties of biocompostable materials help to minimize oxidation reactions, preserving the freshness and quality of the spice. Additionally, these materials can effectively block out UV light, reducing color fading and maintaining the visual appeal of the packaged pepper.

Furthermore, biocompostable packaging materials are often compatible with existing packaging machinery and can be easily processed using standard manufacturing techniques. They exhibit good heat-sealing properties, ensuring proper closure and protection against moisture and air ingress (Chaudhary et al., 2022). The mechanical strength and flexibility of these materials allow for convenient handling, transportation, and storage of pepper products.

Examples of biocompostable packaging solutions for pepper include PLA-based films, cellulose-based coatings, and starch-based laminates (Shlush & Davidovich-Pinhas, 2022). These materials have been extensively researched and developed to meet the specific requirements of pepper packaging, providing an eco-friendly alternative to conventional plastic packaging. Moreover, advancements in biocompostable packaging technology continue to enhance their performance, extending the shelf life of pepper and reducing the environmental impact associated with packaging waste.

In conclusion, biocompostable packaging for pepper offers a promising solution that combines technical functionality, flavor preservation, and environmental sustainability. These materials provide effective barrier properties, moisture resistance, and UV protection, ensuring the quality and shelf life of the packaged pepper. By transitioning to biocompostable packaging, the food industry can contribute to reducing plastic waste and promoting a more sustainable approach to pepper packaging.

13.10 INTELLIGENT PACKAGING

Intelligent packaging, also known as smart packaging or active packaging, incorporates advanced technologies and functionalities to provide additional information or features beyond traditional packaging. Here are some more technical details about intelligent packaging:

- Sensors and Indicators: Intelligent packaging often integrates sensors and indicators to monitor and communicate specific parameters related to the product's condition. For example, temperature sensors can track temperature variations, humidity sensors can measure moisture levels, and freshness indicators can assess product quality based on indicators such as pH or volatile compound emissions.

- Time-Temperature Indicators (TTIs): TTIs are one type of intelligent packaging that provide real-time information about temperature abuse or time-temperature history. They can indicate if a product has been exposed to undesirable conditions that could affect its safety or quality. These indicators use chemical reactions or color changes to visually represent the product's freshness or shelf life (Dainelli et al., 2008).
- RFID and NFC Technology: Radio frequency identification (RFID) and near-field communication (NFC) technologies are commonly used in intelligent packaging. RFID tags or labels contain electronic chips that can store and transmit product information, such as batch numbers, expiration dates, or origin details. NFC allows consumers to interact with the packaging using their smartphones, accessing additional information or verifying product authenticity.
- Gas-Release Systems: Some intelligent packaging systems incorporate gas-release mechanisms to actively modify the atmosphere inside the package. These systems can control the release of specific gases, such as oxygen scavengers or ethylene absorbers, to enhance product freshness, inhibit microbial growth, or delay ripening processes in fruits and vegetables.
- Antitampering Features: Intelligent packaging may include antitampering mechanisms to ensure product integrity and safety. Tamper-evident seals, holographic labels, or QR codes can help consumers verify that the package has not been compromised or opened before purchase (Vanderroost et al., 2014).
- Data Monitoring and Connectivity: Advanced intelligent packaging systems can collect and transmit data regarding various parameters, such as temperature, humidity, or storage conditions. This data can be wirelessly transmitted to monitoring systems or cloud platforms, allowing real-time tracking, quality control, and supply chain optimization.
- Interactive Packaging: Some intelligent packaging incorporates interactive elements to enhance user engagement and experience. For example, packaging with augmented reality (AR) features can provide interactive product information or promotional content when scanned with a smartphone camera (Rani & Ramlie, 2023).

Intelligent packaging offers opportunities for enhanced product safety, freshness monitoring, supply chain optimization, and consumer engagement. The integration of sophisticated sensors, indicators, and connectivity enables better product traceability, quality control, and informed decision-making throughout the packaging life cycle.

13.11   INNOVATIVE PACKAGING DESIGNS

Innovative packaging designs involve the development of new and creative packaging solutions that go beyond traditional formats. Here are some additional technical details about innovative packaging designs:

- Sustainable Materials: Innovative packaging designs focus on the use of sustainable materials that minimize environmental impact. This includes biodegradable or compostable materials, recyclable plastics, and alternative sources like plant-based or bio-based materials (Trajkovska Petkoska et al., 2021).
- Light Weighting: Innovative packaging designs aim to reduce the weight of packaging materials without compromising functionality or protection. This helps minimize material usage, transportation costs, and carbon emissions throughout the supply chain (Elkhattat & Medhat, 2022).
- Active Barrier Technologies: Innovative packaging incorporates advanced barrier technologies to extend the shelf life of products. This can include high-barrier films, coatings, or laminates that provide excellent oxygen, moisture, or light barrier properties to preserve product quality and freshness (Tyagi et al., 2021b).

- Interactive Features: Packaging with interactive features enhances consumer engagement and experience. QR codes, augmented reality (AR), or near-field communication (NFC) technologies enable consumers to access additional product information, recipes, or personalized content through their smartphones (Escobedo et al., 2021).
- Smart Labeling: Innovative packaging designs may include smart labeling solutions that offer functionalities beyond basic information. This can include temperature-sensitive labels, freshness indicators, or interactive elements like touch-responsive labels that change color or texture (Chen et al., 2023).
- Convenience Enhancements: Packaging designs focus on improving convenience for consumers. This includes features like resealable closures, portion control mechanisms, easy-open systems, or ergonomic designs that facilitate handling and usage (Wind Jerry, 2001).
- Active Packaging Systems: Innovative packaging designs can incorporate active packaging systems that actively interact with the product to enhance its quality or functionality. For example, self-heating or self-cooling packages for ready-to-eat meals, moisture-absorbing pads for fresh produce, or controlled-release systems for flavor or fragrance delivery (Dobrucka & Cierpiszewski, 2014).
- Customization and Personalization: Innovative packaging designs offer opportunities for customization and personalization to cater to individual consumer preferences. This can include customizable packaging shapes, personalized labels, or packaging designs that allow consumers to mix and match product components (Wind Jerry, 2001).
- Supply Chain Integration: Innovative packaging designs may include features that improve supply chain efficiency, such as stackable or nestable designs to optimize transportation and storage space, track-and-trace capabilities for inventory management, or tamper-evident features to ensure product integrity (Meherishi et al., 2019).

Innovative packaging designs continuously push the boundaries of traditional packaging to deliver improved functionality, sustainability, consumer engagement, and supply chain efficiency. By leveraging new materials, technologies, and design concepts, they aim to address emerging consumer needs, enhance product differentiation, and reduce environmental impact.

13.12 NANOHYBRID-BASED BIOCOMPOSITES

Nanohybrid-based biocomposites for pepper packaging offer several advantages over traditional packaging materials. These innovative materials combine the benefits of nanotechnology, biopolymers, and other additives to create packaging solutions that are not only environmentally friendly but also provide enhanced functionality and preservation of peppers' quality.

One of the main advantages of nanohybrid-based biocomposites is their improved barrier performance. By incorporating nanoparticles, such as silver nanoparticles or titanium dioxide nanoparticles, into the biocomposite matrix, the packaging material can effectively block the passage of oxygen, moisture, and other contaminants (Ediyilyam et al., 2021). This barrier property helps in maintaining the freshness, flavor, and aroma of pepper for an extended period. Another advantage is the antimicrobial activity provided by nanomaterials. Silver nanoparticles, for example, exhibit strong antimicrobial properties, effectively inhibiting the growth of bacteria, fungi, and other microorganisms that can cause spoilage and contamination of pepper. This antimicrobial activity contributes to the preservation of pepper's quality and extends its shelf life.

Furthermore, nanohybrid-based biocomposites offer improved mechanical strength and durability. The incorporation of natural fibers, such as jute or hemp, as reinforcements enhances the structural integrity and toughness of the packaging material. This increased strength ensures that the packaging can withstand handling, transportation, and storage without compromising the quality of the pepper (Perera et al., 2022).

The use of biopolymers derived from renewable sources, such as starch, cellulose, chitosan, or polylactic acid (PLA), in the biocomposite formulation contributes to the eco-friendliness of the packaging material. These biopolymers are biodegradable and reduce the environmental impact compared to conventional petroleum-based plastics. This aligns with the growing demand for sustainable packaging solutions (Shahbaz et al., 2022).

Examples of nanohybrid-based biocomposites for pepper packaging can include a matrix of starch or PLA reinforced with jute fibers and incorporating silver nanoparticles. This combination provides a strong and eco-friendly packaging material with enhanced barrier properties and antimicrobial activity.

In summary, nanohybrid-based biocomposites offer several advantages for pepper packaging, including improved barrier performance, antimicrobial activity, mechanical strength, and sustainability. By utilizing nanomaterials, biopolymers, and reinforcements, these materials provide a technologically advanced solution that effectively preserves the quality of pepper and extends its shelf life.

13.13 PACKAGING DESIGN AND STRUCTURE

Packaging design refers to the overall visual appearance and layout of the packaging, while packaging structure refers to the physical construction and composition of the packaging materials which is already discussed in the previous section.

- Seal Integrity: Proper seal integrity is essential to prevent the entry of air, moisture, or contaminants. For example, resealable zip-lock bags are commonly used for ground pepper to allow consumers to reseal the package tightly after each use, maintaining freshness and preventing moisture ingress (Ilhan et al., 2021).
- Packaging Size and Format: Packaging size and format should be designed to accommodate the intended quantity of pepper while considering consumer preferences and convenience. Single-serving sachets or portioned spice packets are popular choices for convenience and portion control, ensuring optimal freshness and reducing waste (Eldesouky et al., 2015).
- Labeling: Clear and informative labeling is crucial for consumers to make informed choices. Labels on pepper packaging often include details such as the product name, brand, origin, certification symbols (e.g. organic or fair trade), nutritional information, and storage instructions. This helps consumers identify the product, understand its qualities, and store it correctly.
- Packaging Design for Convenience: User-friendly packaging design enhances convenience and usability. Shaker caps or flip-top dispensers are commonly employed to facilitate easy sprinkling or pouring of ground pepper. Stand-up pouches with tear notches or spouts provide convenience in dispensing and resealing, ensuring ease of use while maintaining product freshness (Eldesouky et al., 2015).

Real-world applications of packaging design and structure for pepper can be found in various commercial products. For instance, a popular brand of ground black pepper offers its product in a resealable stand-up pouch, combining convenience with extended freshness. Another example is the use of glass jars with airtight lids for premium-quality whole peppercorns, providing a visually appealing and protective packaging solution.

13.14 STORAGE DURATION

The storage duration of pepper varies depending on the type of packaging material or advanced technique used. In conventional packaging, such as flexible films, jars, or bottles, it is recommended to consume pepper within 2 to 3 years for optimal flavor and quality. Modified atmosphere packaging

(MAP) allows for an extended storage duration of 3 to 5 years by altering the atmosphere surrounding the pepper, reducing oxygen levels, and controlling humidity. Vacuum packaging, which creates a low-oxygen environment, also extends the recommended storage duration to 3 to 5 years by slowing down oxidation. Active packaging, incorporating technologies like oxygen or moisture absorbers, can further extend the storage duration to 5 years or more by actively controlling the packaging environment. It is important to note that these are general guidelines, and specific recommendations from manufacturers or suppliers should be followed. Proper storage conditions, including suitable temperature, humidity, and light levels, along with appropriate packaging materials or advanced techniques, maximize the shelf life of pepper, preserving its flavor and quality over an extended period.

13.15 CONCLUSIONS

Packaging plays a vital role in preserving the quality, flavor, and shelf life of pepper. The unique characteristics of pepper, such as its composition and susceptibility to flavor loss, aroma degradation, and color changes, necessitate careful consideration when selecting appropriate packaging materials and technologies.

This chapter provides the importance of understanding the composition of pepper and its reactions with the external environment, emphasizing the need for packaging that provides adequate protection against factors like oxidation, photodegradation, heat generation, and volatile component loss.

The chapter also touched upon the significance of storage conditions and shelf life in maintaining pepper's freshness and quality. We emphasized the importance of proper storage duration, optimal temperature and humidity levels, and the role of packaging in minimizing contamination risks during the harvesting, processing, and storage stages.

Different packaging materials commonly used were discussed for pepper, including glass, plastic, and metal containers. We examined their advantages, limitations, and their ability to provide moisture, light, and oxygen barriers. The challenges associated with single-layer packaging and the growing trend towards double-layer packaging for enhanced protection and preservation are described.

Furthermore, we explored recent advances in pepper packaging, such as modified atmosphere packaging (MAP), intelligent packaging, and sustainable packaging solutions. These advancements aim to extend the shelf life, improve quality, enhance safety, and reduce environmental impact.

It is crucial to adhere to packaging regulations and safety standards to ensure consumer protection and maintain product integrity. Compliance with food contact materials regulations, hygiene practices, and the use of eco-friendly packaging materials align with sustainability goals and consumer expectations.

Looking ahead, future trends and innovations in pepper packaging are expected to focus on smart packaging technologies, antimicrobial packaging, and further advancements in sustainable packaging solutions. These developments will continue to drive improvements in packaging efficiency, product quality, and environmental sustainability.

In summary, effective pepper packaging involves careful consideration of factors like composition, reactions with the environment, packaging materials, storage conditions, regulations, and sustainability. By employing appropriate packaging strategies, producers can safeguard the flavor, aroma, and color of pepper, extend its shelf life, ensure food safety, and meet consumer demands for high-quality, sustainable products.

NOTES

1 Glass transition temperature (Tg) in polymers indicates the temperature where the material transitions from a rigid to a flexible state. If Tg is lower, material will transition from glassy state to rubbery state at very low temperature and therefore will increase the movement in polymer chain creating the gaps and thereby the oxygen permeability.

2 Higher humidity disrupts polymer structure, increasing free volume and pathways for oxygen diffusion-Water molecules can hydrogen bond with polar groups in the polymer chains, weakening intermolecular forces. This weakening of forces allows polymer chains to move more freely and increases the overall free volume of the material.

3 Thicker packaging materials create longer pathways for oxygen molecules to traverse, increasing interactions with polymer chains and slowing down diffusion.

REFERENCES

Aggarwal, A., Schmid, M., Patel, M. K., & Langowski, H.-C. (2018). Function-driven investigation of non-renewable energy use and greenhouse gas emissions for material selection in food packaging applications: Case study of yoghurt packaging. *Procedia CIRP*, 69, 728–733. https://doi.org/10.1016/j.procir.2017.11.132

Ahmed, R. S., Khan, M. K., Ahmad, M. H., Arshad, M. S., Ateeq, H., & Rahim, M. A. (2021). Introductory chapter: Herbs and spices-an overview. *Herbs and Spices-New Processing Technologies*, 2–6. https://10.5772/intechopen.100725

Ahvenainen, R., ed. (2003). *Novel Food Packaging Techniques*. Elsevier.

Ajitha, A. R., Aswathi, M. K., Maria, H. J., Izdebska, J., & Thomas, S. (2016). *Multilayer Polymer Films* (pp. 229–258). https://doi.org/10.1007/978-94-017-7324-9_8

Anukiruthika, T., Sethupathy, P., Wilson, A., Kashampur, K., Moses, J. A., & Anandharamakrishnan, C. (2020a). Multilayer packaging: Advances in preparation techniques and emerging food applications. *Comprehensive Reviews in Food Science and Food Safety*, 19(3), 1156–1186. https://doi.org/10.1111/1541-4337.12556

Anukiruthika, T., Sethupathy, P., Wilson, A., Kashampur, K., Moses, J. A., & Anandharamakrishnan, C. (2020b). Multilayer packaging: Advances in preparation techniques and emerging food applications. *Comprehensive Reviews in Food Science and Food Safety*, 19(3), 1156–1186. https://doi.org/10.1111/1541-4337.12556

Anukiruthika, T., Sethupathy, P., Wilson, A., Kashampur, K., Moses, J. A., & Anandharamakrishnan, C. (2020c). Multilayer packaging: Advances in preparation techniques and emerging food applications. *Comprehensive Reviews in Food Science and Food Safety*, 19(3), 1156–1186. https://doi.org/10.1111/1541-4337.12556

Anukiruthika, T., Sethupathy, P., Wilson, A., Kashampur, K., Moses, J. A., & Anandharamakrishnan, C. (2020d). Multilayer packaging: Advances in preparation techniques and emerging food applications. *Comprehensive Reviews in Food Science and Food Safety*, 19(3), 1156–1186. https://doi.org/10.1111/1541-4337.12556

Arvanitoyannis, I. S., & Bosnea, L. A. (2001). Recycling of polymeric materials used for food packaging: Current status and perspectives. *Food Reviews International*, 17(3), 291–346. https://doi.org/10.1081/FRI-100104703

Arvanitoyannis, I. S., & Kotsanopoulos, K. V. (2014). Migration phenomenon in food packaging. food–package interactions, mechanisms, types of migrants, testing and relative legislation—A review. *Food and Bioprocess Technology*, 7(1), 21–36. https://doi.org/10.1007/s11947-013-1106-8

Ayu, R. S., Khalina, A., Harmaen, A. S., Zaman, K., Isma, T., Liu, Q., Ilyas, R. A., & Lee, C. H. (2020). Characterization study of empty fruit bunch (EFB) fibers reinforcement in poly(butylene) succinate (PBS)/starch/glycerol composite sheet. *Polymers*, 12(7), 1571. https://doi.org/10.3390/polym12071571

Azevedo, A. G., Barros, C., Miranda, S., Machado, A. V., Castro, O., Silva, B., Saraiva, M., Silva, A. S., Pastrana, L., Carneiro, O. S., & Cerqueira, M. A. (2022). Active flexible films for food packaging: A review. *Polymers*, 14(12), 2442. https://doi.org/10.3390/polym14122442

Baele, M., Vermeulen, A., Adons, D., Peeters, R., Vandemoortele, A., Devlieghere, F., De Meulenaer, B., & Ragaert, P. (2021). Selecting packaging material for dry food products by trade-off of sustainability and performance: A case study on cookies and milk powder. *Packaging Technology and Science*, 34(5), 303–318. https://doi.org/10.1002/pts.2561

Bamps, B., Buntinx, M., & Peeters, R. (2023). Seal materials in flexible plastic food packaging: A review. *Packaging Technology and Science*, 36(7), 507–532. https://doi.org/10.1002/pts.2732

Baranwal, J., Barse, B., Fais, A., Delogu, G. L., & Kumar, A. (2022). Biopolymer: A sustainable material for food and medical applications. *Polymers*, 14(5), 983. https://doi.org/10.3390/polym14050983

Bayer, I. S. (2021). Biopolymers in multilayer films for long-lasting protective food packaging: A review. In *Sustainable Food Packaging Technology* (pp. 395–426). Wiley. https://doi.org/10.1002/9783527820078.ch15

Camps-Posino, L., Batlle-Bayer, L., Bala, A., Song, G., Qian, H., Aldaco, R., Xifré, R., & Fullana-i-Palmer, P. (2021). Potential climate benefits of reusable packaging in food delivery services. A Chinese case study. *Science of the Total Environment*, 794, 148570. https://doi.org/10.1016/j.scitotenv.2021.148570

Chaudhary, V., Punia Bangar, S., Thakur, N., & Trif, M. (2022). Recent advancements in smart biogenic packaging: Reshaping the future of the food packaging industry. *Polymers*, 14(4), 829. https://doi.org/10.3390/polym14040829

Chen, B., Zhang, M., Chen, H., Mujumdar, A. S., & Guo, Z. (2023). Progress in smart labels for rapid quality detection of fruit and vegetables: A review. *Postharvest Biology and Technology*, 198, 112261. https://doi.org/10.1016/j.postharvbio.2023.112261

Church, I. J., & Parsons, A. L. (1995). Modified atmosphere packaging technology: A review. *Journal of the Science of Food and Agriculture*, 67(2), 143–152. https://doi.org/10.1002/jsfa.2740670202

Coelho, P. M., Corona, B., ten Klooster, R., & Worrell, E. (2020). Sustainability of reusable packaging–Current situation and trends. *Resources, Conservation & Recycling: X*, 6, 100037. https://doi.org/10.1016/j.rcrx.2020.100037

Dainelli, D., Gontard, N., Spyropoulos, D., Zondervan-van den Beuken, E., & Tobback, P. (2008). Active and intelligent food packaging: Legal aspects and safety concerns. *Trends in Food Science & Technology*, 19, S103–S112. https://doi.org/10.1016/j.tifs.2008.09.011

Deshwal, G. K., Tiwari, S., Panjagari, N.R., & Masud S. (2021). Active packaging of fruits and vegetables: Quality preservation and shelf-life enhancement. In *Packaging and Storage of Fruits and Vegetables* (pp. 109–131). Apple Academic Press. https://doi.org/10.1201/9781003161165

Deshwal, G. K., & Panjagari, N. R. (2020). Review on metal packaging: Materials, forms, food applications, safety and recyclability. *Journal of Food Science and Technology*, 57(7), 2377–2392. https://doi.org/10.1007/s13197-019-04172-z

Dey, A., & Neogi, S. (2019). Oxygen scavengers for food packaging applications: A review. *Trends in Food Science & Technology*, 90, 26–34. https://doi.org/10.1016/j.tifs.2019.05.013

Dobrucka, R., & Cierpiszewski, R. (2014). Active and intelligent packaging food-research and development- a review. *Polish Journal of Food and Nutrition Sciences*, 64(1), 7–15. https://doi.org/10.2478/v10222-012-0091-3

Domeño, C., Aznar, M., Nerín, C., Isella, F., Fedeli, M., & Bosetti, O. (2017). Safety by design of printed multilayer materials intended for food packaging. *Food Additives & Contaminants: Part A*, 34(7), 1239–1250. https://doi.org/10.1080/19440049.2017.1322221

Ediyilyam, S., George, B., Shankar, S. S., Dennis, T. T., Wacławek, S., Černík, M., & Padil, V. V. T. (2021). Chitosan/gelatin/silver nanoparticles composites films for biodegradable food packaging applications. *Polymers*, 13(11), 1680. https://doi.org/10.3390/polym13111680

Eldesouky, A., Pulido, A. F., & Mesias, F. J. (2015). The role of packaging and presentation format in consumers' preferences for food: An application of projective techniques. *Journal of Sensory Studies*, 30(5), 360–369. https://doi.org/10.1111/joss.12162

Elkhattat, D., & Medhat, M. (2022). Creativity in packaging design as a competitive promotional tool. *Information Sciences Letters*, 11(1), 135–148. https://doi.org/10.18576/isl/110115

Escobedo, P., Bhattacharjee, M., Nikbakhtnasrabadi, F., & Dahiya, R. (2021). Flexible strain and temperature sensing NFC tag for smart food packaging applications. *IEEE Sensors Journal*, 21(23), 26406–26414. https://doi.org/10.1109/JSEN.2021.3100876

Fávaro, S. L., Rubira, A. F., Muniz, E. C., & Radovanovic, E. (2007). Surface modification of HDPE, PP, and PET films with KMnO4/HCl solutions. *Polymer Degradation and Stability*, 92(7), 1219–1226. https://doi.org/10.1016/j.polymdegradstab.2007.04.005

Gaikwad, K. K., Singh, S., & Ajji, A. (2019). Moisture absorbers for food packaging applications. *Environmental Chemistry Letters*, 17(2), 609–628. https://doi.org/10.1007/s10311-018-0810-z

Gorris, L. G., & Peppelenbos, H. W. (2020). Modified-atmosphere packaging of produce. In *Handbook of Food Preservation* (pp. 349–362). CRC Press.

Grzebieniarz, W., Biswas, D., Roy, S., & Jamróz, E. (2023). Advances in biopolymer-based multi-layer film preparations and food packaging applications. *Food Packaging and Shelf Life*, 35, 101033. https://doi.org/10.1016/j.fpsl.2023.101033

Haghighi, H., Licciardello, F., Fava, P., Siesler, H. W., & Pulvirenti, A. (2020). Recent advances on chitosan-based films for sustainable food packaging applications. *Food Packaging and Shelf Life*, 26, 100551. https://doi.org/10.1016/j.fpsl.2020.100551

Hammouti, B., Dahmani, M., Yahyi, A., Ettouhami, A., Messali, M., Asehraou, A., Bouyanzer, A., Warad, I., & Touzani, R. (2019). Black pepper, the "king of spices": Chemical composition to applications. *Arabian Journal of Chemistry Environment Research*, 6, 12–56.

Hu, S., Wang, H., Han, W., Ma, Y., Shao, Z., & Li, L. (2017). Development of double-layer active films containing pomegranate peel extract for the application of pork packaging. *Journal of Food Process Engineering*, 40(2), e12388. https://doi.org/10.1111/jfpe.12388

Huang, T., Qian, Y., Wei, J., & Zhou, C. (2019). Polymeric antimicrobial food packaging and its applications. *Polymers*, 11(3), 560. https://doi.org/10.3390/polym11030560

Ilhan, I., Turan, D., Gibson, I., & Klooster, R. (2021). Understanding the factors affecting the seal integrity in heat sealed flexible food packages: A review. *Packaging Technology and Science*, 34(6), 321–337. https://doi.org/10.1002/pts.2564

Illeperuma, C. K., & Jayasuriya, P. (2002). Prolonged storage of 'Karuthacolomban' mango by modified atmosphere packaging at low temperature. *The Journal of Horticultural Science and Biotechnology*, 77(2), 153–157. https://doi.org/10.1080/14620316.2002.11511472

Jayakumar, A., Radoor, S., C Nair, I., Siengchin, S., Parameswaranpillai, J., & EK, R. (2021). Polyvinyl alcohol -nanocomposite films incorporated with clay nanoparticles and lipopeptides as active food wraps against food spoilage microbes. *Food Packaging and Shelf Life*, 30, 100727. https://doi.org/10.1016/j.fpsl.2021.100727

Kaiser, K., Schmid, M., & Schlummer, M. (2017). Recycling of polymer-based multilayer packaging: A review. *Recycling*, 3(1), 1. https://doi.org/10.3390/recycling3010001

Kakadellis, S., & Harris, Z. M. (2020). Don't scrap the waste: The need for broader system boundaries in bioplastic food packaging life-cycle assessment – A critical review. *Journal of Cleaner Production*, 274, 122831. https://doi.org/10.1016/j.jclepro.2020.122831

Kausar, A. (2020). A review of high performance polymer nanocomposites for packaging applications in electronics and food industries. *Journal of Plastic Film & Sheeting*, 36(1), 94–112. https://doi.org/10.1177/8756087919849459

Khalaj, M.-J., Ahmadi, H., Lesankhosh, R., & Khalaj, G. (2016). Study of physical and mechanical properties of polypropylene nanocomposites for food packaging application: Nano-clay modified with iron nanoparticles. *Trends in Food Science & Technology*, 51, 41–48. https://doi.org/10.1016/j.tifs.2016.03.007

Khan, M. A. M., and Mittal, A. (2020). Modified atmosphere packaging technique: An overview. *International Conference on Recent Advances in Engineering and Science*.

Kim, H., Panda, P. K., Sadeghi, K., & Seo, J. (2023). Poly (vinyl alcohol)/hydrothermally treated tannic acid composite films as sustainable antioxidant and barrier packaging materials. *Progress in Organic Coatings*, 174, 107305. https://doi.org/10.1016/j.porgcoat.2022.107305

King, K. (2006). Packaging and storage of herbs and spices. In *Handbook of Herbs and Spices* (pp. 86–102). Elsevier. https://doi.org/10.1533/9781845691717.1.86

Kostag, M., & El Seoud, O. A. (2021). Sustainable biomaterials based on cellulose, chitin and chitosan composites – A review. *Carbohydrate Polymer Technologies and Applications*, 2, 100079. https://doi.org/10.1016/j.carpta.2021.100079

Kwon, S., Orsuwan, A., Bumbudsanpharoke, N., Yoon, C., Choi, J, & Ko, S. (2018). A short review of light barrier materials for food and beverage packaging. *Korean Journal of Packaging Science & Technology*, 24, 141–148. https://10.20909/kopast.2018.24.3.141

Lamberti, F. M., Román-Ramírez, L. A., & Wood, J. (2020). Recycling of bioplastics: Routes and benefits. *Journal of Polymers and the Environment*, 28(10), 2551–2571. https://doi.org/10.1007/s10924-020-01795-8

Laria, J. G., Gaggino, R., Kreiker, J., Peisino, L. E., Positieri, M., & Cappelletti, A. (2023). Mechanical and processing properties of recycled PET and LDPE-HDPE composite materials for building components. *Journal of Thermoplastic Composite Materials*, 36(1), 418–431. https://doi.org/10.1177/08927 05720939141

Leelaphiwat, P., Auras, R. A., Burgess, G. J., Harte, J. B., & Chonhenchob, V. (2018). Preliminary quantification of the permeability, solubility and diffusion coefficients of major aroma compounds present in herbs through various plastic packaging materials. *Journal of the Science of Food and Agriculture*, 98(4), 1545–1553. https://doi.org/10.1002/jsfa.8626

Maduwantha, M. I. P., & Jayasinghe, R. A. (2023). Possibilities of development of composite materials from Tetra Pak and metalized film-based packaging waste for non-structural applications. *International Journal of Scientific Engineering and Science*, 7(1), 1–9.

Mandeel, Q. A. (2005). Fungal contamination of some imported spices. *Mycopathologia*, 159(2), 291–298. https://doi.org/10.1007/s11046-004-5496-z

Marangoni Júnior, L., Oliveira, L. M. de, Dantas, F. B. H., Cristianini, M., Padula, M., & Anjos, C. A. R. (2020). Influence of high-pressure processing on morphological, thermal and mechanical properties of retort and metallized flexible packaging. *Journal of Food Engineering*, 273, 109812. https://doi.org/10.1016/j.jfoodeng.2019.109812

Marques, H. M. C. (2010). A review on cyclodextrin encapsulation of essential oils and volatiles. *Flavour and Fragrance Journal*, 25(5), 313–326. https://doi.org/10.1002/ffj.2019

Meherishi, L., Narayana, S. A., & Ranjani, K. S. (2019). Sustainable packaging for supply chain management in the circular economy: A review. *Journal of Cleaner Production*, 237, 117582. https://doi.org/10.1016/j.jclepro.2019.07.057

Meredith, H., Valdramidis, V., Rotabakk, B. T., Sivertsvik, M., McDowell, D., & Bolton, D. J. (2014). Effect of different modified atmospheric packaging (MAP) gaseous combinations on Campylobacter and the shelf-life of chilled poultry fillets. *Food Microbiology*, 44, 196–203. https://doi.org/10.1016/j.fm.2014.06.005

Mujtaba, M., Lipponen, J., Ojanen, M., Puttonen, S., & Vaittinen, H. (2022). Trends and challenges in the development of bio-based barrier coating materials for paper/cardboard food packaging: A review. *Science of the Total Environment*, 851, 158328. https://doi.org/10.1016/j.scitotenv.2022.158328

Müller, P., & Schmid, M. (2019). Intelligent packaging in the food sector: A brief overview. *Foods*, 8(1), 16. https://doi.org/10.3390/foods8010016

Oloyede, O. O., & Lignou, S. (2021). Sustainable paper-based packaging: A consumer's perspective. *Foods*, 10(5), 1035. https://doi.org/10.3390/foods10051035

Ordoñez, R., Atarés, L., & Chiralt, A. (2022). Biodegradable active materials containing phenolic acids for food packaging applications. *Comprehensive Reviews in Food Science and Food Safety*, 21(5), 3910–3930. https://doi.org/10.1111/1541-4337.13011

Oudjedi, K., Manso, S., Nerin, C., Hassissen, N., & Zaidi, F. (2019). New active antioxidant multilayer food packaging films containing Algerian Sage and Bay leaves extracts and their application for oxidative stability of fried potatoes. *Food Control*, 98, 216–226. https://doi.org/10.1016/j.foodcont.2018.11.018

Ozturkoglu-Budak, S. (2017). A model for implementation of HACCP system for prevention and control of mycotoxins during the production of red dried chili pepper. *Food Science and Technology*, 37(suppl 1), 24–29. https://doi.org/10.1590/1678-457x.30316

Paixão, L. C., Lopes, I. A., Barros Filho, A. K. D., & Santana, A. A. (2019). Alginate biofilms plasticized with hydrophilic and hydrophobic plasticizers for application in food packaging. *Journal of Applied Polymer Science*, 136(48), 48263. https://doi.org/10.1002/app.48263

Pal, A. K., & Katiyar, V. (2017). Thermal degradation behaviour of nanoamphiphilic chitosan dispersed poly (lactic acid) bionanocomposite films. *International Journal of Biological Macromolecules*, 95, 1267–1279. https://doi.org/10.1016/j.ijbiomac.2016.11.024

Patel, R., Prajapati, J. P., & Balakrishnan, S. (2015). Recent trends in packaging of dairy and food products. In *Meeting National seminar on Indian Dairy Industry-Opportunities and Challenges* (pp. 118–124). Gujarat, India: Anand Agricultural University, Anand, https://api.semanticscholar.org/CorpusID:4465 9154

Pauer, E., Wohner, B., Heinrich, V., & Tacker, M. (2019). Assessing the environmental sustainability of food packaging: An extended life cycle assessment including packaging-related food losses and waste and circularity assessment. *Sustainability*, 11(3), 925. https://doi.org/10.3390/su11030925

Perera, K. Y., Jaiswal, S., & Jaiswal, A. K. (2022). A review on nanomaterials and nanohybrids based bio-nanocomposites for food packaging. *Food Chemistry*, 376, 131912. https://doi.org/10.1016/j.foodchem.2021.131912

Perumal, A. B., Huang, L., Nambiar, R. B., He, Y., Li, X., & Sellamuthu, P. S. (2022). Application of essential oils in packaging films for the preservation of fruits and vegetables: A review. *Food Chemistry*, 375, 131810. https://doi.org/10.1016/j.foodchem.2021.131810

Piringer, O. G. (1994). Evaluation of plastics for food packaging. *Food Additives and Contaminants*, 11(2), 221–230. https://doi.org/10.1080/02652039409374220

Popescu, P. A., Popa, E. E., Mitelut, A. C., & Popa, M. E. (2020). Development of recyclable and biodegradable food packaging materials – opportunities and risks. *Current Trends in Natural Sciences*, 9(17), 142–146. https://doi.org/10.47068/ctns.2020.v9i17.016

Puligundla, P., Jung, J., & Ko, S. (2012). Carbon dioxide sensors for intelligent food packaging applications. *Food Control*, 25(1), 328–333. https://doi.org/10.1016/j.foodcont.2011.10.043

Qu, P., Zhang, M., Fan, K., & Guo, Z. (2022). Microporous modified atmosphere packaging to extend shelf life of fresh foods: A review. *Critical Reviews in Food Science and Nutrition*, 62(1), 51–65. https://doi.org/10.1080/10408398.2020.1811635

Rani, M. N. H. A., & M. K. Ramlie. (2023). The utilization of augmented reality (AR) applications as packaging design enhancement. *International Journal of Academic Research in Business & Social Sciences*, 13(7), 253–277.

Raźniewska, M. (2022). Compostable packaging waste management—main barriers, reasons, and the potential directions for development. *Sustainability*, 14(7), 3748. https://doi.org/10.3390/su14073748

Salama, H. E., & Abdel Aziz, M. S. (2020). Optimized carboxymethyl cellulose and guanidinylated chitosan enriched with titanium oxide nanoparticles of improved UV-barrier properties for the active packaging of green bell pepper. *International Journal of Biological Macromolecules*, 165, 1187–1197. https://doi.org/10.1016/j.ijbiomac.2020.09.254

Schmidt, J., Grau, L., Auer, M., Maletz, R., & Woidasky, J. (2022). Multilayer packaging in a circular economy. *Polymers*, 14(9), 1825. https://doi.org/10.3390/polym14091825

Shaalan, N. M., Ahmed, F., Saber, O., & Kumar, S. (2022). Gases in food production and monitoring: Recent advances in target chemiresistive gas sensors. *Chemosensors*, 10(8), 338. https://doi.org/10.3390/chemosensors10080338

Shahbaz, A., Hussain, N., Basra, M. A. R., & Bilal, M. (2022). Polysaccharides-based nano-hybrid biomaterial platforms for tissue engineering, drug delivery, and food packaging applications. *Starch – Stärke*, 74(7–8), 2200023. https://doi.org/10.1002/star.202200023

Shin, J., & Selke, S. E. M. (2014). Food packaging. In *Food Processing* (pp. 249–273). John Wiley & Sons, Ltd. https://doi.org/10.1002/9781118846315.ch11

Shlush, E., & Davidovich-Pinhas, M. (2022). Bioplastics for food packaging. *Trends in Food Science & Technology*, 125, 66–80. https://doi.org/10.1016/j.tifs.2022.04.026

Singh, A. K., Ramakanth, D., Kumar, A., Lee, Y. S., & Gaikwad, K. K. (2021). Active packaging technologies for clean label food products: A review. *Journal of Food Measurement and Characterization*, 15(5), 4314–4324. https://doi.org/10.1007/s11694-021-01024-3

Siracusa, V., & Blanco, I. (2020). Bio-Polyethylene (Bio-PE), Bio-Polypropylene (Bio-PP) and Bio-Poly(ethylene terephthalate) (Bio-PET): Recent developments in bio-based polymers analogous to petroleum-derived ones for packaging and engineering applications. *Polymers*, 12(8), 1641. https://doi.org/10.3390/polym12081641

Smith, J. P., Abe, Y., & Hoshino, J. (2018). Modified atmosphere packaging—present and future uses of gas absorbents and generators. In *Principles of Modified-Atmosphere and Sous Vide Product Packaging* (pp. 287–323). Taylor and Francis Group.

Soares, C. T. de M., Ek, M., Östmark, E., Gällstedt, M., & Karlsson, S. (2022). Recycling of multi-material multilayer plastic packaging: Current trends and future scenarios. *Resources, Conservation and Recycling*, 176, 105905. https://doi.org/10.1016/j.resconrec.2021.105905

Subroto, E., Djali, M., Indiarto, R., Lembong, E., & Baiti, N. (2023). Microbiological activity affects postharvest quality of cocoa (Theobroma cacao L.) Beans. *Horticulturae*, 9(7), 805. https://doi.org/10.3390/horticulturae9070805

Syverud, K., & Stenius, P. (2009). Strength and barrier properties of MFC films. *Cellulose*, 16, 75–85. https://doi.org/10.1007/s10570-008-9244-2

Tang, S., Wu, Z., Li, X., Xie, F., Ye, D., Ruiz-Hitzky, E., Wei, L., & Wang, X. (2023). Nacre-inspired biodegradable nanocellulose/MXene/AgNPs films with high strength and superior gas barrier properties. *Carbohydrate Polymers*, 299, 120204. https://doi.org/10.1016/j.carbpol.2022.120204

Trajkovska Petkoska, A., Daniloski, D., D'Cunha, N. M., Naumovski, N., & Broach, A. T. (2021). Edible packaging: Sustainable solutions and novel trends in food packaging. *Food Research International*, 140, 109981. https://doi.org/10.1016/j.foodres.2020.109981

Tyagi, P., Salem, K. S., Hubbe, M. A., & Pal, L. (2021a). Advances in barrier coatings and film technologies for achieving sustainable packaging of food products – A review. *Trends in Food Science & Technology*, 115, 461–485. https://doi.org/10.1016/j.tifs.2021.06.036

Tyagi, P., Salem, K. S., Hubbe, M. A., & Pal, L. (2021b). Advances in barrier coatings and film technologies for achieving sustainable packaging of food products – A review. *Trends in Food Science & Technology*, 115, 461–485. https://doi.org/10.1016/j.tifs.2021.06.036

Tyuftin, A. A., & Kerry, J. P. (2023). The storage and preservation of meat: Storage and packaging. In *Lawrie's Meat Science* (pp. 315–362). Elsevier. https://doi.org/10.1016/B978-0-323-85408-5.00017-0

Usal, T. D., Bektas, C. K., Hasirci, N., & Hasirci, V. (2021). Engineered biopolymers. In *Biological Soft Matter* (pp. 65–88). Wiley. https://doi.org/10.1002/9783527811014.ch3

Vanderroost, M., Ragaert, P., Devlieghere, F., & De Meulenaer, B. (2014). Intelligent food packaging: The next generation. *Trends in Food Science & Technology*, 39(1), 47–62. https://doi.org/10.1016/j.tifs.2014.06.009

Viana Batista, R., Gonçalves Wanzeller, W., & Menezes, V. M. de. (2023). Mathematical model and numerical simulation of oxygen transport in multilayer food packaging: A semi-analytical approach. *Regionem: Revista Interdisciplinar Em Desenvolvimento Sustentável*, 1(1), e12864. https://doi.org/10.36661/2965-3320.2023v1n1.12864

Vilas, C., Mauricio-Iglesias, M., & García, M. R. (2020). Model-based design of smart active packaging systems with antimicrobial activity. *Food Packaging and Shelf Life*, 24, 100446. https://doi.org/10.1016/j.fpsl.2019.100446

Watada, A. E., Kim, S. D., Kim, K. S., & Harris, T. C. (1987). Quality of green beans, bell peppers and spinach stored in polyethylene bags. *Journal of Food Science*, 52(6), 1637–1641. https://doi.org/10.1111/j.1365-2621.1987.tb05894.x

Weber Macena, M., Carvalho, R., Cruz-Lopes, L. P., & Guiné, R. P. F. (2021). Plastic food packaging: Perceptions and attitudes of Portuguese consumers about environmental impact and recycling. *Sustainability*, 13(17), 9953. https://doi.org/10.3390/su13179953

Wind Jerry, R. A. (2001). Customerization: The next revolution in mass customization. *Journal of Interactive Marketing*. Retrieved 7 June 2023, from https://doi.org/10.1002/1520-6653(200124)15:1%3C13::AID-DIR1001%3E3.0.CO;2-%23

Wongsa, P., et al. (2023). Influence of food-packaging materials and shelf-life conditions on dried garlic (Allium sativum L.) concerning quality and stability of allicin/phenolic content. *Food and Bioprocess Technology*, 1–12.

Zabihzadeh Khajavi, M., Ebrahimi, A., Yousefi, M., Ahmadi, S., Farhoodi, M., Mirza Alizadeh, A., & Taslikh, M. (2020a). Strategies for producing improved oxygen barrier materials appropriate for the food packaging sector. *Food Engineering Reviews*, 12(3), 346–363. https://doi.org/10.1007/s12393-020-09235-y

Zabihzadeh Khajavi, M., Ebrahimi, A., Yousefi, M., Ahmadi, S., Farhoodi, M., Mirza Alizadeh, A., & Taslikh, M. (2020b). Strategies for producing improved oxygen barrier materials appropriate for the food packaging sector. *Food Engineering Reviews*, 12(3), 346–363. https://doi.org/10.1007/s12393-020-09235-y

Ajay W. Tumaney, Vallamkondu Manasa,
Palak Daga and Salony Raghunath Vaishnav

14 Pepper Fixed Oil
Novel Source of Oil-Based Nutraceuticals

CONTENTS

14.1 INTRODUCTION

The utilisation of herbs, spices, and seasonings dates back to the inception of human civilization. Traditional ingredients have historically been utilised in culinary practises to augment flavour profiles and offer various health advantages. Traditional medicine is based on historical and natural healthcare practises, such as Ayurveda (Kunwar et al., 2010). According to Devkota and Watanabe (2020), a significant majority, more than 70%, of the global population still depends on medicinal plants and botanicals as their primary healthcare means. The healthcare system is heavily relied upon by individuals due to its cultural acceptability, accessibility, cost, and alignment with religious views.

14.1.1 OVERVIEW OF PEPPER PRODUCTION

Pepper belongs to the genus *Capsicum* and is a member of the *Solanaceae* family. The genus *Capsicum* contains approximately 31 species, 5 of which are cultivated: *C. annuum, C. chinense, C. pubescens, C. baccatum, and C. frutescens*. The average global production of dry and green peppers is estimated to be 3.9 million tonnes and 34.5 million hectares, with dry and green peppers harvested from 1.8 million and 1.9 million hectares, respectively. Despite their immense trait differences, most commercially cultivated pepper cultivars are members of the species *C. annuum*. However, both *C. frutescens* and *C. chinense* are extensively cultivated.

Pepper is a quintessential Indian spice/vegetable and is widely used. India boasts an influential position within spice production, offering diverse spices. The country's climatic diversity, spanning from tropical to sub-tropical and temperate zones, creates an optimal environment for the flourishing growth of numerous spices. This phenomenon extends throughout India's state and union territory, each cultivating various spices. Originating in the elevated terrains of the South Western Ghats in India, black pepper has its roots entrenched in this region. The plant's resilience is evident in its ability to endure temperatures from 10°C to 40°C. Ideal pepper cultivation necessitates an evenly distributed annual rainfall ranging from 125 cm to 200 cm. Given this prerequisite, the climatic circumstances in the southern expanse of India emerge as a paragon, rendering them ideally suited for cultivating this spice as per the Spice Board of India. Compared to bell pepper, Karnataka, Kerala, and Tamil Nadu are major black pepper-producing states in India. The Spice Board of India has reported that in 2021–22, Karnataka produced 39,000 tonnes, Kerala produced 21,000 tonnes, and Tamil Nadu produced 2,000 tonnes of black pepper, as shown in Figure 14.1.

In 1950, India held a significant share of 70% of global pepper cultivation. However, this proportion declined to 46% by 1991, and over the following 60 years, production and export numbers experienced substantial reductions of 66% and 33%, respectively. India's portion of the global market dwindled from 56% to 23% within the same timeframe. Simultaneously, other nations demonstrated notable advancements in their share of the global pepper market, exemplified by the expansion of pepper cultivation in countries like Vietnam. Internationally, it is now grown in Indonesia, Malaysia, Sri Lanka, Thailand, China, Vietnam, Cambodia, Brazil, Mexico, and Guatemala, apart from the country of origin. Today, Vietnam has taken the lead with a 34% share in pepper production, relegating India to the fourth position in this ranking (Ravindran and Kallupurackal, 2012).

14.1.2 Health Benefits of Pepper Oil

The principal nutraceutical components in pepper oils encompass polyphenols, tocopherols, γ-oryzanol, and phytosterols. These nutraceutical sources have demonstrated potential against diabetes, cancer, and disorders related to lipid metabolism (Kowalska et al., 2017).

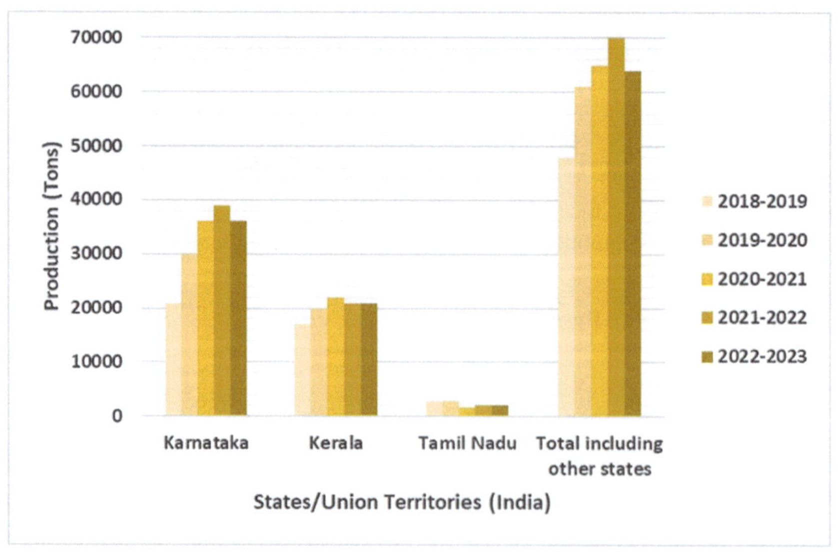

FIGURE 14.1 Production of black pepper in India.

14.1.2.1 Anti-Inflammatory Potential

Reactive oxygen species are accountable for various adverse reactions within the human body, potentially leading to inflammation, cancer, and other complications. Chilli pepper extracts contain anti-inflammatory and antiallergic properties (Lee et al., 2005). Kim et al. (2003) report capsaicin primarily confers anti-inflammatory activity among the many phytochemicals found in chilli pepper. Capsaicin suppressed the obesity-induced inflammatory responses of macrophages in adipose tissue, suggesting its potential use as an analgesic. Capsaicin inhibits the expression of pro-inflammatory cytokines such as tumour necrosis factor- (TNF) and interleukin 1 (IL-1), thereby reducing antigen-induced inflammatory responses (Spiller et al., 2008). The first study on the anti-inflammatory activity of pepper was reported more than 2 decades ago. As highlighted by Mujumdar et al. (1990), there was a proposition that piperine could mitigate acute inflammatory processes through the stimulation of the pituitary-adrenal axis. Subsequent investigations further explored the potential mechanisms underpinning this anti-inflammatory activity. Later, Sabina et al. (2011) reported that piperine (50–100 µg/ml) can suppress the β-glucuronidase and lactate dehydrogenase levels in a dose-dependent manner.

Further, piperine can impede the expression of enzymes such as 5-lipoxygenase and COX-1, which play a role in the biosynthesis of leukotrienes and prostaglandins. These combined effects hold significant potential for mitigating degenerative conditions like rheumatoid arthritis (Stohr et al., 2001). In research conducted by Tasleem et al. (2014), an examination was undertaken to assess the anti-inflammatory potential of pepper and its active constituents by evaluating its effectiveness in treating carrageenan-induced paw edema using a plethysmometer. This study noted that piperine exhibited anti-inflammatory properties at doses of 10 and 15 mg/kg within 30 minutes, the effects of which persisted for 60 minutes. Additionally, hexane and ethanolic fruit extracts demonstrated anti-inflammatory activity at a 10 mg/kg lower dose, extending their impact for 120 minutes (Tasleem et al., 2014).

14.1.2.2 Antimicrobial Activity

Numerous investigations have documented the antimicrobial activity of compounds present in *Capsicum* species. In addition to the pungent phytochemicals, the chilli pepper pigment anthocyanin has antimicrobial activity (Zhao et al., 2009). The antibacterial activities of chilli pepper were reported against the noxious pathogens *Bacillus cereus, B. subtilis, Clostridium sporogenes, C. tetani, Streptococcus pyogenes, Staphylococcus aureus, Escherichia coli, Pseudomonas aeruginosa, Sarcina lutea, Candida albicans*, and *Vibrio cholera* (Chamikara et al., 2016). Capsaicin's minimum inhibitory concentration (MIC) against the gastric pathogen *Helicobacter pylori* was determined to be 10 mg/ml. The total oil (i.e. fixed + essential oil) derived from pepper has demonstrated antimicrobial efficacy against various microorganisms, making it versatile in applications that span from food preservation to infection prevention. Similarly, the essential oil derived from pepper (EO) has demonstrated multiple beneficial properties, including antioxidant, carminative, larvicidal, antibacterial, and antifungal activities (Jeena et al., 2014; Kapoor et al., 2009; Singh et al., 2013; Saleh et al., 2018). However, the oil obtained from *Piper nigrum* exhibited robust antibacterial efficacy against a range of bacteria, such as *Acinetobacter calcoacetica, Alcaligenes faecalis, Bacillus subtilis, Beneckeanatriegens, Brevibacterium linens, Brocothrixthermosphacta, Citrobacter freundii, Clostridium sporogenes, Enterococcus faecalis, Erwinia carotovora, Escherichia coli, Flavobacterium suaveolens, Leuconostoccremoris, Micrococcus luteus, Moraxella sp., Proteus vulgaris, Pseudomonas aeruginosa, Salmonella pullorum, Serratia marcescens, Staphylococcus aureus*, and *Yersinia enterocolitica*, as observed by Dorman and Deans (2000). The study was supported by Zarai et al. (2013). It was established that piperic acid, which is a precursor of piperine (MIC of ~78 mg/ml), displayed more potent antibacterial properties compared to piperine (MIC~312.5 to 625 mg/ml) against various Gram-positive and negative bacteria (Zarai et al., 2013). Furthermore, the extracted essential oil from pepper also holds the potential to safeguard grains against insect-induced damage and manage worm infestations (Dorman and Deans, 2000).

14.1.2.3   Assisting Gastrointestinal Health and Nutrient Absorbance

For ages, pepper and its active constituents have been utilised as traditional solutions for addressing gastric issues. It is widely acknowledged that pepper plays a pivotal role in averting and treating gastrointestinal disorders. By fostering the generation of hydrochloric acid in the stomach, pepper aids in enhancing digestion through the activation of histamine H2 receptors. For example, the active ingredient of black pepper oil, piperine, enhances the operational efficiency of the gastrointestinal tract. Almost 44% to 63% of piperine is absorbed from the mucosal side, and this presence can be detected in both intestinal tissue and serosal fluids, as outlined by Suresh and Srinivasan (2007). The use of pepper substantially elevated the activities of antioxidant enzymes, resulting in the well-being of the mucosal lining, thus providing gastroprotective effects (Prakash and Srinivasan, 2010). Alongside the effects mentioned, pepper and its active components foster the enhancement of digestive enzyme expression and an increase in saliva secretion. Together, these factors lead to an improved digestion process with reduced transit time within the gastrointestinal tract (Bang et al., 2009).

The heightened absorption of nutrients is contingent upon the beneficial alterations in the permeability of the cell membranes within the intestines. Additionally, piperine protects intestinal membranes, guarding them against the corrosive effects of gastric secretions and oxidative damage due to its antioxidant properties. Findings by Okumura et al. (2010) introduced the notion that piperine consumption might contribute to the reduction of body weight and the accumulation of visceral fat by stimulating the expression of the thermogenic protein uncoupling protein-1 in mice (Okumura et al., 2010). Additionally, the juice of pepper amplifies the intestinal absorption of various nutrients, and piperine, in a dose-dependent manner, restrains fatty acid oxidation in rat liver microsomes, as explicated by (Platel et al., 2002).

14.1.2.4   Drug Delivery

Phytochemicals and essential nutrients play a crucial role in upholding human well-being. However, the matter of their bioavailability is a topic that demands the attention of researchers. Certain bioactive compounds display a synergistic interaction, which contributes to the enhancement of mutual absorption. Bell pepper and piperine, for instance, collaborate to heighten the bioavailability of specific phytochemicals such as curcumin and catechins. These presumed effects are underpinned by mechanisms that encompass (1) the facilitation of swift absorption, (2) safeguarding against chemical reactions within the gastrointestinal tract, and (3) defense against oxidative harm, as noted by (Agbor et al., 2006).

Active ingredients of pepper enhance the bioavailability of medications such as carbamazepine and phenytoin (epilepsy drugs) by bolstering their absorption (Pattanaik et al., 2009). Furthermore, in specific investigations, piperine applied black pepper or its active compound, which led to the heightened bioavailability of nimesulide, an anti-inflammatory medication. In terms of acute toxicity in mice, the combination of these substances exhibited a reduced lethal dose (LD_{50}) of 980 mg/kg, compared to the solitary administration of nimesulide (1500 mg/kg), as reported by Gupta et al. (2000). Similarly, drugs like ampicillin and norfloxacin, effective against a variety of bacterial infections, displayed improved potency through co-administration with 20 mg/kg of piperine, as outlined by Janakiraman and Manavalan (2008).

14.1.2.5   Neuroprotective Activity

The recent past has emphasised the potential of functional foods to play a neuroprotective role against various ailments. Pepper has been documented to possess neuroprotective, anticholinesterase, and antioxidant properties, signifying its dual inhibitory action against acetylcholinesterase (AChE), butyrylcholinesterase (BChE), and 2,2-diphenyl-1-picrylhydrazyl (DPPH) free-radical scavenging activity (Tu et al., 2016). Neuroprotective constituents extracted from pepper are chemically characterised as pungent alkaloids, amide alkaloids, and alkamides (Balakrishnan et al., 2023).

In this context, pepper's antioxidant and bioactive constituents enhance brain function (Wattanathorn et al., 2008). Recently, Mostafa et al. (2021) studied the effect of pepper cold-pressed oil (CPO) against scopolamine-induced oxidative stress and memory impairments in rats. The cold-pressed oil resulted in a significant improvement ($p < 0.05$) in the activities of key antioxidant enzymes, including catalase and superoxide dismutase, along with a substantial reduction in malondialdehyde equivalents by ~22%, ~45%, and ~87%, respectively, in comparison to the effects of scopolamine. Moreover, cold-pressed oil administration led to a notable 51% decrease in the elevated acetylcholinesterase levels induced by scopolamine in the hippocampi of rats (Mostafa et al., 2021).

The other study by Li et al. (2007) previously delved into the antidepressant-like effects of piperine, utilising doses of 2.5, 5, and 10 mg/kg. Their findings unveiled that piperine mitigated the effects of chronic mild stress, manifested by alterations in sucrose consumption, plasma corticosterone levels, and open-field activity (Li et al., 2007). In support of the Li et al. (2007) study the Hu et al. (2009) also pointed out that piperine (at doses of 10 and 20 mg/kg) exhibited a positive impact on depression by reducing adrenocorticotropic hormone (ACTH) and corticotropin-releasing hormone (CRH) levels, consequently modulating the hypothalamic-pituitary-adrenal axis (Hu et al., 2009).

14.1.2.6 Anticancer Activity

Applying *C. chinense* fruit extracts to HepG2 cell lines derived from hepatocellular carcinoma inhibited cancer cell proliferation. This inhibitory activity was independently confirmed using methylthiazol tetrazolium (MTT), lactate dehydrogenase leakage, and nitrous oxide (NO) production assays. In malignant cells, the phytochemicals in chilli pepper induce apoptosis. Capsaicin induces apoptosis by impeding the plasma membrane NADH oxidoreductase enzyme in mitochondria. On the other hand, the numerous alkaloids in pepper, piperine, stand out as a significant alkaloid that triggers apoptotic signaling and hinders the progression of the cell cycle. It effectively curtailed tumor growth and metastasis in a mouse 4T1 breast tumor model by activating the intrinsic pathway of apoptosis through caspase 3 and inducing cell cycle arrest in the G2/M phase by reducing the expression of cyclin B1 (Lai et al., 2012). In another study, the ethanolic extract derived from pepper fruit exhibited anticancer properties on three colorectal cancer cell lines, namely HT-29, HCT-116, and HCT-15 (Prashant et al., 2017).

14.2 PEPPER OILS

The oil content of pepper seeds varies between 10.8% and 35.9%, depending on the different types of pepper, such as *Capsicum annuum L* and *Capsicum frutescens L*. Other species include *Capsicum chinense Jacq, Capsicum baccatum L*, and *Capsicum pubescens Ruiz & Pav*. The seed of *C. annuum* had a much higher seed oil content than the other species' seeds. *C. annuum* had the most significant mean (28.1%), maximum (35.9%), and minimum (22.1%) seed oil content among the tested varieties. On the other hand, the seed of *C. pubescens* exhibited notably decreased levels of seed oil concentration in comparison to the remaining farmed species. The seeds of *C. pubescens* exhibited the lowest mean oil content (18.3%) as well as the lowest maximum (21.4%) and minimum (14.6%) oil content values. The observed values within the farmed species exhibited similarity, with an average of 14.9, except for *C. pubescens*, which had a value of 6.9 (Walsh and Hoot, 2001). The pepper seed oil was analysed for the presence of 16 distinct fatty acids, with 4 accounting for most of the composition, which included palmitic, stearic, oleic, and linoleic acids, whose percentages were 11.9%, 4.0%, 11.1%, and 71.1%, respectively. The previous literature has found a high degree of similarity in the fatty acid composition of several pepper seed oils (Embaby and Mokhtar, 2011).

Jarret et al.'s (2013) investigation was carried out on five grown varieties of pepper seeds. The study indicated the presence of four primary fatty acids in these varieties, namely 12.9% for C16:0 (palmitic), 3.4% for C18:0 (stearic), 6.7% for C18:1 (oleic), and 76.0% for C18:2 (linoleic). However, it is important to note that linoleic acid is the predominant fatty acid found in pepper seed oils in all conducted investigations.

The mean percentages for the four primary fatty acids analysed within the taxa were 12.9%, 3.4%, 6.7%, and 76% for C16:0, C18:0, C18:1, and C18:2, respectively. The percentages of C16:0, C18:0, C18:1, and C18:2 exhibited statistically significant increases in *C. annuum var. glabriusculum*, *C. pubescens*, *C. frutescens*, and *C. baccatum var. pendulum*, respectively, compared to the remaining cultivated taxa. The maximum observed values for individual seed samples were as follows: C16:0—*C. annuum var. glabriusculum* (18.8%), C18:0—*C. frutescens* (7.9%), C18:1—*C. baccatum* (12.58%), and C18:2—*C. chinense* (81.1%). The minimum values recorded for each category were as follows: C16:0—*C. pubescens* (9.3%), C18:0—*C. annuum var. glabriusculum* (2.4%), C18:1—*C. baccatum* (4.0%), and C18:2—*C. frutescens* (58.1%). The ranges of individual fatty acid values across the related species exhibited similarities, except for the linolenic acid (C18:2) in *C. frutescens*, which was roughly twice as high (19.4%) compared to the other species.

14.2.1 TOTAL OIL

The oil content of the seeds is reasonably stable, regardless of the geographical location of seed production, within certain boundaries. The seed oil content of various species, except *C. pubescens*, exhibited similarities in their ranges (range = 14.8) but significantly more than that of *C. pubescens* (range = 6.9) (Jarret et al., 2013). On the other hand, the pepper oil yield achieved through cold pressing ranges from 5.12% to 6.65%, which is much lower compared to the oil yield obtained using solvent extraction, which is 13.57%. Solvent extraction is a widely recognised method for conducting industrial-scale oil extractions, typically employed in large-scale operations. Consequently, the crude oils obtained from these extractions are consistently subjected to refinement processes. Indeed, cold pressing is favoured in manufacturing distinctively flavoured, unrefined, and virgin oils (Aydeniz et al., 2014). Based on these, it is recommended to employ solvent extraction to augment the oil output of pepper seeds, mainly if the intention is to refine the oil (Soto et al., 2007).

Similarly, the oil yield in the roasted sample exhibited a slight increase compared to the control sample, and a similar trend was observed for roasted safflower seeds (Aydeniz et al., 2014). The supercritical CO_2 extraction produced an oil yield of 18.4% for pepper seeds (Li et al., 2011). The oil has a specific gravity of 0.918, a refractive index of 1.4738, and Lovibond colour scores of 100 yellow and -46 red (Bosland and Votava, 2000).

Pepper seed oil contains an overall percentage of unsaturated fatty acids that has been reported to range from 81.5% to 87.7%, according to various published literature (Jarret et al., 2013). The recorded percentages of saturated fatty acid in *C. annuum* seed vary from 11.89% to 18.5%. The values were correlated with other investigations, ranging from 10.8% for *C. galapagoense* to 35.8% for *C. annuum*. Previous studies examining the fatty acid content of pepper seed oil have consistently shown a substantial presence of linoleic acid, typically exceeding 70%. Conversely, lower amounts of saturated and unsaturated acids, such as C16:0 and C18:0, have been documented in all the varieties. The significance of linolenic acid improves physical development, the manifestation of skin abnormalities, compromised reproductive function, and the onset of several other clinical manifestations (Xu and Kafkafi, 2003).

Additionally, it offers a safeguard against ischemic heart disease. The results presented in previous studies provide empirical evidence that corroborates the findings mentioned and expands their established boundaries. The absolute values of particular accessions and species should not be regarded as definitive due to the influence of climatic conditions during seed development on the fatty acid composition of seeds despite the relative stability of seed oil content (Demir et al., 2008). Hence, the scientific literature does not contain any available data regarding the fatty acid composition of *Capsicum* spp. seeds towards the influence of climatic conditions.

14.2.2 Essential Oil

Essential oils are acquired through several methods such as hydrodistillation, steam distillation, Soxhlet extraction, infusion, decoction, ultrasonic extraction, cohobation, enfleurage, cold press, or maceration of whole or crushed spices. These entities exhibit volatility and are distinguished by their distinct and potent fragrance. Essential oils find extensive application within the food business, serving as a prominent flavouring agent and playing a pivotal role in meat preparation (Stewart et al., 2016). In addition, they find application within the aromatherapeutic sector to create fragrance compositions, cosmetic goods, and the development of natural insect repellents. Numerous scholarly investigations have documented the antimicrobial, antioxidative, antidiabetic, and anti-inflammatory properties of essential oils derived from several spices (Lee et al., 2020).

The essential oil components of the spices are the secondary metabolites from plants that possess antibacterial, antiviral, antifungal, and insecticidal properties. The extracts and essential oils from aromatic plants with antioxidant and antimicrobial activities for controlling pathogens and toxin-producing microorganisms in foods are gaining interest (Alzoreky and Nakahara, 2003). Essential oil from pepper has a transparent to slightly greenish appearance and is approximately 1% to 3% of the total oil content. It emits a spicy, distinctly peppery fragrance. This oil is characterised by its aroma, reminiscent of its peppery nature.

Recent studies have reported 78 compounds from the essential oil of pepper. The composition of the oils primarily consisted of monoterpene hydrocarbons (~59.2% to 80.1%) and sesquiterpene hydrocarbons (~17.0% to 37.7%). Oxygenated terpenoids constituted ~1.3% to 2.7% of the composition. The predominant constituents were α-pinene (~5.1% to 28.7%), β-caryophyllene (~8.7% to 25.6%), limonene (~15.1% to 19.5%), β-pinene (~9.1% to 15.3%), and δ-3-carene (~9.0% to 12.8%) (Dosoky et al., 2019). In contrast to previous findings, Nie et al. (2023) identified only 25 compounds in pepper essential oil by the GC-MS method. The essential oil extracted from spices is also a complex mélange of various chemical constituents, including aliphatic and aromatic hydrocarbons, aldehydes, ketones, esters, and other secondary metabolites, including terpenoids, polyphenols, etc. (Samfira et al., 2015).

14.2.3 Fixed Oil

Fixed oils are the non-volatile components that remain liquid at room temperature and is typically capable of dissolving in organic solvents. Fixed oil mainly comprises lipophilic constituents, including fats, resins, and waxes. After distilling essential oils from spices, these substances are derived from the residual aromatic spice material. Typically, fixed oils are extracted through organic solvents using Soxhlet extraction or supercritical fluid extraction employing carbon dioxide (CO_2). However, a limited body of research is available regarding the use of fixed oils derived from spices, resulting in a lack of comprehensive understanding in this area. Consequently, these oils have mostly remained unexplored and unexamined.

There is a shortage of scholarly information regarding *Capsicum* species' seed oil content and fatty acid composition, except for *C. annuum*. This presents an analysis of the seed oil content variability in species of *Capsicum* and a closely related genus, ranging from 10.8% to 35.9%. The cultivated species exhibited notable variations in oil content and the composition of specific fatty acids. The seed oil content and fatty acid composition in the wild species exhibited similarities to those observed in the cultivated *Capsicum* species. However, the fixed oil content in the *C. annuum* showed 13.22% after the complete removal of essential oil through steam distillation. It is quite a high percentage of fixed oil compared to the other spices. Whereas the fatty acid profile of the *C. annuum* contains a high amount of polyunsaturated fatty acids, the C18:2 fatty acid was presented as the highest at 61.54%. Next to this, the C16:0 was noticed as high with 20.08%. Similarly, it contains some other fatty acids like C12:0, C14:0, C18:1, and C18:3 in reasonable amounts. The similar fatty acids and their respective percentages were also concrete with the total oil.

14.2.4 OLEORESINS

Oleoresins are naturally occurring blends of essential oils and resins, typically obtained through organic solvents or supercritical fluid extraction methods involving carbon dioxide, either with or without auxiliary solvents like ethanol. According to Wang et al. (2007), oleoresins possess lower volatility and consist of lipophilic constituents, resulting in the consistent scent and flavour of spices. On a global scale, whole spices and their derivatives have been extensively used, including extracts, essential oils, and oleoresins, in various applications such as aromatherapy and producing bioactive components for food, cosmetics, and pharmaceutical industries.

Spice oleoresins have a high degree of concentration and possess significant potential as substitutes. The completed goods exhibit improved dispersion characteristics and require less storage capacity than their related spices. Nevertheless, it should be noted that spice oleoresins display a certain degree of susceptibility to light, heat, and oxygen, resulting in a limited shelf life if appropriate storage conditions are not maintained. The diminished shelf life of pepper oleoresin can be attributed to oxidative and polymeric alterations that occur within its fatty oil constituent and monoterpinic hydrocarbons, such as a-pinene, b-pinene, limonene, sabinene, and D_3-carene. During an extended storage period, chemical and sensory alterations may also manifest in the oleoresin. The degradation of many pigments occurs when they are exposed to oxygen, leading to the conversion of hydroxylic groups into less stable ketones. Subsequently, these substances undergo decomposition into molecules lacking coloration and possessing a reduced carbon framework (Shaikh et al., 2006).

14.3 OIL NUTRACEUTICALS

Plant active compounds have played a crucial role in discovering and developing modern pharmaceuticals (Atanasov et al., 2021). Due to advancements in plant biochemistry research, the investigation and development of an increasing number of chemical constituents derived from plants are now feasible (Bostan et al., 2013). These chemical constituents can maintain health, treat common ailments (they are antiseptic, antimicrobial, anti-inflammatory, decongestants), and enhance immunity (Rani & Prabhu, 2022). Nutraceuticals can be found in various products from various industries, including the food industry, the botanical market, and the market for dietary supplements. Consumption of nutraceuticals in the daily diet has become essential in achieving optimal nutrition and health. This is due to the paucity of evidence for dose, nutraceutical–drug interaction, and their effects on individuals with various diseases and those with normal health (Puri et al., 2022). To improve the efficacy of consuming nutraceuticals, scientists are attempting to improve the nutritional content of these products since excessive food consumption has adverse effects on human health (Chopra et al., 2022). These have been found to have a health-promoting effect and aid in the treatment of a variety of disorders caused by the modern lifestyle, including coronary heart disease, hypertension, diabetes, arthritis and other inflammatory conditions, autoimmune disorders, depression, schizophrenia, and various types of cancers, among others (Opara & Chohan, 2014). These are derived primarily from safflower oil, soybean oil, olive oil, canola oils, other vegetable oils, fish oils, and viscous fishes such as salmon, soybean, flaxseed, canola, pumpkin, and walnuts, among others (Langyan et al., 2022). In addition, it has been observed that regular consumption of fish oils reduces morbidity and mortality hazards associated with cardiovascular diseases, such as ischemic heart disease, nonischemic myocardial heart disease, and hypertension (Chopra et al., 2022). Nutraceuticals could be added to bread, pasta, dairy products, spreads, beverages, and liquid or capsule-based dietary supplements (Nejatian et al., 2022).

Vegetable oils contain various nutraceutical compounds such as polyphenols, tocopherols, phytosterols, oryzanol, and fat-soluble vitamins with numerous health benefits (Deen et al., 2021). These nutraceutical compounds in vegetable oils are crucial in preventing diseases and protecting against various disorders such as obesity, cardiovascular diseases (CVD), and inflammation (Siscovick et al., 2017).

Besides their use in the human diet, vegetable oils are used in various industrial applications such as soaps, detergents, cosmetics, lubricants, ink, varnishes, and paints (Panchal et al., 2017). The major nutraceutical compounds in oils are polyphenols, tocopherols, γ-oryzanol, and phytosterols. Besides this, fixed oils were also a significant source of all these oil-based nutraceuticals (Manasa & Tumaney, 2022).

14.3.1 Polyphenols

Phenolic compounds are by-products of the secondary metabolism in plants. Phenyl propanoids are produced via the shikimate pathway (Vogt, 2010). Phenolic compounds vary in molecular weight, complexity, and distribution. All phenolic compounds have an aromatic arene (phenyl) ring with at least one OH group attached (Croteau et al., 2000). The presence of hydroxyl groups is primarily responsible for the antioxidant properties of polyphenols (Piccolella et al., 2019). These compounds are one of the primary secondary metabolites comprising a diverse pool of biologically active compounds (Usman et al., 2022). The phenolic compounds can be classified based on the number and arrangement of carbon atoms into phenolic acids, flavonoids, lignans, and stilbenes (Gutiérrez-Grijalva et al., 2017). Polyphenols are the most important group of natural antioxidants found in plant sources. A phenolic is created based on compounds with an aromatic ring bearing one hydroxyl group. Polyphenols are associated with a decreased risk of harmful bodily processes induced by oxidative stress. The phenolic-rich sources showed antidiabetic and anticancer properties and activity against lipid metabolism–related disorders (Kowalska et al., 2017). Similarly, they also play an essential role in reducing lipogenesis, increasing lipolysis, and inhibiting pro-inflammatory adipokine secretion (Jeong et al., 2013).

Polyphenols are believed to be associated with the astringent sensation and enhance foods' flavour, colour, and aroma. Due to their cost-effectiveness, simple accessibility, acceptability, prolonged administration, and omnipresence, these compounds have recently attracted considerable interest in diet-based therapy (Imran et al., 2021). As a rich source of antioxidant compounds, these compounds have an outstanding positive effect on overall health and well-being (Ali et al., 2021). In addition, they are well-established cancer-fighting natural therapeutic agents for numerous chronic or non-communicable diseases, including cardiovascular, neurodegenerative, and nephrotoxic diseases (Mark et al., 2019). The pharmacological effects of polyphenols, such as antimicrobial, antioxidant, antidepressant, anti-inflammatory, anticancer, and antilipidemic, are well documented in the scientific literature (Usman et al., 2022). These compounds can also down-regulate the angiogenic growth factors, thereby modifying the signalling involved in tumour progression and inhibiting tumorigenesis (Rani et al., 2021).

The fixed oil of *C. annuum* contains 46.74 mg GAE/100 g oil in terms of phenolic compounds. Among these compounds, myristic acid is the highest quantity at 1.17 mg/100 g of oil. The p-coumaric acid is 0.32 mg/100 g of oil, and trans-Cinnamic acid is 0.14 mg/100 g. These two polyphenols rank as the second and third highest in abundance within the *C. annuum* fixed oil. Likewise, the phenolic compounds derived from the fixed oils of *P. nigrum* were significant, with a content of 275.7 mg GAE/100 g oil. The fixed oil of *P. nigrum* exhibited the highest concentration of rosmarinic acid (78.85 mg/100 g of oil). The second-highest chemical found in the fixed oil was kaempferol, with a concentration of 13.88 mg/100 g of oil. Another species of pepper, specifically *P. longum*, has been found to contain the most significant quantity of total polyphenols, measuring 464.79 mg GAE/100 g oil. This species also contains other compounds such as quercetin, p-coumaric acid, 4-hydroxybenzoic acid, and trans-ferulic acid. It is essential to acknowledge that spice fixed oils generally exhibit higher quantities of polyphenols than palm oil (Neo et al., 2010).

14.3.2 Tocopherols

Tocopherols, the primary form of vitamin E, are fat-soluble compounds with a chromanol ring and a 16-carbon phytyl chain. Depending on the number and position of methyl groups, tocopherols were designated as alpha (α), beta (β), gamma (γ), and delta (δ) (Das Gupta and Suh, 2016). Vitamin E activity is found in a group of eight fat-soluble compounds consisting of α, β, γ, and δ forms of tocopherols and tocotrienols. Among the forms, α-tocopherol is most abundant in animal tissues.

Since their discovery in the past century, tocopherols have garnered much attention from the scientific community. Tocopherols are commonly used as food additives, extracts, or purified synthetic molecules to protect unsaturated lipids (Barouh et al., 2022). Tocopherol is the predominant form of vitamin E in animals and possesses the highest activity in carrying out the vitamin's essential antioxidant functions. Due to the role of oxidative stress in carcinogenesis, extensive research has been conducted on tocopherol's ability to prevent cancer. In some intervention studies, supplementing tocopherol to populations with vitamin E deficiency reduced the risk of malignancy. Recent research has primarily focused on the forms of tocopherols and tocotrienols (T_3). Compared to T_4, these forms have much lower systemic bioavailability, but in numerous animal model and cell-based studies, they have demonstrated more significant cancer-preventive activity. In general, T_3 has even more significant activity than T_4 (Yang et al., 2020).

As natural extracts or pure synthetic molecules, tocopherols are arguably the most widely used antioxidants on an industrial scale (Barouh et al., 2022). Tocopherols play a significant role in activating peroxisome proliferator-activated receptor gamma (PPAR), which is implicated in lipid metabolism-related disorders (Landrier et al., 2009). On an industrial scale, tocopherols are likely the most extensively used antioxidants in combating lipid oxidation in food and cosmetic products, where numerous chemical and physicochemical parameters add to the difficulty. As reactive oxygen species (ROS) are implicated in numerous chronic diseases, including cancer, cardiovascular diseases, and neurodegenerative diseases, the possible preventive effects of the antioxidant vitamin E against these diseases have been intensively studied.

Spices are the richest source of tocopherols, similar to the fixed oils of Indian spices noted for their tocopherols. The effectiveness of tocopherols as antioxidants depends on the concentration of their isomers, namely α-, β-, γ-, and δ-tocopherol. The fixed oils were screened for all α-, β-, γ-, and δ-tocopherol where β- and γ-tocopherol co-eluted.

C. annuum fixed oil is found to contain tocopherols. The α-tocopherol is the most abundant, with a concentration of 99.11 mg per 100 g of oil. This is followed by β+γ-tocopherol, which has a concentration of 17.30 mg per 100 g of oil. The lowest concentration is observed in γ-tocopherol, with just 0.63 mg per 100 g of oil. Nevertheless, the concentration of α-tocopherol in *P. longum* fixed oil was determined to be the lowest at 1.52 mg/100 g oil. The highest concentration of β+γ-tocopherol was found in the fixed oil of *P. longum*, measuring 441.8 mg per 100 g. In a similar vein, the fixed oil derived from *P. nigrum* exhibits the presence of α-, β+γ-, and δ-tocopherols in concentrations of 238.7, 60.38, and 106.08 mg/100 g oil, respectively. Therefore, the observed disparity in composition among the spices was noted.

14.3.3 Phytosterols

Phytosterols, which are plant sterols, exhibit structural similarities to cholesterol and are present within the cellular membranes of plants. According to Moreau et al. (2018), the four prevalent phytosterols include campesterol, sitosterol, stanols, and stigmasterols. Ovesnaâ et al. (2004) reported the efficacy of phytosterols, specifically β-sitosterol, in many therapeutic applications, including anti-inflammatory, antineoplastic, and antipyretic action and colon cancer inhibition. The postulated mechanism of action for β-sitosterol is the inhibition of pre-adipocyte adipogenesis by activating transcriptional factors, specifically peroxisome proliferator-activated receptor gamma (PPARγ), intending to control obesity. According to Awad et al. (2000), utilising β-sitosterols in experimental investigations conducted on 3T3-L1 mouse embryonic fibroblast cell lines demonstrated their ability to regulate pre-adipocyte formation effectively.

Phytosterols, categorised as plant sterols and stanols, are biologically active chemicals naturally present in plant-based food sources. Phytosterols have been suggested to possess numerous pharmacological characteristics, such as the potential to mitigate levels of total and low-density lipoprotein (LDL) cholesterol, thereby diminishing the susceptibility to cardiovascular ailments. Additional health-promoting effects of PSs encompass several beneficial properties such as antiobesity, antidiabetic, antimicrobial, anti-inflammatory, and immunomodulatory activities. Moreover, substantial evidence supports the notion that diets rich in phytosterols may exhibit potent anticancer properties, perhaps leading to a 20% reduction in the chance of developing cancer.

Phytosterols, encompassing a diverse range of more than 200 substances, are primarily present in vegetable oils, nuts, fruits, grains, and other botanical derivatives (Moreau et al., 2002). Consumers consume grain products as part of their daily meals, making them the primary sources of total phytosterol intake (Valsta et al., 2004). According to Carr and Jesch (2006), it has been calculated that the collective consumption of phytosterols from dietary sources amounts to approximately 200 mg to 300 mg per day in several societies. Plant stanols, the saturated counterparts of plant sterols, are present in specific food sources, albeit in lower quantities than plant sterols. Numerous studies have reported the advantageous impacts of phytosterols on various chronic ailments, including cardiovascular diseases (Dash et al., 2021; Hannan et al., 2020) and diabetes (Salehi-Sahlabadi et al., 2020).

Moreover, it has been proposed that diets abundant in phytosterols may potentially mitigate the likelihood of developing cancer. In light of the extensive research conducted in this domain (Dash et al., 2021). Endogenous synthesis of phytosterols does not occur in the human body. According to Scolaro et al. (2019), diets, particularly those consisting of plant-based foods, are the primary sources of nutrition. The average intake of phytosterols in a typical Western diet is approximately 300 mg daily. However, those following a vegetarian diet may consume up to 50% more phytosterols, with an estimated intake ranging from 300 mg to 500 mg daily (Scolaro et al., 2019).

C. annuum fixed oil includes β-sitosterol as the primary phytosterol, with a concentration of 885.82 mg per 100 g. Additionally, a combined amount of 52.93 mg per 100 g of oil is attributed to stigmasterol and campesterol. The presence of β-sitosterol is abundant in spice fixed oil phytosterols, including ergosterol, stigmasterol, campesterol, and β-sitosterol. Additionally, β-sitosterol is widely found in many vegetable oils. Ergosterol was not detected in the sample. Similarly, it is worth noting that *P. nigram* and *P. longum* fixed oils exhibit a significant presence of β-sitosterol as the primary phytosterol, with 1450 and 1439 mg/100 g of oil, respectively. Following closely behind are stigmasterol+ campesterol, with respective quantities of 794.97 and 549.46 mg/100 g of oil. In contrast, it was shown that ergosterol was exclusively detected in the fixed oil of *P. longum*.

14.3.4 ANTIOXIDANTS

The fixed oils derived from spices exhibited significant radical scavenging activity. Analysing the activity of these oils can be challenging due to the complexity of the plant source, resulting in variations in the eradication of distinct radicals by different spice nutraceuticals. Therefore, the *C. annuum* exhibited the highest FRAP activity, measuring 6960.94 mM TE/100 g of oil. It was subsequently followed by the ABTS activity, which measured 86.22 mM TE/100 g of oil. Lastly, the DPPH radicals were eliminated at a rate of 16.71 mg AAE/100 g of oil.

Similarly, it has been observed that *P. nigrum* and *P. longum* fixed oil exhibit the most substantial antioxidant activity, as evidenced by FRAP radical values of 15,945 and 22,733 mM TE/100 g of oil, respectively. In contrast, the ABTS radicals were effectively neutralised at concentrations of 1977 and 5812 mM/100 g of oil, while the DPPH radicals were eliminated at 66.75 and 42.48 mg AAE/100 g of oil. Therefore, it is evident that spice extracts, essential oils, and fixed oils possess significant antiradical scavenging action.

A positive correlation between the total phenolic content in plants and antioxidant activity was established by Saxena et al. (2007). The essential oil extracted from pepper was observed to effectively counteract superoxide and hydroxyl radicals while inhibiting tissue lipid peroxidation. In a study by Jeena et al. (2014), the concentration required for a 50% scavenging effect (IC_{50}) against superoxides and hydroxyl radicals was reported to be higher than 200 µg/ml. The IC_{50} for lipid peroxidation inhibition for pepper was 196 µg/ml. Additionally, the ferric-reducing activity of 50 µg of pepper essential oil was determined to be 1.3 mM, demonstrating a moderate ability to scavenge DPPH radicals (Jeena et al., 2014). In another study, the water and ethanol extracts (at 75 µg/ml) derived from pepper exhibited inhibitory effects of 95.5% and 93.3% on lipid peroxidation, as outlined by Gulcin (2005). Firdos et al. (2017) exploited the antioxidant potential of pepper extract in the stabilisation of sunflower oil in comparison with the synthetic antioxidants butylated hydroxy anisole (BHA) and butylated hydroxytoluene (BHT). They observed that on the 45th day of storage, the stand-alone sunflower oil exhibited an iodine value of 114.01 mg/g. Introducing BHA and BHT into the sunflower oil resulted in iodine values of 123.22 mg/g and 117.99 mg/g on the 45th day of storage.

Interestingly, the iodine value for the stabilised pepper extract at 500 ppm stood at 120.85 mg/g, and the 1000 ppm concentration of pepper stabilised extract measured 155.98 mg/g on the same storage day. Consequently, the iodine values of the stored sunflower oil, treated with synthetic antioxidants and pepper extracts, significantly surpassed those of the standalone sunflower oil sample (Firdos et al., 2017). Kapoor et al. (2009) reported that essential oil and oleoresins displayed superior antioxidant activities to synthetic antioxidants. Similarly, Su et al. (2007) observed that oils hold promise as a viable dietary reservoir of natural antioxidants. Hence, these beneficial constituents position it as a compelling contender for alleviating oxidative stress (Kapoor et al., 2009).

14.4 CONCLUSION

Pepper is a quintessential Indian spice widely used throughout India. India is the fourth-largest producer of pepper. Various hydrophobic extracts from pepper, like oleoresin and essential oil, are widely studied. The oil content of pepper seeds varies between 10.8% and 35.9%. The previously reported nutraceuticals in spice oils mainly comprise therapeutic fatty acids, polyphenols, tocopherols, γ-oryzanol, and phytosterols. Owing to its phytochemical composition, the oil exhibits activity like anti-inflammatory potential, antimicrobial activity, drug delivery, neuroprotective, cancer preventive, etc. Piperine is an alkaloid and a significant component of pepper. Piperine contributes a major part of the biological activity exhibited by pepper oils. The primary fatty acids reported in pepper oils are palmitic, stearic, oleic, and linoleic, with linoleic being the most abundant fatty acid. The significance of linolenic acid improves physical development, the manifestation of skin abnormalities, compromised reproductive function, and the onset of several other clinical manifestations.

Additionally, it offers a safeguard against ischemic heart disease. The pepper is known to have 78 compounds comprising mainly monoterpenes, sesquiterpenes, and oxygenated terpenoids. The major constituents were α-pinene (~5.1% to 28.7%), β-caryophyllene (~8.7% to 25.6%), limonene (~15.1% to 19.5%), β-pinene (~9.1% to 15.3%), and δ-3-carene (~9.0% to 12.8%). Fixed oils are non-volatile oil fractions often obtained by solvent or supercritical CO_2 extraction. Various studies on phenolics, tocopherols, phytosterols, and the antioxidant capacity of oils from pepper are widely studied. Thus, pepper oil is a rich source of various bioactives and confers many health-promoting potentials.

REFERENCES

Agbor, G. A., Vinson, J. A., Oben, J. E., and Ngogang, J. Y. 2006. Comparative analysis of the in vitro antioxidant activity of white and black Pepper. *Nutrition Research* 26(12):659–663.

Ali, A., Wu, H., Ponnampalam, E. N., Cottrell, J. J., Dunshea, F. R., and Suleria, H. A. 2021. Comprehensive profiling of most widely used spices for their phenolic compounds through LC-ESI-QTOF-MS2 and their antioxidant potential. *Antioxidants* 10(5):721.

Alzoreky, N. S., and Nakahara, K. 2003. Antibacterial activity of extracts from some edible plants commonly consumed in Asia. *International Journal of Food Microbiology* 80(3):223–230.

Atanasov, A. G., Zotchev, S. B., Dirsch, V. M., and Supuran, C. T. 2021. Natural products in drug discovery: Advances and opportunities. *Nature Reviews Drug Discovery* 20(3):200–216.

Awad, A. B., Begdache, L. A., and Fink, C. S. 2000. Effect of sterols and fatty acids on growth and triglyceride accumulation in 3T3-L1 cells. *The Journal of Nutritional Biochemistry* 11(3):153–158.

Aydeniz, B., Güneşer, O., and Yılmaz, E. 2014. Physico-chemical, sensory and aromatic properties of cold press produced safflower oil. *Journal of the American Oil Chemists' Society* 91(1):99–110.

Balakrishnan, R., Azam, S., Kim, I. S., and Choi, D. K. 2023. Neuroprotective effects of black pepper and its bioactive compounds in age-related neurological disorders. *Aging and Disease* 14(3):750.

Bang, J. S., Oh, D. H., Choi, H. M., Sur, B. J., Lim, S. J., Kim, J. Y., and Kim, K. S. 2009. Anti-inflammatory and antiarthritic effects of piperine in human interleukin 1 beta-stimulated fibroblast-like synoviocytes and in rat arthritis models. *Arthritis Research & Therapy* 11(2):49.

Barouh, N., Bourlieu-Lacanal, C., Figueroa-Espinoza, M. C., Durand, E., and Villeneuve, P. 2022. Tocopherols as antioxidants in lipid-based systems: The combination of chemical and physicochemical interactions determines their efficiency. *Comprehensive Reviews in Food Science and Food Safety* 21(1):642–688.

Bosland, P. W., and Votava, E. J. 2000. *Peppers: Vegetable and Spice Capsicums.* CABI Publishing, New York.

Bostan, C., Butnariu, M., Butu, M., Ortan, A., Butu, A., Rodino, S., and Parvu, C. 2013. Allelopathic effect of *Festuca rubra* on perennial grasses. *Romanian Biotechnological Letters* 18(2):8190–8196.

Carr, T. P., and Jesch, E. D. 2006. Food components that reduce cholesterol absorption. *Advances in Food and Nutrition Research* 51:165–204.

Chamikara, M. D. M., Dissanayake, D. R. R. P., Ishan, M., and Sooriyapathirana, S. D. S. S. 2016. Dietary, anticancer and medicinal properties of the phytochemicals in chili pepper (Capsicum spp.). *Ceylon Journal of Science* 45(3).

Chopra, A. S., Lordan, R., Horbańczuk, O. K., Atanasov, A. G., Chopra, I., Horbańczuk, J. O., and Arkells, N. 2022. The current use and evolving landscape of nutraceuticals. *Pharmacological Research* 175:106001.

Croteau, R., Kutchan, T. M., and Lewis, N. G. 2000. Natural products (secondary metabolites). *Biochemistry and Molecular Biology of Plants* 24:1250–1319.

Das Gupta, S., and Suh, N. 2016. Tocopherols in cancer: An update. *Molecular Nutrition & Food Research* 60(6):1354–1363.

Dash, R., Mitra, S., Ali, M. C., Oktaviani, D. F., Hannan, M. A., Choi, S. M., and Moon, I. S. 2021. Phytosterols: Targeting neuroinflammation in neurodegeneration. *Current Pharmaceutical Design* 27(3):383–401.

Deen, A., Visvanathan, R., Wickramarachchi, D., Marikkar, N., Nammi, S., Jayawardana, B. C., and Liyanage, R. 2021. Chemical composition and health benefits of coconut oil: An overview. *Journal of the Science of Food and Agriculture* 101(6):2182–2193.

Demir, I., Tekin, A., Ökmen, Z. A., Okcu, G., and Kenanoğlu, B. B. 2008. Seed quality, and fatty acid and sugar contents of pepper seeds (*Capsicum annuum* L.) in relation to seed development and drying temperatures. *Turkish Journal of Agriculture and Forestry* 32(6):529–536.

Devkota, H. P., and Watanabe, M. 2020. Role of medicinal plant gardens in pharmaceutical science education and research: An overview of medicinal plant garden at Kumamoto University, Japan. *Journal of Asian Association of Schools of Pharmacy* 9:44–52.

Dorman, H. J. D., and Deans, S. G. 2000. Antimicrobial agents from plants: Antibacterial activity of plant volatile oils. *Journal of Applied Microbiology* 88(2):308–316.

Dosoky, N. S., Satyal, P., Barata, L. M., da Silva, J. K. R., and Setzer, W. N. 2019. Volatiles of black pepper fruits (*Piper nigrum* L.). *Molecules* 24(23):4244.

Embaby, H. E. S., and Mokhtar, S. M. 2011. Chemical composition and nutritive value of lantana and sweet pepper seeds and nabak seed kernels. *Journal of Food Science* 76(5):C736–C741.

Firdos, A., Tariq, A. R., Imran, M., Niamat, I., Kanwal, F., and Mitu, L. 2017. Antioxidant potential of black pepper extract for the stabilization of sunflower oil. *Bulgarian Chemical Communications* 49(1):31–33.

Gulcin, I. 2005. The antioxidant and radical scavenging activities of black pepper (*Piper nigrum*) seeds. *International Journal of Food Sciences and Nutrition* 56(7):491–499.

Gupta, S. K., Bansal, P., Bhardwaj, R. K., and Velpandian, T. 2000. Comparative anti-nociceptive, anti-inflammatory and toxicity profile of nimesulide vs nimesulide and piperine combination. *Pharmacological Research* 41(6):657–662.

Gutiérrez-Grijalva, E. P., Picos-Salas, M. A., Leyva-López, N., Criollo-Mendoza, M. S., Vazquez-Olivo, G., and Heredia, J. B. 2017. Flavonoids and phenolic acids from oregano: Occurrence, biological activity and health benefits. *Plants* 7(1):2.

Hannan, M. A., Sohag, A. A. M., Dash, R., Haque, M. N., Mohibbullah, M., Oktaviani, D. F., . . . Moon, I. S. 2020. Phytosterols of marine algae: Insights into the potential health benefits and molecular pharmacology. *Phytomedicine* 69:153201.

Hu, Y., Guo, D. H., Liu, P., Rahman, K., Wang, D. X., and Wang, B. 2009. Antioxidant effects of a Rhodobryumroseum extract and its active components in isoproterenol-induced myocardial injury in rats and cardiac myocytes against oxidative stress-triggered damage [Article]. *Pharmazie* 64(1):53–57.

Imran, A., Arshad, M. U., Sherwani, H., Shabir Ahmad, R., Arshad, M. S., Saeed, F., and Anjum, F. M. 2021. Antioxidant capacity and characteristics of theaflavin catechins and ginger freeze-dried extract as affected by extraction techniques. *International Journal of Food Properties* 24(1):1097–1116.

Janakiraman, K., and Manavalan, R. 2008. Studies on effect of piperine on oral bioavailability of Ampicillin and Norfloxacin. *African Journal of Traditional Complementary and Alternative Medicines* 5(3):257–262.

Jarret, R. L., Levy, I. J., Potter, T. L., and Cermak, S. C. 2013. Seed oil and fatty acid composition in Capsicum spp. *Journal of Food Composition and Analysis* 30(2):102–108.

Jeena, K., Liju, V. B., Umadevi, N. P., and Kuttan, R. 2014. Antioxidant, Anti-inflammatory and Antinociceptive Properties of Black Pepper Essential Oil (*Piper nigrum* Linn). *Journal of Essential Oil Bearing Plants* 17(1):1–12.

Jeong, M. Y., Kim, H. L., Park, J., An, H. J., Kim, S. H., Kim, S. J., and Hong, S. H. 2013. Rubi fructus (*Rubus coreanus*) inhibits differentiation to adipocytes in 3T3-L1 cells. *Evidence-based Complementary and Alternative Medicine*:475386.

Kapoor, I. P. S., Singh, B., Singh, G., De Heluani, C. S., De Lampasona, M. P., and Catalan, C. A. N. 2009. Chemistry and in vitro antioxidant activity of volatile oil and oleoresins of Black Pepper (*Piper nigrum*).*Journal of Agricultural and Food Chemistry* 57(12):5358–5364.

Kim, C. S., Kawada, T., Kim, B. S., Han, I. S., Choe, S. Y., Kurata, T., and Yu, R. 2003. Capsaicin exhibits anti-inflammatory property by inhibiting IkB-a degradation in LPS-stimulated peritoneal macrophages. *Cell Signaling* 15:299–306

Kowalska, K., Olejnik, A., Szwajgier, D., and Olkowicz, M. 2017. Inhibitory activity of chokeberry, bilberry, raspberry and cranberry polyphenol-rich extract towards adipogenesis and oxidative stress in differentiated 3T3-L1 adipose cells. *PLoS ONE* 12(11):e0188583.

Kunwar, R. M., Shrestha, K. P., and Bussmann, R. W. 2010. Traditional herbal medicine in Far-west Nepal: A pharmacological appraisal. *Journal of Ethnobiology and Ethnomedicine* 6:1–18.

Lai, L. H., Fu, Q. H., Liu, Y., Jiang, K., Guo, Q. M., Chen, Q. Y., and Shen, J. G. 2012. Piperine suppresses tumor growth and metastasis in vitro and in vivo in a 4T1 murine breast cancer model [Article]. *Acta Pharmacologica Sinica* 33(4):523–530.

Landrier, J. F., Gouranton, E., El Yazidi, C., Malezet, C., Balaguer, P., Borel, P., and Amiot, M. J. 2009. Adiponectin expression is induced by vitamin E via a peroxisome proliferator-activated receptor γ-dependent mechanism. *Endocrinology* 150(12):5318–5325.

Langyan, S., Yadava, P., Sharma, S., Gupta, N. C., Bansal, R., Yadav, R., and Kumar, A. 2022. Food and nutraceutical functions of Sesame oil: An underutilized crop for nutritional and health benefits. *Food Chemistry* 389:132990.

Lee, J. J., Crosby, K. M., Pike, L. M., Yoo, K. S., and Lescobar, D. I. 2005. Impact of genetic and environmental variation of development of flavonoids and carotenoids in pepper (*Capsicum* spp.). *Scientia Horticulturae* 106:341–352

Lee, J. J., Kim, D., Shin, Y., and Kim, Y. 2020. Comparative study of the bioactive compounds, flavours and minerals present in black Pepper before and after removing the outer skin. *LWT—Food Science and Technology* 125:109356

Li, G., Song, C., You, J., Sun, Z., Xia, L., and Suo, Y. 2011. Optimisation of red pepper seed oil extraction using supercritical CO2 and analysis of the composition by reversed-phase HPLC-FLD-MS/MS. *International Journal of Food Science & Technology* 46(1):44–51.

Li, S., Wang, C., Wang, M. W., Li, W., Matsumoto, K., and Tang, Y. Y. 2007. Antidepressant like effects of piperine in chronic mild stress treated mice and its possible mechanisms. *Life Sciences* 80(15):1373–1381.

Manasa, V., and Tumaney, A. W. 2022. Evaluation of the anti-dyslipidemic effect of spice fixed oils in the in vitro assays and the high fat diet-induced dyslipidemic mice. *Food Bioscience* 46:101574.

Mark, R., Lyu, X., Lee, J. J., Parra-Saldívar, R., and Chen, W. N. 2019. Sustainable production of natural phenolics for functional food applications. *Journal of Functional Foods* 57:233–254.

Moreau, R. A., Nyström, L., Whitaker, B. D., Winkler-Moser, J. K., Baer, D. J., Gebauer, S. K., and Hicks, K. B. 2018. Phytosterols and their derivatives: Structural diversity, distribution, metabolism, analysis, and health-promoting uses. *Progress in Lipid Research* 70:35–61.

Moreau, R. A., Whitaker, B. D., and Hicks, K. B. 2002. Phytosterols, phytostanols, and their conjugates in foods: Structural diversity, quantitative analysis, and health-promoting uses. *Progress in Lipid Research* 41(6):457–500.

Mostafa, N. M., Mostafa, A. M., Ashour, M. L., and Elhady, S. S. 2021. Neuroprotective effects of black pepper cold-pressed oil on scopolamine-induced oxidative stress and memory impairment in rats. *Antioxidants* 10(12):15, 1993.

Mujumdar, A. M., Dhuley, J. N., Deshmukh, V. K., Raman, P. H., and Naik, S. R. 1990. Anti-inflammatory activity of piperine. *Japanese Journal of Medical Science and Biology* 43(3):95–100.

Nejatian, M., Darabzadeh, N., Bodbodak, S., Saberian, H., Rafiee, Z., Kharazmi, M. S., and Jafari, S. M. 2022. Practical application of nanoencapsulated nutraceuticals in real food products; a systematic review. *Advances in Colloid and Interface Science* 305:102690.

Neo, Y. P., Ariffin, A., Tan, C. P., and Tan, Y. A. 2010. Phenolic acid analysis and antioxidant activity assessment of oil palm (*E. guineensis*) fruit extracts. *Food Chemistry* 122(1):353–359.

Nie, Y. D., Pan, Y. G., Jiang, Y., Xu, D. D., Yuan, R., Zhu, Y., and Zhang, Z. K. 2023. Stability and bioactivity evaluation of black pepper essential oil nanoemulsion. *Heliyon* 9(4):14730.

Okumura, Y., Narukawa, M., and Watanabe, T. 2010. Adiposity suppression effect in mice due to black pepper and its main pungent component, piperine. *Bioscience Biotechnology and Biochemistry* 74(8):1545–1549.

Opara, E. I., and Chohan, M. 2014. Culinary herbs and spices: Their bioactive properties, the contribution of polyphenols and the challenges in deducing their true health benefits. *International Journal of Molecular Sciences* 15(10):19183–19202.

Ovesnaâ, Z., Lkovaâ, A. V., and Thovaâ, K. H. 2004. Taraxasterol and b-sitosterol: New naturally compounds with chemoprotective/chemopreventive effects Minireview. *Neoplasma* 51(6):407.

Panchal, T. M., Patel, A., Chauhan, D. D., Thomas, M., and Patel, J. V. 2017. A methodological review on bio-lubricants from vegetable oil based resources. *Renewable and Sustainable Energy Reviews* 70:65–70.

Pattanaik, S., Hota, D., Prabhakar, S., Kharbanda, P., and Pandhi, P. 2009. Pharmacokinetic interaction of single dose of piperine with steady-state carbamazepine in epilepsy patients. *Phytotherapy Research* 23(9):1281–1286.

Piccolella, S., Crescente, G., Candela, L., and Pacifico, S. 2019. Nutraceutical polyphenols: New analytical challenges and opportunities. *Journal of Pharmaceutical and Biomedical Analysis* 175:112774.

Platel, K., Rao, A., Saraswathi, G., and Srinivasan, K. 2002. Digestive stimulant action of three Indian spice mixes in experimental rats. *Nahrung-Food* 46(6):394–398.

Prakash, U. N. S., and Srinivasan, K. 2010. Beneficial influence of dietary spices on the ultrastructure and fluidity of the intestinal brush border in rats. *British Journal of Nutrition* 104(1):31–39.

Prashant, A., Rangaswamy, C., Yadav, A. K., Reddy, V., Sowmya, M. N., and Madhunapantula, S. 2017. In vitro anticancer activity of ethanolic extracts of *Piper nigrum* against colorectal carcinoma cell lines. *International Journal of Applied and Basic Medical Research* 7(1):67.

Puri, V., Nagpal, M., Singh, I., Singh, M., Dhingra, G. A., Huanbutta, K., and Sangnim, T. 2022. A comprehensive review on nutraceuticals: Therapy support and formulation challenges. *Nutrients* 14(21):4637.

Rani, R., Kela, A., Dhaniya, G., Arya, K., Tripathi, A. K., and Ahirwar, R. 2021. Circulating microRNAs as biomarkers of environmental exposure to polycyclic aromatic hydrocarbons: Potential and prospects. *Environmental Science and Pollution Research* 28(39):54282–54298.

Rani, V., and Prabhu, A. (2022). Nutraceuticals derived from dietary spices as candidate molecules in targeting glioma signaling pathways. *Revista Brasileira de Farmacognosia* 32(4):514–526.

Ravindran, P. N., and Kallupurackal, J. A. 2012. Black pepper. In *Handbook of Herbs and Spices*, pp. 86–115. Cambridge, England: Woodhead Publishing Series in Food Science, Technology and Nutrition.

Sabina, E. P., Nagar, S., and Rasool, M. 2011. A role of piperine on monosodium urate crystal-induced inflammation-an experimental model of gouty arthritis. *Inflammation* 34:184–192.

Saleh, B. K., Omer, A., and Teweldemedhin, B. J. M. F. P. T. 2018. Medicinal uses and health —benefits of chili pepper (Capsicum spp.): A review. *MOJ Food Process and Technology* 6(4):325–328.

Salehi-Sahlabadi, A., Varkaneh, H. K., Shahdadian, F., Ghaedi, E., Nouri, M., Singh, A., and Mirmiran, P. 2020. Effects of Phytosterols supplementation on blood glucose, glycosylated hemoglobin (HbA1c) and insulin levels in humans: A systematic review and meta-analysis of randomized controlled trials. *Journal of Diabetes & Metabolic Disorders* 19:625–632.

Samfira, I., Rodino, S., Petrache, P., Cristina, R. T., Butu, M., and Butnariu, M. 2015. Characterization and identity confirmation of essential oils by mid infrared absorption spectrophotometry. *Digest Journal of Nanomaterials and Biostructures* 10(2):557–566.

Saxena, R., Venkaiah, K., Anitha, P., Venu, L., and Raghunath, M. 2007. Antioxidant activity of commonly consumed plant foods of India: Contribution of their phenolic content. *International Journal of Food Sciences and Nutrition* 58(4):250–260.

Scolaro, B., Andrade, L. F. D., and Castro, I. A. 2019. Cardiovascular disease prevention: The earlier the better? A review of plant sterol metabolism and implications of childhood supplementation. *International Journal of Molecular Sciences* 21(1):128.

Shaikh, J., Bhosale, R., and Singhal, R. 2006. Microencapsulation of black pepper oleoresin. *Food Chemistry* 94(1):105–110.

Singh, S., Kapoor, I. P. S., Singh, G., Schuff, C., De Lampasona, M. P., and Catalan, C. A. N. 2013. Chemistry, antioxidant and antimicrobial potentials of white pepper (*Piper nigrum* L.) essential oil and oleoresins. *Proceedings of the National Academy of Sciences India Section B-Biological Sciences* 83(3):357–366.

Siscovick, D. S., Barringer, T. A., Fretts, A. M., Wu, J. H., Lichtenstein, A. H., Costello, R. B., and Mozaffarian, D. 2017. Omega-3 polyunsaturated fatty acid (fish oil) supplementation and the prevention of clinical cardiovascular disease: A science advisory from the American Heart Association. *Circulation* 135(15):e867–e884.

Soto, C., Chamy, R., and Zuniga, M. E. 2007. Enzymatic hydrolysis and pressing conditions effect on borage oil extraction by cold pressing. *Food Chemistry* 102(3):834–840.

Spiller, F., Alves, M. K., Vieira, S. M., Carvalho, T. A., Leite, C. E., Lunardelli, A., Polomi, J. A., Cunha, F. Q., and Oliveira, J. R. 2008. Anti-inflammatory effects of red pepper (*Capsicum baccatum*) on carrageenan- and antigen induced inflammation. *Journal of Pharmacy and Pharmocolog* 60:473–478.

Stewart, T. A., Lowe, H. I. C., and Watson, C. T. 2016. (All spice) essential oil extracted via hydrodistillation, solvent and super critical fluid extraction methodologies. *American Journal of Essential Oils and Natural Products* 4(3):27–30.

Stohr, J. R., Xiao, P. G., and Bauer, R. 2001. Constituents of Chinese Piper species and their inhibitory activity on prostaglandin and leukotriene biosynthesis in vitro. *Journal of Ethnopharmacology* 75(2–3):133–139.

Su, L., Yin, J. J., Charles, D., Zhou, K., Moore, J., and Yu, L. L. 2007. Total phenolic contents, chelating capacities, and radical-scavenging properties of black peppercorn, nutmeg, rosehip, cinnamon and oregano leaf. *Food Chemistry* 100(3):990–997.

Suresh, D., and Srinivasan, K. 2007. Studies on the in vitro absorption of spice principles—curcumin, capsaicin and piperine in rat intestines. *Food and Chemical Toxicology* 45(8):1437–1442.

Tasleem, F., Azhar, I., Ali, S. N., Perveen, S., and Mahmood, Z. A. 2014. Analgesic and anti-inflammatory activities of *Piper nigrum* L. *Asian Pacific Journal of Tropical Medicine* 7:461–468.

Tu, Y. B., Zhong, Y. J., Du, H. J., Luo, W., Wen, Y. Y., Li, Q., and Li, Y. F. 2016. Anticholinesterases and antioxidant alkamides from *Piper nigrum* fruits. *Natural Product Research* 30(17):1945–1949.

Usman, I., Hussain, M., Imran, A., Afzaal, M., Saeed, F., Javed, M., and Saewan, S. 2022. Traditional and innovative approaches for the extraction of bioactive compounds. *International Journal of Food Properties* 25(1):1215–1233.

Valsta, L. M., Lemström, A., Ovaskainen, M. L., Lampi, A. M., Toivo, J., Korhonen, T., and Piironen, V. 2004. Estimation of plant sterol and cholesterol intake in Finland: Quality of new values and their effect on intake. *British Journal of Nutrition* 92(4):671–678.

Vogt, T. 2010. Phenylpropanoid biosynthesis. *Molecular Plant* 3(1):2–20.

Walsh, B. M., and Hoot, S. B. 2001. Phylogenetic relationships of Capsicum (Solanaceae) using DNA sequences from two noncoding regions: The chloroplast atpB-rbcL spacer region and nuclear waxy introns. *International Journal of Plant Sciences* 162(6):1409–1418.

Wang, Q., Jiang, L., and Wen, Q. 2007. Effect of three extraction methods on the volatile component of *Illicium verum* Hook. f. analyzed by GC-MS. *Wuhan University Journal of Natural Sciences* 12:529–534.

Wattanathorn, J., Chonpathompikunlert, P., Muchimapura, S., Priprem, A., and Tankamnerdthai, O. 2008. Piperine, the potential functional food for mood and cognitive disorders. *Food and Chemical Toxicology* 46(9):3106–3110.

Xu, G., and Kafkafi, U. 2003. Seasonal differences in mineral content, distribution and leakage of sweet pepper seeds. *Annals of Applied Biology* 143(1):45–52.

Yang, C. S., Luo, P., Zeng, Z., Wang, H., Malafa, M., and Suh, N. 2020. Vitamin E and cancer prevention: Studies with different forms of tocopherols and tocotrienols. *Molecular Carcinogenesis* 59(4):365–389.

Zarai, Z., Boujelbene, E., Ben Salem, N., Gargouri, Y., and Sayari, A. 2013. Antioxidant and antimicrobial activities of various solvent extracts, piperine and piperic acid from *Piper nigrum. LWT-Food Science and Technology* 50(2):634–641.

Zhao, X. Y., Zhang, C., Guigas, C., Ma, Y., Corrales, M., Tauscher, B., and Hu, X. S. 2009. Composition, antimicrobial activity, and antiproliferative active capacity of anthocyanin extracts of purple corn (*Zea mays* L.) from China. *European Food Research and Technology* 228:759–765.

15 Effect of Newer Non-Thermal Processing

Admajith M. Kaimal, Amrutha M. Kaimal and Rekha S. Singhal

15.1 INTRODUCTION

Capsicum pepper is cultivated as a cash crop across the globe (Obayelu et al., 2021). They are of South American origin and were traditionally used for medicinal applications. Gradually, Mexicans introduced pepper into their culinary practices before 8000 BC as a chilli pickle (Wangmo and Dendup, 2021). Thus, *Capsicum* is considered one of the primitive food additives. The name "red pepper" was introduced by Christopher Columbus when he returned to Europe after an expedition (1492–1493 AD). Soon, it became a popular ingredient in several cuisines due to its pungent taste and aroma, which in fact became their quality parameter. Several varieties are available in the market with varying ranges of pungency and colour (green, yellow, orange and red) to cater to various applications (Obayelu et al., 2021). There are 38 species of plants from the *Capsicum* genus, of which 5 varieties are cultivated (*C. annum, C. chinense, C. pubescens, C. baccatum* and *C. frutescens*). In fact, *Capsicum* varieties are differentiated on their attributes and are widely known as bell pepper, sweet pepper and hot pepper (Santos et al., 2008). These peppers are consumed either as whole fruit (chilli) or dried powder (paprika).

Besides culinary applications, they are widely used in nutraceutical and medicinal formulations. The bioactive compounds from peppers such as carotenoids, capsaicinoids and vitamins also demonstrate several biological activities (Montoya-Ballesteros et al., 2014). Capsaicin, a pungent alkaloid, is extensively used as a cardiovascular and gastrointestinal stimulant, as an anti-inflammatory and in anticancer formulations. The carotenoids, *viz.* capsanthin, β-carotene, lutein and β-cryptoxanthin, are utilized in various pharmaceutical formulations for their antioxidant and anti-inflammatory activities.

15.1.1 A Brief Overview of *Capsicum* Production

The increasing demand for ethnic and fusion cuisines augmented the popularity of several indigenous pepper varieties. The culinary market and the nutraceutical potential of pepper constituents have resulted in a massive surge in pepper production globally. According to the Food and Agricultural Organisation (FAO), green chilli and bell pepper production reached 36 million tonnes (MT) in 2021 globally. China (16.7 MT) accounted for approximately 50% of world production, followed by Turkey (3.1 MT), Indonesia (2.7 MT) and Mexico (2.6 MT) (FAO, 2023a). However, dried chilli and pepper are produced more in South Asian countries. The global production of dried chilli in 2021 was 4.8 MT, with India producing 2.05 MT, followed by Bangladesh (0.5 MT) and Thailand (0.34 MT) (FAO, 2023a). Besides fresh and dried pepper, a significant quantity of oleoresin is produced and exported worldwide. India, Spain, Hungary and Morocco are the main exporters of value-added pepper products (Po et al., 2018). The consumption of pepper is greater in tropical countries, with per capita consumption reaching 15 g per person per day in Mexico. Conversely, the consumption is less than 1 g for Nordic countries (Mózsik, 2016). Paprika accounts for 65% of the total pepper consumed, followed by chillies (35%).

15.1.2 Post-Harvest Scenario and Losses

Peppers are generally grown in temperate and tropical humid regions. They are harvested once the chillies reach maturity and ripen to red colour. Depending on the variety, harvesting is done 6 to 7 weeks after flowering. The harvesting is generally done manually and stored/processed depending on the intended use (Po et al., 2018). Chillies incur significant losses after harvest due to unsanitary storage conditions and poor post-harvest operations. Fresh chillies suffer higher losses due to higher moisture content. They typically contain 91% moisture and have a water activity above 0.9. Literature reveals that mesophilic aerobic bacterial load could range from 10^2 to 10^9 cfu/g (Santos et al., 2008). Some studies have also reported the presence of coliforms (2.3×10^5 cfu/g) and *Bacillus cereus* (10^4 cfu/g) (Santos et al., 2008; Ayob et al., 2022). Apart from bacteria, fungal spores of *Aspergillus* and *Penicillium* have also been detected (Santos et al., 2008). The higher water activity and unhygienic post-harvest practices would favour microbial growth, leading to post-harvest losses of up to 4.8% for pimento chilli (FAO, 2023b). However, a report from the government of India on post-harvest profile suggest that 15% to 25% of chillies are lost due to higher moisture (Ministry of Agriculture and Cooperation, 2009; Jalgaonkar et al., 2022).

Chillies are sun-dried and processed for paprika powder in several countries such as India, Mexico and Hungary. They are dried to 10% moisture to yield paprika with water activity below 0.5 (Santos et al., 2008). The lower water activity of paprika would decelerate microbial growth, thereby increasing the shelf life (6 months). Though the water activity is low, some microbes still survive as spores. Thus, storage of dehydrated pepper and paprika powders is critical. These products would absorb moisture during storage and transportation if improperly packaged. This, in turn, alleviates the risk of microbial spoilage, especially mycotoxin contamination. Therefore, several food safety authorities have established regulations for maximum tolerable limits for mycotoxin in *Capsicum* fruits. The European Union regulates the aflatoxin at 10 µg/kg and Indonesia at 20 mg/kg (Santos et al., 2008). Apart from paprika and chilli, several products are prepared from peppers. Smoked pepper or *chipotles* are prepared by the traditional Nahuatl smoking method. Pimento and jalapeño are canned as rings. Peppers are also fermented to prepare pickles. Commercially, paprika oleoresin is prepared from solvent extraction of *Capsicum* (Po et al., 2018). A detailed description of several post-harvest processing methods and various pepper products are collated in Chapters 10 and 12, respectively.

15.2 IMPACT OF PAPRIKA CONSTITUENTS DURING STORAGE

The pungency level and colour of *Capsicum* define the quality of the pepper. Both these parameters are significantly affected during ripening. The harvesting time, ripening status and storage conditions of chilli are critical in ensuring high-quality pepper. Peppers, when attached to the plant, demonstrate climacteric characteristics. On detachment, they show non-climacteric behaviour (Krajayklang et al., 2000). Thus, understanding the changes during ripening and post-harvest chilli storage is imperative (Figure 15.1).

15.2.1 PROXIMATE COMPOSITION

The proximate composition includes moisture, fat, fibre, protein and ash content. The moisture content is one of the significant quality parameters of fresh *Capsicum* and ranges from 85.2% to

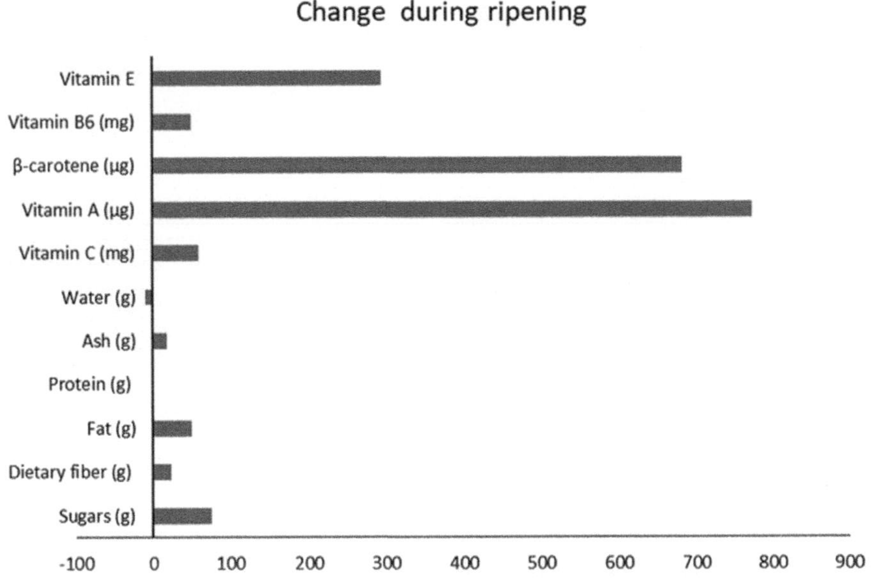

FIGURE 15.1 Biochemical changes during ripening of pepper (Po et al., 2018).

92.5% (Samira et al., 2013). Moisture loss is a major problem experienced during post-harvest, resulting in a softer and flaccid product. Depending on the variety and storage conditions, moisture loss could be up to 9.2% in 28 days (Samira et al., 2013). The moisture loss from chillies could be attributed to the anatomy of the fruit (Figure 15.2). *Capsicum* is a hollow fruit with higher surface area-to-weight ratio than other fruits, such as tomatoes. Besides, the pepper locules are porous and lack an internal water reservoir to regulate the pericarp moisture (O'Donoghue et al., 2018). The moisture loss predominately takes place through the pericarp and the cuticle. However, the integrity of the cuticle layer diminishes during storage. Cell wall permeability increases with the progress of ripening (Samira et al., 2013). Simultaneously, an increase in total soluble solids (TSS) also takes place during the initial stages (4.9 to 6.1°Brix) followed by a reduction after 8 days of storage (6.1° to 5.2° Brix) (Kasampalis et al., 2022; Samira et al., 2013). The initial increase in TSS could correspond to the moisture loss incurred during storage and sugar synthesis via organic acid. Whereas, once the ripening proceeds, the sugar would be utilized for respiration, resulting in the depletion of TSS (Samira et al., 2013). In parallel, the pH and titratable acidity shows a significant difference during storage. Typically, an increase in pH is observed during storage, with specific exceptions with some varieties. The acidity present in *Capsicum* emanates from ascorbic acid, citric acid and malic acid. These acids are precursors in several biochemical reactions such as the Krebs cycle. They also participate in metabolic reactions responsible for cellulose and pectin degradation and pigment synthesis, which are critical during ripening. Therefore, the pH tends to reduce during storage. Meanwhile, a reduction in lipid content also takes place during storage. It is also interesting to note that a higher reduction in lipids could be observed with varieties susceptible to moisture loss. The literature postulates that the cell membrane damage caused due to dehydration could result in lipid loss (Maalekuu et al., 2006). This could be substantiated

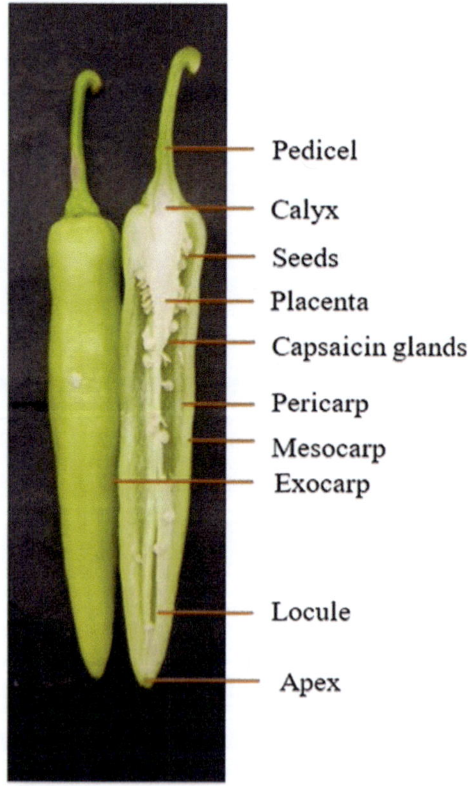

FIGURE 15.2 Anatomy of pepper.

by the loss of phospholipids, which serve as the structural blocks of cell membranes (Maalekuu et al., 2006). However, a nominal difference in protein content could be seen for pepper during storage (Kasampalis et al., 2022). Nevertheless, microbial contamination such as *Aspergillus flavus* infestation could considerably increase the protein content and reduce the fat content (Tripathi and Mishra, 2009).

15.2.2 Vitamins

Capsicum is a good source of ascorbic acid, tocopherol and pro-vitamin A. Vitamin C (ascorbic acid) tends to increase during ripening, and the concentration ranges from 20 to 247 mg/100 g fresh pepper (Samira et al., 2013). Certain varieties such as Arnoia demonstrate a 27% to 30% increase in vitamin C on ripening, while others do not show a significant difference (Martínez et al., 2007). Higher vitamin C content in ripened pepper could correlate to increased glucose levels required for ascorbic acid synthesis (Hamed et al., 2019). However, the level of ascorbic acid reduces gradually after 2 days of storage, and the degradation rate increases drastically with storage temperature (Rahman et al., 2015). Similarly, pepper also contains tocopherol and pro-vitamin A. The concentration of these vitamins increases during ripening (Hernández-Pérez et al., 2020). Especially, β-carotene (pro-vitamin A) is produced during ripening via biosynthesis pathways (Villa-Rivera and Ochoa-Alejo, 2020).

15.2.3 Polyphenols

The polyphenols in *Capsicum* include phenolic acids, tannins and flavonoids. The accumulation of these secondary metabolites is associated with ripening (Hamed et al., 2019). Flavonoids represent a significant proportion of total polyphenols and exhibit higher antioxidant activity. The flavonoids in pepper are derivatives of luteolin and quercetin. They are widely found in the pericarp of pepper (Hernández-Pérez et al., 2020). Recent research suggests that the soluble polyphenols increase during ripening (Kasampalis et al., 2022). However, these polyphenols tend to degrade with time. The degradation could be associated with the polyphenol oxidase, which also increases after harvesting (Barbagallo et al., 2012).

15.2.4 Carotenoids

Carotenoids are phytochemicals composing 8 isoprene units responsible for the characteristic colour of pepper. They are essential for photomorphogenesis and assist in photosynthesis (Villa-Rivera and Ochoa-Alejo, 2020). They also demonstrate various nutraceutical benefits such as antioxidant, anticancer and anti-inflammatory activities. *Capsicum* is a rich source of β-carotene, lutein, β-cryptoxanthin and capsanthin (Kennedy et al., 2021). They are produced by *de novo* synthesis during ripening (Gómez-García and Ochoa-Alejo, 2013). The carotenoid content (48.5 to 3211 mg/100 g) increases up to 60 times on maturity. Green peppers have higher quantities of lutein (41% of all carotenoids), and on maturity, capsanthin (35% of all carotenoids) becomes more predominant (Gómez-García and Ochoa-Alejo, 2013). However, carotenoids are susceptible to degradation on storage. The stability of carotenoids is higher for paprika (dried chilli) than fresh peppers (Arimboor et al., 2015). The degradation of carotenoids in fresh chillies could be mediated by lipoxygenase enzyme and microbial growth (Schweiggert et al., 2007). Moisture content plays a significant role in the stability of carotenoids in paprika powder. Paprika with 10% to 18% moisture showed higher carotenoid stability than low-moisture paprika (< 8%) (Schweiggert et al., 2007). The storage temperature and illumination intensity have a degradative effect on carotenoids. This degradation is primarily caused by free radical-mediated oxidation which could be accelerated by heat and light (Montoya-Ballesteros et al., 2014). Post-harvest pre-treatments such as heat and ozone treatment accelerate the degradation of carotenoids. Modified atmospheric storage and vacuum packaging could improve the stability of carotenoids (Arimboor et al., 2015).

15.2.5 CAPSAICINOIDS

Capsaicinoids are a group of alkaloids present in pepper and are responsible for the pungency thereof. They are concentrated in the placenta, pericarp and seed tissues of pepper. Capsaicin and dihydrocapsaicin are the main capsaicinoids present in pepper, accounting for 90% of all capsaicinoids. Nominal amounts (< 10%) of homodihydrocapsaicin, nordihydrocapsaicin and homocapsaicin are also present (O'Donoghue et al., 2018). The concentration of capsaicinoids is considered to be an important quality parameter for peppers. In fact, peppers are characterized according to the capsaicinoid content, which is expressed in terms of Scoville heat units (SHU). They are produced by a series of complex biochemical pathways in mitochondria and cytoplasm. These biochemical reactions are a conjunction of phenylpropanoid and branched-chain fatty acid pathways (Naves et al., 2019). The concentration of capsaicinoids tends to augment with maturity. However, the ratio of capsaicin to dihydrocapsaicin reduces (4.5 to 1.4) during ripening. The biochemical reactions during ripening could be responsible for the conversion of dihydrocapsaicin to capsaicin (Victoria-Campos et al., 2015). The capsaicinoid content appears to increase during initial days of storage, followed by a reduction. This could correspond to ripening-related biochemical changes during initial days of storage followed by degradation of capsaicinoids through enzyme oxidation. However, the capsaicin content in green and red peppers seems to decrease during initial months of frozen storage, followed by a steady increase till 6 months. On the contrary, dihydrocapsaicin increased during initial months followed by decrease. This could correspond to the biochemical reactions responsible for capsaicin synthesis at frozen conditions (Victoria-Campos et al., 2015). Meanwhile, reports on post-harvest storage (ambient and evaporated cold storage) of chillies suggest an initial increase in capsaicinoid content followed by a decrease (Samira et al., 2013). The warmer temperatures of post-harvest storage (compared to frozen storage) accelerate the ripening process, which then increases the capsaicinoids. The warmer temperature also favours oxidation by peroxidase and capsaicin oxidase, resulting in degradation of capsaicinoids after ripening. The activity of these enzymes could be diminished by reducing water activity (Samira et al., 2013).

15.3 NON-THERMAL PROCESSING OF PAPRIKA

Peppers are a highly perishable horticulture crop with huge demand around the globe and are consumed in different forms like fresh fruit, dried chilies and chilli paste. Post-harvest management is a key challenge in extending shelf life of the produce. The physiological, biochemical changes and microbial activities start soon after harvest. Proper post-harvest handling conditions could reduce these changes (Obayelu et al., 2021). Drying is a common method of processing, in which whole chillies are dried under the sun or with mechanical driers. These drying techniques use heat to remove water under controlled conditions, which can reduce enzymatic and microbial activities. However, thermal treatments accelerate the degradation of bioactive compounds such as tocopherols, vitamin C, carotenoids, flavonoids, nutrients and sensory quality of the produce (Montoya-Ballesteros et al., 2014). In tropical and sub-tropical countries, inadequate drying and humid conditions could enhance fungal growth and contaminate dried chillies with mycotoxins. This urges the need for non-thermal processing techniques to preserve chillies that could, partially or completely, substitute thermal processing techniques to reduce processing time and nutritional loss (Table 15.1). These techniques are reliable for manufacturing microbiologically safe products and with retention of sensory attributes. Yet some of these processes can generate heat, leading to a nominal rise in temperature. However, the impact on food is relatively nominal compared to thermal treatments, where food materials are exposed to higher temperature for a longer time.

15.3.1 HIGH-PRESSURE PROCESSING (HPP)

HPP is a promising non-thermal preservation technique in which the food material is exposed to uniform high pressure, ranging from 300 to 800 MPa, for a specific period. This process follows three principles: Le Chatelier's, microscopic ordering and isostatic principle. Le Chatelier's principle explains the variation in an equilibrated system while applying pressure. It states that a phenomenon accompanied by a decrease in volume is enhanced by pressure. Secondly, the principle of microscopic ordering describes the increase in degree of molecular ordering in a substance with an increase in pressure at a constant temperature. Finally, the isostatic principle portrays that when uniform and instantaneous pressure is applied, the pressure acts equally in all directions of the material. The efficacy of the HPP process is affected by various parameters such as pressure, temperature, moisture content and protein structure of the food. In liquid foods, treatment pressure is the main factor affecting the log reduction of microbial load, whereas water activity governs the lethality in powdered products. It is observed that the temperature of the product is increased by 3°C in every 100MPa increase in pressure. This could correspond to adiabatic heating. The magnitude of temperature rise also depends on the constituents of the food (Singh and Heldman, 2001).

15.3.1.1 Mode of Action

The isostatic high pressure breaks the covalent bonds in cell constitutes including lipids, proteins and nucleic acid, thereby denaturing the enzymes and disintegrating the bacterial cell wall and cell membrane (Figure 15.3). A pressure of 50–100 MPa can inhibit protein synthesis, and pressure above 300 MPa can cause irreversible denaturation of proteins and enzymes (Srinivas et al., 2018). This could alter the permeability of the bacterial cell membrane and disrupt cell integrity, causing morphological changes. Moreover, sudden release during the pressure withdrawal phase could trigger the unfolding of various proteins. Contrary to the fate of microbial cell constituents, the food material constituents such as vitamins, amino acids and phytochemicals seem to be the least affected. HPP is also effective in the inactivation of spores but requires process optimization to increase efficiency (Rifna et al., 2019).

15.3.1.2 Instrumentation

The fundamental parts of an HPP system are a pressure-holding vessel, pressure-transmitting medium/fluid, and pressure pump (Figure 15.3). Solid foods or packed products are placed inside the vessel. Thereafter, transmitting fluid is pumped to fill the space around the product, and pressure is applied to the fluid through a high-pressure pump. Once the pressure reaches a desired value, the product is held for a specific time to achieve the intended lethality. Usually, water is used as a pressure-transmitting fluid and is reused for upcoming batches. In the case of liquid or pumpable semi-solid foods, a container is equipped with a free piston into which liquid foods are introduced. The free piston is adjusted by pumping pressure-transmitting fluid over the free side of the piston (Singh and Heldman, 2001). Certain designs with compact rollers could also be deployed for pressuring solid food. But these designs result in the comminuting of the end product.

15.3.1.3 Application

A recent study deliberated the effect of HPP on red pepper paste (Woldemariam et al., 2022). The study reported a reduction of aerobic mesophilic bacteria by 4.5 log CFU/g at 527 MPa for 517 seconds and yeast and mould count to 1 log CFU/g at 600 MPa for 315 seconds, clearly indicating yeast and mould to be more vulnerable to HPP than aerobic mesophilic bacteria. A noticeable increase in colour was also seen with increased treatment time and pressure. In contrast, Castro et al. (2008)

TABLE 15.1

Application of Various Non-Thermal Treatments on Quality Attributes of Pepper

Process	Product	Variable parameters	Major results	Conclusion	Reference
HPP	Whole red pepper paste	Pressure 100–600 MPa Time -30–600 s	• Reduction of aerobic mesophilic bacteria count by 4.5 log CFU/g at 527 MPa for 517 s • Reduction of yeasts and moulds counts to 1 log CFU/g at 600 MPa for315 s.	Quadratic model was used for optimization of process variable and suggested 536 MPa for 125 s	(Woldemariam et al., 2022)
HPP	Cut green and red bell peppers	100 and 200 MPa (10 and 20 min)	• The reduction of polyphenol oxidase activity, pectin methylesterase activity in green peppers • No reduction on microbial activity	Effectively reduce enzyme activity and microbial load	(Castro et al., 2008)
PEF	Red pepper flakes	Frequencies: 100–180 Hz and Energy: 0.97–17.28 J	• Maximum inactivation rate (64.37%) at 17.28J	Effective in inactivation of aflatoxin with reduced mutagenic and improved quality properties	(Evrendilek et al., 2022)
PEF	Cut red pepper pieces	Electric field strength of 1.0–2.5 kV/cm, pulse width 30µs, time: 1–4 s	• Cell membrane damages, without visible structural changes • Reduced drying time by 34.7%	PEF is an effective pre-treatment in drying and can produce end products with improved quality	(Won et al., 2015)
UV-C	Red pepper	Dosage 1, 3, 7 and 14 kJ/m2, wavelength 254 nm	• Delayed postharvest damages and reduced chilling injury	Effective in reducing spoilage in chillies and chilling injury	(Vicente et al., 2005)
UV-C	Ground hot pepper	Dosage-3.5, 7.0 and 10 kJ/m, wavelength -254 nm	• Completely eliminated *Salmonella* and *E. coli.* • Increased content of total flavonoids, total phenols with increasing treatment time, and thus increased antioxidant activity	Reduced microbial load	(Hassan et al., 2020)
CP	Dried red pepper	Cp with 10, 30, 45, 60 s Voltage 205 240V AC and frequency of 13.56 MHz	• Drying rate increased with increase in treatment time, • Colour retention improved till 30s, followed by decrease for longer treatment time	30 s of pre-treatment is optimum for retention of red pigment and improved drying rate	(Zhang et al., 2019)
CP	Red pepper powder	Power: 300–900 W, time: 20 min (N$_2$, N$_2$-O$_2$)	• Effective in reduction *A. flavus* with no significant colour change	Alternative method for reduction of *A. flavus*	(Abdi et al., 2019)
PAW	Red pepper	PAW generated from argon and oxygen with time of exposure (5 to 60 min)	• Reduced pesticide residue and decontaminated microorganisms	Effective in reducing the pesticide residue to MRL and microbial decontamination	(Sawangrat et al., 2022)
Ozone	Red cayenne pepper	Ozonated water (bubbling of ozone 150 g/150 L)	• Could maintain quality up to 30 days	Ideal for shelf life extension in whole pepper	(Sasmita et al., 2019)
Ozone	Cut red bell peppers	Ozonated water and blanching (blanching at 50 and 60°C; ozone at 0.3 and 2.0 ppm)	• Aqueous ozone showed 0.5 to 1 log reduction	Aqueous ozone effective for microbial reduction and comparable with bleaching at 50°C	(Alexandre et al., 2011)

(Continued)

TABLE 15.1 (*Continued*)

Application of Various Non-Thermal Treatments on Quality Attributes of Pepper

Process	Product	Variable parameters	Major results	Conclusion	Reference
Ozone	Cut red bell peppers	Chlorine (200°µL/L), ozonated water (1°µL/L) and gaseous ozone (0.7°µL/L)	• Samples treated with gaseous ozone (2.3 to 2.6 log reduction) on 14th day of storage showed positive impact • Ozonated and chlorinated water pre-treatment were not effective	Most effective pre-treatment is gaseous ozone in case of cut pieces	(Horvitz and Cantalejo, 2012)
Electron beam and gamma ray irradiation	Dried red pepper	EB: 3 kGy and dose rate of 1 and 5 kGy/s; GR: 3 kGy and dose rate of 1.8–9 kGy/h	• Showed better lethality for electron beam than gamma and X-rays towards total aerobic microbes	Electron beam irradiation was more effective than gamma and X-ray	(Kyung et al., 2019)
PL	Red pepper flakes	Flash 1 to 10 Fluence/flash—1 J/cm2	• 1 log reduction of artificially inoculated *B. Subtillus* on 10 flashes	Effective for inactivation of vegetative cells	(Nicorescu et al., 2013)
PL	Cut pieces of red pepper	Fluence of 4 and 32 J/cm2	• Reduction of microbial load • Marginal increase in product temperature	Effective for inactivation of vegetative cells	(Rybak et al., 2021)
Blue LED	Dried red pepper	25000 mcd radiant powers at 470 nm for 45 min	• Study compared UV, LED and CP, and LED could show 42.1% reduction of total aflatoxin	Effective for inactivation of aflatoxin and gave better results than UV	(Kalathil et al., 2023)
UV-C	Red pepper dried	60 W for 90 min	• Study compared UV, LED and Cp, and UV could show 15.8 % reduction of total aflatoxin	Effective for inactivation of aflatoxin and lower reduction compared to UV and CP	(Kalathil et al., 2023)
Pressured argon	Cut green pepper	2 to 6 MPa	• Delayed microbial growth • Reduces physiological changes • Retention of firmness by controlling water loss	Bacteriostatic activity with lesser impact on bioactives	(Shen et al., 2019)
PL +CP	Red pepper powder and flakes	CP voltage and frequency: 10 kV and 15 kHz, IPL voltage and frequency: 2010 V and 3.8 Hz	• Highest inactivation -2.9 log CFU/g • No change in colour, capsaicin and ascorbic acid concentrations and antioxidant activity • Changes in flavour were observed	Potential technique in microbial inactivation	(Lee et al., 2020a, 2020b)
Ozone + blanching	Cut red bell peppers	Blanching 50–60°C; Ozone 0.3–2.0 ppm	• The microbial reduction was lower than ozone and blanching treatment alone.	No synergistic effect observed	(Alexandre et al., 2011)
UV-C + mild heating	Dried red pepper powder	UV intensity 3.40 mW/cm2, heating 25, 35, 45, 55, and 65°C, time 5, 10 min, dosage UV 20.4 kJ/m2 and 40.8 kJ/m2	• Showed synergistic effect and observed higher log reduction compared to separate process	Potential technique in microbial inactivation	(Cheon et al., 2015)
UV + NIR	Dried red pepper powder	UV intensity 2.62 mW/cm2, NIR power 500 W	• Reduction in microbial load • Nominal effect on colour, capsaicin and dihydrocapsaicin content	Showed synergistic effect	(Ha and Kang, 2013)
Microwave +CP	Red pepper flakes	Microwave power density 0.17 and 0.25 W/m2, CP-power 50–1000 W	• Significant reduction in spores of *Bacillus cereus* and *Aspergillus flavus* • Notable reduction in capsaicin and ascorbic acid	Have potential in inactivation of spores Lower-power microwaves retained sensory quality, capsaicin and dihydrocapsaicin content	(Kim et al., 2019)

HPP: high-pressure processing, CP: cold plasma, PEF: pulse electric field, PL: pulse light, UV: ultraviolet, PAW: plasma-activated water.

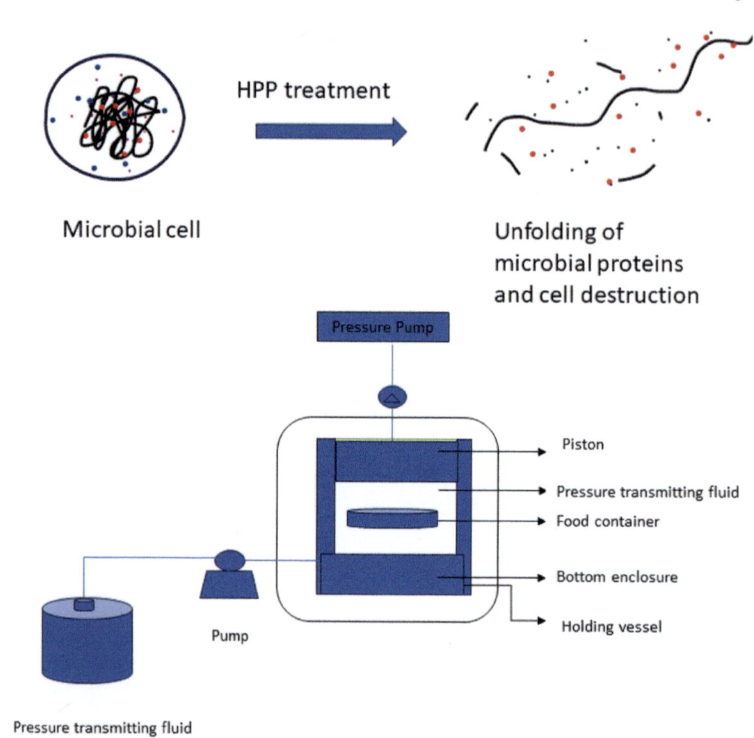

FIGURE 15.3 Schematic representation of high-pressure processing.

observed no reduction in microbial activity on pressure-treated red and green peppers (100 and 200 MPa). Meanwhile, the HPP is also reported to decrease the polyphenol oxidase activity in red and green peppers and increase that of pectin methyl esterase. This could be due to the enhanced extractability of pectin methyl esterase after HPP, owing to higher cell wall permeability. Finally, the peroxidase activity in green and red peppers was also reduced by 70% and 40%, respectively. The pressurized green peppers were firmer than thermally processed.

15.3.2 Pulsed Electric Field Processing (PEF)

PEF processing is a promising food-preservation technology in which food is exposed to a high-intensity short-pulse electric field of 20–80 kV/cm for a fraction of a second (microseconds). Food materials are placed between the two electrodes and an electric field is generated for nanoseconds to microseconds. The main principle behind microbial inactivation is electro-permeabilization (Rifna et al., 2019). The moisture content of the food is an important factor governing the lethality of microorganisms in this process. PEF effectively inactivates microbes in liquid and semi-liquid foods with high moisture content and is unsuitable for low moisture foods such as powders and dry fruits. However, the high-intensity electric field could increase the core temperature of food. This could be overcome by processing the food samples at lower temperatures (Zhang et al., 2019). Other factors that influence the efficiency of processing are the strength of the electric field, treatment time, treatment temperature, wave type, microbial parameters (shape, size, orientation of cell to be inactivated) and dielectric property of food to be treated.

15.3.2.1 Mode of Action

Initially, researchers proposed electroporation as the principle of microbial inactivation, which explained the electrical collapse of cell membranes when the transmembrane electrical potential approaches. Later, the electro-permeabilization principle was proposed, where the changes in monolayer cell membrane functionality were associated with loss of semi-permeability of the cell membrane (Figure 15.4). Free radical oxidation is also presumed to cause microbial reduction, owing to generation of free radicals of hydrogen peroxide in the treated solutions. These free radicals can react with bacterial macromolecules and reduce cell functionality, leading to cell death. Gram-negative bacteria are reported to be more vulnerable to PEF processing than Gram-positive bacteria (Zhang et al., 2019; Rifna et al., 2019).

15.3.2.2   Instrumentation

The basic parts of the PEF unit are an electric field generator and a treatment chamber (Figure 4). The treatment chamber can be considered as space where food is subjected to an electric field. The unit also contains a provision for pumping liquid foods. Usually, foods are pumped into a tube or pipe surrounded by electrodes. A minimum of two electrodes is provided for high-voltage and ground-level electrodes. Different configurations of electrodes can be used such as parallel plate, coaxial or co-linear. Since PEF can increase the food temperature, supplementary cooling systems are added to the unit.

15.3.2.3   Application

PEF has been reported to be a dependable technique for enzyme inactivation and pre-treatment technique for improved drying, osmotic dehydration and extraction (Zhang et al., 2019; Rifna et al., 2019). The use of PEF in fruit juice extraction has also shown a remarkable increase in yield of juice with higher content of bioactive(s), which has been attributed to the disruption of cell membrane breakage (Ade-Omowaye et al., 2002). Won et al. (2015) highlighted that the PEF pre-treated red peppers showed high drying rates with improved colour. The spray-dried juice of red bell pepper of PEF treated samples showed higher vitamin C and non-polar bioactive compounds (Rybak et al., 2020). The applicability of PEF in the inactivation of aflatoxin in red pepper flakes is also reported (Evrendilek et al., 2022).

15.3.3   Pulsed Light Processing (PL)

PL is an effective non-thermal surface decontamination treatment using high-intensity light. The light produced from the source is 20,000 times more intense than sunlight approaching sea level

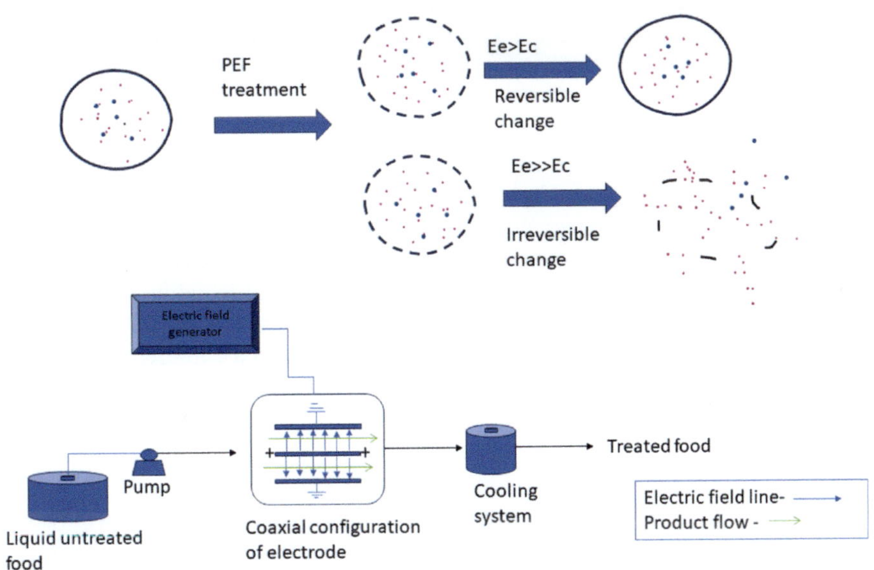

FIGURE 15.4   Schematic representation of pulse electric field.

(Zhang et al., 2019). This technique involves treating the food with short pulses of intense light with a broad-spectrum wavelength of 90 to 1100 nm (UV light: 100–400 nm, visible light: 400–700 nm and infrared light: 700–1100 nm). The light is flashed 1 to 20 times a second with energy density (0.01–50 J/cm^2). The wavelength of light used in treatment varies with the food type (Rybak et al., 2021). In pulsed UV, wavelength varies from 200 to 280 nm (UV C). PL raises the temperature. This is called the photochemical effect and is associated with the absorption of infrared light (Zhang et al., 2019). Researchers suggest PL to be a dependable technique for surface inactivation of microbes in foods and contact surfaces (Rifna et al., 2019; Lee et al., 2020). Besides, the process does not leave any residue after processing. The factors affecting efficiency of PL processing are flashes per second, pulsed energy level, the distance between sample and source, the voltage applied, wavelength, treatment time, the kind of sample treated, type of sample, microorganism and microbial load (Rifna et al., 2019). Several researches refer to PL as pulsed UV, intense short-pulse white light and broad-spectrum light (Rifna et al., 2019; Lee et al., 2020).

15.3.3.1 Mode of Action

The inactivation of microorganisms in PL processing is believed to be due to its effect on DNA replication. When the contaminated food is exposed to a broad spectrum of intense light, DNA of microorganisms absorbs the UV range of light and form photoreactive products, which can interfere with the transcription and translation of DNA. This could alter the genome of microbes, leading to death. In much simpler way, the formation of pyrimidine dimers during processing can stop DNA replication and cause lethality. Decontamination is caused by photophysical and photothermal effects (by absorption UV and IR range from the broad spectrum). Some microorganisms can undergo repair called photoreactivation. However, the reactivation rate is lower for pulsed UV–exposed organisms than for continuous UV treatment. Gram-positive bacteria are more resistant to PL than Gram-negative organisms. The differences in cell composition and inherent repair mechanisms are responsible for this trend (Oms-Oliu et al., 2010).

15.3.3.2 Instrumentation

The main components of the PL unit are an inert gas flash lamp as intense light generator, power supply, a provision for high-energy electric pulse and treatment chamber (Figure 15.5). The electrical energy from power supply is stored as electromagnetic energy in a capacitor and released as high intense electric pulses. When electric pulses are passed through an inert gas lamp, a high-intense broad spectrum of white light is generated. Generally for food applications, a xenon lamp is used as the light source with 1 to 20 flashes per second at an energy ranging from 0.01 to 50 J/cm^2 (Oms-Oliu et al., 2010).

15.3.3.3 Application

Exposure of fresh red pepper to pulsed light (190–1100 nm) reported a significant reduction of aerobic microbial count and fungi with temperature rise in the substrate varying from 12.7 to 23.1°C. A dosage of 16 J/cm^2 is reported to be the most efficient for reduction of microorganisms with minimal tissue damage in pepper (Rybak et al., 2021). Similarly, microbial lethality, aflatoxin and ochratoxin degradation was observed in red pepper powder when exposed to PL (fluence: 1 to 9.1 J/cm^2) (Woldemariam et al., 2022). The PL treatment of 10 J/cm^2 showed a 1 log reduction of artificially inoculated *Bacillus subtilis* bacteria on red pepper flakes (Nicorescu et al., 2013).

15.3.4 Ozone

Ozone is a pale-blue or colourless gas containing three atoms of oxygen. It can quickly decompose to oxygen in air and water with an oxidation potential of 2.07 V. This property of ozone is responsible for higher oxidizing power, which is, in turn, associated with disinfectant, sanitizing and antimicrobial properties. They are widely used for decontamination applications in food industries

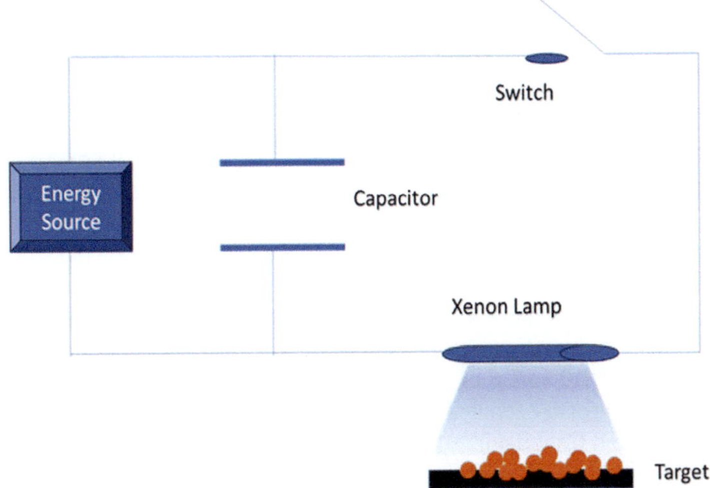

FIGURE 15.5 Schematic representation of pulse light treatment.

such as disinfection of food contact surfaces, equipment and water. Ozone treatment is ideal for the decontamination of broad-spectrum microorganisms in food products. It can even react faster than chlorine with organic matter, leaving no harmful by-products. The efficiency of ozone treatments relies on the medium of application. When using gaseous ozone, temperature, pressure and relative humidity could affect the efficiency. When using aqueous ozone, pH, conductivity and organic matter composition are major factors for its efficacy. In addition, the concentration of ozone, type of microorganisms and nature of substrate are also deciding factors (Epelle et al., 2023).

15.3.4.1 Mode of Action

The inactivation of microorganisms is achieved by direct interaction of ozone with cell components. It can penetrate the cells and oxidize fundamental components of cells such as DNA, RNA, proteins, enzymes and lipids, cumulatively leading to a lethal effect. Several studies identified that ozone initially targets the surface of bacteria and reacts with unsaturated fatty acids in lipids of the cell membrane, which leads to leakage and rupture of the entire cell. Besides direct interaction with ozone, cell inactivation can also be associated with indirect attack by reactive oxidative species formed as a result of interaction of ozone with air or water. $OH\bullet$, $HO_2\bullet$, $O_2^{-}\bullet$, $O_3^{-}\bullet$, $HO_3^{-}\bullet$, H_2, O_2, and O^{-} are a few examples of such reactive species. The formation of these reactive species is governed by the medium (air or water) of application. For example, a hydroxyl radical ($OH\bullet$) is formed when aqueous ozone is used and has an oxidation potential of 2.80 V, which can also initiate destructive reactions. As free radicals are involved in destructive reactions, usually all reactions involved are chain reactions leading to lethality and log reduction in microbial load (Epelle et al., 2023).

15.3.4.2 Instrumentation

Typically, ozone processors using aqueous ozone have oxygen concentrators, ozone generators (corona discharge generators), a bubbling system and reaction vessels. The ozone generated from the ozone generators is supplied to the bubbling system, and the macro bubbles are introduced into the water contained in the reaction chamber. The food products to be decontaminated are immersed in the water for a specific time. Besides, the aqueous ozone and ozone bubble collapsing on the food surface can also initiate the reaction. Certain designs that are documented use nano bubbling technology to increase the concentration of ozone solubilized in water and exhibit positive results (Epelle et al., 2023). In systems using gaseous ozone, ozone gas at specific dosage and operating conditions is pumped to treatment chamber with food and exposed for specific time period. Usually, these systems are equipped with ozone regenerators to remove residual ozone in food (Figure 15.6).

15.3.4.3   Application

Washing of cut red bell pepper with ozonated water (0.30 ppm and 2.00 ppm) has shown 0.5 to 1 log reduction of microbial load (Alexandre et al., 2011). This pre-treatment with ozonated water on red cayenne pepper could maintain quality for up to 30 days on dedicated cold storage compared to 15 days for the control sample (Sasmita et al., 2019). A study evaluating the effectiveness of aqueous ozone, gaseous ozone and chlorine pre-treatment on freshly cut red peppers stored in modified atmospheric packages concluded gaseous ozone to be the most effective method. Gaseous ozone showed a log reduction of 2.3 to 2.6 of aerobic mesophile bacteria than that of the control on the 14th day of storage. In contrast, ozonated water and chlorine-treated samples showed a similar load as the control (Horvitz and Cantalejo, 2012).

15.3.5   Non-Thermal Plasma

Ionization of gaseous matter generates plasma, the fourth state of matter. It contains molecules, positive ions, negative ions, free radicals, UV photons, atoms and nuclei, all equalizing plasma charge to neutral. These active species are responsible for the inactivation of microbes and enzymes. The composition of active species differs with composition of gases, which in turn governs the efficiency of microbial inactivation and other applications. Non-thermal or cold plasma (CP) finds enormous application in enzyme inactivation, pesticide disinfection, sterilization and seed germination (Rifna et al., 2019). Moreover, plasma discharge is used to produce plasma-activated water (PAW). This is formed when plasma or ionized gases are discharged into water, facilitating the dissolution of free radicals and reacting species into water. The secondary reacting species in water comprises peroxide, nitrate ions, nitric acid and nitric oxide radicals. PAW finds application in microbial inactivation and pesticide reduction. The activity of PAW reduces with time. Hence the time between the production of PAW and treatment is critical (Sawangrat et al., 2022).

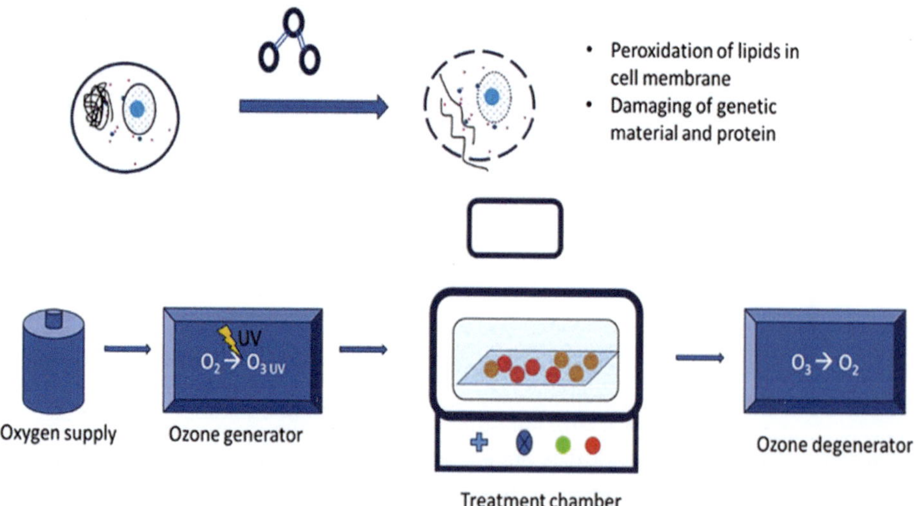

FIGURE 15.6   Schematic representation of gaseous ozone treatment system.

15.3.5.1 Mode of Action

The inactivation of microorganism(s) can happen in various ways. Primarily, the energetic species contained in CP can act on covalent bonds of vital cell components, leading to cell lethality. Secondly, the diffusion of reactive species through the cell membrane can denature proteins and enzymes and initiate lipid peroxidation. At the same time, ions and free radicals can cause localized etching in cell membrane facilitating the active entry of other reactive species, leading to cell damage and rupture. Thirdly, UV radiation present in plasma is absorbed by the DNA of microorganisms, which interferes with cell replication and thereby inhibits the growth of microbes. The reaction of oxygen and nitrogen also contributes to the antimicrobial activity of microbes, as these target unsaturated covalent bonds of lipids contained in cell membranes (Kim et al., 2014). The antimicrobial activity of PAW is contributed by secondary species dissolved in water. The reactive nitrogen and oxygen species rupture the cell membrane by oxidizing lipids and proteins and initiating a free radial chain reaction (Sawangrat et al., 2022).

15.3.5.2 Instrumentation

The main components of a CP treatment unit are power supply, plasma generator, gas supply (gas to be ionized), treatment chamber and cooling system (Figure 15.7). Electrical energy is utilized by plasma generator to ionize gas inside the treatment chamber. Thereafter, the generated plasma is applied to food materials inside the treatment. There are different types of plasma generators: corona discharge, glow discharge, dielectric barrier discharge, pulse discharge, high-frequency discharge and microwave discharge. Plasma generated by the unit can either be directly used on food or transported (indirect treatment) or utilized to produce PAW. The selection of plasma generator is dependent on power supply frequency, alternating or direct current source and shape of electrodes (Sakudo et al., 2020).

15.3.5.3 Application

A study comparing the effect of CP (20 min) and gamma radiation (10 KGy) on microbial reduction, colour and sensory properties of red pepper powder inferred both the processes to be effective in decontamination. However, treated samples showed significant difference in colour and sensory scores (Abdi et al., 2019). Similarly, another study on CP pre-treatment established that CP can replace chemical treatments to increase the drying rate (Zhang et al., 2019). The quality attributes of pepper showed no evident changes with increase in exposure time. However, the retention of red pigment was initially improved till 30 seconds and later showed a negative trend. A combination of CP with nitrogen and mixture of nitrogen and oxygen could also reduce *A. flavus* without

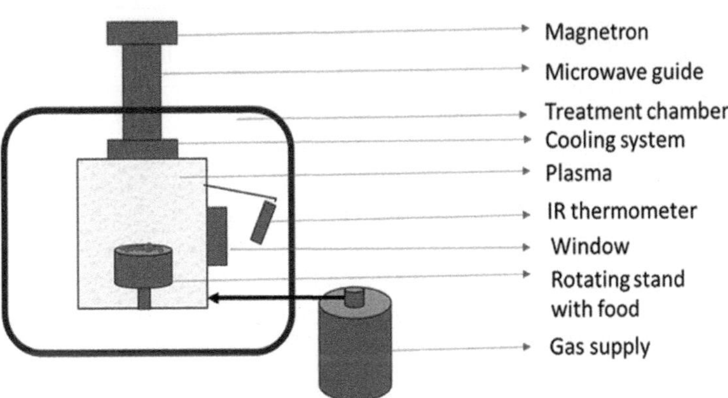

FIGURE 15.7 Schematic representation of cold plasma treatment.

impacting the colour (Abdi et al., 2019). Application of CP can be extended to production of PAW for decontamination of pepper and reduction of pesticide residue (Sawangrat et al., 2022). They could successfully reduce the pesticide residue to the maximum residue limits mandated by regulatory agencies. Microbial decontamination of red pepper could also be performed with PAW generated from an argon and oxygen gas mixture (Sawangrat et al., 2022).

15.3.6 IRRADIATION

Food irradiation has developed wide popularity in the food industry because of the effective decontamination, eco-friendly nature and minimal post-processing residue. Irradiation is the process of exposing food to radiation. The radiation is of two types, ionizing and non-ionizing radiation. Ionizing radiation has the power to generate ions from the atoms or molecules. Ionization happens when electrons are knocked out from neutral atoms or molecules on exposure to radiation. Non-ionizing radiations, on the other hand, do not have sufficient energy to ionize gas molecules.

15.3.6.1 Ionizing Radiation

Gamma rays, X-rays and electron beams are the ionizing radiations which are approved for food application by the Codex general standard for irradiated foods. Gamma rays are produced when high energy photons from ^{60}Co or ^{137}Cs are released. X-rays are formed when accelerated electrons (5MeV $\geq$ Energy level) bombard metal plates, while accelerated electrons (10MeV $\geq$ Energy level) from the machine source are utilized for producing an electron beam.

Gamma radiation from ^{60}Co is most recommended owing to its higher penetration. The impact of ionizing radiation on microorganisms can be explained in two ways: direct and indirect effect. The interaction between radiation and DNA could damage and prevent DNA replication, which could cease microbial growth, called direct effect. The indirect effects result from radiolysis of cellular water within 10^{-6} s leading to the formation of oxidative reactive species. These reactive species can break bonds in vital cell constituents and result in cell lysis (Ayob et al., 2022). Several studies have reported a reduction in aflatoxin and microorganisms in pepper (Ayob et al., 2022; Balakrishnan et al., 2022). Attempts were made by authors to reduce aflatoxin to a "non-detectable limit" in red chilies, red pepper, red pepper powder, red pepper paste with respective maximum dosage of 10 kGy, 5 kGy, 3.5 kGy and 3.5 kGy (Yoon et al., 2014; Kyung et al., 2019; Song et al., 2014).

The red pepper irradiated with gamma rays in the range of 1 to 5 kGy showed increased lethality of *E. coli* O157:H7 and *S. typhimurium* with increased dosage (Song et al., 2014). Gamma irradiation of 10 kGy could disinfect hot pepper contaminated with *Aspergillus flavus* (Ayob et al., 2022). A study comparing effects of electron beam and gamma rays on dried pepper powder revealed higher dosage of gamma (9 kGy) and lower dosage of electron beam (5 kGy) could reduce aerobic plate counts effectively (Kyung et al., 2019). Another study that compared the efficacy of gamma, X-ray and electron beam on dried red pepper powder showed better lethality for electron beam than gamma and X-rays towards total aerobic microbes (Jung et al., 2015). However, Byun et al. (2019) noted that in contrast to gamma ray and electron beam, X-rays are not effective in decontamination of *Aspergillus flavus*.

15.3.6.2 Non-Ionizing Radiation

UV, visible light, infrared and radio waves belong to non-ionizing radiation. However, ultraviolet radiation and visible light are considered non-thermal technologies, as infrared and radio waves could significantly increase the temperature of the food system. The application of UV in surface decontamination is well known, and the treatment ranges [UV-A (315–400 nm), UV-B (280–315 nm), and UV-C (<280 nm)] are selected based on target microorganisms (Figure 15.8). The inactivation of microorganism and related toxins are associated with absorption of UV radiation by DNA and cellular materials, leading to the production of photodegrading components or the resultant

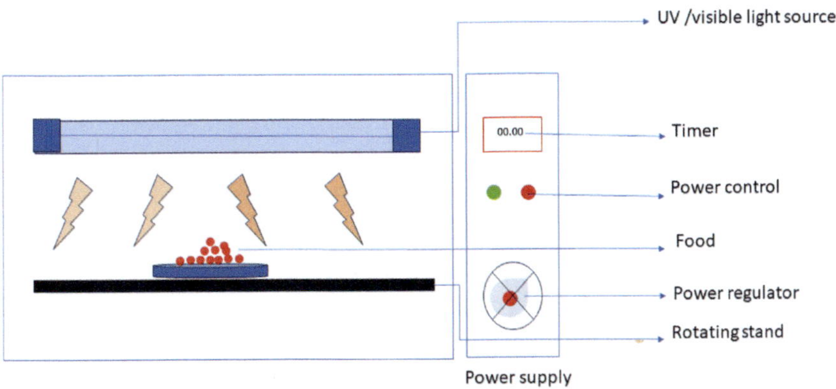

FIGURE 15.8 Schematic representation of ultraviolet and visible light processing.

free radicals. UV can influence post-harvest changes as the nitrogen- and oxygen-reactive species formed during UV irradiation directly impact pathways of physiological changes (Byun et al., 2019). This capability of UV-C is evident from the study, which delayed ripening and associated post-harvest changes on UV-C irradiated pepper samples. Similar work on peppers also highlighted an analogous trend of delayed post-harvest damages and reduced chilling injury (Vicente et al., 2005). The decrease in microbial load on increasing power of exposure to UV-C treatment in hot peppers is also observed (Hassan et al., 2020), and so is the potential of visible light on the effect on the microbial population (Angarano et al., 2020). However, there are limited works available on its application on peppers. The mode of inactivation is linked with oxidative reactive species formation as a result of a broad spectrum of light irradiation. These reactive species can damage the vital cell components, resulting inactivation. Mostly LED sources are used to produce visible light of particular power and wavelength.

The impact of UV-C and blue LED light on reducing aflatoxin in red chillies has been explored (Kalathil et al., 2023). The authors reported a 15.8% and 42.1% reduction in total aflatoxin for UV-C and blue LED (470 nm) treatments, respectively. In recent days, visible light is used in photosensitization treatments in which certain bioactive compounds can interact and inactivate microorganisms. These bioactive compounds (photosensitizers) can produce reactive species on illumination (Rodoni and Lemoine, 2023).

15.3.7 Ultrasound

Ultrasound is a widely used non-thermal technology and finds application in emulsification, degassing, defoaming, extraction, cleaning and sanitization. Ultrasound waves are mechanical energy waves of frequency above 20 kHz. Generally, waves of frequency range 18 to 100 kHz and energy range of 10 to 1000 W/cm² are used. Ultrasonic wave produces a set of compression and rarefactions cycle inside the medium, leading to cavitation (Figure 15.9). This facilitates localized increase in pressure (up to 50 Mpa) and temperature (up to 550°C), which thins the cell membrane and damages cell structure. Alexandre et al. (2013) reported a reduction in artificially inoculated *L. innocua* on red bell pepper after ultrasound treatment (35 kHz and 120 W) using water.

15.3.8 Other Novel Technologies

So far, several non-thermal technologies have been discussed to decontaminate microorganisms. Traditionally, water activity of the produce is reduced to augment the shelf life by dehydration or

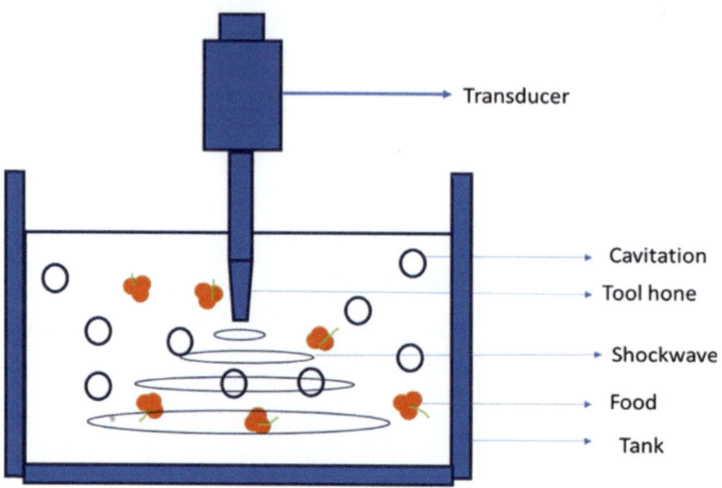

FIGURE 15.9 Schematic diagram of ultrasonic processing.

addition of hydrocolloids. Recently, efforts were made to reduce the water activity by non-thermal means. This can be achieved by locking water molecules and blocking the availability of free water for the growth and survival of microorganisms. The main beneficial part of the technique is the retention of sensory attributes of food, as no additional substance or residue is added or removed. One such technology is the usage of pressurized non-polar gases in the preservation of food. Pressured oxygen, nitrogen or inert gases are generally used to inactivate or control microbial growth (Shen et al., 2019; Fujiwara et al., 2019). Even though this technique finds application in various food preservation, its application is limited to pressured argon in pepper (Meng et al., 2012). The pressurized argon is reported to delay the microbial growth, reduce physiological changes and retain firmness by controlling water loss. All these benefits are achieved by reducing water activity (Wu et al., 2012). The clathrate hydrate of non-polar gases is formed at higher temperature and pressure. Clathrate hydrate is a cagelike molecule which can entrap other molecules. Water based clathrate hydrate assists in restricting the mobility of water and lowering the water activity. This could reduce the microbial growth. Pressurized argon (2 to 6 MPa) on green peppers has shown a significant delay in microbial growth, water loss and enzyme activity. The work also demonstrated 4 MPa of argon gas to be the most effective for reduction in microbial and enzyme activity (Meng et al., 2012).

15.3.9 SYNERGISTIC TECHNOLOGIES

Synergistic technologies combine non-thermal or thermal technologies to achieve desired lethality in lesser time. The challenges of developing ideal combinations, considering the type, matrix (absorbent or non-absorbent), water activity, storage conditions and nutrients, are critical in deciding the effectiveness of the outcome. Besides the target microorganism, microbial load is also considered while developing efficient combination. Pulse light plasma is a combination of PL and plasma and is a synergistic technology that shows promising results in reducing the microbial load with minimal impact on nutrients. It is produced when food is exposed to CP simultaneously under intense pulses of high-energy light. In red pepper flakes, it has been reported to efficiently reduce artificially inoculated spores of *Aspergillus flavus*, *Escherichia coli* O157:H7 and *Bacillus pumilus* without any significant changes in moisture content or colour (Nicorescu et al., 2013). Another study discussed the reduction of indigenous microbial load on red pepper powder with pulsed light plasma (Lee et al., 2020). The study inferred that efficacy of pulse light plasma is influenced by water activity of sample. The treatment also showed no significant reduction in colour, antioxidant activity, ascorbic acid and capsaicin content, with a nominal impact on volatiles. The combination CP and microwave treatment on red pepper flakes has also shown a significant reduction in artificially inoculated *B. cereus spores* with nominal impact on colour. Samples with lower surface-to-volume ratio and higher water activity gave better results. The efficiency of inactivation was also increased by increasing microwave power (Kim et al., 2017). The same group observed similar results when microwave and cold plasma treatment were used to inactivate *B. cereus* and *A. flavus* spores in red pepper flakes. Microwave power positively impacted inactivation efficiency (Kim et al., 2019). A synergistic effect was also observed when CP was combined with mild heating. Results evinced a reduction in artificially inoculated *Bacillus cereus* in red pepper powder (Jeon et al., 2020).

Furthermore, combining UV-C treatment and mild heating (65°C) showed better reduction of artificially inoculated *Escherichia coli* O157:H7 and *Salmonella typhimurium* on red pepper powder than using individual treatments. However, nominal yet significant impact on colour was observed (Cheon et al., 2015). Similarly, lethality was enhanced when near-infrared and UV (Ha and Kang, 2013), gamma irradiation and sodium dichloroisocyanurate (Yoon et al., 2014) were used in combination. In contrast, an antagonist effect was observed when blanching was combined with aqueous ozone treatment, pulsed light plasma and UV radiation (Alexandre et al., 2011).

15.4   EFFECT OF NON-THERMAL PROCESS ON PAPRIKA CONSTITUENTS

15.4.1   PROXIMATE COMPOSITION

The proximate composition of *Capsicum* is nominally affected during non-thermal processing (Table 15.2). Numerous authors substantiated that the moisture loss is insignificant at lower intensities, whereas the loss becomes significant at higher intensities (Woldemariam et al., 2022; Kim et al., 2019; Balakrishnan et al., 2022; Cheon et al., 2015). This loss could be corresponding to the increased permeability of cell membranes caused by non-thermal technologies. It is important to understand that the magnitude of moisture loss during non-thermal treatments is extremely low compared to thermal treatments (Rybak et al., 2022). However, the impact of pressurized argon treatment on moisture content was less than other non-thermal technologies (Meng et al., 2012). In fact, the study evinced that pressurized argon treatment tends to reduce the water loss during storage. This could be ascribed to the change in water distribution in the pepper matrix, thereby confining the water mobility (Meng et al., 2012). Similarly, sugar content was unaffected by various non thermal technologies such as ultrasound and PEF. Nevertheless, the sugar content was drastically reduced by thermal blanching (40% reduction) and combined use of ultrasound and PEF (50% reduction) (Rybak et al., 2020). The extreme decrease in combined treatment could be due to the synergistic action of electroporation and cavitation, which could facilitate sugar degradation (Han et al., 2012). Correspondingly, the pH of pepper was unaffected by UV, blue LED, ultrasonic and ozone treatments (Alexandre et al., 2013). Meanwhile, a reduction in pH was observed for CP-treated pepper. The hydronium ions produced from reactive species of nitrogen contained in plasma could form mineral acids (Kalathil et al., 2023). A similar impact could also be seen with water activity of pepper samples (Kim et al., 2019). Studies also highlighted that the efficacy of several non-thermal treatments is better at higher water activity (Lee et al., 2020; Rifna et al., 2019). Additionally, the protein content of red pepper reduced by 45% on HPP compared to 60% in thermal blanching (Castro et al., 2008). A reduction in volatile oil content was also observed after non-thermal treatment (UV, blue LED, CP) of dried pepper (Kalathil et al., 2023).

TABLE 15.2

Impact of Non-Thermal Treatments on Pepper Constituents

Technology	Parameters	Product	Key findings on pepper constituents	Reference
HPP	100 to 600 MPa and holding time of 30–600 s	Red pepper paste	• Marginal and non-significant increase in carotenoids • Nominal increment in antioxidant activity with increased in pressure • Nominal increment in antioxidant activity with increased in pressure	(Woldemariam et al., 2022)
Thermal blanching and HPP	HPP: 100–200 Mpa for 10–20 min; thermal blanching: 70–98°C for 1–2.5 min	Fresh red chilli	• The protein content reduced up to 60% and 45% for thermal blanching and HPP, respectively • Vitamin C decreased with process time for thermal blanching (45% reduction) as well as HPP (20% reduction)	(Castro et al., 2008)
PEF	Field strength: 0.5–2.5, No. of pulse: 20, pulse duration: 400 μs	Fresh red chilli	• Decrease in vitamin C and carotenoids with temperature	(Ade-Omowaye et al., 2002)
Ultrasonic treatment, PEF, combination of non-thermal with thermal blanching	Pulse no.: 6–12; field density: 1.07 kV/cm, specific energy: 1–3 kJ/kg; ultrasound intensity: 3 W/cm2, frequency: 21 kHz, time: 30 min; thermal blanching: 98°C for 3 min	Red bell pepper	• Carotenoids degraded up to 24% during steam blanching, nominal reduction was observed for ultrasound and PET did not show significant degradation • Vitamin C degraded up to 46% by convention thermal blanching, 15% degradation was observed for ultrasound and PET did not show a significant degradation • 13% reduction in polyphenols by thermal blanching, PET degraded polyphenols by 7% and ultrasound did not degrade • Reduction in sugar content with thermal treatment and no significant difference in for non-thermal samples	(Rybak et al., 2020)
PL	Voltage: 2.1 kV, no. of pulse: 7–53; fluence: 4–32 J/cm2	Fresh cut red bell pepper	• No significant difference in vitamin C • Nominal but insignificant decrease in polyphenol content with fluence • Decrease in carotenoid content with fluence • Antioxidant activity nominally increased with fluence	(Rybak et al., 2021)
CP	750 W @20 kHz for 15–60 s each side	Dried red pepper	• Insignificant colour difference • Initial increase followed by decrease in carotenoids • Increase in antioxidant activity	(Zhang et al., 2019)
Mild Heating and CP	Heat treatment: 60°C for 5–20 min; CP: 1 kV, 43 kHz for 5–20 min	Powdered red pepper	• No significant difference in pH and colour values	(Jeon et al., 2020)

(*Continued*)

TABLE 15.2 (*Continued*)

Impact of Non-Thermal Treatments on Pepper Constituents

Technology	Parameters	Product	Key findings on pepper constituents	Reference
Electron beam and gamma ray irradiation affect	EB: 3 kGy and dose rate of 1 and 5 kGy/s; GR: 3 kGy and dose rate of 1.8–9 kGy/h	Dried red pepper powder	• No significant difference in capsanthin and colour values for ER samples. • A significant change in extractable colour was observed for high dose GR treated samples.	(Kyung et al., 2019)
Gamma Irradiation	2–6 kGy	Dried chilli	• Moisture change was insignificant in lower doses, 1% reduction was observed at higher dose • No decrease in ascorbic acid, capsaicinoids and carotenoids	(Balakrishnan et al., 2022)
Gamma irradiation	Dose rate of 10 kGy/h, 1–5 kGy	Dried chilli	• No significant difference in carotenoids	(Song et al., 2014)
Ultraviolet, ultrasonic, ozone, sanitizer solution treatments	UV: 254 nm, average intensity: 12.36 W m−2; US: 35 kHz and 120 W; ozone: 0.3 mg/l; sanitizer: NaOCl, 200 µg/ml or H_2O_2,: 1%—5% w/w for 2min	Red bell peppers	• Lesser impact on colour for non-thermal treatments than chemical solution • No impact on firmness and pH • Ascorbic acid reduced after treatment (12% for chemical solution and 6% for non-thermal process)	(Alexandre et al., 2013)
Near-infrared radiant heating and UV radiation	NIR: 500 W, UV: 16W for 3 min	Powdered red pepper	• No significant difference in colour	(Ha and Kang, 2013)
UV-C and thermal	UV: Intensity 3.40 mW/cm^2 for 5–10 min at 25–65°C	Powdered red pepper	• No significant difference in colour • Moisture content reduced at higher temperatures	(Cheon et al., 2015)
Ultraviolet, blue LED and CP	UV: 15 W for 30–90 min; CP: 0–100 kV, 0–50 mA for 5–15 min	Dried red pepper	• Reduction in moisture and water activity with exposure time • Reduction in pH for cold plasma–treated samples • Reduction in volatile oil content • Inverse effect on colour and carotenoid content • Initial increase followed by decrease in capsaicinoids content for UV and CP • No impact of blue LED on the capsaicinoid content	(Kalathil et al., 2023)
Pressurized argon treatment	4–6 MPa for 1h	Green peppers	• Reduced the water loss during storage • Reduction in ascorbic acid and chlorophyll with pressure	(Meng et al., 2012)
Pulsed light plasma	CP: 0–10 kV and 6.4–40.0 kHz; PL: 30 kV for 1–6.3 min	Powdered red pepper	• No reduction in colour, ascorbic acid and carotenoid content • Significant difference in flavour profile	(Lee et al., 2020)
Pulsed light plasma	CP: 8–10 kV and 15kHz; IPL: 2.0 kV, 0.8–3.8 Hz, and 6.3 min)	Red pepper flakes	Lower efficiency at lower a_w	(Lee et al., 2020a, b)

(Continued)

TABLE 15.2 (*Continued*)
Impact of Non-Thermal Treatments on Pepper Constituents

Technology	Parameters	Product	Key findings on pepper constituents	Reference
Microwave-combined CP	MW: 900 W, 677 Pa, 20 min; CP: 50–1000 W, He: O_2 = 99.8:0.2	Red pepper flakes	• No significant difference in capsaicinoids in low and medium MW • Capsaicinoids reduced significantly with storage • Reduction in ascorbic acid up to 25% • No significant difference in water activity and colour • Reduced oxidation of carotenoids during storage	(Kim et al., 2019)

HPP: high-pressure processing, CP: cold plasma, PEF: pulse electric field, PL: pulse light, UV: ultraviolet

15.4.1.1 Vitamins

Vitamin C is highly unstable and is prone to oxidation under thermal as well as non-thermal conditions. A reduction of up to 45% was observed during thermal blanching of fresh pepper (Castro et al., 2008). Chemical treatments such as sodium hypochlorite and hydrogen peroxide treatments could limit the degradation to 12% (Alexandre et al., 2013). Recently, several non-thermal technologies such as CP, ultrasound (Kim et al., 2019; Rybak et al., 2020), PEF (Ade-Omowaye et al., 2002), HPP (Castro et al., 2008), ozone (Alexandre et al., 2013) and pressurized argon treatments (Meng et al., 2012) evinced a reduction in vitamin C up to 25%, 20%, 15% and 6% for CP, HPP, ultrasound and ozone treatments, respectively. PEF (Rybak et al., 2020) and pressurized argon–treated (Meng et al., 2012) samples showed a marginal reduction in vitamin C. CP-, ultrasound- and PEF-treated samples did show an initial increase in ascorbic acid followed by degradation. The initial increase could correspond to increased extractability of vitamin C from the samples. Once the ascorbic acid is displaced from the cellular matrix, they undergo rapid oxidation (Rybak et al., 2020; Ade-Omowaye et al., 2002). Nevertheless, certain non-thermal technologies including PL (Rybak et al., 2021), pulsed-light plasma (Lee et al., 2020) and gamma irradiation treatment (Balakrishnan et al., 2022) showed no significant impact on vitamin C.

Apart from vitamin C, red pepper is also rich in vitamins A, B and E. However, literature on the impact of non-thermal treatments on these vitamins is scant.

15.4.2 POLYPHENOLS

Polyphenols are water-soluble and heat-labile phytochemicals that are known for their antioxidant activity. Thermal treatment and storage cause a detrimental effect on polyphenols due to thermal, enzyme-induced oxidation and leaching (blanching). Thermally blanched red peppers experience a 12% degradation in total polyphenol content, while ultrasound treatment could limit the same to 7% (Rybak et al., 2020). Equipping non-thermal technologies such as PL and PEF would ensure minimal degradation of polyphenols (Rybak et al., 2021, 2020). On the contrary, HPP increased the polyphenol content in red pepper paste (Woldemariam et al., 2022); this has been explained to be due to increased extractability following improved cell permeability. Correspondingly, an increase in antioxidant activity could be observed for HPP- and PL-treated samples (Rybak et al., 2021; Woldemariam et al., 2022). These non-thermal technologies also significantly reduced the polyphenol degradation during storage (Kim et al., 2019; Meng et al., 2012), which is due to denaturation of polyphenol oxidase and peroxidase enzymes.

15.4.3 Carotenoids

Carotenoids are photosensitive and thermally labile components responsible for the colour of pepper. Steam blanching of red bell pepper causes 24% of carotenoids to degrade (Rybak et al., 2020). Colour is one of the important quality parameters which determine the price of pepper. Thus, degradation of carotenoids during processing needs to be controlled, and several non-thermal technologies are used as alternatives. However, a minimal yet significant reduction could be observed while using UV, blue LED and PL treatment due to photoinduced oxidation of carotenoids (Kalathil et al., 2023; Cheon et al., 2015; Alexandre et al., 2013). Carotenoid degradation could be further reduced by using CP, PEF, UV and ozone treatment. A nominal reduction could be observed while using these technologies (Rybak et al., 2020; Ade-Omowaye et al., 2002; Zhang et al., 2019). These studies unanimously reported an initial increase in carotenoid content followed by a decrease. This could be due to the higher extractability followed by oxidation of polyphenols by free radicals produced during CP and cavitation, wherein HPP, gamma and electron beam irradiation, pulse light plasma treatments did not influence the carotenoid content (Woldemariam et al., 2022; Kyung, et al., 2019). Besides the reduction in oxidation during processing, these non-thermal technologies are beneficial in reducing the degradation during storage (Kim et al., 2019).

15.4.4 Capsaicinoids

Capsaicinoids are alkaloids responsible for pungency. They undergo thermal degradation and follow first-order kinetics. Results from a study evaluating the impact of various cooking methods on capsaicinoids reveal up to 30% reduction (Bustamante et al., 2022). Non-thermal technologies Such as UV (Kalathil et al., 2023) and CP (Kim et al., 2019) are known to reduce the loss of capsaicinoids. The capsaicinoid degradation was nominal and insignificant at lower and moderate intensities but did reduce at higher intensities of CP and UV radiation. Meanwhile, gamma, electron beam irradiation and blue LED treatment did not affect the capsaicin content (Kim et al., 2019; Balakrishnan et al., 2022).

15.5 FOOD SAFETY AND REGULATORY CHALLENGES

Aflatoxins are carcinogenic fungal metabolites that have chronic adverse effects such as immune-suppressing effects and impairing child growth (Ayob et al., 2022). Because chillies are susceptible to fungal growth, aflatoxin is a major contaminant. Several incidents of higher aflatoxin content in chillies urged regulatory authorities around the globe to have multiple checks and strict regulations. The European Commission has set maximum permissible limit for aflatoxin B1 and total aflatoxin to 5 and 10 µg/kg for chillies (whole or ground) and chilli powder (Petrini, 2009). The maximum residual limit of total aflatoxin limit for India, Singapore, Pakistan, Switzerland, Sri Lanka, Japan, Thailand, Turkey and the United States is 30 µg/kg, 5 µg/kg, 30 µg/kg, 10 µg/kg, c30 µg/kg, 10 µg/kg, 20 µg/kg, 10 µg/kg and 20 µg/kg, respectively (FAO/WHO, 2017).

The chief advantages of non-thermal technologies are their potential to reduce toxins below the permissible levels. However, regulatory approval from different national authorities is required for the commercial use of these technologies. Several countries have initiated the process. One such initiative from the Indian regulatory authority was implementing food safety and standards (approval for non-specific food and food ingredients) regulation, 2017 (FSSAI, 2017). This regulation details provision for manufacturing and marketing of products manufactured using novel technology through a separate process of approval from regulatory authority under non-specified food and food ingredient category.

A widely accepted non-thermal processing for microbial decontamination is irradiation. The Codex general standard for irradiated food specifies the requirement of license for establishments involved in irradiation of food, labelling requirements [facility details, dosage, treatment date, Radura symbol (optional) and lot identification] and the restriction on maximum absorbed dosage to 10 kGy for food (Codex Alimentarius Commission, 2003). In India, all irradiated foods shall carry the Radura symbol along with the purpose of irradiation. The permissible dosage for microbial decontamination in spices is 6 to 14 kGy (FSSAI, 2017).

15.6 FUTURE PROSPECTS

The quality and nutraceutical value of chillies are dependent on the bioactive components. These components are highly unstable, and appropriate considerations should be taken for selection and optimization of post-harvest operations. The non-thermal treatments are capable of a similar lethality as thermal treatments with minimal impact on these bioactives. This notion could be substantiated by voluminous literature on non-thermal processing of chillies. However, certain aspects are yet to be explored, and further studies would enable a better understanding of the viability of these technologies at industrial level. Further research on areas identified as follows would fill the research gap and would provide a conclusive overview of non-thermal pepper processing:

- The impact of non-thermal technologies on vitamins A, B and E during pepper processing
- The effect on chillies' essential oil composition and the losses associated with non-thermal processing
- Pilot scale studies on non-thermal technologies (PET, PL, CP)
- Further investigations on high-intensity visible lights including blue LED for disinfection of chillies
- Exploring the use of photosensitizers to enhance the efficiency of non-ionic irradiation and PL during chilli processing
- Extensive toxicological evaluation on the by-product(s) formed during non-thermal processing
- Techno-economic evaluation of non-thermal technologies for pepper preservation
- Pressurized nitrogen and oxygen treatments for increasing the shelf life of peppers
- Extensive evaluation on application of non-thermal technologies as pre-treatment before extraction and drying
- Studies on preservation of indigenous pepper species by non-thermal treatments

15.7 CONCLUSIONS

Post-harvest handling is the foremost area where controls can be applied to narrow down the losses. Understanding the chemistry, biochemistry and microbiology of peppers is critical in identifying the best preservation technique. It could be inferred from literature that non-thermal technologies are superior to thermal technologies in conserving the bioactive compounds while delivering a similar lethality. Technologies such as gamma irradiation, PL, combined thermal and non-thermal treatments, blue LED and pulse light plasma shows promising results. CP, PEF, ultrasonic, HPP and pressurized argon treatments showed nominal impact on pepper constituents. The degradation of bioactives is limited at lower and moderate intensities, wherein high-intensity treatments have detrimental effects. Therefore, optimization of process parameters is required for better results, thereby assuring a high-quality product with extended shelf life. Besides high initial investment, delay in regulatory approval and product-/variety-based optimization for commercialization are hindering the applications of non-thermal technologies at commercial scale.

REFERENCES

Abdi, S., Hosseini, A., Moslehishad M., & Dorranian, D. (2019). Decontamination of red pepper using cold atmospheric pressure plasma as alternative technique. *Applied Food Biotechnology*, 6 (4), 247–254. https://doi.org/10.22037/afb.v6i4.26002.

Ade-Omowaye, B. I. O., Rastogi, N. K., Angersbach, A., & Knorr, D. (2002). Osmotic dehydration of bell peppers: Influence of high intensity electric field pulses and elevated temperature treatment. *Journal of Food Engineering*, 54 (1), 35–43. https://doi.org/10.1016/S0260-8774(01)00183-2.

Alexandre, E. M. C., Brandão, T. R. S., & Silva, C. L. M. (2013). Impact of non-thermal technologies and sanitizer solutions on microbial load reduction and quality factor retention of frozen red bell peppers. *Innovative Food Science & Emerging Technologies*, 17, 99–105. https://doi.org/10.1016/j.ifset.2012.11.009.

Alexandre, E. M. C., Santos-Pedro, D. M., Brandão, T. R. S., & Silva, C. L. M. (2011). Influence of aqueous ozone, blanching and combined treatments on microbial load of red bell peppers, strawberries and watercress. *Journal of Food Engineering*, 105 (2), 277–282. https://doi.org/10.1016/j.jfoodeng.2011.02.032.

Angarano, V., Smet, C., Akkermans, S., Watt, C., Chieffi, A., & Van Impe, J. F. M. (2020). Visible light as an antimicrobial strategy for inactivation of *Pseudomonas fluorescens* and *Staphylococcus epidermidis* biofilms. *Antibiotics*, 9 (4), 171–193. https://doi.org/10.3390/antibiotics9040171.

Arimboor, R., Natarajan, R. B., Menon, K. R., Chandrasekhar, L. P., & Moorkoth, V. (2015). Red pepper (*Capsicum annuum*) carotenoids as a source of natural food colors: Analysis and stability—a review. *Journal of Food Science and Technology*, 52 (3), 1258–1271. https://doi.org/10.1007/s13197-014-1260-7.

Ayob, O., Hussain, P. R., Naqash, F., Riyaz, L., Kausar, T., Joshi, S., & Azad, Z. R. A. A. (2022). Aflatoxins: Occurrence in red chilli and control by gamma irradiation. *International Journal of Food Science and Technology*, 57 (4), 2149–2158. https://doi.org/10.1111/ijfs.15088.

Balakrishnan, N., Yusop, S. M., Rahman, I. A., Dauqan, E., & Abdullah, A. (2022). Efficacy of gamma irradiation in improving the microbial and physical quality properties of dried chillies (*Capsicum annuum* L.): A review. *Foods*, 11 (1), 1–18. https://doi.org/10.3390/foods11010091.

Barbagallo, R. N., Chisari, M., & Patané, C. (2012). Polyphenol oxidase, total phenolics and ascorbic acid changes during storage of minimally processed 'California Wonder' and 'Quadrato d'Asti' sweet peppers. *LWT—Food Science and Technology*, 49 (2), 192–196. https://doi.org/10.1016/j.lwt.2012.06.023.

Bustamante, K., Guajardo, J. I. A., & Cahill, T. (2022). Thermal degradation of capsaicin and dihydrocapsaicin during cooking. *Journal of the Arizona-Nevada Academy of Science*, 49 (2), 99–108. https://doi.org/10.2181/036.049.0207.

Byun, K.-H., Cho, M.-J., Park, S.-Y., Chun, H.-S., & Ha, S.-D. (2019). Effects of gamma ray, electron beam, and X-ray on the reduction of *Aspergillus flavus* on red pepper powder (*Capsicum annuum* L.) and *gochujang* (red pepper paste). *Food Science and Technology International*, 25 (8), 649–658. https://doi.org/10.1177/1082013219857019.

Castro, S. M., Saraiva, J. A., Lopes-da-Silva, J. A., Delgadillo, I., Loey, A. V., Smout, C., & Hendrickx, M. (2008). Effect of thermal blanching and of high pressure treatments on sweet green and red bell pepper fruits (*Capsicum annuum* L.). *Food Chemistry*, 107 (4), 1436–1449. https://doi.org/10.1016/j.foodchem.2007.09.074.

Cheon, H.-L., Shin, J.-Y., Park, K.-H., Chung, M.-S., & Kang, D.-H. (2015). Inactivation of food borne pathogens in powdered red pepper (*Capsicum annuum* L.) using combined UV-C irradiation and mild heat treatment. *Food Control*, 50, 441–445. https://doi.org/10.1016/j.foodcont.2014.08.025.

Codex Alimentarius Commission. (2003). Codex general standard for irradiated foods. *Codex Standard*, 1–3.

Epelle, E. I., Macfarlane, A., Cusack, M., Burns, A., Okolie, J. A., Mackay, W., Rateb, M., & Yaseen, M. (2023). Ozone application in different industries: A review of recent developments. *Chemical Engineering Journal*, 454, 140188–140208. https://doi.org/10.1016/j.cej.2022.140188.

Evrendilek, G. A., Bulut, N., Atmaca, B., & Uzuner, S. (2022). Prediction of *Aspergillus parasiticus* inhibition and aflatoxin mitigation in red pepper flakes treated by pulsed electric field treatment using machine learning and neural networks. *Food Research International*, 162, 111954–111963. https://doi.org/10.1016/j.foodres.2022.111954.

FAO. (2023a). FAOSTAT. *Crops and Livestock Products*. https://www.fao.org/faostat/en/#data/QCL/visualize.

FAO. (2023b). FAOSTAT. *Food Balances*. https://www.fao.org/faostat/en/#data/FBS.

FAO/WHO. (2017). Discussion paper on the establishment maximum levels for mycotoxins in spices. *Joint FAO/WHO Food Standards Programme: Codex Committee on Contaminants in Foods*, 1–24. https://www.fao.org/fao-who-codexalimentarius/sh-proxy/en/?lnk=1&url=https%253A%252F%252Fworkspace.fao.org%252F52Fsites%252Fcodex%252FMeetings%252FCX-735-11%252FREPORT%252FREP17_CFe.pdf.

FSSAI. (2017). Approval of non-specified food and food ingredients. *The Gazette of India*. https://fssai.gov.in/upload/uploadfiles/files/Gazette_Notification_NonSpecified_Food_Ingredients_15_09_2017.pdf.

Fujiwara, A., Hatayama, N., Matsuura, N., Yokota, N., Fukushige, K., Yakura, T., & Yokomise, H. (2019). High-pressure carbon monoxide and oxygen mixture is effective for lung preservation. *International Journal of Molecular Sciences*, 20 (11), 2719–2730. https://doi.org/10.3390/ijms20112719.

Gómez-García, M., & Ochoa-Alejo, N. (2013). Biochemistry and molecular biology of carotenoid biosynthesis in chili peppers (*Capsicum* spp.). *International Journal of Molecular Sciences*, 14 (9), 19025–19053. https://doi.org/10.3390/ijms140919025.

Ha, J.-W., & Kang, D.-H. (2013). Simultaneous near-infrared radiant heating and UV radiation for inactivating *Escherichia coli* O157: H7 and *Salmonella enterica* serovar typhimuriumin powdered red pepper (*Capsicum annuum* L.). *Applied and Environmental Microbiology*, 79 (21), 6568–6575. https://doi.org/10.1128/AEM.02249-13.

Hamed, M., Kalita, D., Bartolo, M. E., & Jayanty, S. S. (2019). Capsaicinoids, polyphenols and antioxidant activities of *Capsicum annuum*: Comparative study of the effect of ripening stage and cooking methods. *Antioxidants*, 8 (9), 364–382. https://doi.org/10.3390/antiox8090364.

Han, Z., Zeng, X. A., Fu, N., Yu, S. J., Chen, X. D., & Kennedy, J. F. (2012). Effects of pulsed electric field treatments on some properties of tapioca starch. *Carbohydrate Polymers*, 89 (4), 1012–1017. https://doi.org/10.1016/j.carbpol.2012.02.053.

Hassan, A. B., Al Maiman, S. A., Sir Elkhatim, K. A., Elbadr, N. A., Alsulaim, S., Osman, M. A., & Mohamed Ahmed, I. A. (2020). Effect of UV-C radiation treatment on microbial load and antioxidant capacity in hot pepper, fennel and coriander. *LWT—Food Science and Technology*, 134, 109946–109953. https://doi.org/10.1016/j.lwt.2020.109946.

Hernández-Pérez, T., del Rocío Gómez-García, M., Valverde, M. E., & Paredes-López, O. (2020). *Capsicum annuum* (hot pepper): An ancient Latin-American crop with outstanding bioactive compounds and nutraceutical potential. A review. *Comprehensive Reviews in Food Science and Food Safety*, 19 (6), 2972–2793. https://doi.org/10.1111/1541-4337.12634.

Horvitz, S., & Cantalejo, M. J. (2012). Effects of ozone and chlorine postharvest treatments on quality of fresh-cut red bell peppers. *International Journal of Food Science & Technology*, 47 (9), 1935–1943. https://doi.org/10.1111/j.1365-2621.2012.03053.x.

Jalgaonkar, K., Mahawar, M. K., Girijal, S., & Geeta, H. P. (2022). Post-harvest profile, processing and value addition of dried red chillies (*Capsicum annum* L.). *Journal of Food Science and Technology*, no. 0123456789, 1–19. https://doi.org/10.1007/s13197-022-05656-1.

Jeon, E. B., Choi, M.-S., Kim, J. Y., & Park, S. Y. (2020). Synergistic effects of mild heating and dielectric barrier discharge plasma on the reduction of *Bacillus cereus* in red pepper powder. *Foods*, 9 (2), 171–181. https://doi.org/10.3390/foods9020171.

Jung, K., Song, B.-S., Kim, M. J., Moon, B.-G., Go, S.-M., Kim, J.-K., Lee, Y.-J., & Park, J.-H. (2015). Effect of X-ray, gamma ray, and electron beam irradiation on the hygienic and physicochemical qualities of red pepper powder. *LWT—Food Science and Technology*, 63 (2), 846–851. https://doi.org/10.1016/j.lwt.2015.04.030.

Kalathil, N., Thirunavookarasu, N., Lakshmipathy, K., Chidanand, D. V., Radhakrishnan, M., & Baskaran, N. (2023). Application of light based, non-thermal techniques to determine physico-chemical characteristics, pungency and aflatoxin levels of dried red chilli pods (*Capsicum annuum*). *Journal of Agriculture and Food Research*, 13, 100648–100657. https://doi.org/10.1016/j.jafr.2023.100648.

Kasampalis, D. S., Tsouvaltzis, P., Ntouros, K., Gertsis, A., Gitas, I., Moshou, D., & Siomos, A. S. (2022). Nutritional composition changes in bell pepper as affected by the ripening stage of fruits at harvest or postharvest storage and assessed non-destructively. *Journal of the Science of Food and Agriculture*, 102 (1), 445–454. https://doi.org/10.1002/jsfa.11375.

Kennedy, L. E., Abraham, A., Kulkarni, G., Shettigar, N., Dave, T., & Kulkarni, M. (2021). Capsanthin, a plant-derived xanthophyll: A review of pharmacology and delivery strategies. *AAPS PharmSciTech*, 22 (5), 203. https://doi.org/10.1208/s12249-021-02065-z.

Kim, J. E., Choi, H. S., Lee, D.-U., & Min, S. C. (2017). Effects of processing parameters on the inactivation of *Bacillus cereus* spores on red pepper (*Capsicum annum* L.) flakes by microwave-combined cold plasma treatment. *International Journal of Food Microbiology*, 263, 61–66. https://doi.org/10.1016/j.ijfoodmicro.2017.09.014.

Kim, J. E., Lee, D.-U., & Min, S. C. (2014). Microbial decontamination of red pepper powder by cold plasma. *Food Microbiology*, 38, 128–136. https://doi.org/10.1016/j.fm.2013.08.019.

Kim, J. E., Oh, Y. J., Song, A. Y., & Min, S. C. (2019). Preservation of red pepper flakes using microwave-combined cold plasma treatment. *Journal of the Science of Food and Agriculture*, 99 (4), 1577–1585. https://doi.org/10.1002/jsfa.9336.

Krajayklang, M., Klieber, A., & Dry, P. R. (2000). Colour at harvest and post-harvest behaviour influence paprika and chilli spice quality. *Postharvest Biology and Technology*, 20 (3), 269–278. https://doi.org/10.1016/S0925-5214(00)00141-1.

Kyung, H. K., Ramakrishnan, S. R., & Kwon, J. H. (2019). Dose rates of electron beam and gamma ray irradiation affect microbial decontamination and quality changes in dried red pepper (*Capsicum annuum* L.) powder. *Journal of the Science of Food and Agriculture*, 99 (2), 632–638. https://doi.org/10.1002/jsfa.9225.

Lee, H. S., Park, H. H., & Min, S. C. (2020a). Microbial decontamination of red pepper powder using pulsed light plasma. *Journal of Food Engineering*, 284, 110075–110084. https://doi.org/10.1016/j.jfoodeng.2020.110075.

Lee, S. Y., Park, H. H., & Min, S. C. (2020b). Pulsed light plasma treatment for the inactivation of *Aspergillus flavus* spores, *Bacillus pumilus* spores, and *Escherichia coli*O157:H7 in red pepper flakes. *Food Control*, 118, 107401. https://doi.org/10.1016/j.foodcont.2020.107401.

Maalekuu, K., Elkind, Y., Leikin-Frenkel, A., Lurie, S., & Fallik, E. (2006). The relationship between water loss, lipid content, membrane integrity and LOX activity in ripe pepper fruit after storage. *Postharvest Biology and Technology*, 42 (3), 248–255. https://doi.org/10.1016/j.postharvbio.2006.06.012.

Martínez, S., Curros, A., Bermúdez, J., Carballo, J., & Franco, I. (2007). The composition of Arnoia peppers (*Capsicum annuum* L.) at different stages of maturity. *International Journal of Food Sciences and Nutrition*, 58 (2), 150–161. https://doi.org/10.1080/09637480601154095.

Meng, X., Zhang, M., & Adhikari, B. (2012). Extending shelf-life of fresh-cut green peppers using pressurized argon treatment. *Postharvest Biology and Technology*, 71, 13–20. https://doi.org/10.1016/j.postharvbio.2012.04.006.

Ministry of Agriculture and Cooperation. (2009). *Post Harvest Profile of Chilli*. Government of India. http://www.credall.org.in/images/npvol07.pdf.

Montoya-Ballesteros, L. C., González-León, A., García-Alvarado, M. A., & Rodríguez-Jimenes, G. C. (2014). Bioactive compounds during drying of chili peppers. *Drying Technology*, 32 (12), 1486–1499. https://doi.org/10.1080/07373937.2014.902381.

Mózsik, G. (2016). It remains a mystery why people living in hot climates consume spicier food. *Temperature*, 3 (1), 50–51. https://doi.org/10.1080/23328940.2015.1131033.

Naves, E. R., de Ávila Silva, L., Sulpice, R., Araújo, W. L., Nunes-Nesi, A., Peres, L. E. P., & Zsögön, A. (2019). Capsaicinoids: Pungency beyond *Capsicum*. *Trends in Plant Science* 24 (2), 109–120. https://doi.org/10.1016/j.tplants.2018.11.001.

Nicorescu, I., Nguyen, B., Moreau-Ferret, M., Agoulon, A., Chevalier, S., & Orange, N. (2013). Pulsed light inactivation of *Bacillus subtilis* vegetative cells in suspensions and spices. *Food Control*, 31 (1), 151–157. https://doi.org/10.1016/j.foodcont.2012.09.047.

Obayelu, O. A., Adegboyega, O. M., Sowunmi, F. A., & Idiaye, C. O. (2021). Factors explaining postharvest loss of hot pepper under tropical conditions. *International Journal of Vegetable Science*, 27 (6), 526–535. https://doi.org/10.1080/19315260.2021.1879342.

O'Donoghue, E. M., Brummell, D. A., McKenzie, M. J., Hunter, D. A., & Lill, R. E. (2018). Sweet *Capsicum*: Postharvest physiology and technologies. *New Zealand Journal of Crop and Horticultural Science*, 46 (4), 269–297. https://doi.org/10.1080/01140671.2017.1395349.

Oms-Oliu, G., Martín-Belloso, O., & Soliva-Fortuny, R. (2010). Pulsed light treatments for food preservation. A review. *Food and Bioprocess Technology*, 3 (1), 13–23. https://doi.org/10.1007/s11947-008-0147-x.

Petrini, O. (2009). Scientific reasons behind European commission legislation no. 1881/2006 on heavy metals in food. *Pagine Di Micologia*, 32, 119–127.

Po, L. G., Siddiq, M., & Shahzad, T. (2018). Chili, peppers, and paprika. In *Handbook of Vegetables and Vegetable Processing* (pp. 633–660). John Wiley & Sons, Ltd. https://doi.org/10.1002/9781119098935.ch27.

Rahman, M. S., Al-Rizeiqi, M. H., Guizani, N., Al-Ruzaiqi, M. S., Al-Aamri, A. H., & Zainab, S. (2015). Stability of vitamin C in fresh and freeze-dried *Capsicum* stored at different temperatures. *Journal of Food Science and Technology*, 52 (3), 1691–1697. https://doi.org/10.1007/s13197-013-1173-x.

Rifna, E. J., Singh, S. K., Chakraborty, S., & Dwivedi, M. (2019). Effect of thermal and non-thermal techniques for microbial safety in food powder: Recent advances. *Food Research International*, 126, 108654–108674. https://doi.org/10.1016/j.foodres.2019.108654.

Rodoni, L. M., & Lemoine, M. L. (2023). Shedding light about the use of photosensitizers and photodynamic treatments during postharvest of fruit and vegetables: Current and future state. *Postharvest Biology and Technology*, 204, 112463. https://doi.org/10.1016/j.postharvbio.2023.112463.

Rybak, K., Wiktor, A., Kaveh, M., Dadan, M., Witrowa-Rajchert, D., & Nowacka, M. (2022). Effect of thermal and non-thermal technologies on kinetics and the main quality parameters of red bell pepper dried with convective and microwave–convective methods. *Molecules*, 27 (7), 2164–2185. https://doi.org/10.3390/molecules27072164.

Rybak, K., Wiktor, A., Pobiega, K., Witrowa-Rajchert, D., & Nowacka, M. (2021). Impact of pulsed light treatment on the quality properties and microbiological aspects of red bell pepper fresh-cuts. *LWT—Food Science and Technology*, 149, 111906–111916. https://doi.org/10.1016/j.lwt.2021.111906.

Rybak, K., Wiktor, A., Witrowa-Rajchert, D., Parniakov, O., & Nowacka, M. (2020). The effect of traditional and non-thermal treatments on the bioactive compounds and sugars content of red bell pepper. *Molecules*, 25 (18), 1–17. https://doi.org/10.3390/molecules25184287.

Sakudo, A., Misawa, T., & Yagyu, Y. (2020). Equipment design for cold plasma disinfection of food products. In *Advances in Cold Plasma Applications for Food Safety and Preservation* (pp. 289–307). Elsevier. https://doi.org/10.1016/B978-0-12-814921-8.00010-4.

Samira, A., Woldetsadik, K., & Workneh, T. S. (2013). Postharvest quality and shelf life of some hot pepper varieties. *Journal of Food Science and Technology*, 50 (5), 842–855. https://doi.org/10.1007/s13197-011-0405-1.

Santos, L., Marin, S., Sanchis, V., & Ramos, A. J. (2008). *Capsicum* and mycotoxin contamination: State of the art in a global context. *Food Science and Technology International*, 14 (1), 5–20. https://doi.org/10.1177/1082013208090175.

Sasmita, E., Restiwijaya, M., Yulianto, E., Arianto, Y. F., Kinandana, A. W., & Nur, M. (2019). Effect of ozone technology applications on physical characteristics of red cayenne pepper (*Capsicum frutescens* L.) preservation. *Journal of Physics: Conference Series*, 1217 (1), 012007–012013. https://doi.org/10.1088/1742-6596/1217/1/012007.

Sawangrat, C., Phimolsiripol, Y., Leksakul, K., Thanapornpoonpong, S., Sojithamporn, P., Lavilla, M., & Boonyawan, D. (2022). Application of pinhole plasma jet activated water against *Escherichia coli, Colletotrichum gloeosporioides*, and decontamination of pesticide residues on chili (*Capsicum annuum* L.). *Foods*, 11 (18), 2859–2875. https://doi.org/10.3390/foods11182859.

Schweiggert, U., Kurz, C., Schieber, A., & Carle, R. (2007). Effects of processing and storage on the stability of free and esterified carotenoids of red peppers (*Capsicum annuum* L.) and hot chilli peppers (*Capsicum frutescens* L.). *European Food Research and Technology*, 225 (2), 261–270. https://doi.org/10.1007/s00217-006-0413-y.

Shen, X., Zhang, M., Devahastin, S., & Guo, Z. (2019). Effects of pressurized argon and nitrogen treatments in combination with modified atmosphere on quality characteristics of fresh-cut potatoes. *Postharvest Biology and Technology*, 149, 159–165. https://doi.org/10.1016/j.postharvbio.2018.11.023.

Singh, R. P., & Heldman, D. R. 2001. *Introduction to Food Engineering*. Gulf Professional Publishing.

Song, W. J., Sung, H. J., Kim, S. Y., Kim, K. P., Ryu, S., & Kang, D. H. (2014). Inactivation of *Escherichia coli* O157: H7 and *Salmonella typhimurium* in black pepper and red pepper by gamma irradiation. *International Journal of Food Microbiology*, 172, 125–129. https://doi.org/10.1016/j.ijfoodmicro.2013.11.017.

Srinivas, M. S., Madhu, B., Srinivas G., & Jain, S. K. (2018). High pressure processing of foods: A review. *The Andhra Agricultural Journal*, 65, 467–476, https://www.researchgate.net/publication/328652367.

Tripathi, S., & Mishra, H. N. (2009). Nutritional changes in powdered red pepper upon *in vitro* infection of *Aspergillus flavus*. *Brazilian Journal of Microbiology*, 40 (1), 139–44. https://doi.org/10.1590/S1517-83822009000100024.

Vicente, A. R., Pineda, C., Lemoine, L., Civello, P. M., Martinez, G. A., & Chaves, A. R. (2005). UV-C treatments reduce decay, retain quality and alleviate chilling injury in pepper. *Postharvest Biology and Technology*, 35 (1), 69–78. https://doi.org/10.1016/j.postharvbio.2004.06.001.

Victoria-Campos, C. I., De Jesús Ornelas-Paz, J., Ramos-Aguilar, O. P., Failla, M. L., Chitchumroonchokchai, C., Ibarra-Junquera, V., & Pérez-Martínez, J. D. (2015). The effect of ripening, heat processing and frozen storage on the *in vitro* bioaccessibility of capsaicin and dihydrocapsaicin from Jalapeño peppers in absence and presence of two dietary fat types. *Food Chemistry*, 181, 325–332. https://doi.org/10.1016/j.foodchem.2015.02.119.

Villa-Rivera, M. G., & Ochoa-Alejo, N. (2020). Chili pepper carotenoids: Nutraceutical properties and mechanisms of action. *Molecules*, 25 (23), 5573–5595. https://doi.org/10.3390/molecules25235573.

Wangmo, C., & Dendup, T. (2021). Post-harvest handling and losses of green chilies: A case study from Bhutan. *Indonesian Journal of Social and Environmental Issues (IJSEI)*, 2 (3), 284–292. https://doi.org/10.47540/ijsei.v2i3.329.

Woldemariam, H. W., Emire, S. A., Teshome, P. G., Töpfl, S., & Aganovic, K. (2022). Microbial inactivation and quality impact assessment of red pepper paste treated by high pressure processing. *Heliyon*, 8 (12), e12441–12450. https://doi.org/10.1016/j.heliyon.2022.e12441.

Won, Y. C., Min, S. C., & Lee, D.-U. (2015). Accelerated drying and improved color properties of red pepper by pretreatment of pulsed electric fields. *Drying Technology*, 33 (8), 926–932. https://doi.org/10.1080/07373937.2014.999371.

Wu, Z. S., Zhang, M., & Wang, S. (2012). Effects of high pressure argon treatments on the quality of fresh-cut apples at cold storage. *Food Control*, 23 (1), 120–127. https://doi.org/10.1016/j.foodcont.2011.06.021.

Yoon, M., Jung, K., Lee, K.-Y., Jeong, J.-Y., Lee, J.-W., & Park, H.-J. (2014). Synergistic effect of the combined treatment with gamma irradiation and sodium dichloroisocyanurate to control gray mold (*Botrytis cinerea*) on paprika. *Radiation Physics and Chemistry*, 98, 103–108. https://doi.org/10.1016/j.radphyschem.2013.12.039.

Zhang, X.-L., Zhong, C.-S., Mujumdar, A. S., Yang, X.-H., Deng, L.-Z., Wang, J., & Xiao, H.-W. (2019). Cold plasma pretreatment enhances drying kinetics and quality attributes of chili pepper (*Capsicum annuum* L.). *Journal of Food Engineering*, 241, 51–57. https://doi.org/10.1016/j.jfoodeng.2018.08.002.

16 Adulteration and Authenticity Testing of Chilli

B. Sasikumar

CONTENTS

16.1 INTRODUCTION

Chilli (*Capsicum* spp.) is an indispensable culinary spice used all over the world and is valued mainly for its pungency and colour. Chilli is also used in beverages, sauces, defence, and cosmetic and medicinal preparations (Berke and Shieh 2001). The commodity is traded as dry whole fruits, crushed chilli, chilli powder, and value-added products like fermented chilli, chilli oil, chilli paste, oleoresin, sauces besides green chilli, etc. Among the various forms, chilli powder and paste are more vulnerable to adulteration (Chakrabarti and Roy 2003). Chilli powder (also spelled chile, chilli, or powdered chili) is the dried, pulverized fruit of one or more types of chilli pepper, sometimes with the addition of other spices, making it chilli powder blend or chilli seasoning mix (Farrell 1998).

Good-quality chilli is very relevant for the perceived biological efficiency of the commodity and their pungency, flavour, or aroma. The health-conscious public all over the world are increasingly asking for quality products, be it for health, culinary, or cosmetic uses.

Regulatory agencies, food processors, and consumers are all involved in detecting adulterants and authenticating raw materials/value-added products of food products in order to adhere to food quality and safety standards (Che Man et al. 2005). Standards and conformity assessment of commodities assumes high global relevance in herbs and spices as well as other food products due to the non-tariff barriers to trade such as sanitary and phytosanitary (SPS) and pre-shipment inspection (PSI) agreements, which mandate that the product is safe, free from adulterants/contaminants and has the desired quality.

Adulteration detection and authenticity testing of food and agricultural commodities such as grains, pulses, beverages, olive oil, horticultural products including spices, and medicinal herbs are important for value/quality assessment besides restricting unfair competition and unethical trade practices and protecting consumers against defrauding practices commonly observed in unscrupulous trade. Further, deceitful adulteration of food and agricultural products containing undeclared constituents or parts inconsistent with the label claims may cause health issues in sensitive men and women (Asensio et al. 2008; Marcus and Grollman 2002).

DOI: 10.1201/9781003378259-16

A report to assess the contamination data of processed chilli pepper and tomatoes over the past four decades since the establishment of the Rapid Alert System for Food and Feed (RASFF) attests that out of 887 notifications assessed, 446 were of chilli and tomato contamination (Figure 16.1), and India emerged as the country of origin with the highest number of reported cases relating to chilli pepper adulteration/contamination, including pesticides and mycotoxins. The study also revealed that almost all unauthorized dyes in the notifications were more than the range (0.5 to 1 mg/kg) of the detection limit of Sudan dyes and other related dyes, as per the European Union standards (Essuman et al. 2023).

Hazards like pesticide residues and mycotoxins (contaminants) as well as heavy metals in chilli/chilli products are out of the scope of this review.

16.2 ADULTERANTS IN CHILLI

Major additives in chilli/chilli products are presented in Table 16.1.

Naked-eye visualization fails to discriminate an original chilli sample from an adulterated one, even at very high concentration, be it any kind of adulterants (Figure 16.2, Dhanya 2009).

16.3 TECHNIQUES FOR ADULTERANT DETECTION

Many techniques have been developed to detect adulteration in chilli and chilli products.

16.3.1 PHYSICAL AND SENSOR BASED METHODS

Physical methods involved in adulteration determination and authentication are microscopic gross structural study or histological evaluation and other techniques based on solubility, bulk density, texture, imaging, colour reaction, etc. Though physical methods are simple to use, their regulatory/commercial value is limited. Sensing methods involving customized or non-customised sensors to detect Sudan dyes and other dyes or inert additives, a relatively new approach, are more sensitive and accurate.

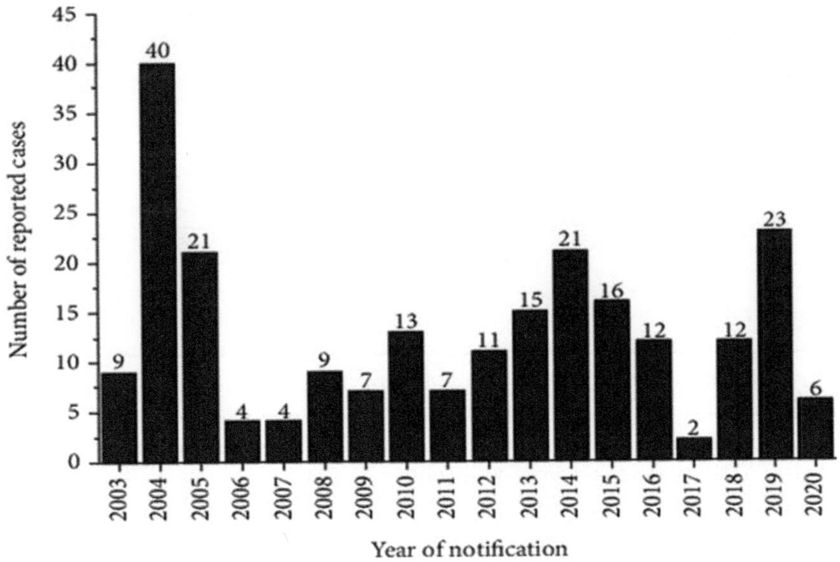

FIGURE 16.1 Yearwise notification reports for chilli pepper adulteration/fraud during the period 2003 to 2020 (Source: Essuman et al. 2023).

TABLE 16.1

Major Adulterants in Chilli/Chilli Products

Item	Chemical and inert adulterant	Biological adulterant	Reference (pooled)
Chilli whole	Dyes, mineral oil	—	Dhanya and Sasikumar 2010, Rohaeti et al. 2019
Chilli powder/ Paprika powder	Sudan dyes, Ponceau 4R (E124), coal tar, para red, n-vanilylnonamide, rhodamine B (RhB), fast garnet, oil orange, mineral oil, talc powder, brick powder, soap stone, salt powder	Powdered fruits of 'Choti ber' *Ziziphus nummularia* (Burm.f.) Wight & Arn. (Figure 16.1), red beet pulp, almond shell dust, peanut, rice flour and bran, extra amounts of bleached pericarp, seeds, calyx and peduncles of chilli, plant straw, foliage, starch of cheap source, tomato wastes, onion, garlic, Sumac (*Rhus coriaria* Linn.) fruit powder, potato starch, acacia gum, Bixin (E160b), and annatto	Mitra et al. 1961, Banerjee et al. 1974, Berke and Shieh 2001, Mazzetti et al. 2004, Dhanya and Sasikumar 2010, Hong Mei and Meng 2014, Swetha et al. 2016, Sasikumar et al. 2016, Kim and Baik 2016, Osman et al. 2019, Périat et al. 2019, Sannino and Savini 2021, Horn et al. 2021, Lopez et al. 2022, Oliveira et al. 2022, Essuman et al. 2023
Chilli oil	rhodamine B (RhB), Sudan dyes	—	Wang et al. 2014, Sha et al. 2022

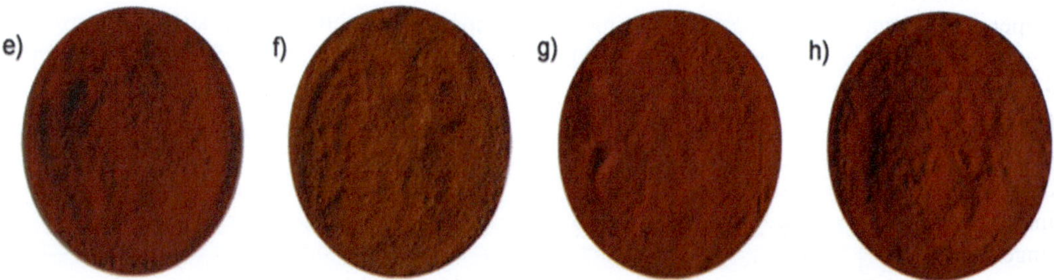

FIGURE 16.2 Adulteration of chilli powder with biological adulterants (e) genuine chilli, (f) *Ziziphus nummularia* fruit powder, (g) simulated sample [chilli: *Ziziphus nummularia* (80:20)], (h) market sample (Source: Dhanya 2009).

Microscopic evaluation was used to detect the existence of plant parts such as straws, leaves, and stalk besides red beet pulp, tomato waste, and added starch in chilli (Schwein and Miller 1967; Pruthi 1980; Chakrabarti and Roy 2003; Hong Mei and Meng 2014).

Sen et al. 2017 used simple water floatation to detect brick powder presence in chilli, while Nitika et al. 2022 developed antenna based sensors to detect brick powder adulteration in chilli powder. Desai and Patel (2021) used a novel method involving digital imaging coupled with logistic regression to detect brick powder adulteration of chilli powder. Artificial intelligence–aided adulteration detection and quantification is the most recent approach in detecting brick powder adulteration in chilli powder. Sarkar et al. (2023) demonstrated machine learning–based algorithms to detect brick powder adulteration in red chilli powder for the first time.

Multiwall carbon nanotubes (MWCNTs) as well as treated pencil graphite electrode with DNA, o-phenylenediamine, and gold nanoparticle bioimprinted polymer have been used for the detection of Sudan I and Sudan II adulteration, respectively, in chilli powder (Yang et al. 2010; Rezaei et al. 2016). Li et al. (2015a) developed a new method for the detection of Sudan I based on a conducting poly(p-aminobenzene sulphonic acid) (p-ABSA) -film-modified electrode, in hot chilli and ketchup samples with strong accumulation ability and excellent electrocatalytic activity. A highly sensitive electrochemical sensor based on La3+-doped Co3O4 nanocubes for determination of Sudan I in chilli powder and paste is described by Moghaddam et al. (2019).

Another recent sensing technique developed to detect Sudan dyes in chilli powder involves the use of a highly selective and sensitive benzimidazole-based bifunctional sensor (Abbasi et al. 2021). Simultaneous ultra low detection limits for RhB (0.0072 ng/mL), Sudan I (0.0040 ng/mL), and Sudan II (0.0260 ng/mL) in chilli seasoning were achieved by a single-emission, dual-enzyme immunofluorometric magnetosensor. Two enzymatic reactions catalyzed by horseradish peroxidase-labeled rhodamine (RhB) antibody and glucose oxidase-labeled Sudan dyes (SuDs) antibody were carried out within a functional microfluidic chip, leading to production of strongly fluorescent reso-rufin. The signals emitted were captured and quantified by a compact analyzer with the help of a smartphone (Guan et al. 2022).

A luminescent chemosensor probe based on Ru(II) bipyridine complex for detection of Sudan I through inner filter effect between luminophore, $[Ru(bpy)_2(CIP)]^{2+}$, and quencher is perfected to detect the dye in commercial chilli powder samples (Sheth et al. 2022).

Khan et al. 2020 utilized hyperspectral camera imaging for the detection of oil (solvent) and Sudan dye in red chilli.

A simple household colour test involving adding hexane to the sample, followed by mixing of acetonitrale to the supernatant and recording the colour change of the solution, was developed to detect Sudan dye in chilli (Jaiswal et al. 2016).

16.3.2 Analytical Methods

Depending on the basic principle, the techniques can be grouped into different types as follows.

16.3.2.1 Chromatographic Techniques

Paper chromatography, thin-layer chromatography (TLC), high-performance thin-layer chroma-tography (HPTLC), high-performance liquid chromatography (HPLC), HPLC fingerprinting with chemometrics, gel permeation chromatography-liquid chromatography-tandem mass spectrometry interfaced with electro spray ionization (GPC–LC–ESI-MS/MS), multiwavelength chromatographic finger print comparison, or their modifications are some of the different chromatographic methods in vogue to detect the synthetic adulterants (dyes) and mineral oil in chilli and chilli products (Mitra et al. 1961; Stelzer 1963; Navarao et al. 1965; Schwein and Miller 1967; Sen et al. 1973; Marshall 1977; Tateo and Bononi 2004; Daood and Biacs 2005; Sun et al. 2007; Ertaş et al. 2007; Mustafa et al. 2013; Rani et al. 2015; Zhu et al. 2016; Gandhi and Mashru 2019; Sebaei et al. 2019; Esteki et al. 2019; Ullah et al. 2023; Sahu et al. 2023).

A straightforward thin-layer chromatographic (TLC) analysis of 70 samples of ground chilli from Karachi, Pakistan, revealed that 50% of the samples are adulterated with Sudan I and IV (Pervezi et al. 2017).

However, literature review reveals hybrid protocols involving one or the other chromatographic techniques coupled with superior extraction methods, duration of extraction, effective dispersants and detectors, improved separation systems, spectroscopy, chemometry, etc. as the current trends.

Simultaneous extraction and/or detection of various azo dyes in chilli/chilli products are not only cost-effective but also of high throughput. Simultaneous extraction and detection of water-soluble and fat-soluble synthetic dyes in chilli and other food stuffs have been obtained using dimethyl sulfoxide (DMSO) as the extraction solvent followed by high-performance liquid chromatography–diode array detection–electrospray mass spectrometry (Ma et al. 2006). Various other methods such as the two-directional high-performance thin-layer chromatographic method for the simultaneous determination of metanil yellow and Sudan dyes in chilli besides curcumin in turmeric, and curry powders; liquid chromatography–atmospheric pressure photoionization–tandem mass spectrometry for the simultaneous quantitative determination of Sudan dyes (I–IV) in chilli powder and simulated tomato sauce; miniaturized method termed 'in-line micro-matrix solid-phase dispersion' (in-line MMSPD) coupled with high performance liquid chromatography (HPLC) for the concurrent extraction and detection of Sudan dyes (Sudan I–IV, Sudan orange G, Sudan black B, and Sudan red G) in chilli; liquid chromatography–diode-array detection for simultaneously detecting eight oil-soluble and 10 water-soluble prohibited dyes in chilli sauce and paprika powder; a monolithic polymerized high internal phase emulsion (PolyHIPE) column combined with high-performance liquid chromatography (HPLC) for the simultaneous extraction and sensitive determination of Para red and Sudan dyes in chilli samples have been reported (Dixit et al. 2008; Murty et al. 2009; Li et al. 2013; Rajabi et al. 2015; Uematsu et al. 2017; Du et al. 2017). Experimenting with better separation systems, Schwack et al. 2018 described a new HPTLC method involving caffeine-impregnated silica gel plates for simultaneous detection of the eight most frequently found azo dyes in chilli, paprika and chilli sauce besides turmeric powder, curry paste, and palm oil. Ultra-high-performance liquid chromatography (UHPLC) accompanied by high-resolution mass spectrometry with an Orbitrap analyzer and Ultrafast liquid chromatography along with ultra-violet detector (UFLC-UV) analysis are other two recent techniques developed for the simultaneous determination of 12 banned azo dyes (Sudan I–IV, Sudan black B, Sudan orange G, Sudan red 7B, Sudan red B, Sudan red G, Butter yellow, Para red, and rhodamine B) in chilli and other traded spices (Adjei et al. 2020; Sannino and Savini 2021).

A pressurised liquid extraction (PLE) with acetone along with gel permeation chromatography (GPC) clean-up followed by liquid chromatography (LC) conjoined with electrospray ionization in positive mode tandem mass spectrometry (ESI-MS–MS) to detect Sudan (I–IV), Sudan Orange G, Sudan Red 7B and Para red in hot chilli food samples was previously described (Pardo et al. 2009).

Electrolyte-assisted microemulsion breaking in vortex-agitated solidified floating organic drop microextraction (VA-SFODME-EAMB), a simple and efficient liquid phase microextraction technique, linked with high-performance liquid chromatography, was developed for analysing Sudan I-IV in chilli powder, chilli oil, chilli sauce, and chilli paste (Sricharoen et al. 2017; Sricharoen et al. 2019).

As an alternative to the traditional chromatographic approach, a simple and rapid extraction method based on supramolecular solvent-based microextraction of Sudan dyes in chilli-containing food items accompanied by liquid chromatography-photodiode array was proposed (Jiménez et al. 2010). Mustafa et al. 2013 also perfected a quick and simple HPLC method coupled with UV-Vis detector for the determination of Sudan and Para red dyes in red chilli.

Fukuji et al. 2011 developed a partial filling micellar electrokinetic chromatography (MEKC) for the quantitative determination of Sudan dyes (I–IV) in chilli sauces, while a more sensitive and selective micellar electrokinetic chromatography-MS/MS method using ammonium perfluoro octanoate (APFO) as volatile surfactant for the determination of Sudan dyes in chilli products is described by González et al. 2020.

Dispersive liquid-phase microextraction with solidification of floating organic drop (SFO–DLPME) is one of the most interesting sample preparation techniques developed in recent years to detect adulterants in food materials. A rapid and efficient SFO–DLPME combined with high-performance liquid chromatography (HPLC) was demonstrated for the extraction and sensitive detection of Sudan I–IV in chilli sauce and chilli oil after trial and error involving variables such as the type and volume of extractants and dispersants, pH and volume of sample solution, extraction time and temperature, sample preparation technique, ion strength, and humic acid concentration (Chen and Huang 2014; Sulaiman et al. 2022).

Various parameters such as extraction efficiency of the adulterants, including the type of desorption solvent, sample loading rate, sample volume, and pH, have also been studied by different authors. Liquid–liquid extraction with ultra-performance liquid chromatography—tandem mass spectrometer (UPLC-MS/MS) analysis, a fast and reliable method, was developed for the estimation and quantification of very low levels of Sudan I–IV (3.3–8.709 μg/kg) in chilli powder (Schummer et al. 2013). Organic solvent-free air-assisted liquid–liquid microextraction for optimized extraction of illegal azo dyes and their main metabolites from spices, cosmetics, etc. is another advancement in this regard (Barfi et al. 2015). Ji et al. 2017 described a molecularly imprinted solid-phase extraction (MISPE) method based on SH-Au modified silica gel for Sudan dyes in red chilli powder.

Rhodamine B adulteration in chilli oil has been ascertained using thin-layer chromatography (TLC) in combination with surface-enhanced Raman spectroscopy (SERS) as well as deep eutectic solvent extraction followed by an ultra-high-performance liquid chromatograph equipped with a fluorescence detector (Wang et al. 2014; Wang et al. 2017).

Rao et al. 2021 perfected a simple and precise ultra-high-performance liquid chromatography with photodiode array detector–based analytical method for detection and quantification of metanil yellow and Sudan I from powders of chilli and turmeric.

Though chromatographic methods are generally used to detect the synthetic adulterants, intertype and interregional variation (adulteration) of paprikas using HPLC-UV fingerprints was described recently (Cetó et al. 2020).

Chemometric algorithms–assisted high-performance liquid chromatography with diode array detection (HPLC-DAD), a more recent approach, was proposed for screening and quantifying 12 azo dyes illegally added into different food products including chilli (Dong et al. 2021).

A screening method employing thin-layer chromatography (TLC) as a separation method linked to the retardation factor, excitation, and fluorescence spectra besides fluorescence lifetime exhibited by RhB when excited with ultraviolet or green light, followed by direct fluorescent measurement of RhB on the TLC plate using a fibre optic probe coupled to a commercial spectrofluorometer or to an instrumental set up for laser-induced fluorescence measurements, is described to detect rhodamine B in chilli powder (Knecht 2023).

16.3.2.2 Spectroscopic Analysis

Spectroscopic methods like UV, visible, mid- or near-infrared (MIR, NIR), Raman, fluorescence, and nuclear magnetic resonance (NMR) have been deployed to detect various adulterants in traded spices and other food items since long (Calbiani et al. 2004; Meuren 2010; Bharathi et al. 2018; Sasikumar 2019; Kaavya et al. 2020). Spectroscopy alone or in conjunction with other analytical methods including chemometry has been widely used in adulteration detection and authentication of chilli/chilli products (Table 16.2).

16.3.2.3 PCR-Based Molecular Methods

DNA-based methods are the best option for biological adulterant detection in spices and other food products as the strategy is independent of the physical form of the sample (whole or powdered), age, environmental factors, storage, and processing conditions (Balachandran et al. 2015). Molecular methods involve the amplification of one or more regions of DNA and have good scope in food authentication due to their sensitivity, rapidity, specificity, and simplicity (Sasikumar et al. 2016). Random amplified polymorphic DNA (RAPD) (Williams et al. 1990; Mane et al. 2006; Dhanya et al. 2008), arbitrarily primed PCR (AP-PCR) (Welsh and McClelland 1990), DNA amplification fingerprinting (DAF) (Anolles et al. 1991), intersimple sequence repeat (ISSR) (Zietkiewicz et al. 1994), directed amplification of minisatellite-region DNA (DAMD) (Heath et al. 1993), sequence characterized amplified regions (SCAR) (Paran and Michelmore 1993; Dhanya et al. 2011[a]), amplification refractory mutation system (ARMS) (Newton et al. 1989), simple sequence repeat (SSR) analysis (Litt and Lutty 1989; Melchiade et al. 2007), species-specific PCR (Sasaki et al. 2004), single nucleotide polymorphism (SNP) (Johnston et al. 2013), and real-time PCR (Lockley and Bardsley 2000), etc. are the most frequently used DNA based approaches besides the relatively latest DNA barcoding, credited with universality and reproducibility (Swetha et al. 2016).

Lekha et al. 2001 used inter simple sequence repeats polymerase chain reaction (ISSR-PCR) and fluorescent inter simple sequence polymerase chain reaction (FISSR- PCR) markers for differentiating

TABLE 16.2

Spectroscopic Methods in Adulteration Detection of Chilli Products

Item	Adulterant	Method	Reference
Chilli products	Sudan dyes (I–IV)	Liquid chromatography-electrospray-tandem mass spectroscopy	Calbiani et al. 2004
Chilli powder and chilli-containing food products	1-phenylazo-2-naphthol (Sudan I)	GPC cleanup and HPLC with LC/MS	Mazetti et al. 2004
Hot chilli, spices, and oven-baked foods	Sudan I	HPLC/APCL-MS	Tateo and Bononi 2004
Chilli powder and other food products	Sudan azo dyes	HPLC-APCI-ion trap MS/MS	Rovellini 2005
Chilli	Para red, Sudan orange G, Sudan I–IV, Sudan red 7B, and rhodamine B	Reversed phase HPLC coupled with mass spectrometry (tandem in time—ion trap mass analyser)	Botek et al. 2007
Paprika	Sudan dyes I–IV	UV–visible spectroscopy and multivariate classification	Di Anibal et al. 2009
Chilli powder	Sudan I	Surface Enhanced Raman Scattering with Multivariate Chemometrics	Cheung et al. 2010
Paprika	Sudan dyes	High-resolution ^{1}H nuclear magnetic resonance and chemometric treatment	Di Anibal et al. 2011
Paprika	Sudan I	Surface-enhanced Raman spectroscopy (SERS) and multivariate analysis	Di Anibal et al. 2012
Chilli powder	Sudan I–IV, Sudan red G, Sudan red 7B, Sudan black B, and Sudan yellow	Ultra-high performance liquid chromatography-MS/MS (UHPLC-MS/MS)	Wang et al. 2013
Chilli powder samples and tomato sauce	Para red	Simple spectrophotmetric method followed by thin-layer chromatography	Mustafa et al. 2013
Chilli powder	Sudan I dye	Infrared and Raman spectroscopy	Haughey et al. 2015
Paprika powder	Sudan I	Biosensor combining molecularly imprinted polymers (MIPs), thin-layer chromatography (TLC), and surface-enhanced Raman spectroscopy (SERS)	Gao et al. 2015
Chilli powder	Sudan dyes	Real time-mass spectrometry	Li et al. 2015b
Paprika powder	Sudan dye	FTIR spectroscopy	Lohumi et al. 2017
Paprika powder	Sudan I	Nuclear magnetic resonance spectroscopy	Hu et al. 2017
Chilli powder	Sudan I	Thin-layer chromatography (TLC) and highly specific sensing using surface-enhanced Raman scattering (SERS) spectroscopy	Kong et al. 2017
Chilli flakes	Sudan dyes	Surface-enhanced Raman spectroscopy coupled with Au–Ag core-shell nanospheres	Ou et al. 2017
Paprika powder	Azo dyes	Gas chromatography-mass spectrometry	Otero et al. 2017
Chilli powder	Peanut	Liquid chromatography-mass spectrometry (LC-MS/MS)	Vandekerckhove et al. 2017
Chilli powder	Sudan III–IV	Artificial neural network (ANN) –aided UV-visible spectroscopy	Islam et al. 2018

(Continued)

TABLE 16.2 *(Continued)*

Spectroscopic Methods in Adulteration Detection of Chilli Products

Item	Adulterant	Method	Reference
Chilli powder	Azo dyes	Matrix-assisted laser desorption ionization time-of-flight mass spectrometry (MALDI-TOF MS)	Negrete et al. 2019
Chilli, paprika, and other spices	Sudan I, Sudan IV, Bixin (E160b), and Ponceau 4R (E124)	Ultra-high-performance liquid chromatography hyphenated with a quadrupole/time-of-flight mass spectrometry (UHPLC-QTOF-MS) with sequential window acquisition of all theoretical fragment-ion spectra (SWATH)	Périat et al. 2019
Paprika	Sudan I	Non-destructive Raman spectroscopy	Maraña et al. 2019
Chilli sauce	Sudan II	Zinc oxide nanoparticles (ZnONPs)/carbon paste electrode (CPE) characterised by x-ray diffraction spectroscopy and scanning electron microscope technique	Heydari et al. 2019
Paprika	Sudan dyes and Para red	NIR spectroscopy with chemometrics.	Hansen et al. 2019
Chilli oil	Azo dyes	High-performance thin-layer chromatography-mass spectroscopy (HPTLC-MS)	Madhukar and Vinayak 2020
Chilli products	Sudan dyes	Micellar electrokinetic chromatography-MS/MS using a volatile surfactant	González et al. 2020
Paprika powder	Sudan I	Fluorescence spectroscopy combined with second-order calibration	Maraña et al. 2020
Chilli powder	Sudan (III and IV), Sudan orange G	Liquid chromatography/tandem mass spectrometry	Mohamed et al. 2021
Paprika powder	Azorubine and Ponceau 4R, beetroot and sumac powder	^{1}H nuclear magnetic resonance spectroscopy combined with one-class classification coupled with chemometrics	Horn et al. 2021
Red chillies	Sudan red	^{1}H-nuclear magnetic resonance (NMR) relaxometry	Shomaji et al. 2021
Chilli products	Sudan I–IV, Sudan orange, Sudan red 7B, Para red, rhodamine B	Ultra-high-performance liquid chromatography-tandem mass spectrometry (UHPLC–MS/MS)	Larrañaga et al. 2021
Paprika powder	Potato starch, acacia gum, and annatto	Portable near-infrared spectroscopy	Oliveira et al. 2022
Chilli oil	rhodamine B and Sudan I	Thin-layer chromatography-surface-enhanced Raman spectroscopy (TLC-SERS).	Sha et al. 2022
Chilli and cumin powders	23 synthetic dyes	Dilute-and-shoot–based SWATH-MS	He et al. 2023

four disputed chilli seed samples, a case of adulteration, and marketing spurious seeds of inferior chilli varieties in the guise of an elite variety. Out of 17 ISSR-anchored primers used, 9 primers could clearly differentiate all the four disputed samples.

Dhanya et al. 2008 demonstrated the utility of RAPD markers for the detection of bioadditives viz., dried and powdered fruits of 'Choti ber', dried red beet pulp and almond shell dust in commercial samples of chilli powder. Out of six market samples analysed along with genuine chilli samples, one market sample (C5) amplified a 'Choti ber'(*Ziziphus nummularia*) band of approximately 400 bp with primer OPA 10, while none of the market samples amplify any red beet pulp or almond shell dust–specific bands.

Dhanya et al. 2011[b] further developed random amplified polymorphic DNA sequence characterised amplified region (RAPD SCAR) to detect *Ziziphus nummularia* adulteration in market samples of chilli. Out of six market samples analysed along with four genuine chilli varietal samples, one market sample (C5) amplified a *Zizhiphus nummularia* specific band (~389 bp) (Figure 16.3).

RAPD markers are also found useful in detecting smoked paprika adulteration with paprika extracted from foreign varieties of pepper. Two RAPD primers, S13 and S22, generated two smoked paprika specific bands of 641 and 704 bp, respectively (Hernández et al. 2010).

Economically motivated adulteration (EMA) of red chilli powder with a mixture of garlic (*Allium sativum* L.) and onion (*Allium cepa* L.), as little as 0.05%, could be detected using species-specific PCR of nuclear rDNA-ITS regions (Kim and Baik 2016).

Use of RT-PCR and enzyme linked immunoassay (ELISA) have been demonstrated to detect low levels of peanut in chilli powder (Vandekerckhove et al. 2017).

DNA barcoding loci TS2 and *psbA-trnH* have been used to detect adulteration of chilli with related species, similar-looking cheaper substitutes, and other crop-based products such as rice, corn, or wheat flour (Zhang et al. 2019).

16.3.3 Other Methods

Sudan I, II, III, and IV in hot chilli powder are successfully determined by rapid fluorescence assay using polyethyleneimine-coated copper nanoclusters or $CsPbBr_3$ perovskite quantum dots (QDs) (Ling et al. 2014; Wu et al. 2018). A new quantitative way for detecting Sudan I in chilli powder using 1,8-diamino naphthalene–copper (II) system as a turn-on fluorescence probe was perfected by Liu et al. 2017.

For the extraction and removal of Sudan dyes from complicated matrixes, magnetic molecularly imprinted polymers (MMIPs) were developed by the surface-initiated reversible addition fragmentation chain transfer (RAFT) polymerization using Sudan I as the template followed by characterization using transmission electron microscope, Fourier transform infrared spectroscopy, X-ray photoelectron spectroscopy, vibrating sample magnetometer, and x-ray diffraction. Benefiting from the controlled/living property of the RAFT strategy, the uniform MIP layer was successfully grafted on the surface of RAFT agent-modified $Fe_3O_4@SiO_2$ nanoparticles, favoring the fast mass transfer and rapid binding kinetics. The developed MMIPs were used as the solid-phase extraction

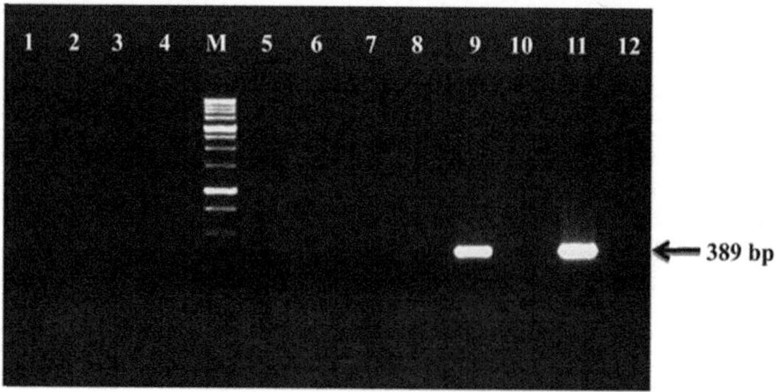

FIGURE 16.3 PCR amplification of *Ziziphus nummularia* specific SCAR primer pair Z1 (Z1 F, Z1 R) in pure chilli samples, commercial samples of chilli powder and *Z. nummularia*. Lane 1- Chilli var.'Sannam', Lane 2-Chilli var.'Bydagi', Lane 3- Chilli var. 'Pant-C-1', Lane 4-Chilli var. 'KtPL-19', Lane 5- Market sample C1, Lane 6- Market sample C2, Lane 7- Market sample C3, Lane 8- Market sample C4, Lane 9- Market sample C5, Lane 10- Market sample C6, Lane 11- *Z. nummularia*, Lane 12- Negative control and M- 1 Kb DNA ladder (Biogene, USA) (Source: Dhanya 2009).

sorbents to selectively extract four Sudan dyes (Sudan I–IV) from chilli powder samples to a level ranging from 74.1% to 93.3% with relative standard deviation lower than 6.4% and the relative standard uncertainty lower than 0.029% (Xie et al. 2015).

A new method involving silver-coated cellophane paper-based surface-enhanced Raman scattering (SERS) substrates for detecting ultra low levels of rhodamine 6G in chilli powder was perfected by Wei et al. 2019.

Magnetic nanoparticles, ferroferric oxide nanoparticles coated with polystyrene, were synthesised to effectively and efficiently extract Sudan dyes from chilli powders. The extraction procedure comprised liquid–solid extraction and magnetic solid phase extraction. The conditions were optimised to achieve efficient magnetic solid phase extraction, including extraction and desorption time, type and volume of the desorption solvent, and the mass of the adsorbents. Repeatability tests showed satisfactory recovery rates of 80.2% to 115.8%, with a relative standard deviation < 3.8% (Yu et al. 2019).

An up-conversion molecularly imprinted ratiometric fluorescent probe with a monodisperse nuclear-satellite structure and its test strip are designed that can avoid fluorescent background interference to detect Sudan I in chilli powder. The detection mechanism is based on the selective recognition of Sudan I by imprinted cavities on the surface of a ratiometric fluorescent probe and the inner filter effect between Sudan I molecules and the emission of up-conversion materials ($NaYF_4$:Yb,Tm) (Shi et al. 2023).

16.4 CHILLI—AN ADULTERANT

While on the one hand chilli is adulterated with many biological additives apart from other additives, chilli itself is an adulterant in black pepper.

Using the DNA barcoding locus *trnH-psbA*, complemented by HPLC analysis, the presence of chilli (*Capsicum annuum* L.) adulteration in traded black pepper (*Piper nigrum* L.) powder was demonstrated for the first time (Parvathy et al. 2014). *Piper nigrum* samples gave amplicon of size 350 bp, while adulterated samples (with chilli) gave amplicons of 650 bp and 350 bp (Figure 16.4), respectively, for chilli and black pepper. One of the market samples of black pepper proved positive for the chilli-specific band (650 bp).

Subsequently, Dissanayake et al. 2016, using the same DNA barcoding locus *trnH-psbA*, confirmed the adulteration of black pepper powder with chilli. Further, DNA barcoding coupled with high-resolution melting analysis (BarHRM) and backed by Principal Component Analysis (PCA)

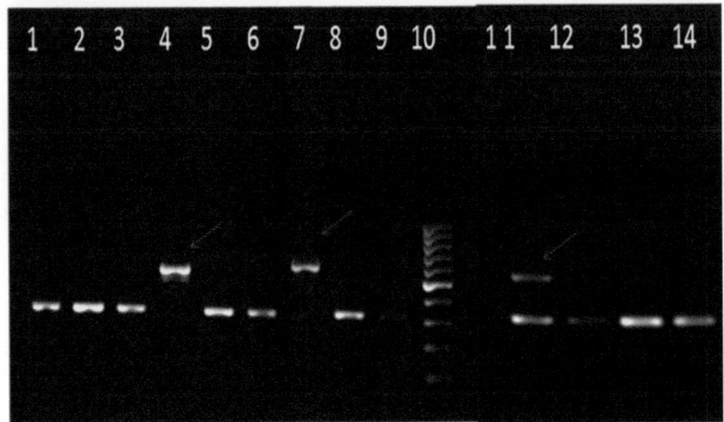

FIGURE 16.4 Amplification of *trnH-psbA* locus (Lanes 1–3 -*Piper nigrum*, lane 4- *Capsicum annuum*, lane 5- Market sample1, lane 6- Market sample2, lane 7- Market sample3, lane 8- Market sample4, lane 9- Market sample5, lane 10- 100bp ladder, lane 11- Market sample6, lane 12- Market sample7, lane 13- Market sample8, lane 14- Market sample 9) (Source: Parvathy et al. 2014).

has been effectively used to authenticate black pepper adulteration with papaya seeds and chilli (Herath and Wijesinghe 2020). Wilde et al. 2019 demonstrated the application of near and Fourier-transform infrared spectroscopy with chemometrics to screen black pepper powder adulterated with chilli, papaya seeds besides with pericarps and other plant parts of black pepper.

Chilli waste powder or other toxic wastes from chilli may be recycled as an additive to reinforce spent black pepper powder by traders.

16.5 FUTURE PERSPECTIVES AND CONCLUSION

Red chilli is one of the most popular spices to flavor dishes owing to its distinct taste and color. The demand for red chilli is increasing in the global market, the major drivers being rising awareness about health benefits of the commodity and steady growth in food, pharmaceutical, and cosmetics industries, coupled with increased purchasing power and advancement in processing and packaging technologies. The global market for red chilli was valued at $1 billion in 2021. Unfortunately, this high-value, low-volume commodity has been subjected to deliberate, economically-motivated adulteration, leading to reduced perceived biological value and posing health risks besides eroding public faith. Adulteration is also a major economic fraud involving public health. Reliable, easy, sensitive, and high-throughput traceability and authentication methods coupled with quality standards thus assume significance. Adoption of a state-of-the-art supply chain practice in the chilli industry coupled with stringent quality control at various levels and quality linked pricing in addition to cutting edge commercial adulteration diagnostic kits, if they can be developed and deployed, will be a sure shot to ensure the quality of the traded produce.

REFERENCES

Abbasi, A., Ansari, I.I. and M. Shaki. 2021. Highly selective and sensitive benzimidazole based bifunctional sensor for targeting inedible azo dyes in red chilli, red food color, turmeric powder, and Cu(Ii) in coconut water. *Journal of Fluorescence* 31:1353–1361.

Adjei, J.K., Ahormegah, V., Boateng, A.K., Megbenu, H.K. and S. Owusu. 2020. Fast, easy, cheap, robust and safe method of analysis of Sudan dyes in chilli pepper powder. *Heliyon* 6:e05243.

Anolles, G., Bassam, B.J. and P.M. Gresshoff. 1991. DNA amplification fingerprinting using very short arbitrary oligonucleotide primers. *Biotechnology* 9:553–557.

Asensio, L., Gonzalez, I., Garcya, T. and R. Martyn. 2008. Determination of food authenticity by enzyme linked immunosorbent assay (ELISA). *Food Control* 19:1–8.

Balachandran, K.R.S., Mohanasundaram, S. and S. Ramalingam. 2015. DNA barcoding: A genomic based tool for authentication of phytochemicals and its products. *Botanics: Targets and Therapy* 5:77–84.

Banerjee, T.S., Guha, K.C., Saha, A. and B.R. Roy. 1974. Examination of oil soluble colours from food by solvent partitioning and chromatography. *Journal of Food Science and Technology* 11:230–232.

Barfi, B., Asghari, A., Rajabi, M. and S. Sabzalian. 2015. Organic solvent-free air-assisted liquid–liquid microextraction for optimized extraction of illegal azo-based dyes and their main metabolite from spices, cosmetics and human bio-fluid samples in one step. *Journal of Chromatography B* 998–999:15–25.

Berke, I.G. and S.C. Shieh. 2001. Capsicum, chillies, paprika, bird's eye chilli. In *Handbook of Herbs and Spices*, ed. Peter, K.V., 111–122. Woodhead Publishing Limited, Cambridge.

Bharathi, S.K.V., Sukitha, A., Moses, J.A. and C. Anandharamakrishnan. 2018. Instrument—based detectin methods for adulteration in spice and spice products: A review. *Journal of Spices and Aromatic Crops* 27:106–118.

Botek, P., Poustka, J. and J. Hajšlo. 2007. Determination of banned dyes in spices by liquid chromatography–mass spectrometry. *Czech Journal of Food Science* 25:17–24.

Calbiani, F., Careri, M., Elviri, L., Mangia, A., Pistara, L. and I. Zagnoni. 2004. Development and in- house validation of a liquid chromatography-electrospray-tandem mass spectroscopy methods for the simultaneous detection of Sudan I, Sudan II, Sudan II and Sudan IV in hot chilli products. *Journal of Chromatography A* 1042:123–130.

Cetó, X., Sánchez, C., Serrano, N., Díaz-Cruz, J.M. and O. Núñez. 2020. Authentication of paprika using HPLC-UV fingerprints. *LWT* 124:109153.

Chakrabarti, J. and B.R. Roy. 2003. Adulterants, contaminants and pollutants in Capsicum products. In *Capsicum: The Genus Capsicum. Medicinal and Aromatic Plants-Industrial Profiles*, ed. De, A.K., 231–235. Taylor & Francis, London.

Che Man, Y.B., Syahariza, Z.A., Mirghani, M.E.S., Jinap, S. and J. Baker. 2005. Analysis of potential lard adulteration in chocolate and chocolate products using Fourier transform infrared spectroscopy. *Food Chemistry* 90:815–819.

Chen, B. and Y. Huang. 2014. Dispersive liquid-phase microextraction with solidification of floating organic droplet coupled with high-performance liquid chromatography for the determination of Sudan dyes in foodstuffs and water samples. *Journal of Agriculture and Food Chemistry* 62:5818–5826.

Cheung, W., Shadi, I., Xu, Y. and R. Goodacre. 2010. Quantitative analysis of the banned food dye Sudan-1 using surface enhanced Raman scattering with multivariate chemometrics. *Journal of Physical Chemistry C* 114:7285–7290.

Daood, H. and P.A. Biacs. 2005. Simultaneous determination of Sudan dyes and carotenoids in red pepper and tomato products by HPLC. *Journal of Chromatographic Science* 43:461–465.

Desai, N. and D. Patel. 2021. Identification of adulterants in household chilli powder from its image using logistic regression methods. *Reliability: Theory and Application* (Special issue) 16:384–391.

Dhanya, K. 2009. *Detection of Probable Plant Based Adulterants in Selected Powdered Market Samples of Spices Using Molecular Techniques.* PhD Thesis, Manglore University, Manglore, Karnataka, India, p. 251.

Dhanya, K. and B. Sasikumar. 2010. Molecular marker based adulteration detection in traded food and agricultural commodities of plant origin with special reference to spices. *Current Trends in Biotechnology and Pharmacy* 4:454–489.

Dhanya, K., Syamkumar, S., Jaleel, K. and B. Sasikumar. 2008. Random amplified polymorphic DNA techniques for the detection of plant based adulterants in chilli powder (Capsicum annuum). *Journal of Spices and Aromatic Crops* 17:75–81.

Dhanya, K., Syamkumar, S., Siju, S. and B. Sasikumar. 2011a. Sequence characterized amplified region markers: A reliable tool for adulterant detection in turmeric powder. *Food Research International* 44:2889–2895.

Dhanya, K., Syamkumar, S., Siju, S. and B. Sasikumar. 2011b. SCAR markers for adulteration detection in ground chilli. *British Food Journal* 113:656–668.

Di Anibal, C.V., Marsal, F.M., Callao, M.P. and I. Ruisánchez. 2012. Surface Enhanced Raman Spectroscopy (SERS) and multivariate analysis as a screening tool for detecting Sudan I dye in culinary spices. *Spectrochimica Acta, Part A: Molecular and Biomolecular Spectroscopy* (SAA) 87:135–141.

Di Anibal, C.V., Odena, M., Ruisánchez, I. and M.P. Callao. 2009. Determining the adulteration of spices with Sudan I-II-II-IV dyes by UV–visible spectroscopy and multivariate classification techniques. *Talanta* 79:887–892.

Di Anibal, C.V., Ruisánchez, I. and M.P. Callao. 2011. High-resolution 1H Nuclear Magnetic Resonance spectrometry combined with chemometric treatment to identify adulteration of culinary spices with Sudan dyes. *Food Chemistry* 124:1139–1145.

Dissanayake, D.R.R.P., Herath, H.P.M.D., Dissanayake, M.D.M.I.M., Chamikara, M.D.M., Jayakody, M.M., Amarasekara, S.S.C., Kularathna, K.W.T.R., Karannagoda, N.N.H., Ishan, M. and S.D.S.S. Sooriyapathirana. 2016. The length polymorphism of the locus psbA-trnH is idyllic to detect the adulterations of black pepper with papaya seeds and chilli. *The Journal of Agricultural Sciences* 11:74–87.

Dixit, S., Khanna, S.K. and M. Das. 2008. Method using simple extraction and 2-directional high-performance thin-layer chromatography (HPTLC) was developed for the simultaneous determination of curcumin, metanil yellow, and Sudan dyes in turmeric, chili, and curry powder formulations. *Journal of AOAC International* 91:1387–1396.

Dong, M.Y., Wu, H.L., Long, W.J., Wang, T. and R.Q. Yu. 2021. Simultaneous and rapid screening and determination of twelve azo dyes illegally added into food products by using chemometrics-assisted HPLC-DAD strategy. *Microchemical Journal* 171:106775.

Du, F., Zheng, Z., Deng, J., Zou, J., Zeng, Q., Li, J. and G. Ruan. 2017. High internal phase emulsion polymeric monolith extraction coupling with high-performance liquid chromatography for the determination of para red and Sudan dyes in chilli samples. *Food Analytical Methods* 10:2018–2026.

Ertaş, E., Özer, H. and C. Alasalvar. 2007. A rapid HPLC method for determination of Sudan dyes and Para Red in red chilli pepper. *Food Chemistry* 105:756–760.

Essuman, E.K., Teye, E., Dadzie, R.G. and L.K.S. Amoah. 2023. Pesticide residues and unauthorized dyes as adulteration markers in chilli pepper and tomato. *International Journal of Food Science* 2023. http://doi.org/10.1155/2023/5337150

Esteki, M., Shahsavari, Z. and J.S. Gandara. 2019. Food identification by high performance liquid chromatography fingerprinting and mathematical processing. *Food Research International* 122:323–327.

Farrell, K.T. 1998. Spices, condiments and seasonings. In *Chapman & Hall Food Science Book*, 215–217, 550 Brookshire Rd Suite A, Greer, SC 29651, Springer, USA.

Fukuji, T.S., Puyana, M.C., Tavares, M.F.M. and A. Cifuentes. 2011. Fast determination of Sudan dyes in chilli tomato sauces using partial filling micellar electrokinetic chromatography. *Journal of Agricultural and Food Chemistry* 59:11903–11909.

Gandhi, M. and R. Mashru. 2019. Detection of adulterants in red chili powder with special emphasis on qualitative and quantitative estimation of Sudan I dye in red chili powder. *International Journal of Research & Review* 6:107–112.

Gao, F., Hu, Y., Chen, D., Li-Chan, E.C.Y., Grant, E. and X. Lu. 2015. Determination of Sudan I in Paprika powder by molecularly imprinted polymers–thin layer chromatography–surface enhanced Raman spectroscopic biosensor. *Talanta* 143:344–352.

González, D.M., Jáč, P., Švec, F. and L. Nováková. 2020. Determination of Sudan dyes in chili products by micellar electrokinetic chromatography-MS/MS using a volatile surfactant. *Food Chemistry* 310:125963.

Guan, T., Jiang, Z., Liang, Z., Liu, Y., Huang, W., Li, X., Shen, X., Li, M., Xu, Z. and H. Lei. 2022. Single-emission dual-enzyme magnetosensor for multiplex immunofluorometric assay of adulterated colorants in chili seasoning. *Food Chemistry* 366:130594.

Hansen, G.J.T., Almonacid, J., Albertengo, L., Rodriguez, M.S., Di Anibal, C. and C. Delrieux. 2019. NIR-based Sudan I to IV and Para-Red food adulterants screening. *Food Additives & Contaminants: Part A* 36:1163–1172.

Haughey, A., Galvin-King, P., Ho, Y.C., Bell, S.E.J. and C.T. Elliot. 2015. The feasibility of using near infrared and Raman spectroscopic techniques to detect fraudulent adulteration of chili powders with Sudan dye. *Food Control* 48:75–83.

He, G., Hou, X., Han, H., Qiu, S., Li, Y., Qin, S., Qiu, B. and M. Liang. 2023. A dilute-and-shoot based SWATH-MS approach for rapid analysis of 23 synthetic dyes in spices. *Journal of Food Composition and Analysis* 115:104878.

Heath, D.D., Iwama, G.K. and R.H. Devlin. 1993. PCR primed with VNTR core sequences yields species specific patterns and hypervariable probes. *Nucleic Acids Research* 21:5782–5785.

Herath, H.M.P. and W.R.P. Wijesinghe. 2020. High-resolution melting traceability of black pepper adulteration with papaya seeds, chili and/or other potential plants. In *International Conference on Applied and Pure Sciences, 2020. Faculty of Science*, 13. University of Kelaniya, Kelaniya, Sri Lanka.

Hernández, A., Aranda, E., Martín, A., Benito, M.J., Bartolomé, T. and M. de G. Córdoba. 2010. Efficiency of DNA typing methods for detection of smoked paprika "Pimenton de la Vera" adulteration used in the elaboration of dry-cured Iberian pork sausages. *Journal of Agricultural and Food Chemistry* 58:11688–11694.

Heydari, M., Ghoreishi, S.M. and S. Khoobi. 2019. Novel electrochemical procedure for sensitive determination of Sudan II based on nanostructured modified electrode and multivariate optimization. *Measurement* 142:105–112.

Hong Mei, M. and Z. Meng. 2014. Study on the microscopic identification of adulterated plant origin powdered seasonings. *Discourse Journal of Agriculture and Food Sciences* 2:264–268.

Horn, B., Esslinger, S., Hasseek, C.F. and J. Riedi. 2021. 1H NMR spectroscopy, one-class classification and outlier diagnosis: A powerful combination for adulteration detection in paprika powder. *Food Control* 128:108205.

Hu, Y., Wang, S., Wang, S. and X. Lu. 2017. Application of nuclear magnetic resonance spectroscopy in food adulteration determination: The example of Sudan dye I in paprika powder. *Scientific Reports* 7:2637. https://doi.org/10.1038/s41598-017-02921-8.

Islam, M.F., Uddin, M.N., Rana, A.A. and M.M. Karim. 2018. Development of a chemometric method for the analysis of Sudan III-IV dyes adulteration in chili powder using UV-visible spectroscopy data. *Journal of Scientific and Innovative Research* 7:30–35.

Jaiswal, S., Yadav, D.S., Mishra, M.K. and A.K. Gupta. 2016. Detection of adulterants in spices through chemical method and thin layer chromatography for forensic consideration. *International Journal of Development Research* 6:8824–8827.

Ji, W., Zhang, M., Wang, T., Wang, X., Zheng, Z. and G. Junjie. 2017. Molecularly imprinted solid-phase extraction method based on SH-Au modified silica gel for the detection of six Sudan dyes in chili powder samples. *Talanta* 165:18–26.

Jiménez, F.J.L., Rubio, S. and D.P. Bendito. 2010. Supramolecular solvent-based microextraction of Sudan dyes in chilli-containing foodstuffs prior to their liquid chromatography-photodiode array determination. *Food Chemistry* 121:763–769.

Johnston, S.E., Lindqvist, M., Niemelä, E., Orell, P., Erkinaro, J., Kent, M.P., Lien, S., Vaha, J.P., Vasemagi, A. and C.R. Primmer. 2013. Fish scales and SNP chips: SNP genotyping and allele frequency estimation in individual and pooled DNA from historical samples of Atlantic salmon (Salmo salar). *BMC Genomics* 14:439.

Kaavya, R., Pandiselvam, R., Mohammed, M., Dakshayani, R., Kothakota, A., Ramesh, S.V., Cozzolino, D. and C. Ashok Kumar. 2020. Application of infrared spectroscopy techniques for the assessment of quality and safety in spices: A review. *Applied Spectroscopy Reviews* 55:593–611.

Khan, M.H., Saleem, Z., Ahmad, M., Sohaib, A., Ayaz, M. and M. Mazzara. 2020. Hyperspectral imaging for color adulteration detection in red chili. *Applied Science* 10:5955.

Kim, J.H. and S.H. Baik. 2016. Molecular identification of economically motivated adulteration of red pepper powder by species-specific PCR of nuclear rDNA-ITS regions in garlic and onion. *Food Analytical Methods* 9:3287–3297.

Knecht, G.T. 2023. *Development of a Screening Methodology for the Analysis of Rhodamine B in Foodstuffs*. Honours Undergraduate Thesis, University of Central Florida, Orlando, p. 43.

Kong, X., Squire, K., Chong, X. and A.X. Wang. 2017. Ultra-sensitive lab-on-a-chip detection of Sudan I in food using plasmonics-enhanced diatomaceous thin film. *Food Control* 79:258–265.

Larrañaga, A., Chamú, S.E., Santos, F.J. and E. Moyano. 2021. Determination of banned dyes in red spices by ultra-high-performance liquid chromatography-atmospheric pressure ionization-tandem mass spectrometry. *Analytica Chimica Acta* 1164:338519.

Lekha, D.K., Kathirvel, M., Rao, G.V. and J. Nagaraju. 2001. DNA profiling of disputed chilli samples (Capsicum annuum) using ISSR-PCR and FISSR-PCR marker analysis. *Forensic Science International* 116:63–68.

Li, B.L., Luo, J.H., Luo, H.Q. and N.B. Li. 2015a. A novel conducting poly(p-aminobenzene sulphonic acid)-based electrochemical sensor for sensitive determination of Sudan I and its application for detection in food stuffs. *Food Chemistry* 173:594–599.

Li, J., Ding, X.M., Liu, D.D., Guo, F., Chen, Y., Zhang, Y.B. and H.M. Liu. 2013. Simultaneous determination of eight illegal dyes in chili products by liquid chromatography–tandem mass spectrometry. *Journal of Chromatography B* 942–943:46–52.

Li, Z., Zhang, Y.W., Zhang, Y.D., Bai, Y. and H.W. Liu. 2015b. Rapid analysis of four Sudan dyes using direct analysis in real time-mass spectrometry. *Analytical Methods* 7:86–90.

Ling, Y., Li, J.X., Qu, F., Li, N.B. and H.Q. Luo. 2014. Rapid fluorescence assay for Sudan dyes using polyethyleneimine-coated copper nanoclusters. *Microchimica Acta* 181:1069–1075.

Litt, M. and J.A. Lutty. 1989. A hyper variable microsatellite revealed by in vitro amplification of a dinucleotide repeat within the cardiac muscle acting gene. *American Journal of Human Genetics* 44:397–401.

Liu, J., Zhou, Z., Wang, L. and M. Xu. 2017. Using a 1,8-diamino naphthalene–copper (II) system as a turn-on fluorescence probe for detecting Sudan I. *Spectroscopy Letters* 50:411–416.

Lockley, A.K. and R.G. Bardsley. 2000. DNA based methods for food authentication. *Trends in Food Science Technology* 11:67–77.

Lohumi, S., Joshi, R., Kandpal, L.M., Lee, H., Kim, M.S., Cho, H., Mo, C., Seo, Y.W., Rhaman, A. and B.K. Cho. 2017. Quantitative analysis of Sudan dye adulteration in paprika powder using FTIR spectroscopy. *Food Additives & Contaminants: Part A* 34:678–686.

Lopez, A.M., Nicolini, C.M., Aeppli, M., Luby, S.P., Fendorf, S. and J.E. Forsyth. 2022. Assessing analytical methods for the rapid detection of lead adulteration in the global spice market. *Environmental Science and Technology* 56(23):16996–17006.

Ma, M., Luo, X., Chen, B., Su, S. and S. Yao. 2006. Simultaneous determination of water-soluble and fat-soluble synthetic colorants in foodstuff by high-performance liquid chromatography–diode array detection–electrospray mass spectrometry. *Journal of Chromatography A* 1103:170–176.

Madhukar, N.S. and M.S. Vinayak. 2020. A novel digitally optimized rapid quantification of carcinogenic aryl azo amines from various food matrices by HPTLC-MS. *Journal of Liquid Chromatography & Related Technologies* 43:445–454.

Mane, B.G., Tanwar, V.K., Girish, P.S. and V.P. Dixit. 2006. Identification of species origin of meat by RAPD–PCR technique. *Journal of Veterinary Public Health* 4:87–90.

Maraña, O.M., Eskildsen, C.E., Afseth, N.K., Díaz, T.G., de la Peña, A.M. and J.P. Wold. 2019. Non-destructive Raman spectroscopy as a tool for measuring ASTA color values and Sudan I content in paprika powder. *Food Chemistry* 274:187–193.

Maraña, O.M., Eskildsen, C.E., de la Peña, A.M., Diaz, T.G. and J.P. Wold. 2020. Non-destructive fluorescence spectroscopy combined with second-order calibration as a new strategy for the analysis of the illegal Sudan I dye in paprika powder. *Microchemical Journal* 154:104539.

Marcus, D.M. and A.P. Grollman. 2002. Botanical medicines: The need for new regulations. *The New England Journal of Medicine* 347:2073–2076.

Marshall, P.N. 1977. Thin layer chromatography of sudan dyes. *Journal of Chromatography A* 136:1835–1840.

Mazzetti, M., Fascioli, R., Mazzoncini, I., Spinelli, G., Morelli, I. and A. Bertoli. 2004. Determination of 1-phenyl azo-2-naphthol (Sudan 1) in chilli powder and in chilli containing food products by GPC cleanup and HPLC with LC/MS confirmation. *Food Additives and Contaminants A* 21:935–941.

Melchiade, D., Foroni, I., Corrado, G., Santangelo, I. and R. Rao. 2007. Authentication of the 'Annurca' apple in agro-food chain by amplification of microsatellite loci. *Food Biotechnology* 21:33–43.

Meuren, M. 2010. Spectrophotometric techniques. In *Food Authentication and Traceability*, ed. Lees, M., 184–185. Wood Head Publishing, Cambridge.

Mitra, S.N., Sengupta, P.N. and B.R. Roy. 1961. The detection of oil soluble cpal-tar dyes in chilli (Capsicum). *Journal and Proceedings of the Institute of Chemistry* 33:69.

Moghaddam, M., Tajik, S. and H. Beitollahi. 2019. Highly sensitive electrochemical sensor based on La3+-doped Co3O4 nanocubes for determination of sudan I content in food samples. *Food Chemistry* 286:191–196.

Mohamed, S.H., Salim, A.I., Issa, Y.M., Atwa, M.A. and R.H. Nassar. 2021. Evaluation of different Sudan dyes in Egyptian food samples utilizing liquid chromatography/tandem mass spectrometry. *Food Analytical Methods* 14:2038–2050.

Murty, R.V.S., Chary, N.S., Prabhakar, S., Raju, P.N. and M. Virmani. 2009. Simultaneous quantitative determination of Sudan dyes using liquid chromatography–atmospheric pressure photoionization–tandem mass spectrometry. *Food Chemistry* 115:1555–1562.

Mustafa, S., Khan, R.A., Sultana, I. and N. Nasir. 2013. Estimation of para red dye in chilli powder and tomato sauces by a simple spectrophotmetric method followed by thin layer chromatography. *Journal of Applied Science and Environmental Management* 17:177–184.

Navarao, S., Ortuno, A. and F. Ooasta. 1965. Thin layer chromatographic determination of synthetic dyes in foods.1. Fat soluble azo dyes in paprika. *Anales de Bromatologia* 17:269.

Negrete, M.A.A., Wrobel, K., Barrientos, E.Y., Escobosa, A.R.C., Amguliar, F.J.A. and K. Wrobel. 2019. Determination of sulfonated azo dyes in chili powders by MALDI-TOF MS. *Analytical and Bioanalytical Chemistry* 411:5833–5843.

Newton, C.R., Graham, A., Heptinstall, L.E., Powell, S.J., Summers, C., Kalsheker, N., Smith, J.C. and A.F. Markham. 1989. Analysis of any point mutation in DNA. The amplification refractory mutation system (ARMS). *Nucleic Acids Research* 17:2503–2516.

Nitika, K.J. and R. Khanna. 2022. Exploration of adulteration in common raw spices using antenna-based sensor. *International Journal of Microwave and Wireless Technologies*:1–13. https://doi.org/10.1017/S175907872200112X.

Oliveira, M.M., Cruz-Triado, J.P., Roque, J.V., Teófilo, R.F. and D.F. Barbin. 2022. Portable near-infrared spectroscopy for rapid authentication of adulterated paprika powder. *Journal of Food Composition and Analysis* 87:103403.

Osman, A.G., Raman, V., Haider, S., Zulfiqar, A. and A.G. Chittiboyina. 2019. Overview of analytical tools for the identification of adulterants in commonly traded herbs and spices. *Journal of AOAC International* 102:376–385.

Otero, P., Saha, S.K., Hussein, A., Barron, J. and P. Murray. 2017. Simultaneous determination of 23 azo dyes in paprika by gas chromatography-mass spectrometry. *Food Analytical Methods* 10:876–884.

Ou, Y., Pei, L., Lai, K., Huang, Y., Rasco, B.A., Wang, X. and Y. Fan. 2017. Rapid analysis of multiple sudan dyes in chili flakes using surface-enhanced Raman spectroscopy coupled with Au–Ag core-shell nanospheres. *Food Analytical Methods* 10:565–574.

Paran, I. and R.W. Michelmore. 1993. Development of reliable PCR-based markers linked to downy mildew resistance genes in lettuce. *Theoretical and Applied Genetics* 85:985–993.

Pardo, O., Yusà, V., Leon, N. and A. Pastor. 2009. Development of a method for the analysis of seven banned azo-dyes in chilli and hot chilli food samples by pressurised liquid extraction and liquid chromatography. *Talanta* 78:178–186.

Parvathy, V.A., Swetha, V.P., Sheeja, T.E., Leela, N.K., Chempakam, B. and B. Sasikumar. 2014. DNA barcoding to detect chilli adulteration in traded black pepper powder. *Food Biotechnology* 28:25–40.

Périat, A., Bieri, S. and N. Mottier. 2019. SWATH-MS screening strategy for the determination of food dyes in spices by UHPLC-HRMS. *Food Chemistry X* 1:100009.

Pervezi, K., Ahmed, F., Dewani, R., Ayazi, T., Mehboobi, S.J. and S.A. Soomro. 2017. Qualitative investigation of prohibited food colours in red hot chilli & curry powder collected from Karachi city. *Pakistan Journal of Pharmacology* 34:17–22.

Pruthi, J.S. 1980. *Spices and Condiments-Chemistry, Microbiology, Technology.* Academic Press Inc., New York, p. 449.

Rajabi, M., Sabzalian, S., Barfi, B., Beydokhti, S.A. and A. Asghari. 2015. In-line micro-matrix solid-phase dispersion extraction for simultaneous separation and extraction of Sudan dyes in different spices. *Journal of Chromatography* 1425:42–50.

Rani, R., Medhe, S. and M. Srivastav. 2015. HPTLC–MS based method development and validation for the detection of adulterants in spices. *Journal of Food Measurement and Characterization* 9:186–194.

Rao, S.N., Reddy, V.C.S., Patel, S., Meti, S. and B. Huggar. 2021. Determination of banned adulterants in turmeric and chilli powders using ultra-high-performance liquid chromatography. *Journal of Liquid Chromatography & Related Technologies* 44:235–243.

Rezaei, B., Boroujeni, M.K. and A.A. Ensafi. 2016. Development of Sudan II sensor based on modified treated pencil graphite electrode with DNA, o-phenylenediamine, and gold nanoparticle bioimprinted polymer. *Sensor Actuat B-Chem* 222:849–856.

Rohaeti, E., Muzayanah, K., Septaningsih, A.D. and M. Rafi. 2019. Fast analytical method for authentication of chili powder from synthetic dyes using UV-Vis spectroscopy in combination with chemometrics. *Indonesian Journal of Chemistry* 19:668–674.

Rovellini, P. 2005. Determination of Sudan azo dyes: HPLC-APCI-io trap MS/MS method. *Rivista Italiana delle Sostanze Grasse* 82:299–303.

Sahu, P.K., Purohit, S., Tripathy, S., Mishra, D.P. and B. Acharya. 2023. Current trends in HPLC for quality control of spices. In *High Performance Liquid Chromatography—Recent Advances and Applications*, eds. Núñez, O., Sentellas, S., Saurina, J. and M. Granados. Intech Open. http://doi.org/10.5772/intechopen.110897.

Sannino, A. and S. Savini. 2021. A fast and simple method for the determination of 12 synthetic dyes in spicy foods by UHPLC-HRMS. *ACS Food Science and Technology* 1:107–112.

Sarkar, T., Choudhury, T., Bansal, N., Arunachaleshwaran, V.R., Khayrullin, Shariati, M.A. and J.M. Lorenz. 2023. Artificial intelligence aided adulteration detection and quantification for red chilli powder. *Food Analytical Methods* 16:721–748.

Sasaki, Y., Fushimi, H. and K. Komatsu. 2004. Application of single-nucleotide polymorphism analysis of the trnK gene to the identification of Curcuma plants. *Biological and Pharmaceutical Bulletin* 27:144–146.

Sasikumar, B. 2019. Advances in adulteration and authenticity testing of turmeric (Curcuma longa L.). *Journal of Spices and Aromatic Crops* 28:96–105.

Sasikumar, B., Swetha, V.P., Parvathy, V.A. and T.E. Sheeja. 2016. Advances in adulteration and authenticity testing of herbs and spices. In *Advances in Food Authenticity Testing: Improving Quality Throughout the Food Chain*, ed. Downey, G., 585–562. Wood Head Publishing/Elsevier, Sawston, Cambridge.

Schummer, C., Sassel, J., Bonenberger, P. and G. Moris. 2013. Low-level detections of sudan I, II, III and IV in spices and chili-containing foodstuffs using UPLC-ESI-MS/MS. *Journal of Agricultural and Food Chemistry* 61(9):2284–2289.

Schwack, W., Pellissier, E. and G. Morlock. 2018. Analysis of unauthorized Sudan dyes in food by high-performance thin-layer chromatography. *Analytical and Bioanalytical Chemistry* 410:5641–5651.

Schwein, W.G. and B.J. Miller. 1967. Detection and identification of dehydrated red beets in Capsicum spices. *Journal of Official Analytical Chemists International* 50:223.

Sebaei, A.S., Youssif, M.A. and A.A.-M. Ghazi. 2019. Determination of seven illegal dyes in Egyptian spices by HPLC with gel permeation chromatography clean up. *Journal of Food Composition and Analysis* 84:103304. https://doi.org/10.1080/10408347.2021.1905503.

Sen, A.R., Sengupta, P., Ghosh Dastidar, N. and T.V. Mathew. 1973. Chromatographic determination of small amount of mineral oil in chillies. *Research and Industry* 18:97–99.

Sen, S., Mohanty, P. and V. Sneetha. 2017. Detection of food adulterants in chilli, turmeric and coriander powders by physical and chemical methods. *Research Journal of Pharmacy and Technology* 10:3057–3060.

Sha, X., Han, S., Fang, G., Li, N., Lin, D. and W. Hasi. 2022. A novel suitable TLC-SERS assembly strategy for detection of Rhodamine B and Sudan I in chili oil. *Food Control* 138:109040.

Sheth, S., Li, M. and Q. Song. 2022. Luminescent chemosensor based on Ru(II) bipyridine complex for detection of Sudan I through inner filter effect. *Journal of Fluorescence* 30:1543–1551.

Shi, T., Liu, T., Zhang, J., Cai, D. and Y. Zhang. 2023. A test strip constructed by molecular imprinting for ratiometric fluorescence with ultra-low limit of detection for selective monitoring of Sudan I in chili powder. *Microchimica Acta* 190:263.

Shomaji, S., Masna, N.V.R., Ariando, D., Paul, S.D., Herron, K.H., Forte, D., Mandal, S. and S. Bhunia. 2021. Detecting dye-contaminated vegetables using low-field NMR relaxometry. *Foods* 10:2232.

Sricharoen, P., Limchoowong, N., Sripakdee, T., Nuengmatcha, P. and S. Chanthai. 2017. Electrolyte-assisted microemulsion breaking in vortex-agitated solidified floating organic drop microextraction for preconcentration and analysis of Sudan dyes in chili products. *Analytical Methods* 9:3810–3818.

Sricharoen, P., Limchoowong, N., Techawongstien, S. and S. Chanthai. 2019. Ultrasound-assisted emulsification microextraction coupled with salt-induced demulsification based on solidified floating organic drop prior to HPLC determination of Sudan dyes in chilli products. *Arabian Journal of Chemistry* 12:5223–5233.

Stelzer, H. 1963. Identification of synthetic colouring in paprika. *Nutricion, Bromatologia, Toxicologia* 2:177.

Sulaiman, U., Shah, F. and R.A. Khan. 2022. QuEChERS sample preparation integrated to dispersive liquid-liquid microextraction based on solidified floating organic droplet for spectrometric determination of sudan dyes: A synergistic approach. *Food and Chemical Toxicology* 159:112742.

Sun, H.W., Wang, F.C. and L.F. Ai. 2007. Determination of banned 10 azo dyes in hot chilli products by gel permeation chromatography-liquid chromatography-elctrospray ionization-tandem mass spectrometry. *Journal of Chromatography A* 1164:120–128.

Swetha, V.P., Sheeja, T.E. and B. Sasikumar. 2016. DNA Barcoding as an authentication tool for food and agricultural commodities. *Current Trends in Biotechnology and Pharmacy* 10:384–402.

Tateo, F. and M. Bononi. 2004. Fast determination of Sudan I by HPLC/APCI-MS in hot chilli, spices, and oven-baked foods. *Journal of Agricultural and Food Chemistry* 52:655–658.

Uematsu, Y., Mizumachi, T. and K. Monma. 2017. Simultaneous analysis of oil-soluble, basic, and acidic illegal dyes in foods using liquid chromatography–diode-array detection. *Journal of AOAC International* 100:1102–1109.

Ullah, A., Chan, M.W.H., Aslam, S., Khan A., Abbas, Q., Ali, S., Ali, M., Hussain, A., Mirani, Z.H., Hassan, S.S., Kazmi, M.R., Ali, S., Hussain, S. and A.M. Khan. 2023. Banned Sudan dyes in spices available at markets in Karachi, Pakistan. *Food Additives & Contaminants: Part B* 16:69–76.

Vandekerckhove, M., Droogenbroeck, B.V., Loose, M.D., Taverniers, I., Daeseleire, E., Gevaert, P., Lapeere, H. and C.V. Poucke. 2017. Development of an LC-MS/MS method for the detection of traces of peanut allergens in chili pepper. *Analytical and Bioanalytical Chemistry* 409:5201–5207.

Wang, C., Cheng, F., Wang, Y., Gong, Z., Fan, M., and J. Hu. 2014. Single point calibration for semi-quantitative screening based on an internal reference in thin layer chromatography-SERS: The case of Rhodamine B in chili oil. *Analytical Methods* 6:218–7223.

Wang, L., Du, Y., Xiao, Z., Li, Y., Li, B. and G. Yang. 2017. Determination of trace rhodamine B in chili oil by deep eutectic solvent extraction and an ultra high-performance liquid chromatograph equipped with a fluorescence detector. *Analytical Science* 33:715–717.

Wang, L., Zheng, J., Wang, T. and B. Che. 2013. Determination of eight Sudan dyes in chili powder by UPLC-MS/MS. *Engineering* 5:154–157.

Wei, W., Gong, X., Sun, J., Pan, D., Huang, Q. and C. Wang. 2019. Cellophane paper-based surface-enhanced Raman scattering (SERS) substrates for detecting Rhodamine 6G in water and chilli powder. *Vibrational Spectroscopy* 102:52–56.

Welsh, J. and M. McClelland. 1990. Fingerprinting genomes using PCR with arbitrary primers. *Nucleic Acids Research* 18:7213–7218.

Wilde, A.S., Haughey, S.A., King, P.G. and C.T. Elliott. 2019. The feasibility of applying NIR and FT-IR fingerprinting to detect adulteration in black pepper. *Food Control* 100:1–7.

Williams, J.G., Kubelik, A.R., Livak, K.J., Rafalski, J.A. and S.V. Tingey. 1990. DNA polymorphisms amplified by arbitrary primers are useful as genetic markers. *Nucleic Acids Research* 18:6531–6535.

Wu, C., Lu, Q., Miu, X., Fang, A., Li, H. and Y. Zhang. 2018. A simple assay platform for sensitive detection of Sudan I-IV in chilli powder based on CsPbBr3 quantum dots. *Journal of Food Science and Technology* 55:2497–2503.

Xie, X., Chen, L., Pan, X. and S. Wang. 2015. Synthesis of magnetic molecularly imprinted polymers by reversible addition fragmentation chain transfer strategy and its application in the Sudan dyes residue analysis. *Food Chemistry* 1405:32–39.

Yang, D., Zhu, L. and X. Jiang. 2010. Electrochemical reaction mechanism and determination of Sudan I at a multi wall carbon nanotubes modified glassy carbon electrode. *Journal of Electroanal Chemistry* 640:17–22.

Yu, X., Lee, J.K., Liu, H. and H. Yang. 2019. Synthesis of magnetic nanoparticles to detect Sudan dye adulteration in chilli powders. *Food Chemistry* 299:125144.

Zhang, M., Shi, Y., Sun, W., Wu, L., Xiong, C., Zhu, Z., Zhao, H., Zhang, B., Wang, C. and X. Liu. 2019. An efficient DNA barcoding based method for the authentication and adulteration detection of the powdered natural spices. *Food Control* 106:106745.

Zhu, Y., Wu, Y., Zhou, C., Zhao, B., Yun, W., Huang, S., Tao, P., Tu, D. and S. Chen. 2016. A screening method of oil-soluble synthetic dyes in chilli products based on multi-wavelength chromatographic fingerprints comparison. *Food Chemistry* 192:441–451.

Zietkiewicz, E., Rafalski, A. and D. Labuda. 1994. Genome fingerprinting by simple sequencerepeat (SSR)-anchored polymerase chain reaction amplification. *Genomics* 20:176–183.